ıd Scope

stry, the purposeful growing of trees and crops in interacting ions, began to attain prominence in the late 1970s, when the onal scientific community embraced its potentials in the tropics gnized it as a practice in search of science. During the 1990s, the of agroforestry for solving problems related to deterioration of rms, increased soil erosion, surface and ground water pollution, eased biodiversity was recognized in the industrialized nations s, agroforestry is now receiving increasing attention as a le land-management option the world over because of its l, economic, and social attributes. Consequently, the ge-base of agroforestry is being expanded at a rapid rate as d by the increasing number and quality of scientific publications s forms on different aspects of agroforestry.

g full and efficient use of this upsurge in scientific agroforestry challenge and an opportunity to the agroforestry scientific ity. In order to help prepare themselves better for facing the and seizing the opportunity, agroforestry scientists need access esized information on multi-dimensional aspects of scientific ty.

m of this new book-series, *Advances in Agroforestry*, is to offer he art synthesis of research results and evaluations relating to aspects of agroforestry. Its scope is broad enough to encompass all aspects of agroforestry research and development. tions are welcome as well as solicited from competent authors aspect of agroforestry. Volumes in the series will consist of books, subject-specific monographs, peer-reviewed publica- of conferences, comprehensive evaluations of specific projects, r book-length compilations of scientific and professional merit ance to the science and practice of agroforestry worldwide.

Advances in Agrofores

Volume 10

Series Editor

P. K. Ramachandran Nair, Gainesville, USA

For further volumes:
http://www.springer.com/series/6904

Aims

Agrofo
combi
interna
and re
relevar
family
and de
too. T
sustain
ecologi
knowle
illustra
of vari
Mal
is both
commu
challen
to syn
agrofo
The
state-o
differer
any a
Contri
on any
referen
tions o
and otl
and rel

Jagdish Chander Dagar
Anil Kumar Singh
Ayyanadar Arunachalam
Editors

Agroforestry Systems in India: Livelihood Security & Ecosystem Services

Springer

Editors
Jagdish Chander Dagar
Anil Kumar Singh
Ayyanadar Arunachalam
Natural Resources Management
Indian Council of Agricultural Research
New Delhi, Delhi
India

ISSN 1875-1199 ISSN 1875-1202 (electronic)
ISBN 978-81-322-1661-2 ISBN 978-81-322-1662-9 (eBook)
DOI 10.1007/978-81-322-1662-9
Springer New Delhi Heidelberg New York Dordrecht London

Library of Congress Control Number: 2013951504

Printed on acid-free paper

Springer is part of Springer Science+Business Media (www.springer.com)

Foreword

Agroforestry has come a long way since its "reinvention" about 35 years ago. Today, it ranks high among the significant land-management initiatives that have been undertaken the world over during the past few decades. Indeed, agroforestry is now recognized as an important approach to ensuring food security and rebuilding resilient rural environments.

India has been an all-time leader in agroforestry. The South and Southeast Asia region comprising India is often described as the cradle of agroforestry. Almost all forms of agroforestry systems exist across India in ecozones ranging from humid tropical lowlands to high-altitude and temperate biomes, and perhumid rainforest zones to parched drylands. The country ranks foremost among the community of nations not only in terms of this enormous diversity and long tradition of the practice of agroforestry, but also in fostering scientific developments in the subject. Following the "First National Seminar on Agroforestry" in 1979, a head-start in agroforestry research was made in the early 1980s with the launching of an All India Coordinated Research Project on Agroforestry with research and development centers in different parts of the country.

Despite this rich and varied experience, the information-base on agroforestry in India is scattered and poorly documented. Other than my respected colleague (the late) Dr. K. G. Tejwani's 1994 book "Agroforestry in India," and a scatter of articles, book chapters, and reports, there is no comprehensive database or documentation of agroforestry in India. Little wonder then that agroforestry is not receiving its deserving level of attention and prominence in national development efforts in India, and the breadth and wealth of agroforestry experience in India is little known even within—let alone outside—India. This book containing authoritative accounts of agroforestry in different ecoregions of India represents a bold step forward in fulfilling the long-felt need to fill that void.

Having been in constant contact with Dr. J. C. Dagar, the book's Principal Editor, during various stages of the book's progression, I am aware of the extent of patient and persistent efforts that have gone into the production of such a multi-authored volume. I congratulate all editors, chapter authors, and others who have been involved in the project on their splendid accomplishment, and express the global agroforestry community's gratitude to them for providing such a valuable contribution to the agroforestry literature.

Gainesville, Florida, USA, July 2013 P. K. R. Nair

Preface

Agroforestry has been the way of life since time immemorial, but the subject is relatively new as a science. Still, India has the distinction of having the longest history of formal agroforestry research in the world. It has come of age during the past three decades when activities and interest in agroforestry research, education, and training have increased tremendously. Its importance has been felt more during the last one decade in the scenario of climate change when perennial farming systems are considered more appropriate and sustainable for livelihood security in general and poor, marginal, and landless farmers in particular.

Today, agroforestry is taught at the senior undergraduate and postgraduate levels in many institutes and universities around the world, either as a separate subject or as a part of the regular curricula of agriculture, forestry, ecology, and other related programs. Although several books on the subject have been published during the past few years, there is still no single publication that may be recognized giving detailed description covering all the aspects on the subject. This publication is an effort to make up for this deficiency.

The introductory chapter traces the brief history of the development of agroforestry underlying the concepts of the subject, and the very purpose of publication. It is followed by 12 chapters dealing with the majorly practiced systems found in the tropics and temperate regions, especially different agro-climatic regions of India and the recent developments in each of them are discussed in detail. The site-specific systems/practices include edaphic and climatic parameters, ecology, vegetation, site-specificity, and structure and livelihood security of different stakeholders. Important success stories have been highlighted, research gaps have been identified, and the way forward has also been suggested. Finally, in the Synthesis Chapter the gist of the book is given. The important initiatives taken at the level of researchers and policy makers and the way forward for agroforestry research in India have been highlighted.

Recent concerns such as agroforestry for ecosystem services, mitigating climate change, carbon sequestration, biodiversity conservation, value addition, and role in health services through medicinal and aromatic plants have been addressed separately. There is a global consensus that integration of trees/perennials into farms, grazing lands, and other production landscapes helps to capture social, economic, cultural, ecological, and environmental benefits. With these developments in agroforestry science, we are now in a better position than ever before to capitalize the promise of multi-functional agriculture to make a difference in the lives of millions of people across the globe.

Overall, agricultural ecosystem can be further improved through agroforestry to ensure environmental restoration (including defended habitats), enhanced farm productivity, nutrient security, and realization of ecological services including climate change mitigation and adaptation for improved rural livelihoods. Nonetheless, this calls for a policy to help strengthen agroforestry research and tree development at the national level. This volume is a holistic effort to consummate the flagged research in agroforestry that addresses socio-biological issues of the very practice, apart from mundane ecological and climate change-related issues.

Most of the contributors are experts in their relevant fields and have added their field experiences to the value of this publication. We sincerely thank all the contributors and reviewers, who contributed enormously and cooperated so splendidly under strict and difficult time schedules. We also acknowledge the critical overview given by Prof. Dr. P. K. R. Nair, Florida University, Gainesville, USA, which helped us in improving the contents of various chapters and integrating them such that the document appears to be whole rather than the sum of its components. He was also kind enough to write the Foreword. We hope that the publication will be very useful for scientists, policymakers, environmentalists, educationists, and researchers for shaping this very important field of present-day science.

New Delhi, India

J. C. Dagar
A. K. Singh
A. Arunachalam

Contents

Introduction

1

Jagdish Chander Dagar, Anil Kumar Singh,
and A. Arunachalam

Abstract

Agroforestry is an age-old land use system as the process of human evolution has been from forests when man learnt the art of domesticating plants and animals after leaving the hunting and gathering habit. The hunting and food gathering system gradually gave way to food producing systems. Incidentally, some stray references related to tree plantations occur in different texts of the Vedic literature. Archeological excavations corroborate early tree domestication around the settlements in South Asia. In India, organized research in agroforestry was initiated in early 1970s through industry participation in plantation of commercial tree species through Indian Council of Agricultural Research (ICAR) institutes. As follow up, the All India Coordinated Research Project (AICRP) on Agroforestry was established in 1983 by the ICAR through which research work was carried out through Agricultural Universities situated in different agroclimatic zones. Forest Survey of India has reported that about 25 million ha area in the country (8.2 % of the total reported geographical area) is under agroforestry in both irrigated and rainfed agriculture which also includes trees outside forests and scattered trees on and off the agricultural fields. In this publication, the various chapters are compiled in such a way that a clear picture of various agroforestry systems both traditional and improved found in different agroecological regions is presented. Some systems are present across a number of climatic regions (for example salt-affected and waterlogged areas); but the problems are of different nature in different regions. To deal with such cases, separate chapters are included to present agroforestry approaches dealing with specific problems. Agroforestry systems provide excellent opportunities for carbon sequestration and mitigating climate change, hence it needs a special strategy in policy initiatives.

J. C. Dagar (✉) · A. K. Singh · A. Arunachalam
Natural Resource Management Division, Indian Council of Agricultural Research, Krishi Anusandhan Bhavan-II, New Delhi 110012, India
e-mail: dagarjc@gmail.com

J. C. Dagar et al. (eds.), *Agroforestry Systems in India: Livelihood Security & Ecosystem Services*, Advances in Agroforestry 10,
DOI: 10.1007/978-81-322-1662-9_1, © Springer India 2014

Early History of Agroforestry in India

Agroforestry is a collective name for land use systems and technologies, where woody perennials are deliberately grown on the same land-management units as agricultural crops and/or animals, either in some form of spatial arrangement or on a temporal sequence. In agroforestry systems, there are both ecological and economical interactions between the different components (Lundgren and Raintree 1982). Historically agroforestry is an age-old land use system as the process of human evolution has been from forests when man learnt the art of domesticating plant and animals after leaving the hunting and gathering habit. In about 700 BC, the hunting and food gathering system gradually gave way to food producing systems. Horticulture as co-existent with agriculture is found to have been prevalent in India from early historic period (500 BC to First century AD) when a certain amount of share in garden crops started to have been enjoyed by the King for providing irrigation. Incidentally, some stray references occur in different texts of the Vedic literature. For example, the cultivation of date palm (*Phoenix dactylifera*), banana (*Musa* x *paradisiaca*), pomegranate (*Punica granatum*), coconut (*Cocos nucifera*), jujube (*Ziziphus mauritiana*), *aonla* (*Emblica officinalis*), *bael* (*Aegle marmelos*), lemon (*Citrus limon*), and many other fruit trees and requirement of livestock in agriculture and mixed economy of agriculture and cattle breeding may be traced in proto-history chalcolithic periods of civilization (Raychaudhuri and Roy 1993). The role of many common trees such as *khejri* or *sami* (*Prosopis cineraria*), *aswattha* (*Ficus religiosa*), *palasa* (*Butea monosperma*), and *varana* (*Crataeva nurvala*) in Indian folk life has been mentioned in ancient literature of Rig Veda, Atharva Veda, and other Indian scriptures (Mann and Saxena 1980).

Archeological excavations corroborate early tree domestication around the settlements in South Asia. The evidences of this dates back to the Mesolithic period (10,000–4000 BC) when fruits of 63 plants including *bael, aonla, ber* (*Ziziphus* spp.), figs (*Ficus* spp.), *mahua* (*Madhuca indica)*, mango (*Mangifera indica*), etc., were reportedly consumed in one or the other form and were domesticated near the habitats (Randhawa 1980). Puri and Nair (2004) mentioned that rearing of silkworm (*Bombyx mori*) and lac insect (*Kerria lacca*) was practiced in the Indian sub-continent during the Epic era of Ramayana and Mahabharata (7000 and 4000 BC, respectively). The *Rishi*, *Muni,* and *Guru* (teachers of that time) were residing in forest and derived food in the form of vegetables, dry fruits, sweetening medium, medicine, and gum besides fuel from the forest. Further, they developed *Ashrams* (Institutions of learning) and started growing fruit trees like mango, guava (*Psidium guajava*), *mahua*, and others along with flowers, vegetables, and *Kutoo* (food grains) to meet their needs. Emperor Ashoka, a great Indian ruler (273–232 BC), encouraged a system of arbori-horticulture of banana, mango, jackfruit (*Artocarpus heterophyllus*), and grapes (*Vitis vinifera*). Further, the travelog of Ibn Battuta (Persian traveler, 1325–1354 AD) provides the earliest literary evidence of intensively cultivated landscapes of Malabar coast with coconut (*Cocos nucifera*) along with black pepper (*Piper nigrum*) around the habitats (Randhawa 1980). Warriar (1995) while describing Wayanad in Western Ghats, now in Kerala State, as Green Paradise mentioned that plough agriculture was prevalent in Wayanad as early as in the Megalithic Age (between 400 BC and 400 AD), and spices like black pepper, ginger (*Zingiber officinale*), and cardamom (*Elettaria cardamomum*) were often grown in association with woody perennials—as support or shade trees, since the early Middle Ages (500–1400 AD). The contents of *Krishi Gita*—over 300 year-old book of agricultural verses in Malayalam—also reflect on the need to maintain tree cover on the land scape, plant fruit trees on cleared forests, gardens, and other leftover lands, avenue planting as well as leaving vestiges of forests in the midst of cultivated landscapes (Kumar 2008). Natural history studies during the two previous centuries (Mateer 1883; Logan

1906) also signify that the people in the southern parts of peninsular India traditionally used their homesteads for a variety of needs such as food, energy, shelter, medicines, and other purposes. These evidences show that agroforestry was at the central stage in meeting the livelihood requirements since ancient times in India. Agroforestry practices were also prevalent in other parts of the world as described by Conklin (1957), Hailey (1957), Raghavan (1960), King (1987) and Nair (1993).

In India, organized research in agroforestry was initiated in early 1970s through industry participation in plantation of commercial tree species through Indian Council of Agricultural Research (ICAR) institutes. In 1979, first agroforestry Seminar was conducted at Imphal with participation of International Centre for Research in Agroforestry (ICRAF). As follow up, the All India Coordinated Research Project (AICRP) on Agroforestry was established in April, 1983 by the ICAR. The Council took lead in conducting systematic research in agroforestry initially through several coordinated projects and later by establishing National Research Center for Agroforestry (NRCAF) in 1988 at Jhansi. The Center was established to cater basic, strategic, and applied research needs in the field of agroforestry. The AICRP on Agroforestry—a large agroforestry network operational since 1983 was transferred to NRCAF in 1997 by empowering Director, NRCAF as the Project Coordinator of this big program. At present, the AICRP is being operated at 25 State Agricultural Universities (SAUs), 11 ICAR institutes and one Indian Council of Forestry Research and Education (ICFRE) institute. Since the early 1980s several agroforestry systems have been developed which have gone to farmers' fields and provided livelihood support to poor farmers.

Presently, the NRCAF is recognized worldwide for its research and development capabilities, agroforestry database, and information repository and natural resource management on watershed basis. The center conducts basic and strategic agroforestry research involving forest and fruit trees, shrubs, bamboos, arable crops, pastures, livestock, fish, and their interactions. Equal emphasis is given on tree improvement and human resource development (HRD) research programs.

Major Agroforestry Systems and Practices in India

In agroforestry literature the words "systems" and "practices" are often used as synonyms. In real sense agroforestry system is a specific local example of a practice, characterized by environment, plant species, and their arrangement, management, and socio-economic functioning. An agroforestry practice in turn denotes a distinctive arrangement of components in space and time (Nair 1993). Although hundreds of agroforestry systems have been recorded, they all consist of about 20 distinct agroforestry practices. In other words, the same or similar practices are found in various systems in different situations.

Nair (1993) classified agroforestry systems based on nature and arrangement of components (structural classification), function of the systems, ecological classification, and socio-economic criteria. In actual sense none of the system can be followed as such because most of the categories of agroforestry systems are found in all agroecological zones. In Indian literature, several terms for different systems such as agri(o)silviculture, agri(o)silvipasture, agri(o)horticulture, agri-silvi-horticulture, agri-silvi-hortipasture, hortiagriculture, horti-silvi(o)agriculture, horti-silvipasture, silvipasture, sylvopastoral, etc., are frequently used but the rationale and criteria for using or defining such terms have seldom been explained. While classifying agroforestry systems based on the type of components Nair (1989, 1993) classified agroforestry systems as agrisilviculture, silvopastoral, and agrosilvopastoral. He preferred using term agrisilviculture (rather than agrosilviculture) to denote the combination of trees and crops, whereas agrosilvopastoral (rather than agrisilvipastoral) was used for crops + animals/pasture + trees. His intention was to limit the use of the word agrisilviculture only to those combinations involving agricultural crops and trees. He has argued that the word agrosilviculture can encompass all forms of agriculture (including animal husbandry) with trees, and would thus be another word for agroforestry. That again was the reasoning behind the use of all-inclusive "agro" prefix agrosilvipastoral.

In this classification, the word tree was probably commonly used for forest as well as fruit trees, hence later the word "horti" was used as prefix for fruit trees. In this publication, more common classifications adopted broadly in agroforestry literature involving all categories explained by Nair (1993) has been followed; furthermore, when the tree included is a fruit tree, it is referred to as a "horti"-system, and the term tree is used to encompass all trees other than fruit trees.

With the above background, agroforestry systems in India are categorized/classified as follows:

(a) Trees with arable crops or tree-based systems—includes scattered trees on farm land, boundary plantations, shelter belts and wind breaks, and wood lots.
(b) Trees on pasture lands (silvopastoral systems)—trees + pastures/grasses + animals.
(c) Fruit tree-based cropping systems—arable crops or grasses as intercrops with fruit trees.
(d) Commercial tree-based cropping systems—including commercial trees like *Populus* and *Eucalyptus* along arable crops.
(e) Plantations on rangeland or grazing/pasture land—plantations such as coconut, rubber, red oil palm, etc. + grasses/legume fodders + animals.
(f) Plantation based multi-storey cropping system—plantations including clove, cardamom, coconut, coffee, etc., + forest trees/shrubs + pine apple + shade tolerant crops + grasses + vines + vegetables, etc.
(g) Others—such as home gardens, multi-enterprise farming systems, biodrainage plantations, aquaforestry, apiculture with forestry, etc.

Some of these systems are traditional (Pathak and Dagar 2000) while in many research efforts have been made. In some modern systems such as alley cropping (hedgerow intercropping), sand dune stabilization, reclamation of degraded lands including salt-affected lands, biodrainage plantations, even home gardens, improved fallows, watershed management, and multi-enterprise farming system much research inputs (technologies) have been incorporated in the recent past. The major agroforestry systems/practices found/identified in different regions of India are mentioned in Table 1.1.

Description of the systems and practices in different agroecological regions are included in specific chapters in this book. Some location-specific aspects of general nature of agroforestry systems that are not described in the specific chapters for different agroecological regions are briefly described here.

Shifting Cultivation

Shifting cultivation, one of the most primitive traditional agroforestry practices, refers to farming system in tropics and subtropics in which land under natural vegetation (usually forests) is cleared by the slash and burn method, cropped with common arable crops for a few years, and then left unattended while the natural vegetation regenerates. Traditionally, the fallow period used to be 10–20 years but now it is reduced to 3–5 years. In India, about 600 thousand families in 48 districts are cultivating 2.27 million ha area (Table 1.2) as shifting cultivation (FSI 1997; North Eastern Council 1997).

It is evident that shifting cultivation has become unsustainable primarily due to reduced *jhum* cycle owing to the increase in population pressure. Many workers (Borthakur 1992; Ramakrishnan 1992; Tripathy and Barik 2003; Tomar et al. 2012) suggested alternatives or improvements to shifting cultivation.

Nair (1993) has included about 44 species of perennial legumes used in Asian farming systems which may help in improving the fallow. These include species of *Acacia, Albizia, Alnus, Cajanus, Calliandra, Casuarina, Erythrina, Faidherbia, Flemingia, Gliricidia, Inga, Leucaena, Parkinsonia, Pithecelobium, Prosopis, Robinia,*

Table 1.1 Common agroforestry systems/practices found in India

Agroforestry practice	Arrangement of major components	Agroecological adaptations/distribution and special remarks
1. Shifting cultivation (Traditional as well as improved fallows)	Fast growing trees planted after clearing forest and left to grow in fallow period + arable crops	Tribal areas of Northeastern states, Orissa, Andhra Pradesh (enough scope of improvement)
2. Taungya (Traditional)	Combined stand of woody and arable crops during early stages of establishment	In all regions mainly tribal belts (enough scope of improvement)
3. Multi-purpose trees on crop lands (Traditional and improved)	Trees scattered on fields, boundary plantations, live hedges (forest and fruit trees + arable crops)	All regions especially in subsistence farming, also integrated with animals mainly in dry regions (Many improved MPTs are available)
4. Plantation-based cropping systems (Improved)	(i) Integrated multi-storey mixture of plantation crops	Plantations like coffee, cacao, coconut + fruit trees + spices like clove, cardamom + vines (black pepper) + shade crops, etc., in humid and sub-humid tropical regions
	(ii) Commercial plantation crops with shade trees scattered on field or dense in boundaries	Tea plantations + shade trees in hilly regions
	(iii) Tree plantations with arable crops	Commercial plantations (Poplar, Eucalyptus, bamboo) in Indo-Gangetic Plains on crop fields
	(iv) Fruit orchards and arable intercrops or fodder crops	All regions including temperate hilly regions
	(v) Plantations on pastures/grazing lands	In humid and sub-humid tropical regions, mainly coastal regions
5. Alley cropping (Hedgerow intercropping) (Improved systems)	Leguminous trees/shrubs trained in rows as alleys + arable crops/grasses	In humid and sub-humid regions, mostly on sloping lands to conserve soil
6. Home gardens (Improved home gardens also available)	Multi-storey combination of MPTs (mainly fruit & plantation trees, shrubs), vines, shade crops, vegetables, spices, etc., poultry, ducks, and livestock around homesteads	All regions especially in hilly regions and coastal highly populated regions
7. Trees on grazing/pasture lands	Scattered trees/shrubs (used as fodder + shade) on grazing lands (involves livestock)	In all hilly and dry regions; also nearby forests
8. Afforestation for land reclamation and soil conservation (Improved)	Trees, shrubs, and grasses on highly degraded lands including eroded, mined, salt-affected, and waterlogged area (improved systems)	In all arid and semi-arid regions to rehabilitate ravine lands, mined areas, salt-affected lands, biodrainage plantations along canals and crop fields, and sand dunes stabilization

(continued)

Table 1.1 (continued)

Agroforestry practice	Arrangement of major components	Agroecological adaptations/distribution and special remarks
9. Shelter belts, wind breaks, and live hedges	Trees and shrubs sometimes succulent cactus, etc., as live hedges	In all regions
10. Woodlots	Mostly on common property lands, along roads, railway lines, avenue trees, sacred groves, etc.	In all regions
11. Cut and carry systems (protein banks)	Usually leguminous fodder and fuel-wood trees/shrubs and fodder grasses and legumes on *Panchayat* lands	In many progressive villages throughout country
12. Multi-enterprise farming systems (Improved)	Forest and fruit trees + food crops + vegetables + fodder crops + live stock + poultry + duckry + piggery + fish in pond + floriculture + apiculture + natural gas/solar energy	Improved and highly remunerative; size of enterprise depends on the size of land holding of the farmer; being adopted in all regions
13. Others (Improved)	Aquaforestry, apiculture, and cultivation of medicinal and aromatic plants with forest and fruit trees; fish culture with mangroves; domestication of ornamental plants from forests and wild places, grazing in forests, etc.	New adventures but being followed at many places

Table 1.2 Area under shifting cultivation in India

States	Total area (000 ha)	Families involved (No.)	Districts (No.)
Arunachal Pradesh	261	54,000	10
Assam	310	58,000	3
Manipur	360	70,000	5
Meghalaya	265	52,290	5
Mizoram	45	50,000	3
Nagaland	633	1,16,046	7
Tripura	108	43,000	3
Andhra Pradesh	103	23,200	4
Orissa	184	1,41,000	8
All India	2,269	6,07,536	48

Source FSI (1997), North Eastern Council (1997)

and *Sesbania.* The intercropping between the fast growing leguminous trees during fallow phase is one of the approaches while finding alternative to shifting cultivation. Besides technical inputs socio-economic considerations are also important in solving the problems related to shifting cultivation. There is urgent need of settling the land tenureship issue educating the people about the adverse impacts of short *jhum* cycles. There is need to provide employment opportunities and regular income through proper utilization of natural resources and by equitable distribution of wasteland among the tribal people. The cooperative efforts should be encouraged for carrying out forest-based activities such as basket making (bamboo-based cottage industry), rope making, cane furniture, processing of minor forest produce, honey collection, etc. For these activities, efforts should be made to create viable market for sailing the products on remunerative prices. The local administration must ensure the implementation of total literacy campaign, which due to remoteness and un-supportive attitude of tribal people has not been successful so far. For awareness and educating specially the women and children, services of various non-governmental organizations and voluntary agencies, besides the regular government machinery, are required on regular basis. Eco-development plans for areas under shifting cultivation should be developed on priority on site-specific basis involving sustainable agroforestry practices. Determining the population supporting capacity of a *jhum* stand may be one of the major aspects for checking the degradation of the environment and depletion of the resources. Overall strategy should be developed which ensures improving livelihood of people by efficient utilization of natural resources including land, water, biodiversity, and external input in a practical and profitable manner enhancing the environmental safety. Integrated approach involving arable crops, fruits, animal husbandry, fishery, and forestry with appropriate conservation measures for natural resources would be most effective in overall development of the shifting cultivation areas.

Taungya

The *taungya* system in the tropics is like an organized and scientifically managed shifting cultivation, a forerunner to agroforestry. The word is reported (Blanford 1958) to have originated in Myanmar (earlier Burma); *tauang* means hill and *ya* means cultivation. Earlier it was a local name for shifting cultivation and later subsequently used to describe afforestation as well. Today the system is known by different names such as *Tumpangsari* in Indonesia; *Kainginning* in the Philippines; *Lading* in Malaysia; *Chena* in Sri Lanka; *Kumri, Jhooming, Poonam, Taila* and *Tackle* in different parts of India; *Shamba* in East Africa; *Parcelero* in Puerto Rico; and *Consorciarcao* in Brazil (Nair 1993).

Essentially, the system consists of growing annual agricultural crops along with the forestry species during the early years of establishment of the forestry plantation. The important tree species grown in this system include *Shorea robusta, Tectona grandis, Dalbergia sissoo, Acacia catechu, Eucalyptus globulus, Populus deltoides*, and *Pinus patula*. The land basically belongs to the forestry department and upon their large-scale lease, allowed the subsistence farmers to raise their crops and in turn the farmers protect the tree saplings. It can be considered a step in the process of transformation from shifting cultivation. It is not merely the temporary use of a piece of land and a poverty level wage, but a chance to participate equitably in diversified and sustainable agroforestry economy. There are numerous reports describing *taungya* practices of different regions but research data on changes of the soil fertility and management aspects are, however, scarce. Alexander et al. (1980) based on 2 years data on the Oxisols of Kerala mentioned disadvantage of *taungya* causing erosion hazard caused by soil preparation during cultivation for the agricultural crops. The surface horizons became partly eroded and subsurface horizons were gradually exposed. The addition of crop residues to the soil surface was found to be a very effective way of minimizing soil loss and exposure. The farmers may manage the system in more sustainable way as is the case in watersheds if they are leased the land for longer period.

Home Gardens

Home gardens depict a transition stage between tropical forest ecosystem and arable cropping that mutually supports the sustainable agriculture and forest ecosystems. It also preserves the biodiversity. Much has been written about home gardens and numerous terms have been used by various workers. These include mixed-garden horticulture, house garden, Javanese home gardens, compound farm, kitchen garden, household garden, and homestead agroforestry (Nair 1993). Plantation crops such as coconut (*Cocos nucifera*), cacao (*Theobroma cacao*), coffee (*Coffea arabica*), arecanut (*Areca catechu*), and vine black pepper (*Piper nigrum*) often are the dominant components of many home gardens of humid tropics. Fruits such as banana (*Musa paradisiaca*), papaya (*Carica papaya*), mango (*Mangifera indica*), guava (*Psidium guajava*), custard apple (*Annona squamosa*), and jackfruit (*Artocarpus heterophyllus*) are the major components of some tropical home gardens including coastal regions of India. Dagar (1995), Kumar (2006, 2010), Pandey et al. (2007) and Kumar and Kunhamu (2011) have given comprehensive account of home gardens in coastal and Island regions of India.

Plantation Based Cropping Systems

During the recent past in India a sizeable data has been generated from coconut-based cropping systems and is related to intercropping under coconut plantations and fruit tree orchards, integrated mixed farming in small holdings, grazing under coconut, factors favoring intensification of land use with coconut, and multi-storey tree gardens. Important crops—plantation combinations include cereals (rice-*Oryza sativa*, finger millet-*Eleusine coracana,* and maize-*Zea mays*); pulses (pigeon pea-*Cajanus cajan*, green gram-*Vigna radiata*, black gram-*Vigna mungo*, gram-*Cicer arietinum*, soybean-*Glycine max*, cowpea-*Vigna unguiculata*, etc.); oil seeds (groundnut- *Arachis hypogaea*, mustard-*Brassica juncea, B. nicra, B. napus*, sesame-*Sesamum indicum*, safflower-*Carthamus tinctorius*, sunflower-*Helianthus annuus*, etc.); root crops (sweet potato-*Ipomoea batatas*, tapioca-*Manihot esculenta*, yams-*Dioscorea* spp., and taro-*Colocasia esculenta*, etc.); spices and condiments (ginger-*Zingiber officinale*, turmeric-*Curcuma domestica*, cardamom-*Elettaria cardamomum*, cinnamon-*Cinnamomum zeylanicum*, clove-*Syzygium aromaticum*, nut meg-*Myristica fragrans*, chilies-*Capsicum acuminatum, C. annuum*, and black pepper-*Piper nigrum*); fruits like pineapple-*Ananas comosus*, mango-*Mangifera indica*, banana-*Musa* x *paradisiaca*, papaya-*Carica papaya*, and bread fruit-*Artocarpus altilis*); other crops (cotton-*Gossypium herbaceum, G. arboretum, G. hirsutum*, sugarcane- *Saccharum officinarum*, potato-*Solanum tuberosum*, abaca-*Musa textilis* and several vegetables); tree crops (areca nut-*Areca catechu*, cacao (*Theobroma cacao*, tea-*Camellia sinensis*, and coffee-*Coffea arabica*); improved pasture grasses include species of *Brachiaria, Dichanthium, Panicum, Setaria, Paspalum,* and *Pennisetum*; while improved forage legumes include species of *Stylosanthes, Desmodium, Glycine, Pueraria, Phseolus, Leucaena,* and *Macroptilium*. Many trees such as species of *Populus, Eucalyptus, Moringa, Erythrina, Ficus, Tamarindus, Gliricidia, Ceiba,* and *Cordia* also find the place in these systems.

In North-eastern Himalayan regions domestication of large cardamom (*Ammomum subulatum*) plantations under *alder* (*Alnus nepalensis)* and its collection from the natural forests by indigenous Lepcha and Limbu tribes is an age-old agroforestry practice. Besides *Alnus nepalensis*, there are 29 other tree species, supporting this plantation crop. Tree management practices by farmers involve harvesting trees above 16 cm basal diameter to assist natural regeneration of younger tree seedlings and open canopy to regulate light at the ground. This tree management system provides continuous supply of fodder and fuel wood. The nitrogen fixing trees help site improvement and better growth of cardamom. Alder is also grown to enhance soil fertility particularly by the tribal people for growing maize, millet, potato, chilies, barley, vegetables, etc. This tree is main component of coffee and cardamom at lower and higher altitudes as a shade tree.

In Punjab and Haryana farmers are growing poplar (*Populus deltoides*) and *Eucalyptus* (particularly in areas with high water table) in rows on their cultivated fields having rice-wheat cropping system or sugarcane. This is more prevalent in irrigated situations.

Scattered Trees on Farmlands

The practice of growing agricultural crops under scattered trees on farmlands is quite old and seems to have scarcely changed for centuries. In ancient India, trees were given more importance than crops in tree-crop mixed cropping. Today also the trees are found grown scattered in agricultural fields for many uses such as shade, fodder, fuel wood, fruit, small timber, vegetables, and medicinal uses. Some of the practices are very extensive and highly developed. For example, growing of *Prosopis cineraria* and *Ziziphus nummularia* in arid areas; *Dalbergia sissoo, Acacia nilotica*, and mango in Indo-Gangetic plains; *Grewia optiva, Quercus* spp, and many other tree species in the Himachal Pradesh; *Eucalyptus globulus* in the southern hills of Tamil Nadu; and *Borassus flabellifer* in peninsular coastal regions. Nair and Dagar (1991) documented a profile of numerous tree species found growing in different agroclimatic regions of India. There are strong convictions for the acceptance of these trees on agricultural fields since time immemorial. The very fact that Khejri (*Prosopis cineraria*) is omnipresent in dry regions and its occurrence is encouraged in all the cultivated fields and village grazing grounds. It shows that its usefulness is generally and widely accepted by land owners who have a strong conviction that the tree does not hinder crop productivity in the adjoining areas. Moreover, studies conducted have shown that the soil under *P. cineraria* has more organic matter, total nitrogen, total and available phosphorus and potassium, and micronutrients (Zn, Mn, Cu, and Fe). Similarly *Ziziphus nummularia* is preferred and is a favorite bush in arid Rajasthan.

In the foot hills of Shivaliks, Grewal (1992) reported that growing Bhabar grass (*Eulaliopsis binata*) with *Eucalyptus* and *Acacia catechu* is highly economical system. In Indo-Gangetic plains the farmers retain trees of *Acacia nilotica, Acacia catechu, Azadirachta indica, Butea monosperma, Dalbergia sissoo, Gmelina arborea, Morus alba, Mangifera indica, Syzygium cuminii*, and *Ziziphus mauritiana*. Farmers in sub-humid *terai* region of Indo-Gangetic plains, prefer *Dalbergia sissoo*, *Psidium guajava*, *Mangifera indica*, *Morus alba*, *Syzyzium cuminii*, and *Grewia nudiflora*. In Bihar *Dalbergia sissoo*, *Litchi chinensis*, and *Mangifera indica* are frequently grown on fields. Farmers in northeastern region prefer *Alnus nepalensis*, *Artocarpus chaplasha*, and species of *Bambusa*, *Dendrocalamus*, and fruit trees like *Mangifera indica*, *Emblica officinalis*, and *Parkia roxburghii*. In coastal areas of peninsular India, *Borrasus flabellifer* is found scattered in the fields of groundnut, rice, and green gram. Other most common trees found on farmers' fields are *Moringa oleifera*, *Tamarindus indica*, *Ceiba pentendra*, *Anacardium occidentale*, *Cocos nucifera* palm, and fruits like banana, custard apple, guava, and pomegranate.

Trees on Farm Boundaries

Trees, which are grown in agricultural fields or on field bunds, are also often and usually grown on farm boundaries. In northern parts of India particularly in Haryana and Punjab, both *Eucalyptus* and *Populus* are commonly grown along field boundaries or bunds of paddy fields. Other trees, which are found grown as boundary plantations or live-hedge, include *Acacia nilotica*, *Dalbergia sissoo*, and *Prosopis juliflora*. Farmers of Sikkim grow bamboo (*Dendrocalamus*) all along irrigation channels. In coastal areas of Andhra Pradesh, *Borassus* is most frequent palm. In Andamans, *Gliricidia sepium*, *Jatropha* spp, *Ficus sp.*, *Ceiba pentendra*, *Vitex trifoliata*, and *Erythrina indica* are grown frequently as live-hedges. At many places succulents like *Agave* and many cactoids are grown as common live fence.

Many of the boundary plantations also help as shelterbelts and wind breaks particularly in fruit orchards. In Bihar, *Dalbergia sissoo* and *Wendlandia exserta* are most common plantations. *Casuarina equisetifolia* and *Acacia auriculaeformis* are extensively planted on field bunds and along sandy coastal areas in Orissa.

Woodlots

In many parts, farmers grow trees in separate blocks as woodlots along with agricultural fields. Now the practice is expanding fast due to shortage of fuel wood and demand of poles or pulpwood in industry. For example, bamboo poles are in great demand for orange orchards in Nagpur area and *Eucalyptus* and *Populus* for WIMCO Industries. Woodlots are being raised mostly on large farms due to the increase of labor costs and labor management, lack of irrigation facilities, and risk of crop investments. Woodlots of *Casuarina equesitifolia*, bamboo, *Populus deltoides*, *Eucalyptus*, red sanders (*Pterocarpus santalinus*), *Dalbergia sissoo*, *Grewia nudiflora*, and many others have become popular in many parts of the country. In recent times, these woodlots have shrunk at a fast rate due to increase in population pressure.

Systems for Soil Conservation/Amelioration/Reclamation

About 121 million ha of land in India is suffering from different degradation problems (NAAS 2010) including serious wind and water erosion and salinity. The deep and narrow gullies are best controlled by putting them to permanent vegetation after closure to grazing. Afforestation with suitable tree species like *Acacia nilotica*, *A. eburnea*, *Azadirachta indica*, *Butea monosperma*, *Prosopis juliflora*, *Dalbergia sissoo*, *Tectona grandis*, species of *Bambusa* and *Dendrocalamus* helps in stabilizing the gullies and ravines. Other adaptable species include grasses like Dichant*hium annulatum*, *Bothriochloa pertusa*, *Iseilema laxum*, *Cynodon dactylon*, *Dactyloctenium aegyptium*, *D. sindicum*, *Heteropogon contortus*, *Themeda quadrivalvis*, *Sehima nervosum*, and *Cenchrus ciliaris* which will besides providing fodder will also help in checking soil erosion. Grasses such as *Vetiveria zizanioides* and *Saccharum benghalensis* along with trees like *Terminalia arjuna*, *Eucalyptus*, and *Casuarina equisetifolia* help in stabilizing the coastal sand dunes.

Tree species such as *Acacia tortilis, A. nilotica, A. senegal, Prosopis cineraria, P. juliflora, Azadirachta indica, Tecomella undulata, Salvadora oleoides, Capparis decidua, Cordia rothii, Albizia lebbeck*, and *Cassia siamea* are suitable for growing on sand dunes in arid regions along with shrubs like *Calligonium polygonoides, Acacia jacquimontii,* and *Ziziphus nummularia*; and grasses like *Cenchrus ciliaris, Lasiurus scindicus,* and *Saccharum benghalensis*. These have been found effective in arresting drifting of soil/sand, as high as 1,450 t ha^{-1} (Hirekerur et al. 1991). Comprehensive results on afforestation of the salt-affected highly degraded soils with suitable species have been published by Dagar and Singh (1993), Singh et al. (1993), Tomar et al. (2003a, b, 2010) and Dagar (2012). The salt tolerant tree species are used for fodder, fuel wood, timber, and shade. Tree plantation techniques have been evolved for planting trees on highly sodic soils using auger-hole technique (Singh et al. 1993; Dagar et al. 2001a, b); on saline soils by furrow method (Tomar et al. 1998); on calcareous degraded lands using saline water (Tomar et al. 2003b); and on waterlogged farmers' fields (Ram et al. 2011). Several species have been identified for these situations. Tree species such as *Acacia nilotica, A. farnesiana, Parkinsonia aculeata, Prosopis juliflora,* and *Tamarix articulata* are the ideal for both alkali and saline soils while *Casuarina glauca* is an additional tool for saline waterlogged situations. *Azadirachta indica, Salvadora persica, Acacia nilotica*, and *Prosopis juliflora* are successful plants in black cotton saline vertisols. *Salicornia bigonia, Terminalia catappa, Pandanus* spp., and *Salvadora persica* (all oil yielding plants) have been assessed as the ideal choice for planting along with sandy coastal regions. In canal command areas where salinity and waterlogging are created due to seepage *Eucalyptus camaldulensis, E. tereticornis, Casuarina equisetifolia, C. glauca, Acacia nilotica*, and *Acacia ampleceps* are grown with success to check the seepage and salinity.

Shelter Belts

Arid regions witness very high wind velocity throughout the year and sand can initiate movement of particles even at 12–14 km h^{-1} wind velocity. Farmers build kind of obstacles to stop sand movements called *kana bandi* (e.g., in Rajasthan) either by using pieces of small dead wood or local vegetation to check wind velocity within safer limits (Mathur 1995). *Crotalaria burhia, Leptadenia pyrotechnica,* and A*erva javanica* are planted in 20–25 m wide rows across the wind direction. Between the lines of these under shrubs, grasses such as *Cenchrus ciliaris, C. setigerus*, and *Lasiurus scindicus* are planted on leeward side of each break. This permanent vegetation helps accumulating sand near them which is again spread in the field. This also helps increased crop yields along the lines.

Trees on Rangelands

As pointed earlier *Salvadora oleoides, S. persica, Capparis decidua, Tecomella undulata, Acacia nilotica, A. leucophloea, Prosopis cineraria*, and now *Prosopis juliflora* are the most frequent trees on common community grazing lands in dry regions. In coastal areas, coconut is most common tree on pasture lands. Sometimes trees like *Trema tomentosa, Moringa oleifera, Morinda citrifolia, Gliricidia sepium, Albizia lebbeck, Pongamia pinnata*, and *Ficus* spp. are grown and lopped for fodder. A cattle rearing usually involves grazing on these pastures. An organized form of silvopastoral system assures 10 t $ha^{-1}year^{-1}$ forage biomass production (as against 1 t $ha^{-1}year^{-1}$ from natural stands) at 10 year rotation in dry zones (Pathak et al. 1995) besides assuring soil conservation, carbon sequestration, and employment generation. While explaining the nature of grassland dynamics and their management Dagar and Pathak (2005) have cited several examples of trees playing crucial role in management of grazing lands in different agroecological regions. Based on long-term studies Rai (2012) reported the importance of *Ailanthus excelsa, Acacia tortilis, Acacia nilotica, Hardwickia binata,* and *Leucaena leucocephala* based silvopastoral systems for live stock production.

Aquaforestry

Throughout the coastal regions particularly along the Andhra coast farmers are cultivating fish and prawn in saline water and growing coconut and other trees on bunds of ponds. These trees help in producing feed in the form of litter to fishery and generating extra income to the farmer. Now the fish culture in association of the mangroves is also advocated which are proved to be rich source of nutrition to the aquatic life and breeding ground for juvenile fish, prawn, muscles, turtles and variety of other animals (Dagar et al. 1991; Dagar 1995, 2003). Poultry is another adventure in some of these regions. A well-balanced system of animal husbandry including goat farming, poultry, ducky, turtles, and fishes in the small ponds in home gardens make a balanced system of high moisture, energy, and nutrient use efficiency per unit area The leaves of many trees such as *Gliricidia sepium, Leucaena leucocephala*, and *Moringa oliefera* have been found to serve as fish feed when offered as pallets and improved the productivity of fish pond. In many parts of the country, farmers are raising forest and fruit trees and vegetables on the dykes of fish ponds on their farms in multi-enterprise mode and are generating nutrition rich food for the family and good income.

Apiculture with Trees

The coastal population has been harvesting honey from the bees-making-hives on trees. Apiculture is now considered a profitable profession throughout the country. Many farmers are interested to go for this business by selecting different kinds of plantations. Eventually, the scopes of bee keeping are more prominent in the agroforestry systems due to high floral diversity. Apiculture in farming system mode is a sustainable component for generating regular income.

Improved Agroforestry Practices and Systems in India

It has been possible to conduct a Diagnosis and Design (D&D) exercise for existing agroforestry practices in India, generate valuable information, and identify important agroforestry practices from different parts of the country through the (AICRP) on Agroforestry. The collection and evaluation of multi-purpose trees (MPTs) resulted in establishment of arboretum in each coordinating center of AICRP. A collection of 184 species was made by these coordinating centers. This was followed by identification of important tree species for agroforestry research for various agroclimatic conditions. Each center was allocated two tree species for seed/germplasm collection and conducting provenance trials. A significant contribution of the project was tree selection and improvement of species such as *Populus deltoides, Eucalyptus tereticornis, Dalbergia sissoo, Azadirachta indica, Acacia nilotica, Leucaena leucocephala, Ailanthus excelsa, Pongamia pinnata*, and *Casuarina equisetifolia*. Clonal orchards for *Dalbergia sissoo, Acacia nilotica*, and *Azadirachta indica* have been established. Under National Agricultural Technology Project a program "Agroforestry BASE"—an online database has been developed at National Center for Agroforestry Research (NCAF), Jhansi which is being updated periodically.

Agronomic practices such as planting methods, irrigation, filling mixture composition, fertilization, spacing, and pruning schedules for raising some of the promising MPTs in association with annual crops have been developed and standardized. The suitable crops and cropping sequences which can be grown successfully (without significant reduction in yield) through agronomic manipulations and tree canopy management practices in combination with different forest and fruit trees have been identified. *Morus alba* and *Grewia optiva*-based agroforestry systems for western Himalayas, *Alnus nepalensis*-based system for North Eastern Hill region, *Populus deltoides*-based system for Indo-Gangetic region, *Emblica officinalis*, and *Prosopis cineraria*-based systems for semi-arid and arid regions, *Tectona grandis*-based system for tropical region, and *Gmelina* and *Acacia*-based systems for humid and sub-humid regions have been developed. Packages have also been developed for rehabilitation of different wastelands through suitable agroforestry models.

The AICRP on Agroforestry initiated systematic work on biofuel research in 2003 with major emphasis on *Jatropha* and *Pongamia*. A network project on bamboo-based agroforestry systems has also been initiated in 2007 at six centers. Keeping in view the present day challenges, the project is now focusing on role of agroforestry in meeting the environmental challenges, value addition for creating livelihood opportunities, and application of modern tools and technologies in agroforestry research.

In recent times, substantial research inputs have been put into alley cropping (hedge row intercropping) system in which usually arable crops are grown in alleys formed by hedge rows of trees or shrubs, particularly in high rainfall areas. The hedge rows are cut back at crop planting time and kept pruned during the cropping season to prevent shading and to reduce competition with food crops. The hedge rows are allowed to grow when there are no crops and normally pruned during the season and the pruned material is either used as mulch, fodder, or source of nitrogen for crops. Tree species such as *Leucaena*, *Gliricidia*, *Sesbania,* and *Cassia* have widely tested in alley cropping system. Short duration rainy crops such as pearl millet and sorghum were found to be compatible with *Leucaena leucocephala* and *Gliricidia sepium.* In high rainfall areas like Andamans *Gliricidia sepium* has been found very successful on sloping lands and forage grasses such as hybrid napier (*Pennisetum purpureum*), thin napier (*Pennisetum polystachyon*), *Setaria anceps*, and legume *Stylosanthes guianensis* could be grown for fodder as alley crops which in turn also helped in checking erosion. Other crops such as turmeric (*Curcuma domestica*), ginger (*Zingiber officinale*), *Colocasia esculenta*, etc., also can be grown with hedge row crops. Wider alleys and low cutting heights were found to give higher intercrop yields in *Leucaena* system in semi-arid conditions.

Recently, farmers are adopting multi-enterprise farming systems to ensure perennial and sustainable round the year income from the farm involving many components such as fruit and forest trees, cereal crops, vegetables, fish (if there is pond), live stock, fodder, poultry birds, mushroom, and bee keeping. Live stock component is very important in this enterprise. Due to climate aberration if some component fails then the farmer can get some returns from remaining components. This farming system is most viable and remunerative.

Area Under Agroforestry in India

India has a large number of agroforestry systems and practices of various forms and types so it is very difficult to assess or estimate the extent of the area under agroforestry in the country. However, the Forest Survey of India (FSI) has been engaged in assessing the Trees Outside Forest (TOF) wealth of the country since 1991. In general, trees outside forests mean the trees available on agricultural land, along road side, railways, canals, ponds, orchards, parks, gardens, and homestead. These categories broadly fall under agroforestry. From 1991 onwards, the organization has been trying to improve the methodology of TOF assessment which has so far been based on field inventory methods, by incorporating satellite data of IRS LISS III, PAN, and fused image from these two for editing and refinement of classified images (Rawat et al. 2004; FSI 2011). At the national level, an attempt is being made in the National Carbon Project (NCP) to estimate the total phytomass and carbon density for plants/trees inside and outside the forest through a project taken up by the Indian Space Research Organization (ISRO), Government of India under its ISRO-Geosphere and Biosphere Program in the 11th-Five-Year Plan. Inventories of non-forest areas were also designed to generate information at the state level. For example, the Kerala Forest Research Institute carried out a systematic survey using small villages as sample unit and estimated that the home gardens of Kerala had a total of 440 million trees equivalent to an estimated growing stock of 134.13 million m^3, of which 8.15 million m^3 meet the substantial portion of the total fuel wood requirement in the state (Singh and Chand 2012). Similarly a systematic study was conducted in the state of Madhya Pradesh by the Indian Institute of Forest Management, Bhopal, and highlighted the special distribution and importance of TOF in relation to livelihood of local people. In arid and semi-arid areas of less forest cover such as Rajasthan, Haryana, Punjab, Maharashtra, and Central India, nearly the entire total fuel wood requirements of rural people are met from non-forest resources like TOF. Recently, Singh and Chand (2012) developed a methodology and reported that in Haryana aboveground TOF phytomass varied from 1.26 t/ha in the scattered trees in the rural/urban area to 91.5 t/ha in the dense linear TOF along the irrigation canals. The total aboveground TOF phytomass and carbon content was calculated as 367.04 and 107.34 t/ha, respectively. These studies concluded that the classification of TOF and estimation of phytomass and carbon content in TOF can be successfully achieved through the combined approach of Remote Sensing and GIS-based spatial technique with the supplement of field data.

The forest cover assessment carried out by FSI using satellite data included all lands comprising area of one hectare and more, with a tree canopy density of more than 10 %, irrespective of their land use and ownership. The small patches of trees which are less than one hectare in extent, such as trees in village woodlots, homestead, urban areas, scattered trees, trees along roads, canals, railway lines and trees in block, and linear formation are excluded from the forest cover due to technical limitations of the procedures used. The contribution of such trees/patches to overall cover is estimated statistically using a sampling based methodology and is termed as tree cover. Thus, tree cover comprises tree patches outside the recorded forest area which are not captured by remote sensing satellite during forest cover assessment having area less than the minimum worth mapping area of one hectare. These blocks or linear patches between 0.1 and 1.0 ha are usually scattered trees.

Based on combined approach, the FSI has estimated the total forest and tree cover of India to be 78.3 million ha (including 4,662 km^2 mangrove area), which is 23.81 % of geographical area (FSI 2011). Of this, Trees Outside-Forest consist of about 5.07 % of total forest cover, i.e., 3.51 million ha (Mha) area. Thus, total TOF and tree cover make 12.59 Mha (Table 1.3), which may be considered as under agroforestry systems. FSI (FSI 2011) has reported that 25.31 Mha area in the country (8.2 % of the total reported geographical area of the country) is under agroforestry in both irrigated and rainfed agriculture (Table 1.4) which also includes TOF, scattered trees on and off the agricultural fields, 2.27 Mha area under shifting cultivation (North Eastern Council 1997), and 2.42 Mha home gardens (Kumar 2006).

As per this estimation, 7.0 Mha area is under irrigated agroforestry with commercial/industrial plantations under agrisilviculture (2.63 Mha), fruit orchards or fruit-based cropping systems (2.79 Mha), and trees on boundaries or bunds of agricultural fields (1.58 Mha) through social forestry. About 10.6 Mha area in rainfed agriculture is under agrisilviculture (2.4 Mha), agrihorticulture or fruit trees based cropping systems (1.86 Mha), and trees on grazing/range lands and field boundaries (6.32 Mha). Besides this, about 3 Mha area is under tree cover due to rehabilitation of salt-affected lands, mined wastelands, and trees on common property/community lands.

Considering the importance of agroforestry in India, the Planning Commission of India, the apex planning body for the country, in its report stated that there is a scope for bringing 10 Mha irrigated and 18 Mha rainfed additional area under agroforestry (Planning Commission 2001). Out of this a total of about 7.73 Mha (3.80 Mha

Table 1.3 Area under trees outside forest and tree cover (scattered trees) of India in 2011

Category	Area (000 ha)	Percent of geographical area
Trees outside forest	3,509	1.07
Tree cover (scattered trees)	9,084	2.76
Total	12,593	3.83

Source FSI (2011)

Table 1.4 Estimated agroforestry area in the country

Category	Area (million ha)	Remarks
Agroforestry in irrigated areas		
Agrisilviculture	2.63	Industrial use
Agrihorticulture	2.79	Fruit orchards/fruit trees based cropping systems
Trees on field boundary or bunds	1.58	Social forestry, live fences, etc.
Sub-total (A)	**7.00**	
Agroforestry in rainfed areas		
Agrisilviculture	2.40	Scattered trees on fields, bunds, boundaries
Agrihorticulture	1.86	Fruit orchards/plantation-based cropping systems
Trees on field boundary/bunds	0.74	Social forestry, deliberate live fences, etc.
Silvo-pastoral	5.58	Trees on grazing/range lands
Sub-total (B)	**10.58**	
Other land uses		
Home gardens	2.42[a]	Mostly in coastal areas and North-Eastern states
Shifting cultivation	2.27[b]	Mostly in NEH States, Orissa, Andhra Pradesh
Afforestation of problem soils	2.12	Plantations on salty soils, mine areas, etc.
Trees on community/common lands	0.92	On Panchayat lands, along roads, railways, etc.
Sub-total (C)	**7.73**	
Total agroforestry area **(A + B + C)**	**25.31**[c]	

Source NRCAF (2013)
[a] Kumar (2006)
[b] North Eastern Council (1997)
[c] FSI (2011)

irrigated and 3.93 Mha rainfed land) has been brought under agroforestry through various schemes during the past 10 years. There is further scope of increasing the area under agroforestry in future by another 28.0 Mha which may include the remaining 20.27 Mha area earlier earmarked by the Planning Commission (2001). The major share of the land to be brought under agroforestry will be from fallows, cultivable fallows, pastures, groves, and rehabilitation of problem soils. Thus, a total of 53.31 Mha area (about 17.5 % of the total reported geographical area-TRGA), could potentially be under agroforestry in the near future (NRCAF 2013), making the agroforestry a major activity in India in terms of area occupied after agriculture with 141 Mha, 46 % of the TRGA (Ministry of Agriculture 2010), and forestry with 69.63 Mha, 22.8 % of the TRGA (FSI 2011).

Need and Scope of Agroforestry in India

It is worth mentioning here that agroforestry systems are probably the only means for getting the desired tree cover in the country, especially in states that have low forest area. Agroforestry is not only a technique for growing food crops in association with woody perennials and livestock to optimize the use of natural resources but also a source of renewable energy resource and a means of reducing the risk of environmental degradation and climate change. The livelihood security through agroforestry and its potential in meeting basic needs viz., food, fuel, fodder, and employment generation are well known. By virtue of diversity of the components of the agroforestry systems like food grains, vegetables and fruits, and the nutritional security to the communities could be ensured. The fodder cultivation under agroforestry land use will ensure production of milk, meat, and animal products. Wide range of food crops, pulses, and oil seeds can meet diverse needs of the society. Tree domestication and the commercial processing and marketing of tree products and services, is a new frontier for agroforestry research and or development. Multi-enterprise farming systems have to go long way for bringing sustainability and livelihood security among small and marginal stakeholders. A major role is also emerging in the domain of environmental services, particularly the development of mechanisms to reward the rural poor for the watershed protection, biodiversity conservation, and carbon sequestration that they provide to the society. Agroforestry has the greatest potential to mitigate climate change through adaptation. It is the tested mechanism to tolerate diversified climate through adaptations toward variety of climate aberrations.

Some of the main socio-economic and environmental outcomes/successes of agroforestry development programs in the country have been summarized in a separate chapter dealing with utilization and benefits of agroforestry. The economic analysis of more than 30 agroforestry models practiced in different agroclimatic conditions of the country reveal a high benefit: cost ratio (1.5–3) and Internal Rate of Return (15–40 %). Thus, agroforestry has to play a decisive role not only in supply of timber products, thus saving the forest, but also to be extremely effective in meeting the other requirements including food security and requirements of day today life and also to mitigate climate change.

Agroforestry Research and Technology Development

India has been at the forefront of agroforestry research but still there is a long way to go. Agroforestry research in the ICAR system and other Indian institutes and universities has been in progress since early 1950s. As mentioned earlier, the ICAR launched the AICRP on agroforestry in 1983 representing all agroclimatic zones in the country. Besides ICAR Institutes and SAUs, the ICFRE also supports agroforestry research in education in various parts of the country. Private sector initiatives in agroforestry such as those by WIMCO, Bharatia Agro-Industries Foundation (BAIF), Indian Farms Forestry Development Cooperative (IFFDC), and Imperial Tobacco Company (ITC) are also worth mentioning. Some of the important research and development outcomes in agroforestry in India since organized research in agroforestry began in the early 1980s are listed below:

- ***Characterization of agroforestry systems*** in different agroclimatic zones of the country: By developing a D&D approach and using them for survey, a bench mark information base for major agroforestry systems has been prepared.
- ***Collection and evaluation of multipurpose tree*** species: Creation of arboretums in different agroclimatic zones.
- ***Tree selection and improvement***: Tree species on which tree improvement has focused include poplar (*Populus deltoides*), shisham (*Dalbergia sissoo*), neem (*Azadirechta indica*), semal (*Bombax ceiba*), subabul (*Leucaena leucocephala*), and species of *Casuarina* and *Eucalyptus*.
- Identification of preferred agroforestry tree species for different agroclimatic regions.
- Management practices for different agroforestry systems; Packages of practices have been standardized and their efficacies and economic returns (B:C ratios) worked out to show the

usefulness of the systems under specified agroecological and socio-economic settings.

- ***Development of site-specific agroforestry technologies***: Tree species must suited for rehabilitating problem areas such as degraded lands including salt—affected, waterlogged, and eroded lands, have been identified and management technologies for each developed.
- Design and testing of multi-enterprise farming system models for different agroclimatic situations for livelihood security of small and marginal farmers.
- Identification and development of climate resilient agroforestry systems involving diversified cropping systems and livestock population directly on farmers' fields.

The major thrust area on which agroforestry has focused today in India includes:

- Increased production of wood resources to meet the growing needs of households and industries.
- Environmental amelioration through the C-sequestration, bioremediation, and resource conservation.
- Enhancement of livelihood and employment avenues.
- Promotion of agroforestry adoption through developing decision support systems and other appropriate means.
- Exploration of new and under exploited species of high economic value, such as medicinal, aromatic, oil-yielding, etc.
- Testing and evaluation of marine algae or seaweeds for food, medicine and green manure in identified agroforestry systems.
- Rehabilitation of degraded salt-affected areas using good quality forages, MPTs, plantation crops, and plants of industrial application with due attention to quality assessment of the products obtained from the saline habitats.
- Development of halophytic crops involving fish, shrimp culture, and poultry with agroforestry systems, genetic improvement of salt-tolerant plants, raising nursery with saline water, and multiplication and conservation of useful genetic material.
- Development of sustainable and remunerative multi-enterprise farming systems for marginal and small farmers of different agroclimatic regions working under different stresses, through value addition and policy initiatives.
- Promotion of agroforestry-based industries.
- Transfer of technology transfer.
- Capacity building/human resource development.
- Support and linkages among national and international collaborative research programs.

Based on three decades of experience with agroforestry research and development we have also identified major constraints to the promotion of agroforestry in India which include:

- Lack of an enabling Agroforestry Policy.
- Long gestation periods of tree species.
- Lack of high yielding genetic planting material and silvicultural practices for indigenous tree species.
- Lack of access to credit against agroforestry plantations.
- Absence of marketing institutions.
- Absence of any organized agroforestry extension service.
- Restrictions in harvesting of some timber trees; the cumbersome procedures involved in getting permission for harvesting trees on private lands.

In this publication, the various chapters are compiled in such a way that a clear picture of various agroforestry systems found in different agroecological regions is presented. The regional system descriptions in the chapters present both little studied traditional practices as well as improved systems. Some systems are present across a number of climatic regions; for example salt-affected and waterlogged areas are present in all agroecological regions but the problems are of different nature in different regions. To deal with such cases, a separate chapter is included to present agroforestry approaches to dealing with salinity problems. Similarly, other problematic areas such as mines and ravine lands covered in separate chapter. Most of the degraded lands including salt-affected and waterlogged areas can be brought under vegetation through agroforetsry practices which control erosion, reduce runoff, improve *in situ* soil-moisture conservation, increase water-table, and also improve productivity and profitability. Agroforestry systems that provide excellent opportunities for carbon sequestration are a valuable climate change mitigation strategy, which needs to be incorporated into major policy documents. Concerted efforts are required to identify practical and cost-effective means through changes in land use practices using agroforestry as an option. These issues are discussed in a separate chapter on policy initiatives. Finally a synthesis chapter highlights the important initiatives taken at the level of researchers and policy makers, constraints faced, and directions for future efforts in agroforestry research.

This publication may prove quite useful for different stakeholders interested in agroforestry at global level in general and at country level in particular.

References

Alexander TG, Sobhana K, Balagopalan M, Mary MV (1980) Taungya in relation to soil properties, soil degradation and soil management. Research report 4. Kerala Forest Research Institute, Peechi

Blanford HR (1958) Highlights of one hundred years of forestry in Burma. Empire For Rev 37(1):33–42

Borthakur DN (1992) Agriculture of the North Eastern region with special reference to hill agriculture. Beece Prakashan, Guwahati

Conklin HC (1957) Hununoo agriculture. FAO Forestry Development Paper No.12. Rome, Italy

Dagar JC (1995) Agroforestry systems for the Andaman & Nicobar Islands. Int Tree Crops J 8:107–128

Dagar JC (2003) Biodiversity of Indian saline habitats and management and utilization of high salinity tolerant plants with industrial application for rehabilitation of saline areas. In: Alsharhan AA, Wood WW, Gouie AS, Fowler A, Abdellatif EM (eds) Desertification in the third millennium. Swets and Zeitlinger Publishers, Lisse, pp 151–172

Dagar JC (2012) Utilization of degraded lands/habitats and poor quality water for livelihood security and mitigating climate change. Indian J Agrofor 14(1):1–16

Dagar JC, Pathak PS (2005) Grassland dynamics and their management. Range Manag Agrofor 26(1):7–31

Dagar JC, Mongia AD, Bandyopadhyaya AK (1991) Mangrooves of Andaman and Nicobar Islands. Oxford and IBH Publishing Co. Pvt. Ltd, New Delhi, pp 166

Dagar JC, Sharma HB, Shukla YK (2001a) Raised and sunken bed technique for agroforestry on alkali soils of northwest India. Land Degrad Dev 12:107–118

Dagar JC, Singh G (1993) Afforestration and agroforestry for salt affected soils. Indian Rev Life Sci 13:21–24

Dagar JC, Singh G, Singh NT (2001b) Evaluation of forest and fruit trees used for rehabilitation of semi-arid alkali/sodic soils in India. Arid Land Res Manag 15:115–133

FSI (1997) State of Forest Report. Forest Survey of India, Dehradun, Ministry of Environment and Forests, Government of India

FSI (2011) State of Forest Report. Forest Survey of India, Dehradun, Ministry of Environment and Forests, Government of India

Grewal SS (1992) Watersheds management project, Relmajra (Punjab). Annual Report, 1991–92, CSWCRTI, Dehra Dun, pp 142–143

Hailey L (1957) An African survey. Oxford University Press, Oxford

Hirekerur LR, Lal S, Pofali RM (1991) Approaches to development planning for degraded lands. In: Abrol IP, Dhruvanarayana VV (eds) Technologies for wasteland development ICAR, New Delhi, pp 43–58

King KFS (1987) The history of agroforestry. In: Nair PKR, Stepler HA (eds) Agroforestry: a decade of development. ICRAF, Nairobi, pp 1–11

Kumar BM (2006) Carbon sequestration potential of tropical homegardens. In: Kumar BM, Nair PKR (eds) Tropical homegardens: a time tested example of sustainable agroforestry. Springer, The Netherlands, pp 185–204

Kumar BM (2008) Krishi gita (agricultural verses – trans) A treatise on indigenous farming practices with special reference to malayalam desam (Kerala). Asian Agri-History Foundation, Secunderabad, p 111

Kumar BM (2010) Species richness and aboveground carbon stocks in the homegardens of central Kerala, India. Agric Ecosyst Environ 140:430–440

Kumar BM, Kunhamu TK (eds) (2011) Quarter century of agroforestry research in Kerala: a compendium of research publication. Kerala Agricultural University, Thrissur, p 404

Logan W (1906) Molabar Mannual, vol 2, 2nd edn. Asian Educational Services, Madras, p 772

Lundgren BO, Raintree JB (1982) Sustained agroforestry. In: Nestel B (ed) Agricultural research for development: potentials and challenges in Asia. ISNAR, The Hague, pp 37–49

Mann HS, Saxena SK (1980) Khejri (*Prosopis cineraria*) in the Indian desert: its role in agroforestry. CAZRI, Jodhpur

Mateer (1883) Native life in Travancore. WH Allen, London, pp 450

Mathur BL (1995) Kana bundi indigenous method of controlling wind erosion, Honey Bee (12):7

Ministry of Agriculture (2010) Agricultural statistics at a glance. Directorate of Economics and Statistics, Ministry of Agriculture, Government of India

NAAS (2010) Degraded and wastelands of India: status and spatial distribution. National Academy of Agricultural Sciences (NAAS), New Delhi, pp 158
Nair PKR (ed) (1989) Agroforestry systems in the tropics. Kluwer, Dordrecht
Nair PKR (1993) An introduction to agroforestry. Kluwer Academic Publishers, The Netherlands
Nair PKR, Dagar JC (1991) An approach to developing methodologies for evaluating agroforestry systems in India. Agrofor Syst 16:55–81
North Eastern Council (1997) Basic statistics for north eastern region. North Eastern Council, Shillong, Meghalaya, India
NRCAF (2013) NRCAF Vision 2050. National Research Centre for Agroforestry, Jhansi
Pandey CB, Rai BR, Singh L, Singh A (2007) Homegardens of Andaman and Nicobar, India. Agric Syst 92:1–22
Pathak PS, Dagar JC (2000). Traditional agroforestry sytems: an overview. In:Soni P, Srivastava AK (eds) Landmarks of botany in India. Suriya International Pub., Dehradun, pp 26–49
Pathak PS, Gupta SK, Singh P (1995) IGFRI approaches: rehabilitation of degraded lands. Bulletin Indian Grassland and Fodder Research Institute, Jhansi, pp 1–23
Planning Commission (2001) Report of the Task Force on Greening India for Livelihood Security and Sustainable Development. Planning Commission, Government of India, New Delhi, pp 254
Puri S, Nair PKR (2004) Agroforestry research for development in India: 25 years of experiences of a national program. Agrofor Syst 61:437–452
Raghavan MS (1960) Genesis and history of the kumari system of cultivation. In: Proceedings of the 9th silviculture conference 1956 Forestry Research Insitute (FRI), Dehradun
Rai P (2012) Sylvipastoral system for livestock production: performance of small ruminants. Indian J Agrofor 14(1):23–28
Ram J, Dagar JC, Lal K, Singh G, Toky OP, Tanwar RS, Dar SR, Mukesh K (2011) Biodrainage to combat water logging, increase farm productivity and sequester carbon in canal command area of north-west India. Curr Sci 100 (11): 1673–1680
Ramakrishnan PS (1992) Shifting agriculture and sustainable development: an interdisciplinary study from north eastern India. UNESCO MAB series, Parthenon Publ, Carnforth, Lanes, Paris (Republished by Oxford Univ Press, New Delhi in 1993)
Randhawa MS (1980) A history of Indian agriculture: eighth to eighteenth century, vol 2. Indian Council of Agricultural Research, New Delhi, p 358
Rawat JK, Dasgupta SS, Kumar RS (2004) Assessment of trees outside forest based on remote sensing satellite data. Forest Survey of India, Dehradun, p 9
Raychaudhuri SP, Roy M (1993) Agriculture in ancient India. ICAR, New Delhi
Singh G, Singh NT, Tomar OS (1993). Agroforestry in salt affected soils. Technical Bulletin No.17, CSSRI, Karnal
Singh K, Chand P (2012) Above-ground outside forest (TOF) phytomass and carbon estimation in the semi-arid region of southern Haryana: a synthesis approach of remote sensing and field data. J Earth Syst Sci 121(6):1469–1482
Tomar OS, Gupta RK, Dagar JC (1998) Afforestation techniques and evaluation of different tree species for waterlogged saline soils in semiarid tropics. Arid Soil Res Rehabil 12(4):301–316
Tomar OS, Minhas PS, Sharma VK, Gupta RK (2003a) Response of nine forage grasses to saline irrigation and its schedules in a semi-arid climate of north-west India. J Arid Environ 55:533–544
Tomar OS, Minhas PS, Sharma VK, Singh YP, Gupta RK (2003b) Performance of 31 tree species and soil condition in a plantation established with saline irrigation. For Ecol Manag 177:333–346
Tomar OS, Dagar JC, Minhas PS (2010) Evaluation of sowing methods, irrigation schedules, chemical fertilizer doses and varieties of *Plantago ovata* Forsk to rehabilitate degraded calcareous lands irrigated with saline water in dry regions of north western India. Arid Land Res Manag 24:133–151
Tomar JMS, Das A, Puni L, Chaturvedi OP, Munda GC (2012) Shifting cultivation in northeastern region of India—status and strategies for sustainable development. Indian For 138(1):52–62
Tripathy RS, Barik SK (2003) Shifting cultivation in North East India. In: Bhat BP, Bujarbaruah KM, Sharma YP, Patiram (eds) Proceeding of approaches for increasing agricultural productivity in hill and mountain eco system, ICAR Resaerch Complex for NEH Region, Umiam, Meghalaya, pp 317–322
Warriar MRR (1995) Wayanad in the middle ages. In:Johny KP (ed) Discover Wayanad – the green paradise. District Tourism Promotion Council, Kalpetta, pp 68–69

Soil Conservation and Ecosystem Stability: Natural Resource Management through Agroforestry in Northwestern Himalayan Region

2

S. D. Kashyap, J. C. Dagar, K. S. Pant, and A. G. Yewale

Abstract

Northwestern Himalayan region is spread between 28°43′-37°05′ N latitude and 72°40′-81°02′ E longitude covering an approximate area of 33 million ha. The major natural resources of Western Himalayas are water, forests, floral, and faunal biodiversity. Forests constitute the major share in the land use of the region covering more than 65 % of the total geographical area of the region. The estimated annual soil loss from northwest Himalayas is approximately 35 million tons, which is estimated to cost around US $32.20 million. The rural population in Himachal Pradesh, Jammu and Kashmir, and Uttarakhand constitutes 90.2, 75.2, and 74.3 %, respectively as compared to the national average of 72.2 %. The livestock population in the region has increased tremendously during last three decades and is 21.33 million against human population of 29.53 million (1:1.38). The agriculture including livestock continues to be the dominant sector despite the fact that the area is exposed to adverse and harsh geophysical and agri-silviculture conditions. Strategies by planting fodder trees or grasses in the waste/degraded lands (representing 7.9, 9.8 and 11.5 % of the geographical area in Himachal Pradesh, Jammu and Kashmir, and Uttrakhand, respectively), is needed for enhancing the fodder production. In addition, farm spaces on terrace risers and improved crop production technology coupled with integration of agroforestry will help in bridging the gap

S. D. Kashyap (✉) · K. S. Pant · A. G. Yewale
Department of Silviculture and Agroforestry,
College of Forestry, Dr. Y S Parmar University of
Horticulture and Forestry, Nauni, Solan,
HP 173 230, India
e-mail: sdkforestry@gmail.com

J. C. Dagar
Natural Resource Management Division,
Indian Council of Agricultural Research,
Krishi Anusandhan Bhawan—II, Pusa,
New Delhi 110012, India

J. C. Dagar et al. (eds.), *Agroforestry Systems in India: Livelihood Security & Ecosystem Services*, Advances in Agroforestry 10,
DOI: 10.1007/978-81-322-1662-9_2, © Springer India 2014

between demand and supply of the fodder. The indigenous agroforestry systems such as homestead (*kyaroo*), plantation crop combinations, scattered trees on farm lands/field bunds and bamboo grove, etc., are practiced by the farming community. The land management operations are predominated by different indigenous agroforestry practices which have proven potential and hold promise in alleviating the poverty among rural masses of this hilly region. The agroforestry systems provide unique opportunity for integration of different components in the farming systems, which help to optimize the ecosystem functioning and better management of land, water, and biological resources. These systems need to be further improved with suitable technological interventions considering the local population need, so that the socioeconomic status of the farming communities uplifted. Experiences gained in managing natural resources through well-tested agroforestry systems have been shared in this chapter.

Introduction

The Northwestern Himalayan Region (NWHR) comprises of three states of the Indian Republic viz, Jammu and Kashmir, Himachal Pradesh, and Uttrakhand. Geographically it spreads between 28°43′-37°05′ N latitude and 72°40′-81°02′ E longitude covering an approximate area of 33 million ha contributing about 10 % of total geographical area of the country. The region occupies the strategic position in the northern boundary of the nation and touches the international borders of Nepal, China, and Pakistan. Most of the area is covered by snow-clad peaks, glaciers of higher Himalaya and dense forest covers of mid-Himalaya. The region comparatively shows a thin and dispersed human population due to its physiographic conditions and poor infrastructure development. The Himalayas exhibit great diversity in climate, landforms, ethnicity, resource availability, and agricultural practices.

The rural population in Himachal Pradesh, Jammu and Kashmir, and Uttrakhand constitutes 90.2, 75.2, and 74.3 %, respectively, as compared to the national average of 72.2 % (Census of India 2006). Agriculture, including livestock continues to be the dominant sector despite the fact that the area is exposed to adverse and harsh geophysical and agro-climatic conditions. Climate of the region is conducive for growth of a large variety of plants ranging from tropical to temperate due to different altitudinal ranges varying from 100 m above mean sea level (amsl) to more than 4000 m amsl, i.e., subtropical to cold temperate alpine zone. The region is the natural abode of large number of medicinal and aromatic plants and the value of medicinal herbs from forests is enormous. These medicinal resources are harvested as raw material from wild sources and majority comes from Himalayan region, i.e., temperate, subalpine, and alpine zones.

The Himalayan region has unique advantage and competitive edge over the adjoining states in the plains, i.e., Punjab, Haryana, and Uttar Pradesh due to diverse agro-climatic conditions for cultivation of off-season vegetables, temperate fruits, aromatic rice, and medicinal and aromatic plants, besides the huge potential for organic farming. The Himalayas have also contributed toward the formation of fertile plains. The estimated annual soil loss from northwest Himalaya is approximately 35 million tons, which is estimated to cost around US $32.20 million (VPKAS 2011). The land management operations are predominated by different indigenous agroforestry practices which have proven potential and hold promise in alleviating the poverty of hill people. The region is bestowed with rich natural resources including biodiversity; therefore, an attempt has been made in this chapter to justify the management of these resources through agroforestry which is a viable and most sustainable option for the fragile ecology of the region.

Ecology and General Features

The Western Himalayan region is a part of Hindu Kush Himalayas and is characterized not only by ecological fragility but also by a deep and historical geopolitical sensitivity (Stone 1992). There is considerable variation in climate, physiography, soil, and vegetation between the outer and inner Himalayas. Vegetation largely is controlled by altitude. The Western Himalayas are classified into Lesser Himalayas and the Greater Himalayas. The Lesser Himalayas lie in the north of Shiwalik hills. The mountain ranges in this region are usually 50–100 km wide and 1,000–5,000 m high. Dhauladhar range in Himachal Pradesh, Pir Panjal in Jammu and Kashmir, and Mussoorie in Uttrakhand are some of the important hill ranges. The Western Himalayan region has been divided into four agro-ecological zones (Fig. 2.1) as described in Table 2.1.

Climate

The Western Himalayan Region mainly experiences two seasons namely winter and summer. The average summer temperature in the southern foothills is about 30 °C and the average winter temperature is around 18 °C. In the middle Himalayan valleys, the average summer temperature remains around 25 °C while the winters are really cold. On the higher regions of the middle Himalayas, the summer temperature is recorded at around 15–18 °C while the winters are below freezing point. The region above 4,880 m amsl is below freezing point and is permanently covered with snow. During winters, the snowfall is heavy while the summers are much more mild and soothing. The Himalayan alpine climate varies according to the altitude. The climatic conditions change very quickly in the Himalayan region due to change in the altitude. The climate here is very unpredictable and dangerous too. The people in regions of Ladakh and Zanskar situated in the north of main Himalayan range are unaware of the monsoon season as the average annual rainfall is only a few centimeters (in the form of snow precipitation) resulting in very low humidity

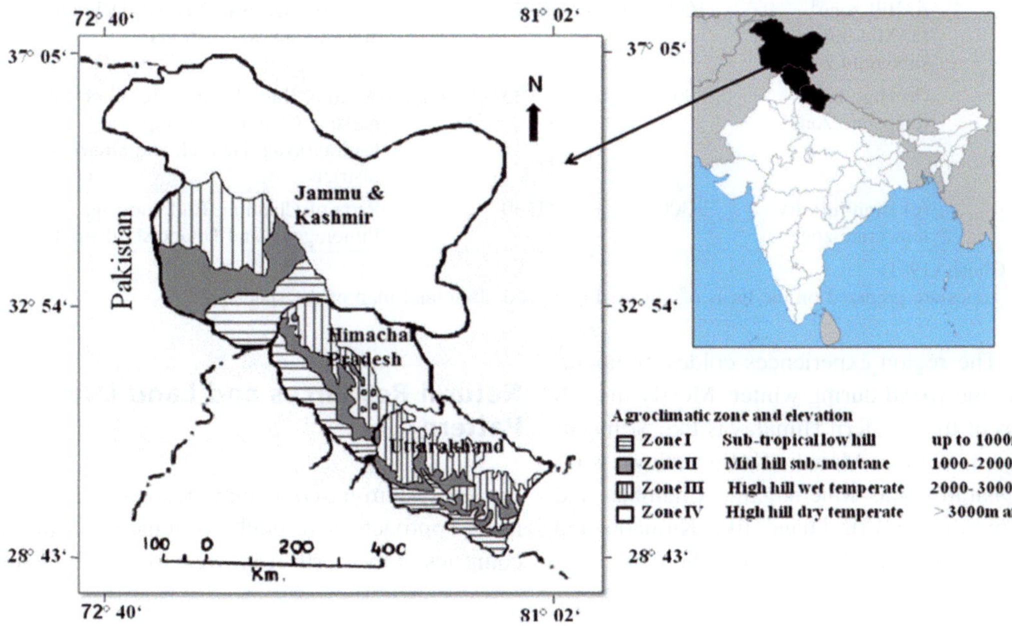

Fig. 2.1 Northwestern Himalayan region with agro-climatic zones

Table 2.1 Area under different agro-climatic zones of Northwestern Himalayan states

States	Agro-climatic zones	Tentative altitudinal ranges (m amsl)	Approximate area (000 ha)	Districts/parts of the state
Jammu and Kashmir	i. Low altitude subtropical zone	≤1000	1551	Jammu, Kathua, Udhampur, and parts of Doda and Rajouri districts
	ii. Mid to high altitude intermediate zone	1000–2000	1482	Poonch and major part of Rajouri, Anantnag, Pulwama, part of Doda district
	iii. Valley temperate zone and mid to high altitude zone	2000–3000 (<1200 Kashmir valleys)	1472	Srinagar, Baramulla, Kupwara and part of Anantnag, Pulwama, Budan, and Poonch
	iv. Cold arid zone (High hills dry temperate zone)	>3000	17719	Kargil, Leh, upper riches of Doda, Anantnag, Baramulla. Kupwara districts
Himachal Pradesh	i. Low hills subtropical zone	≤1000	924	Una, Hamirpur, Bilaspur and parts of Kangra, Chamba, Mandi, Sirmour, and Solan districts
	ii. Mid-hills subhumid zone	1000–2000	573	Part of Chamba, Solan, Shimla, Mandi, Sirmour districts
	iii. High hills temperate wet zone	2000–3000	1821	Parts of Chmaba, Kangra, Mandi, Sirmour, Shimla, and Kullu districts
	iv. High hills dry temperate zone	>3000	2249	Kinnaur, Lahaul Spiti, and parts of Kangra and Chamba districts
Uttrakhand	i. Babhar and tarai zone and subtropical low hills zone	<1000	1216	Haridwar, Udham Singh Nagar, and parts of Nanital, Dehradun, and Paurigarhwal districts
	ii. Hill zone[a] (a) Mid-hill subhumid zone	1000–2000	667	Parts of Garhwal, Nanital, Dehradun, and Champawat districts
	(b) High hills temperate zone	2000–3000	2935	Almora, Bageshwar, Tehri Garhwal, parts of Chamoli, Champawat, Rudraprayag, Nanital, and Uttarkashi districts
	(c) High hill dry temperate zone	>3000	1230	Parts of Chamoli, Rudraprayag, Pithoragarh, and Uttarkashi districts

Source Ghosh (1981)

[a] These zones are prepared on the basis of physiographic and altitudinal map of the states

levels. The region experiences coldest temperatures in the world during winter. Mostly the hill stations of the Western Himalayas like Srinagar, Pahalgam, Shimla, Manali (Kullu valley), Kangra, Dharamshala, Maclodganj, Chamba, and some regions in Uttrakhand like Kumaon and Garhwal experience the monsoon showers.

Natural Resources and Land Use Pattern

An ecosystem-based natural resource management approach is difficult to achieve as many countries in the north from west to east share the resources of this mountain system; whereas Indian Western Himalayan region resources are shared by many Indian states. The Himalayas are full of natural wealth, both renewable and nonrenewable resources. The major natural resources of Western Himalayas are water, forests, floral, and faunal biodiversity; and climate, soil, and ecology are the major determinants of the hill farming systems in different agro-climatic zones for livelihood subsistence.

The conservation of natural resources is not only essential for the food and livelihood security of the Himalayan states but also for the entire country. The hydrological potential of these states consists of vast and rich water resources as glaciers, rivers, and lakes. The high altitude areas of Lesser and Greater Himalayas are covered with glaciers and snow fields and are the origin of number of perennial rivers which heavily drain into Indus and Gangetic basin and form a most fertile Indo-Gangetic plain region of the country known as "food bowl of India." The hydropower and irrigation potential of these two important rivers is given in Table 2.2.

In Uttrakhand 31 natural lakes covering an area of about 300 ha and eight large-sized manmade reservoirs in Tehri and Udham Singh Nagar districts are covering an area of 20,075 ha. The Tehri dam is the largest dam in Uttrakhand followed by Sharda reservoir with 6,880 ha water area and Nanak Sagar reservoir with water area of 4,084 ha is the third largest in the state. These all reservoirs are owned by irrigation department and are used for irrigation purpose to enhance agricultural production. In Himachal Pradesh, the water from Beas and Sutlej rivers has been stored in Pong dam and Bhakra Govind Sagar reservoirs having capacity of 7,290 and 9,621 million cubic meters, respectively for irrigation and power generation. It is major source of irrigation to Punjab, Haryana, and Rajasthan. The catchment area of Ganga in India is approximately 863 thousand km^2 which covers about 26.2 % of the total geographical area of the country particularly of Northern States of Indian Territory and is considered as most fertile region of the world (State Profile Himachal Pradesh 2010).

Forests constitute the major share in the land use of Northwestern Himalayan region covering an area of about 1101, 2023, and 3486 thousand ha in Himachal Pradesh, Jammu and Kashmir, and Uttarakhand, respectively (Ministry of Agriculture 2009). Forests are the second largest natural renewable resources after water. The forest cover and canopy densities has a major role to maintain the hydrological regime in the region as well as to feed the adjoining plain areas for agricultural production. The very dense forests having canopy density more than 70 %

Table 2.2 Catchment area (km^2) and hydropower potential of different rivers in Himachal Pradesh[a]

Major basin	Tributary	Area (km^2)	Hydropower potential (MW)
A. Indus			
	Chenab	7,500	3,032
	Ravi	5,451	2,159
	Beas	20,402	4,604
	Sutlaj	20,000	10,355
	Total A	53,353	20,150
B. Ganga			
	Yamuna	2,320	592
	Total A + B	55,673	20,742
C. Mini-micro projects			
			750
Grand Total		55,673	21,492

Sources State Profile Himachal Pradesh (2010)

[a] Data for Jammu and Kashmir and Uttrakhand not available

and moderately dense forests having canopy density 40–70 % are generally found in high hill temperate wet and mid-hill subhumid zones of the region making it more humid with large number of natural springs due to deep soil percolation and interception of rain water into the soils of these forests. It also maintains optimum temperature for fruit and off-season vegetable production, conservation of biodiversity in the region and to maintain the hydrological potential of the perennial rivers round the year.

The growing stock of trees outside the forest land (ToF) under agroforestry or social forestry has played significant role to enhance the GDP of the country from 1 to 1.70 % (ICFRE 2010). The tree cover in the region increased significantly during last three decades when ICAR initiated All India Coordinated Research Project on Agroforestry (AICRP–AF) in collaboration with International Council for Research in Agroforestry (ICRAF), Nairobi, Kenya during 1982–1983 and farmers were encouraged to grow fodder trees on the bunds to meet the top-fodder demand during the lean period and timber trees to enhance their farm income and to meet their domestic demand. At present the total area under ToF in the region is 7,815 km^2 and total growing stock is 190 million cum (Table 2.3).

The research under AICRP-AF during the last three decades; number of indigenous agroforestry practices and multipurpose trees species (MPTs) are identified in different agro-climatic zones of the states and briefly discussed under various agroforestry systems, and research and development initiatives taken in agroforestry.

The livestock population in the region has increased tremendously during last three decades and is 21.33 million against human population of 29.53 million. In order to feed 21.33 million cattle in Northwestern Himalayan states, 1.35 million tons of fodder are required. The region does not produce adequate fodder and therefore, faces 54 % deficit in green fodder and 34 % deficit in dry fodder. Strategies by planting fodder trees or grasses in the waste/degraded lands which represent 7.9, 9.8, and 11.5 % of geographical area in Himachal Pradesh, Jammu and Kashmir, and Uttrakhand, respectively are needed for enhancing the fodder production. It is also needed to plant grasses, fodder bushes, and wild fruits on the forest floors to feed the wildlife and migratory shepherd (sheep and goats) during their journey from alpines to the lower hills (VPKAS, 2011). In addition, farm spaces on terrace risers and improved crop production technology coupled with integration of agroforestry, will help in

Table 2.3 Estimated production of forest resources in northwestern Himalayan region

Parameters	Jammu and Kashmir	Himachal Pradesh	Uttarakhand	Total
Annual estimated production of timber from forests (000 cum)	55	231	251	537
Annual production of fuel wood from forests (000 tons)	20	0.05	50	70.05
Growing stock TOF (million cum)	147.75 (6550)	21.15 (623)	20.92 (642)	189.82 (7815)
Annual production of timber from TOF (000 cum)	830	939	697	2466
Annual availability of fuelwood from TOF (000 tons)	365	290	297	952
Total live stock (million)	10.99	5.22	5.12	21.33
Livestock dependent on forests (million)	3.02	5.18	4.88	13.08

Source FSI (2011), Census of India (2011), HPFS (2011)
TOF Trees outside forest area in the form of social forestry, agroforestry, homestead, etc
Value in parenthesis represents the area under TOF in km^2

bridging the gap between demand and supply of the fodder.

Amongst other nonrenewable resources are deposits of boron, lead, lithium, coal, chromium, ores of iron, copper, tungsten, zinc, and deposits of building material like limestone, dolomite, and marble. These deposits occur across the length and breadth of the Himalayas cutting across international boundaries. The Himalayas have substantial mineral wealth due to which numbers of cement industries are coming up during last three decades in the region. The Himalayas present a storehouse of biodiversity, where flora and fauna vary extensively with climate diversity from one region to the other and this biodiversity is used for developing new varieties/hybrids in agriculture and horticultural crops to enhance the productivity.

Agriculture

There are several important valleys where intensive agriculture is practiced. These include Kangra and Kullu in Himachal Pradesh and Kashmir valley in Jammu and Kashmir and Doon valley and Babhar and Tarai region in Uttrakhand. In these low hills, agricultural fields are terraced in some parts except plain areas and fruit plantations are raised along with several arable crops, such as paddy, maize, pulses, wheat, oilseeds, potatoes, vegetables, etc. Cultivation is practiced up to 2500 m elevation. The contribution of agriculture and allied sectors to net state domestic product at factor cost (at current prices) in the region ranged between 16.8 and 28.9 % in the region (RBI 2010). The average operational holding in Jammu and Kashmir, Himachal Pradesh and Uttrakhand are 0.67, 1.07, and 0.95 ha, respectively, against the national average of 1.33 ha. The irrigated area in respective states is 42, 19, and 45 %, respectively of the net area sown (FAI 2010). If we consider only hilly region, these figures are much lower than the plains (Ramakrishna et al. 2000).

The crop production systems prevailing in Northwestern Himalayas are based on cereal crops, vegetables, horticulture under different agroforestry practices. Livestock is the integral part of farming in almost all three western Himalayan states. Wheat (*Triticum aestivum/ durum*), paddy (*Oryza sativa*), maize (*Zea mays*), *Hordeum vulgare* (barley), *Elusine coracana* (mandua/ragi), *Pennisetum typhoides* (pearl millet), barnyard millet (*Echinochloa crus-galli*), oats (*Avena sativa*), *Amaranthus caudatus* (ramdana), rice-bean (*Vigna umbellata*), *Fagopyrum esculentum* (buckwheat), lentil (*Lens culinaris*), and soybean (*Glycine max*) are the major field crops. The major vegetable crops are Knol-khol (*Brassica oleracea* var. *gongylodes*), cabbage (*B. oleracea* var *capitata*), cauliflower (*B. oleracea* var. *botrytis*), turnip *Brassica rapa*), radish (*Raphanus sativus*), carrot (*Daucus carota* var. *sativa*), onion (*Allium cepa*), pea (*Pisum sativum*), spinach (*Spinacia oleracea*), garlic (*Allium sativum*), tomato (*Solanum lycopersicum*), chillies (*Capsicum annuum var. acuminatum*), capsicum (*Capsicum annuum*), French bean (*Phaseolus vulgaris*), brinjal (*Solanum melongena*), bottle gourd (*Lagenaria siceraria*), cucurbits (*Cucumis melo, C. sativus*), bitter gourd (*Momordica charantia*), pumpkin (*Cucurbita maxima, C. pepo*), *Brassica campestris, B. oleracea)* and potato (*Solanum tuberosum*). Fruits such as apple (*Malus pumila*), peach (*Prunus persica*), apricot (*P. armeniaca*), plum (*P. domestica*), almond (*P. amygdalus*), common plums (*Spondias* spp), etc., are the major fruit crops. Floriculture is also fast emerging as an important cash generating activity of the production systems in certain areas. Fruit orchards of several species are found in the hills of Himachal Pradesh, Jammu & Kashmir, and Uttrakhand. The fruit trees like plum, peach, apricot, etc., are grown up to 1500 m elevation. Apple orchards are found above 1500 m elevation. The terraced agricultural fields are in the form of narrow strips whose width varies from 2 to 5 m. Plantations of trees on agricultural croplands were not common in the past as enough forests were available in the vicinity, however, after clearing the forests for plantation of fruit trees as orchards in the hills during the last 4–5 decades has created acute shortage of firewood and fodder and has compelled the farmers to grow trees on their farmlands as a part of their farming systems to meet their daily need of fuel, fodder, and timber.

Indigenous Agroforestry Systems

Various traditional agroforestry system occurring in different agro-climatic zones of North-western Himalayan hill states along with their functional units are given in the Table 2.4.

Some Important Indigenous/Traditional Systems of the Region

Homesteads (*Kyaroo*)

A homestead (*kyaroo)* is an operational farm unit in which a number of tree species for fodder, timber, and fuelwood are raised along with livestock, poultry, and/or fish mainly for the purpose of satisfying the farmers' basic needs. It is locally called as *kyaroo* in Kangra, Hamirpur, and Jammu areas, but in general it is developed and managed by every farmer in the region to meet their day-to-day requirements. In a *kyaroo*, multiple crops are present in a multitier canopy configuration. The fodder trees such as *Celtis australis* (Khirak), *Bahaunia variegata* (Kachnar), *Grewia optiva* (Beul), and bamboo species particularly major bamboo (*Dendrocalamus hamiltonii*) and *lathi* bamboo (*D. strictus*) for both timber and fodder are managed in the upper storey, whereas middle storey is constituted of bushes like medicinal *Adhatoda vasica*, *Vitex negundo*, etc. The fruit trees such as pear, plum, lemon and citrus, etc., are grown for domestic use.

During the rainy season cucurbits (vines) are grown along with taro (*Colocasia esculanta*), elephant foot yam (*Dioscoria* spp), and turmeric (*Curcuma domestica*). However, wide variation in the intensity of tree cropping is noticeable in different places. This is generally attributed to the differences in socioeconomic conditions of the households and their response to externally determined changes, particularly prices of inputs and products, dependence on land and tenurial conditions, etc. (Verma 1998).

Plantation Crop Combination

Plantation crops play a major role in national economics because these generate value added goods for the international markets. The important plantation crop of the Himalayan region is tea. Traditionally, tea is grown on waste and marginal lands in association with indigenous forest tree species. In Himachal Pradesh, tea gardens in Kangra, Palampur, and Baijnath valleys are managed under the canopy of *Albizia chinensis*, which not only nurses the tea plants by fixing the atmospheric nitrogen in its roots but also provides shade for the development and maintenance of new tenders. The leaf litter of *Albizia* trees also adds nutrients to the soil and during lean period the trees are used for fodder.

Bamboo Groves

The cultivation of bamboo (*Dendrocalamus hamiltonii, D. structus,* and *Bambusa nutans*) is a common practice in agricultural holdings alongside the streams and irrigation/drainage channels and on the agriculture field along *Nalas* and *Choes.* Bamboos are extensively used in building small farmhouses, cow sheds, piggery enclosures, baskets, mates, fishing rods, hookah pipes, and various household items for daily use (Fig. 2.2), string making, and also used as water conveyers for irrigation/drainage system. Bamboo leaves serve as an excellent winter fodder for cattle. Bamboo stumps/culms also protect the water channels from erosion. This system is limited only to the high rainfall subtropical and mid-hill moist areas where sufficient water is available to grow bamboo, e.g., Palampur, Hamirpur, Una, and Bilaspur areas of Himachal Pradesh; Jammu area of Jammu and Kashmir; and Dehradun area of Uttrakhand are well known for bamboo groves (Verma 1998).

Table 2.4 Traditional agroforestry systems in North western Himalayan region

Zone	System	Functional units
(A) Jammu and Kashmir		
(i) Kashmir valley		
Zone I 1000–2000 m amsl	Agri-silviculture in plains (irrigated)	Forest trees: *Populus deltoides, Populus nigra, Salix alba, Robinia pseudoacacia, Ailanthus altissima, Aesculus indica,* etc.
		Crops: Knol-khol, cabbage, cauliflower, turnip, radish, carrot, onion, pea, spinach, garlic, tomato, chillies, capsicum, French bean, bottle gourd, cucurbits, bitter gourd, pumpkin, potato, etc.
Zone II 2000–2500 m amsl	Agri-silviculture on slopes (mid-hills or foot hills of mountains)	Forest trees: *Robinia pseudoacacia, Ailanthus altissima, Aesculus indica, Populus nigra, Salix alba,* etc.
		Crops: Maize, mustard, carrot, turnip, radish, etc.
Zone III >2500 m amsl	Boundary plantation (Low lying area)	Forest trees: *Populus deltoides, Populus nigra, Salix alba, Aesculus indica,* etc.
		Crops: Paddy, oats, mustard, wheat, etc.
	Forest & fruit trees- based cropping systems (High altitude, rainfed)	Forest trees: *Populus ciliata, Populus nigra, Salix alba, Robinia pseudoacacia, Ailanthus altissima, Aesculus indica,* etc.
		Horticultural trees: Pomegranate (*Punica granatum),* walnut, apricot, peach, pear, apple, almond, etc.
		Crops: Tomato, brinjal, chillies, pea, brassica, cauliflower, cabbage, radish, turnip, onion, oats, wheat, etc.
	Fruit and forest trees based pastoral systems (on the slopes, rainfed)	Forest trees: *Populus deltoides, Populus nigra, Salix alba, Robinia pseudoacacia, Ailanthus altissima, Aesculus indica,* etc.
		Horticultural trees: *Punica granatum*, walnut, apricot, peach, pear, apple, almond, etc.
		Grasses: Oats, *Festuca pratense, Trifolium* spp., *Dactylis glomerata,* etc.
	Homesteads	Forest trees : *Populus deltoides, Populus nigra, Platanus orientalis, Salix alba, Robinia pseudoacacia, Ailanthus altissima, Aesculus indica,* etc.
		Fruit trees: *Malus domestica* (apple), *Juglans regia* (walnut), *Pyrus communis* (pear), *Prunus armeniaca* (apricot), *Prunus amygdalus* (almond), peach (*P. persica*), Cherry (*Prunus avium*) Pistachio nut *(Pistacia vera),* pomegranate (*Punica granatum*), etc.
		Vegetable crops: Cabbage, turnip (*Brassica rapa*), radish, spinach, carrot, onion, peas, etc.

(continued)

Table 2.4 (continued)

Zone	System	Functional units
(ii) Jammu		
Zone I <1000 m amsl Subhumid	Agri-silviculture (subhumid)	Forest trees : *Acacia nilotica, Ailanthus excelsa, Albizia lebbek, Grewia optiva, Bauhinia variegata, Ziziphus mauritiana, Dalbergia sissoo, Dendrocalamus strictus, D. hamiltonii, Melia Azedirach, Morus alba, Toona ciliata, Populus deltoides, Sesbania aegyptica, Leucaena leucocephala*
		Crops: Wheat, maize, pulses, millets, medicinal, and aromatic plants
Zone II and III 1000–2500 m amsl High hill temperate zone	Agri-silviculture	Forest Trees: *Populus ciliata, Salix alba, Robinia pseudoacacia, Alnus nepalensis, Quercus leucotrichophora, Ulmus wallichiana, Celtis australis,* etc.
		Crops: Wheat, oat, maize, medicinal and aromatic plants, *Dioscorea* spp.
	Fruit and forest trees-based cropping systems as well as pastoral systems	Horticulture trees : *Mangifera indica* (mango), *Litchi chinensis* (litchi), *Psidium guajava* (guava), *Carica papaya* (papaya), *Emblica officinalis* (aonla), *Citrus* spp. (lemon, kagzi lime, orange, chakotra, galgal, mosumbi), etc.
		Forest trees : *Populus deltoides* (poplar), *Dalbergia sissoo* (shisham), *Albizia lebbek* (siris), *A. chinensis* (black siris), *Ulmus villosa* (Elm), *Toona ciliata* (toon), *Acacia catechu* (khair), *Olea ferruginea, Anogeissus latifolia* (dhawda), *Grewia optiva* (beul), *Morus leavigata* (Toot), *Bauhinia variegata* (kachnar), *Santalum album* (Chandan), *Pinus roxburghii* (Chir pine), *Terminalia* spp., *Emblica officinalis* (aonla), *Sapindus mukorossi* (soap nut), Bambusa species, etc.
		Crops: Wheat, maize, soybean, *Cicer arietinum* (gram), *Lens esculenta* (lentil), *Brassica* spp. (mustard), *Vigna mungo* (urd), *Vigna umbellata* (rice-bean), *Cajanus cajan* (arhar), *Pisum sativum* (pea), *Abelmoschus esculentus* (okra), *Capsicum frutescens* (capsicum), *Phaseolus vulgaris* (French bean), *Brassica oleracea* (cauliflower), *Brassica capitata* (cabbage), etc.
		Grasses: *Setaria, Panicum,* Napier, etc.

(continued)

Table 2.4 (continued)

Zone	System	Functional units
(B) Himachal Pradesh		
Low hills subtropical zone <1000 m amsl	Fruit and forest trees-based cropping systems as well as pastoral systems	Forest trees: *Aesculus indica, Alnus nitida, Betula, Salix, Quercus leucotrichophora (*Ban oak*)*, Q. *semecarpifolia* (Kharsu oak), *Q. dilatata* (Moru oak), *Rhododendron arboretum* (Buransh), *Abies pindrow* (Himalayan fir), *Picea smithiana* (Himalayan spruce), *Pinus wallichiana, Pinus roxburghii, Cedrus deodara* (Deodar), *Ulmus wallichiana*, *Myrica nagi* (Kafal), Hill bamboo, etc.
		Horticulture trees: *Malus domestica* (apple), *Juglans regia* (walnut), *Pyrus communis* (pear), *Prunus armeniaca* (apricot), *Prunus amygdalus* (almond), Pistachio nut, etc.
		Crops: Wheat, maize, barley, *Elusine coracana* (mandua), pearl millet, *Pisum sativum* (pea), *Phaseolus vulgaris* (French bean), cauliflower, tomato, cabbage, *Amaranthus caudatus* (ramdana), potato, rice-bean, *Fagopyrum esculentum* (buckwheat), etc.
		Grasses: *Chrysopogon, Eragrostis, Apluda,* etc.
Subtemperate mid-hills (1000–1500 m amsl)	Fruit and forest trees-based cropping systems as well as pastoral systems	Forest Trees: Populus species, Juniperus spp., *Pinus gerardiana* (chilgoza pine), *Hippophae, Betula, Robinia pseudoacacia, Salix,* etc.
		Horticulture trees: *Malus pumila* (apple), *Prunus armeniaca* (apricot), almond, Pistachio nut, etc. Crops: Wheat, barley, *Amaranthus* spp, potato, buckwheat, etc.
		Grasses:, *Agropyron semicostatum, Eragrostis, Deschampsia cespitosa, Helictotrichon, Poa,* etc.
Temperate high hills (1500–2500 m amsl)	Fruit and forest trees-based cropping systems as well as pastoral systems	Forest Trees: Populus species, *Juniperus* spp., *Pinus gerardiana* (chilgoza pine), *Hippophae, Betula, Robinia pseudoacacia, Salix,* etc.
		Horticulture trees: *Malus pumila* (apple), *Prunus armeniaca* (apricot), almond (*P. amygdalus*), Pistachio nut (*Pistacia vera*), etc.
		Crops: Wheat, barley, *Amaranthus,* potato, buckwheat, etc.
		Grasses:, *Agropyron semicostatum, Eragrostis, Deschampsia cespitosa, Helictotrichon, Poa,* etc.

(continued)

Table 2.4 (continued)

Zone	System	Functional units
Dry temperate zone (>2500 mamsl)	Fruit and forest trees based cropping systems	Forest trees: *Eucalyptus*, *Populus deltoides* (poplar), *Anthocephalus cadamba* (kadam), *Dalbergia sissoo* (shisham), *Tectona grandis* (teak), *Toona ciliata* (toon), *Acacia catechu* (khair), *Holoptelea integrifolia* (kanju), *Adina cordifolia* (haldu), etc.
		Horticulture trees: *Mangifera indica* (mango), *Litchi chinensis* (litchi), *Psidium guajava* (guava), *Artocarpus integerifolia* (jack fruit), *Carica papaya* (papaya), *Emblica officinalis* (aonla), *Citrus* sp. (lemon, kagzi lime, orange, chakotra, galgal, mosumbi), etc.
		Crops: Wheat, maize, rice, *Glycine max* (soybean), *Cicer arietinum* (gram), *Lens esculenta* (lentil), mustard, *Saccharum officinarum* (sugarcane), *Curcuma domestica* (turmeric), *Phaseolus mungo* (urd), *P. aureus* (mung), *Vigna umbellata* (rice-bean), *Cajanus cajan* (arhar), *Pisum sativum* (pea), *Abelmoschus esculentus* (okra), garlic, onion, *Capsicum frutescens* (capsicum), *Phaseolus vulgaris* (French bean), cauliflower, *Brassica oleracea* var. *capitata* (cabbage), etc.
(C) Uttrakhand		
Tarai and Bhabar, submountain and low hills (up to 1000 m amsl)	Fruit and forest trees-based cropping systems as well as pastoral systems	Forest trees: *Aesculus indica*, *Alnus nitida*, *Betula*, *Salix*, *Quercus leucotrichophora*, *Q. semecarpifolia*, *Q. dilatata*, *Fraxinus micrantha*, *Rhododendron arboretum*, *Populus ciliata*, *Taxus baccata*, *Abies pindrow*, *Cedrus deodara*, *Myrica nagi*
		Horticulture trees: *Malus pumila* (apple), *Juglans regia* (walnut), *Pyrus communis* (pear), *Prunus armeniaca* (apricot), *Prunus amygdalus* (almond), Pistachio nut, etc.
		Crops: Wheat, maize, barley, *Elusine coracana* (mandua), *Paspalum* spp. (Jhingora), pearl millet, pea, French bean, cabbage, *Amaranthus caudatus* (ramdana), potato, rice-bean, buckwheat, etc.
		Grasses: *Eragrostis* spp., *Bromus inermis*, *Festuca arundinacea*, *Dactylis glomerata*, *Koeleria cristata* , *Cymbopogon distans*, *Chrysopogon royleanus*, *Danthonia cachemyriana*, *Digitaria decumbense*, *Festuca arundinacea*, *F. pratense*, *Phleum repens*, *Setaria anceps*, *Chrysopogon montanus*, *Lolium multiflorum*, *Panicum clandestinum*, *Chloris gayana*, *Lolium rigidum*, *Arundinella nepalensis*, *Phleum pratense*, *Poa annua*, etc.

(continued)

Table 2.4 (continued)

Zone	System	Functional units
Subtemperate mid-hills (1000–1500 m amsl)	Fruit and forest trees-based cropping systems as well as pastoral systems	Forest trees: *Hippophae* spp., *Betula, Salix,* etc.
		Horticulture trees: *Malus pumila* (apple), *Prunus armeniaca* (apricot), almond (*P. amygdalus*), Pistachio nut, etc.
		Crops: Wheat, barley, potato, buckwheat etc.
		Grasses: *Dichanthium* spp., *Koeleria cristata, Calamagrostis emodensis, Festuca lucida, Brachypodium sylvaticum, Trisetum spicatum, Andropogon tristis, Phleum repense, Agropyron semicostatum, Eragrostis* spp., *Deschampsia cespitosa, Helictotrichon virescens* and *Dactylis glomerata*, etc.
Temperate high hills (1500–2500 m amsl)	Forest and fruit trees-based cropping and pastoral systems	Forest trees: *Aesculus indica* (pangar), *Alnus nitida, Betula, Salix, Quercus leucotrichophora* (Banj), *Q. semecarpifolia* (kharsu), *Q. dilatata* (moru), *Fraxinus micrantha, Rhododendron arboretum* (Buransh), *Populus ciliata* (Poplar), *Taxus baccata* (thuner), *Abies pindrow* (raga), *Cedrus deodara* (deodar), *Myrica esculenta* (kaiphal)
		Arable crops: Wheat (Kaiphal), maize, barley, *Elusine coracana* (mandua), *Paspalum* sp (Jhingora), pearl millet, pea, French bean, cabbage, *Amaranthus caudatus* (ramdana), potato, rice-bean, *Fagopyrum esculentum* (buckwheat)
		Horticultural trees: *Malus pumila* (apple), *Juglans regia* (walnut), *Pyrus communis* (pear), *Prunus armeniaca* (apricot), *Prunus amygdalus* (almond), Pistachio nut Grasses: *Aragrostis* spp, *Bromus inermis, Festuca arundinacea, Dactylis glomerata,, Kobretia* sp, *Cymbopogon distans, Chrysopogon royleanus, Danthonia cachemyriana, Digetaria decumbense, Festuca arundinacea, F. pratense, Phleum repens, Setaria anceps, Chrysopogon orientalis, Lolium multiflorum, Panicum cladystymum, Chloris gayana, Lolium vigidum, Arundinella nepalensis, Phleum pratense, Poa annua*
Dry temperate zone (>2500 m amsl)	Agri-silviculture Horti-silviculture Silvi-pastoral	Forest trees: Species of *Hippophae, Betula, Salix*, etc.
		Arable crops: Wheat, barley, amaranthus, potato, buckwheat
		Horticultural trees: *Malus pumila* (apple), *Prunus armeniaca* (apricot), almond, Pistachio nut Grasses: *Dichanthium annulatum, Koeleria cristata, Calamagrostis emodensis, Festuca lucida, Brachypodium sylvaticum, Trisetum spicatum, Andropogon tristis, Phleum repense, Agropyron semicostatum, Agrostis munroana, Deschampsia caespitosa, Helictotrichon virescens*, and *Dactylis glomerata*

Sources Tewari et al. (2007), Verma et al. (2007), Mughal and Khan (2007), Saleem and Gupta (2007)
For scientific names of crops see text just before the table

Fig. 2.2 Handmade bamboo artifacts for sale in the market of Himachal Pradesh

Sea-Buck-Thorn Based Agroforestry Systems in Cold Desert Areas

Sea-buck-thorn (*Hippophae rhamnoides*) is well known for its environmental benefits, desertification control, and land reclamation in fragile cold arid ecosystems. It fixes nitrogen by symbiotic association with microorganisms, e.g., *Frankia* to the tune of about 180 kg $ha^{-1}yr^{-1}$. Its plantation serves as windbreaks and also checks pedestrian traffic. Traditionally, it is planted around agricultural fields for protection of crops against stray animals and as a fuel wood because it is a potential energy plant in the region. The calorific value of dry sea-buck-thorn wood is 4,785 calories kg^{-1}. It is fast growing shrub and can stump every 3–5 years and hence reduce pressure on other native woody plants. In Ladakh region, sea-buck-thorn (*Hippophae*) is harvested from wild at large scale (Fig. 2.3) and fruit pulp was sold worth of INR 14 million in 2007 for making fruit juices as a trade name "Leh berry" which has medicinal value. Its plantation/cover area accounts for less than 5 % of its potential of the region and if fully utilized the shrub can change the entire economy of the region (Stobdan et al. 2008).

Research and Development Initiatives in Agroforestry

Identification of Multipurpose Tree Species

Surveys were conducted in different agro-climatic zones of the states for the identification of different multipurpose tree species (MPTs) grown on farmers fields and adjoining forest areas and more than 60 % land is found under the jurisdiction of forest department for their livelihood subsistence. The farmers keep the trees for fuel, fodder, and timber. The important MPTs with their climatic zones and altitudinal distribution alongwith method of planting and uses are given in Table 2.5.

Many trees such as *Grewia optiva* are frequently used in cropping systems by the farmers. Progressive farmers grow vegetables such as tomato (*Solanum lycopersicum*), brinjal (*Solanum melongena*), chillies (*Capsicum annuum*), etc., as remunerative crops along with trees (Fig. 2.4)

Fig. 2.3 Seabuck-thorn (*Hippophae rhamnoides*) fruits harvested by the local people for sale and domestic purposes. Inset showing the fruit branch of seabuckthorn

Fodder Values of Important Trees and Grasses

The fodder values of promising fodder trees and grasses have been studied for meeting farmers demand as well as to understand their preference to grow specific tree on their farm lands and presented in Table 2.6.

Soil Amelioration Potential Through Agroforestry

Northwestern Himalayan region is most important ecosystem of the country to meet the irrigation potential of the Indo-Gangetic plain region which is called as "Food Ball of India" as already stated. They are characterized by undulating barren hill slopes, undulating agriculture fields and erratic rainfall pattern. The hilly terrain is subjected to runoff and soil loss of varying degrees. The geological formation of Himalayas is young in age and very weak, having scanty vegetation in some of the areas as a result of which the area is subjected to high erosion. Biotic pressure is also important reason for causing soil degradation. The dispersion ratio is used to assess the erosion behavior of the soils. Kumar et al. (2002) reported that dispersion ratio was comparatively higher under cultivated lands (18.50–21.82) followed by orchards (13.25–14.46) and lowest under forests (11.64–12.93). The erosion ratio also followed the same trend (Table 2.7) for which the vegetative cover could be the reason.

More the vegetation higher the organic carbon content which results in higher water stable aggregates. Agroforestry practices consist of at least one woody component so it results in more addition of organic matter and increase in the organic carbon content. The canopy cover which increases over the unit area plays important role to limit the soil erosion by varying agents like water, wind, etc.

Barren lands recorded very high soil loss (86.05 t ha^{-1} yr^{-1}) followed by intensively cultivated areas (58.87 t ha^{-1} yr^{-1}), and fruit tree—based cropping system (18.27 t ha^{-1} yr^{-1}), degraded forests (17.76 t ha^{-1} yr^{-1}), and lowest soil loss (3.46 t ha^{-1} yr^{-1}) was recorded in dense forests (Table 2.8). The vegetation protects the soil against impact of falling rain drops, increases the roughness of soil surface, reduces the speed of runoff, binds the soil mechanically, and improves the physical, chemical, and biological properties of soil. Alternative land uses such as forest and fruit trees—based cropping systems on unstable slopes reduce losses and generate additional income to the farmers and have sufficient biomass for nutrient recycling (Sharma et al. 2002).

The silvopastoral and agri-silviculture systems provided best mechanism for conservation in Shiwalik hills. The soil loss and runoff under

Table 2.5 Multipurpose tree species (MPTs) of Northwestern Himalayan region

S. No.	Name of species	Climatic zone Altitudinal range (m amsl)	Method of planting	Uses
1.	*Abies pindrow*	2000–3500	SL	Timber, packing cases, plywood, pulp, and paper
2.	*Acacia auriculaeformis*	<1000	DS, SL	Timber, fuel, ornamental
3.	*A. catechu*	<1200	DS, SL	Fodder, timber, fuel, katha and cutch, gum, soil conservation
4.	*A. nilotica*	<700	DS, SL	Fuel, timber, tannin, gum, fodder
5.	*Acer acuminatum*	2400–2800	SL	Fuel, timber
6.	*A. caesium*	2400–2800	SL	Fuel, timber
7.	*A. pictum*	2400–2800	SL	Fuel, timber
8.	*A. oblongum*	1000—1800	DS,SL	Timber, agricultural implements
9.	*Aegle marmelos*	<1000	DS,SL	Fuel, fruit, gum, bark, and fruit medicine
10.	*Aesculus indica*	1200–2400	SL,DS	Timber, fodder, fruit, ornamental
11.	*Ailanthus altissima*	1500–2400		Timber, packaging cases, fodder
12.	*Albizia chinensis*	700–1200	DS, SL	Timber, fodder, fuel and shaded tree in tea gardens
13.	*A. lebbek*	700–1200	DS, SL	Timber, fodder, medicinal, fuel
14.	*A. procera*	110–700	DS, SL	Fodder, fuel, and timber
15.	*Alnus nepalensis*	700–1200	DS, SL	Timber, fuel, soil conservation
16.	*A. nitida*	700–1200	DS, SL	Timber, fuel, soil conservation
17.	*Anogeissus latifolia*	200–700	SL	Fuel and fodder
18.	*Artocarpus lakoocha*	200–500	DS, SL	Timber, fruit, fodder, Vegetables
19.	*Azadirachta indica*	<500	DS, SL	Timber, bark and seed—medicine, insecticidal, fertilizer
20.	*Bambusa balcoa*	<1000	Rhizomes, offsets	Constructional purposes, paper and pulp, cottage industry
21.	*B. tulda*	<1000	Rhizomes, offsets	Constructional purposes, mats, basket, young shoot pickled, and eaten
22.	*Bauhinia variegata*	<1200	SL, DS	Cherry gum, bark-dye and medicine, fodder, fuel, flower buds vegetables
23.	*Betula alnoides*	>3000	SL, DS	Plywood, furniture, tool handles
24.	*B. utilis*	>3000	SL	Bark papery (Bhojpatra), timber, fodder
25.	*Bombax ceiba*	<1200	DS, SL, BC	Match industry, plywood
26.	*Butea monosperma*	<500	DS, SL	Fodder, fuel, lac and dye making
27.	*Carpinus wiminea*	>2000	SL	Ornamental, bobbin manufacture
28.	*Cassia fistula*	<800	DS, SL	Fuel, timber
29.	*Celtis australis*	>700	SL, BC, DS	Timber, fuel, fodder, fruit-edible and also of medicinal value, sports goods, utensils
30.	*Cordia myxa*	<700	DS, SL	Kernel—medicine, timber for agriculture implements, pickle
31.	*Coriaria nepalensis*	1200–2500	SL	Boxwood, tannin, food for silk worm

(continued)

Table 2.5 (continued)

S. No.	Name of species	Climatic zone Altitudinal range (m amsl)	Method of planting	Uses
32.	*Dalbergia sissoo*	100–1200	DS, SL	Timber, furniture, sleepers, plywood, fuel, fodder
33.	*Dendrocalamus hamiltonii*	500–1800	DS, Rhizomes	Paper pulp, vegetable—young shoots, constructional
34.	*D. strictus*	300–1200	DS, SL, Rhizomes	Paper making, cottage industries, mathematical instruments
35.	*Emblica officinalis*	100–1000	DS,SL	Fruit, hair dyes, fodder, tannin, timber, fuel, myrobalan, medicine, hair oil
36.	*Erythrina suberosa*	400–1000	BC, SL	Light timber, cork, packing cases, hosts for black pepper, ornamental
37.	*Eucalyptus camaldulensis*	<1800	SL	Timber, charcoal, fuel, gum-medicinal and apiculture (Honey bee) flora
38.	*E. tereticornis*	<1200	SL	Timber, paper pulp, plywood, fuel, essential oil, honey bee flora
39.	*Ficus palmata*	<1000	SL, BC	Ornamental, fodder, edible fruits
40.	*Grewia elastica*	<500	SL	Ornamental, timber, toy making, fuel, shade tree
41.	*G. optiva*	500–1500	SL	Timber-cot frames, fiber, fodder, fuel wood
42.	*Hippophae rhamnoides*	>3000	DS, BC	Medicinal, fuel, soil conservation, and fodder
43.	*Juglans regia*	>2000	DS, SL	Timber—furniture and carving, gun-stocks, fruits
44.	*Juniperus communis*	>3000	DS, SL	Volatile oil, medicinal, resin, wad, esters
45.	*J. polycarpus*	>3000	DS, SL	Timber, walking sticks, drinking cups, fuel and charcoal, medicinal, essential oil
46.	*Lagerstroemia speciosa*	<1000	SL	Timber-constructional purposes, furniture, agricultural implements, telegraph poles, railway sleepers, medicinal, fodder
47.	*Lannea coromandelica*	<800	DS,SL, BC	Matchwood, pulpwood, gum and brushwood, soil conservation
48.	*Leucaena leucocephala*	<1200	DS, SL	Paper pulp, fodder, light timber, fuel
49.	*Mangifera indica*	<600	SL	Edible fruits, fatty oil, plywood, shoe heels
50.	*Melia azedarach*	<2000	DS, BC,SL	Box planks, fuel wood, paper pulp, medicinal
51.	*Moringa oleifera*	<500	BC, DS	Fruits-edible, medicinal, fodder
52.	*Morus alba*	<1200	BC, DS, SL	Fruits-edible, timber, sports goods, fodder
53.	*M. serrata*	1500–2500	BC,DS	Furniture, toys, sports goods, fodder
54.	*Ougeinia oojeinensis*	<800	BC,SL, DS	Cart, carriage, building construction, fodder agricultural implements, and edible fruits
55.	*Picea smithiana*	>2200–3000	SL	Planking, general filling and joinery, packing cases

(continued)

Table 2.5 (continued)

S. No.	Name of species	Climatic zone Altitudinal range (m amsl)	Method of planting	Uses
56.	*Pinus roxburghii*	<1500	SL	Timber, resin, pulpwood
57.	*P. wallichiana*	<1500–2500	SL	Timber, resin, pulpwood
58.	*Pithecellobium dulce*	<400	BC,DS	Packing cases, fuel, agricultural implements, fruits-edible, yield fatty oil, fodder
59.	*Platanus orientalis*	1500–2500	SL, BC	Ornamental, fuel, cheap timber
60.	*Populus alba*	>2500	BC,SL	Matchwood, pulpwood, light timber, fuel
61.	*P. ciliata*	1500–2500	RS, RC,BC	Matchwood, pulpwood, light timber, fuel, fodder
62.	*P. deltoides*	<1200	BC	Matchwood, pulpwood, light timber, fuel, packaging cases
63.	*P.euphratica*	>2500	B0C	Matchwood, pulpwood, light timber, fuel, and fodder
64.	*P. nigra*	>2500	BC	Matchwood, pulpwood, light timber, fuel, and fodder
65.	*Prunus domestica*	>1500	SL	Fruits-edible, fodder, timber, fuel
66.	*P. padus*	>1200	SL	Fruits-edible, fodder, timber, fuel
67.	*Psidium guajava*	<700	SL	Fruits-edible, wood engraving, spear handles, instruments, and lac turnery, medicinal
68.	*Punica granatum*	600–1200	BC, SL	Fruits-edible, ornamental, fuel wood, fodder
69.	*Quercus dilatata*	2500–3000	BC	Fuel, fodder
70.	*Q. leucotrichophora*	1500–2500	DS	Fuel, fodder
71.	*Q. semecarpifolia*	2500–3500	BC	Fuel, fodder
72.	*Salix alba*	<700	BC	Cricket bats, matchwood, tool handles, fuel, fodder
73.	*S. flabellaria*	>3000	BC	Fuel, fodder
74.	*S. fragilis*	>3000	BC	Light timber, charcoal for gun powder, fuel, soap (Saponine)
75.	*Santalum album*	<1000	DS,RS, SL	Fragrant heartwood, sandal oil, carving work, medicinal, agarbatties, perfumery
76.	*Shorea robusta*	<600	DS, SL	Timber, constructional purposes, plywood, paper pulp in mixture
77.	*Syzygium cuminii*	<600	DS, SL, ST, BC	Fruit-edible and medicinal, timber plywood, tools and implements, fuel, fodder
78.	*Terminalia bellirica*	<700	SL, DS, ST	Timber, pulp, plywood, fruit (myrobalan) medicinal, fodder
79.	*T. chebula*	<700	SL, DS	Fruits (myrobalan), medicinal, timber
80.	*Toona ciliata*	<2000	DS, SL	Timber, plywood, ornamental, fodder
81.	*Ziziphus muritiana*	<500	DS,SL, BC	Fruits, fuel, tannin, medicinal, fodder

DS Direct sowing, *SL* seedling (entire planting), *BC* Branch cutting *RC* Root cutting, *RS* Root suckers, *ST* Stump planting

Source UHF (1987), Bhatt et al. (2010), Dalvi and Ghosh (1982)

Fig. 2.4 Tomato (*Solanum lycopersicum*) as inter-crop with *Grewia optiva* trees

Eucalyptus tereticornis with bhabar grass (*Eulaliopsis binata*) reduced to 0.07 t ha^{-1} and 0.05 % in comparison to 5.65 t ha^{-1} and 23.0 % under cultivated fallow and 2.69 t ha^{-1} and 20.50 % under *Sesamum indicum*—*Brassica campestris* systems, respectively. Thus, tree-based systems were found to be very useful in Shiwalik region where erosion is a major cause of soil degradation and nutrient depletion (Grewal 1993). During the 9-year-study period, the average annual monsoon rainfall was about 1000 mm and it caused 347 mm runoff and 39t ha^{-1} soil loss due to erosion every year from fallow plots. The runoff and soil loss were reduced by 27 and 45 %, respectively by contour cultivation of maize (*Zea mays*). Contour tree-rows of *Leucaena leucocephala* hedges reduced the runoff and soil loss by 40 % and 48 %, respectively over the maize plot (reducing soil loss to 12.5 t ha^{-1}). This reduction in erosion was primarily due to the barrier effect of trees or hedgerows and micro-terraces formed through sediment deposition along the contour barriers. Such vegetative measures, that are productive while being protective, offer viable alternative for erosion control in areas with gentile slopes of the valley region. High density block plantations of eucalyptus and leucaena almost completely controlled the erosion losses and can be recommended for steeper slopes that are vulnerable to heavy erosion (Narain et al. 1998).

The soils were analyzed for different nutrient status in the 20 years old plantation of fodder trees in subtropical humid zone of Himachal Pradesh and it was observed that organic carbon (OC) and available N, P, K, and Ca increased significantly in each type of block plantation. The highest OC contents (2.75 %) were observed under *Ulmus villosa* and *Albizia stipulata* syn. *A. chinensis* (2.74 %). The highest available nitrogen kg ha^{-1} was under *Albizia stipulata* and *Dalbergia sissoo*, i.e., 458 and 459 kg ha^{-1}, respectively; available P in *Grewia optiva* (459 kg ha^{-1}) and exchangeable Ca in *Dalbergia sissoo* plantation (5880 kg ha^{-1}), whereas pH was observed near to neutral and EC was almost same as in the control in all the plantations (Table 2.9). *Albizia stipulata, Bauhinia variegata, Bombax ceiba, Celtis australis, Dalbergia sissoo, Grewia optiva, Robinia pseudoacacia, Sapindus mukorossi, Toona ciliata, Quercus luecotricophora,* and *Ulmus villosa,* etc., are the important tree species which the farmers of the western Himalayan region have grown on their farmlands to supplement their fodder and fuelwood needs. There is no doubt in some of these species also render benefit to the farming community indirectly, i.e., through atmospheric nitrogen fixation, addition of organic matter in the form of litter fall and also conservation of soil and water, the most important natural resources for the livelihood.

Table 2.6 Fodder quality of promising agroforestry fodder trees and grasses of Northwestern Himalayan region

Fodder trees	CP %	CF %	Ether extract %	Ash %	Ca %	P %
(A) Lower and mid-hill (900–1800 m amsl)						
Albizia chinensis	15.08	31.64	4.39	5.50	1.02	0.20
Albizia lebbeck	16.81–26.50	26.47–37.59	2.85–4.68	7.11–11.54	2.02	0.14
Anogeissus latifolia	7.45–11.48	16.38–24.15	2.68–4.41	8.68–10.93	3.03	0.34
Bamboos	12.00–19.00	12.00–20.00	2.50–4.00	11.98–19.04	0.50–1.12	0.05–0.26
Bauhinia variegata	10.73–15.91	25.28–32.97	1.33–3.93	6.37–12.31	2.40	0.22
Celtis australis	14.47–15.33	19.45–21.45	2.54–5.62	11.66–17.81	4.87	0.18
Ficus roxburghii	12.28–13.35	7.71–17.79	4.49–4.65	9.42–15.52	1.31–2.19	0.17–0.22
Grewia optiva	15.60–19.05	18.88–22.12	2.17–6.70	9.60–14.30	3.20	0.21
Leucaena leucocephala	24.20	13.30	4.40	10.80	1.98	0.27
Morus alba	15.00–27.64	9.07–15.27	2.30–8.04	14.32–22.87	2.43	0.24
Pittosporum floribundum[a]	14.70	10.25	3.81	9.54	2.80	0.20
Grasses						
Avena fatua	6.50	35.70	1.80	8.90	0.80	0.10
Dichanthium annulatum	2.08	39.60	–	0.50	0.50	–
Euchlaena maxicana	4.47–11.99	19.57–32.20	1.20–2.34	–	0.65–0.91	0.16–0.28
Medicago sativa	19.90	29.51	1.81	14.10	2.80	0.74
Panicum maximum	2.8–16.1	41–57	28–34	–	0.32–0.76	0.35–0.80
Pennisetum hybrid (Napier)	10.20	30.50	2.10	16.20	0.50	0.40
Pennisetum pedicellatum	6.50	35.80	3.20	14.40	0.40	0.30
Sorghum bicolor	7.75	32.36	1.73	8.55	-	–
Trifolium alexandrinum	15.80–26.70	14.90–28.50	1.40–3.00	12.90–16.00	1.48–2.58	0.18–0.31
Zea mays	6.74	85.95	2.09	8.15	–	–
(B) Higher Hills (1800 m amsl and above)						
Fodder trees						
Celtis australis	14.47–15.33	19.45–21.45	2.54–5.62	11.66–17.81	3.47–4.87	0.18
Morus alba	15.00–27.67	9.07–15.27	2.30-8.04	14.32-22.87	2.42-4.71	0.23-0.97
Morus serrata	14.00–22.00	9.07–14.27	2.30–6.00	13.32–22.87	2.40–4.60	0.20–0.95
Populus ciliata	10.00–12.00	22–26	4.00-8.00	8.00-11.00	–	–
Quercus dilatata	9.56	29.06	4.52	5.11	0.17	0.35
Q. glauca	9.62	29.04	4.14	7.6	1.87	0.23
Q. leucotrichophora	10.20–11.42	31.34	3.53–4.84	5.13–6.43	0.90–1.65	0.11–0.15
Robinia pseudoacacia	14–25.5	46.50	3.30	6.09	1.17	–
Salix alba	12.00–17.00	33.35	4.00-5.00	7.00–11.00	–	–
Grasses						
Dactylis glomerata	18.80	33.20	–	8.96	–	–

(continued)

Table 2.6 (continued)

Fodder trees	CP %	CF %	Ether extract %	Ash %	Ca %	P %
Festuca arundinacea	12.10–22.10	–	–	–	0.41–0.50	0.30–0.37
Festuca pratense	10.92–23.07	20.37–34.09	2.12–4.72	7.19–14.39	–	–

Source UHF (1988); [a] Mhaiskar (2012)
CP crude protein, CF crude fibre

Table 2.7 Erodibility characteristics of soils under different land uses under mid-hill conditions

Location (Land use)	Dispersion ratio	Erosion ratio	WSA (%) (>0.25 mm dia)	MWD (mm)
Nauni (C)	18.92	17.78	72.11	2.70
Nauni (O)	14.46	13.42	89.17	3.86
Lavighat (C)	20.42	21.14	68.36	1.73
Mukhari (C)	21.82	20.80	66.24	1.25
Randon Ghonron (C)	18.50	19.92	79.32	2.80
Majhat (F)	12.93	14.38	93.27	5.49
Kandhaghat (O)	13.25	15.50	92.89	4.26
Ranno (F)	11.64	13.24	95.32	5.22

C Cultivation, *F* Forest, *O* Orchard, *WSA* Water stable aggregates; *MWD* Mean Weight Diameter
Source Kumar et al. (2002)

Table 2.8 Effect of land use and slope on soil loss (t ha^{-1}) under mid-hill conditions

Percent slope → Land use category ↓	5–10	10–15	15–35	>35	Total	Area (ha)	Weighted soil loss (t ha^{-1} yr^{-1})
Dense forests	0.01	5.43	2.31	1.49	9.24	2,674.15	3.46
Mixed forests	0.05	1.81	3.85	6.38	12.09	1,366.29	8.85
Moderately dense forests	0.02	2.92	21.25	9.01	33.20	3,461.14	9.59
Degraded forests	0.00	5.44	7.66	6.88	19.98	1,124.86	17.76
Fruit-based cropping system	0.07	8.86	13.00	26.66	48.59	2,659.33	18.27
Intensive cultivation (sole)	1.73	59.77	38.12	54.47	154.09	2,617.33	58.87
Barren lands	0.00	6.14	11.16	12.76	30.06	349.34	86.05
Soil loss (1000 tonnes)	1.88	98.24	100.80	137.39	338.31		
Area (ha)	177.40	5,737.00	5,250.40	535.15			

Source Sharma et al. (2002)

Identification of Agroforestry Systems Under Different Land Holding Categories

During diagnostic surveys, different indigenous agroforestry systems were identified and their prevailing intensity on different land holdings was also surveyed and is presented in Table 2.10.

The marginal and small farmers usually maintain and develop all types of agroforestry systems on their agriculture field bunds and also in the homesteads/kitchen gardens. They grow trees and vegetables to meet their own requirement as well as retain fodder trees to feed the cattle during lean periods in all the agro-climatic zones. They also keep a unit of their land under fruit trees, vegetable crops, and grasses (cut and

Table 2.9 Effect of block plantation of agroforestry species on physico-chemical properties of the soil

Name of the species	OC (%)	pH	EC_2 (dSm^{-1})	Avail.N (kg ha^{-1})	Avail.P (kg ha^{-1})	Avail. K (kg ha^{-1})	Ex. Ca (kg ha^{-1})
Albizia stipulata	2.74	7.1	0.33	458	39	437	5337
Acer oblongum	2.13	7.1	0.36	415	30	422	5285
Bauhinia retusa	2.27	7.1	0.34	459	40	477	5477
B. variegata	2.55	7.2	0.34	386	29	370	5275
Bombax ceiba	2.29	7.0	0.35	317	38	438	5127
Celtis australis	2.01	7.1	0.35	307	33	448	5369
Dalbergia sissoo	2.51	7.1	0.34	459	32	409	5880
Grevillea robusta	2.27	7.0	0.30	329	31	337	5516
Grewia optiva	2.30	7.1	0.34	273	41	459	5409
Melia composita	2.21	7.2	0.36	250	40	357	5191
Robinia pseudoacacia	2.46	7.2	0.34	474	39	420	5673
Sapindus mukorossi	2.34	7.0	0.33	304	30	437	5275
Terminalia arjuna	2.38	7.0	0.35	275	28	388	4175
Toona ciliata	1.90	6.9	0.35	361	30	336	4873
Prunus armeniaca	2.06	7.1	0.36	329	30	443	5093
Punica granatum	2.38	7.1	0.33	292	29	392	47.60
Paulownia fortunii	2.67	7.0	0.29	279	30	423	5182
Quercus leucotrichophora	1.93	7.0	0.36	319	27	349	5384
Ulmus villosa	2.75	7.0	0.35	321	29	431	4984
Control	1.85	7.1	0.36	251	24	315	3802

Source UHF (2010)

carry) to feed the livestock. The pasture land for grazing is generally state-owned common *Panchayat* or a community land or the state forest department owned land. Every farmer keeps a small unit of land for hay making (*ghasnies*) to feed their cattle in the winter season.

Improved and Managed Agroforestry Systems

There are many possibilities of interventions into the traditional agroforestry systems due to accumulated knowledge through research and have other technological opportunities. Based on research and experiences of farmers the following agroforestry systems in Western Himalayan regions are suggested for adaptations in different regions:

Subtropical Low Hill Zone

The dominant, successful, and remunerative agroforestry systems in low hill subtropical zone include Kinnow (*Citrus*) or mango (*Mangifera indica*)- based cropping systems; poplar (*Populus deltoides*) and *Eucalyptus* based agri-silvicultural systems; multipurpose (mainly fodder) trees in the *ghasnies* (grass lands); and trees on field bunds and along slopping lands which also support growing of fodder grasses, etc., for generating additional farm income along with meeting their own domestic requirements.

Sometime block plantations of poplar and eucalyptus are also carried out by the big or absentee farmers to supply commercial raw material to the wood-based industries or to the local saw mills/furniture manufacturers to enhance their income. These days Poplar based cropping systems are quite frequently found

Table 2.10 Occurrence of agroforestry systems in Himachal Pradesh

Agroforestry systems	Famers categories					
	Uplands			Low land		
	Marginal (<1 ha)	Small (1–2 ha)	Medium (2–5 ha)	Marginal (<1 ha)	Small (1–2 ha)	Medium (2–5 ha)
i. Low hill subtropical region (<1000 m amsl)						
Agri-silviculture	–[c]	–[c]	–[a]	–[c]	–[c]	–[a]
Agri-silvi-horticulture	–[b]	–[b]	–[b]	–[b]	–[b]	–[b]
Pastoral- silviculture	–[a]	–[a]	–[a]	–[a]	–[a]	–[a]
Pastoral -silvi-horticulture	–[b]	–[a]	–[a]	–[b]	–[a]	–[a]
Pastoral-horticulture	–[b]	–[a]	–[a]	–[a]	–[a]	–[a]

Agroforestry systems	Famers categories			
	Marginal (<1 ha)	Small (1–2 ha)	Medium (2–5 ha)	Large (>5 ha)
ii. Mid-hill moist temperate region at Nauni (1275 m amsl) in Solan district				
Agri-silviculture	–[b]	–[b]	–[a]	–[a]
Agri-silvi-horticulture	–[b]	–[a]	–[a]	–[a]
Silvopastoral	–[c]	–[b]	–[b]	–[a]
Silvo-horti-pastoral	–[b]	–[b]	–[a]	–[b]
Horti-pastoral	–[b]	–[b]	–[b]	–[c]

Agroforestry systems	Famers categories					
	Uplands			Low land		
	Marginal (<1 ha)	Small (1–2 ha)	Medium (2–5 ha)	Marginal (<1 ha)	Small (1–2 ha)	Medium (2–5 ha)
iii. Moist temperate high hills (2000–3000 m amsl)						
Agri-silviculture	–[c]	–[b]	–[b]	–[b]	–[b]	–[b]
Agri-silvi-horticulture	–[b]	–[b]	–[b]	–[b]	–[b]	–[b]
Silvopastoral	–[b]	–[b]	–[c]	–[b]	–[b]	–[b]
Horti-pastoral	–[c]	–[c]	–[c]	–[b]	–[b]	–[b]
Horti-silvopastoral	–[b]	–[b]	–[b]	–[a]	–[a]	–[a]

Agroforestry systems	Famers categories					
	Uplands			Low land		
	Marginal (<1 ha)	Small (2–5 ha)	Medium (>5 ha)	Marginal (<1 ha)	Small (2–5 ha)	Medium (>5 ha)
iv. High hill dry temperate (devoid vegetation) and cold desert (>3000 m amsl)						
Agri-silviculture	–[b]	–[b]	–[b]	–[c]	–[c]	–[c]
Agri-horticulture	–[c]	–[c]	–[c]	–[b]	–[b]	–[b]
Agri-silvo-horticulture	–[c]	–[c]	–[c]	–[b]	–[b]	–[b]
Silvopastoral	–[b]	–[b]	–[c]	–[b]	–[b]	–[b]
Silvo-horti-pastoral	–[a]	–[b]	–[b]	–[a]	–[a]	–[b]

[a] Absent/Partial; [b] Supportive; [c] Major

Source Modified from Verma et al. (2007)

Table 2.11 Economics of agri-horticultural system (Kinnow + wheat and Kinnow + mustard)

Crops	Cost (INR ha^{-1})						Total cost (INR ha^{-1})	Returns (INR ha^{-1})	
	Input			Labor				Gross	Net
	Seed	Ferti	Total	Bullock	Human	Total			
Wheat	1,500	5,165	6,665	6,400	6,200	12,600	19,265	29,868	10,603
Mustard	240	3,234	3,474	2,200	3,500	5,700	9,174	24,030	14,845
Kinnow (275 plants/ha)	....	.....	10,196	.....	12,287	12,287	36,335[a]	83,505	47,170
Wheat+ Kinnow	1,205	4,151	15,523	5,143	4,983	22,412	51,787	1,04,857	53,070
Mustard + Kinnow	193	2,599	12,988	1,768	2,813	16,868	43,708	1,00,115	56,408

[a] INR 13,852 for orchard establishment cost is also included; Av. Yield 45 kg per plant of Kinnow (12 year age) has been considered in the analysis. (US \$ 1 = INR 43.41, Dec, 2008)
Source Sharma et al. (2008)

Fig. 2.5 Poplar (*Populus deltoides*) with turmeric (*Curcuma domestica*) as inter-crop)

particularly with commercial crops (Fig. 2.5) like turmeric (*Curcuma domestica*) and ginger (*Zingiber officinale*).

Success Story Related to Fruit Trees-Based Cropping System in the Region

In the subtropical low hill zone valley the farmers of Nurpur (Kangra), Nadaun, Hamirpur, and Una districts of Himachal Pradesh are getting net return of INR 10,603 and 14,845 per hectare from wheat (*Triticum aestivum*) and mustard (*Brassica juncea*) crops, respectively when these are grown on slopes along with Kinnow (*Citrus* sp) fruit trees. The net return from Kinnow (12 years old plantation) as sole crop was INR 47,170 per hectare but when wheat and mustard were integrated with Kinnow, the net return increased to INR 56,048 (Table 2.11). The analysis revealed that cultivation of mustard is more profitable as compared to wheat with Kinnow. However, the net profits in both the cases were higher as compared to the sole Kinnow. Economics of agri-horticultural systems revealed that both gross as well as the net returns increased on per hectare basis. Agroforestry models are thus important to enhance the land use efficiency as well as productivity.

Forest and Fruit Trees-Based Cropping System

A cropping system was developed which is consisted of Kinnow-mandarin (*Citrus nobilis* × *C. deliciosa*) 400 plants per hectare, subabul

(*Leucaena leucocephala*) in quincunx method at varying plant density (0, 100, 166, and 277 plants ha^{-1}) such as woody perennials and agricultural crops viz., wheat (*Triticum aestivum*) and mash (*Lens culinaris*) grown under irrigated conditions. Various yield parameters are shown in Table 2.12.

The maximum number of fruits (174 thousands ha^{-1}) which is the only economical part of the system for sale, fodder yield (2504 kg ha^{-1}), and fuel-wood yield (1593 kg ha^{-1}) were obtained when Kinnow was spaced at 5 m x 5 m and *Leucaena* trees were planted at a distance of 10 m × 5 m in between the rows of Kinnow. It is evident from the table that there is decrease in agriculture crop yield in the system when density of *Leucaena* trees is increased; on the other hand there is a significant increase of Kinnow fruits by incorporating *Leucaena* plants. The *Leucaena* plants fix the atmospheric nitrogen and improved the soil fertility which is available to the Kinnow trees and also provide additional benefits in terms of fuel wood and fodder for livelihood subsistence of the farmers. It is, therefore, very important that horticultural crops planted at standard distance should be incorporated with leguminous fodder trees for enhancing the income as well as livelihood subsistence in the region.

Agri-silvicultural System

In subtropical low hills, i.e., Ponta valley of Himachal Pradesh, different varieties of sugar cane (*Saccharum officinarum*), i.e., Co 88, Co 767, Coj 64, and Co 7717 were cultivated under different clones of poplar (*Populus deltoides*), i.e., PD 3294, PD G3, PD G48, and PD 1/56; which were planted at 5 × 5 m spacing. The bole girth of the poplar trees varies significantly and was maximum in clone PD G48 and maximum average cane yield was recorded to be 117 t ha^{-1} in clone Co 7717 followed by 116.5 t ha^{-1} in Coj 64 under poplar plantation as compare to Co 767 (112 t ha^{-1}) and Co 88 (107 t ha^{-1}). Hence, Co 7717 and Coj 64 of sugarcane were found suitable for successful growing under poplar based agroforestry system (Table 2.13).

Mid-hill Subhumid Zone

Intercropping of Aromatic and Medicinal Plants Under High Density Peach Plantation

The yield of fruits increased when aromatic and medicinal plants namely Tulsi (*Ocimum sanctum*), Ashwgandha (*Withania somifera*), and Kalmeghh (*Andrographis paniculata*) were grown between the rows of high density (4.5 m row-to-row and 2 m plant-to-plant) of 5-year-old peach (*Prunus persica*) plantations in the mid-hill Himalayas. There was a significant increase in yield of peach (4.5 × 2.5 m) by 7.77, 11.8, and 9.02 % with Tulsi, Ashwagandha, and Kalmegh, respectively over sole peach (Table 2.14; Fig. 2.6). It was also observed that under peach plantation, there was a significant increase in biomass of all these medicinal plants showing that partial shade is helping the growth of these plants.

The growing of aromatic and medicinal plants under fruit trees is highly profitable. Thakur et al. (2010) and Verma et al. (2010) also recommended growing of *Digitalis lanata, Matricarea chamomilla, Salvia sclaria,* and *Ocimum basilicum* with poplar (*Populus deltoides*) as shade tolerant intercrops and were proved to be good option for farm diversification. Verma and Thakur (2010) recommended the cultivation of *Withania somnifera* (Ashwagandha) with peach, and also with fruit tree *Morus alba* and grass *Setaria* for high productivity and profitability.

Carbon Sequestration in Peach, Aromatic, and Medicinal Plants Based Agroforestry Systems

The rate of carbon sequestration in the aromatic and medicinal plants was comparatively higher than the sole crops, i.e., 1.21, 0.87, and 1.14 t ha^{-1} yr^{-1} in Tulsi, Ashwagandha, and Kalmeghh under high density peach plantation in comparison to 1.00, 0.60, and 0.89 t $ha^{-1}year^{-1}$in the open sole crops, respectively. The rate of change in 1 year was highest in sole peach plantation being a perennial crop, and due to increase in growth with the age, i.e., 0.20 t ha^{-1} $year^{-1}$. In Ashwagandha under peach, the rate of change in

Table 2.12 Fruit, fodder, fuelwood, and grain yield under forest and fruit tree-based system

AF system Δ	*Number of fruits (000 ha^{-1})	Leucaena dry fodder yield (kg ha^{-1})	Fuelwood yield (kg ha^{-1})	Wheat yield (kg ha^{-1})	Mash yield (kg ha^{-1})
K (400) + A	143	–	–	890	450
K (400) + L (100) + A	159	15	898	855	426
	(+11.3)			(−3.9)	(−5.6)
K (400) + L (166) + A	174	25	1593	795	3.6
	(+21.8)			(−10.7)	(−19.3)
K (400) + L (277) + A	132	38	2339	695	321
	(−7.6)			(−21.9)	(−28.7)

Parenthesis represents the percentage increase (+) or decrease (−); *Number of Kinnow per kg = 8–10
K Kinnow-mandarin, *L Leucaena* and *A* Agriculture crops-wheat in *rabi* and mash in *kharif*
Δ Figure in brackets shows number of trees; *Source* Chauhan et al. (1997)

Table 2.13 Growth parameters of trees and cane yield (tons ha^{-1}) of sugarcane varieties grown under different poplar clones

Poplar clone	Growth		Sugarcane yield (tons ha^{-1})				
	Height (m)	DBH (cm)	Co 88	Co 767	Co j64	Co 7717	Mean
PD 3294	13.13	13.58	106.60	112.40	115.56	118.77	113.33
PD G3	14.27	14.34	105.97	111.90	116.90	116.33	112.77
PD G48	14.94	14.60	102.50	111.20	115.66	115.66	111.25
PD 1/56	14.45	14.44	111.50	112.50	117.97	116.93	114.72
Mean			106.64	112.00	116.52	116.92	

Source Chauhan and Dhiman (2003) *DBH* diameter at breast height

Table 2.14 Yield of peach fruits and biomass production of three medicinal plants when grown in different agroforestry systems in mid-hill Himalayas

Agroforestry components		Fruit yield (t ha^{-1})	Above ground biomass of medicinal plants (t ha^{-1})	Below ground biomass of medicinal plants (t ha^{-1})
(i)	Peach (sole) at 4.5 m x 2.5 m	3.99	–	–
(ii)	Tulsi with peach	4.30	2.30	0.37
		(7.77)	(21.06)	(12.12)
(iii)	Ashwagandha with peach	4.46	1.01	0.79
		(11.78)	(46.38)	(21.54)
(iv)	Kalmeghh with peach	4.35	2.31	0.17
		(9.02)	(26.23)	(13.13)
(v)	Tulsi (sole)	–	1.90	0.33
(vi)	Ashwagandha (Sole)	–	0.69	0.65
(vii)	Kalmegh (sole)	–	1.83	0.15

Parenthesis represents the percentage increase
Source Tripathi (2012)

consecutive year was maximum, i.e., 0.12 t $ha^{-1}yr^{-1}$ and minimum was −0.15 t ha^{-1} yr^{-1} in Tulsi, i.e., productivity reduced because it is a light demanding species. Similarly, rate of carbon emission and mitigation t $ha^{-1}year^{-1}$ in aromatic and medicinal plants was higher under peach than

Fig. 2.6 Tulsi (*Ocimum sanctum*) cultivated as intercrop with peach (*Prunus persica*)

in the open due to more biomass production (Table 2.15). Therefore, agroforestry with aromatic and medicinal plants in the mid-hill subhumid zone of the Himalayan region is a viable option for climate change for CO_2 mitigation.

Biomass Production in Improved Silvopastoral System

A silvopastoral system was established on undulating farm land having slope more than 50 % to meet the fodder requirement of the cattle as one of the livelihood subsistence. The plantation of four promising fodder trees, i.e., *Celtis australis, Morus alba, Grewia optiva*, and *Leucaena leucocephala* was done in *gradonies* (continuous contour trenches) in alleys, i.e., at 1 m (plant-to-plant) and 4 m (row-to-row) apart. In between the rows, the grass *Setaria anceps* var. Kanachangula was planted intensively to have a complete land coverage. The single row fodder trees were pollarded at 0.5 m height, in other single row fodder tree pollarded at 1.5 m height and double row of fodder trees (0.5 m apart) pollarded as single row at 0.5 m and second row at 1.5 m height. The data for leaf fodder, branches as fuel wood, and grass planted underneath were recorded consecutively for 2 years after 12 years of plantation. It was observed that leaf fodder production was comparatively higher in *Morus alba* in all pollarding heights and varied from 1680 kg ha^{-1} at 1.5 m pollarding height to 2726 kg ha^{-1} pollarding at 0.5 m and 1.5 m height together. The mean leaf fodder production was 1897 kg ha^{-1} followed by 1767 kg ha^{-1} in *Leucaena leucocephala,* 699 kg ha^{-1} in *Grewia optiva*, and 263 kg ha^{-1} in *Celtis australis*. The overall productivity including grass grown under the trees between the rows was maximum in the *Morus alba* based agroforestry systems, i.e., 15.42 t ha^{-1} followed by 12.4 t ha^{-1} in *Leucaena leucocephala,* 9.91 t ha^{-1} in *Grewia optiva,* and minimum (9.5 t ha^{-1}) in *Celtis australis*. The available nitrogen increased from 280 to 307 kg ha^{-1}, available P from 1.7 to 20.7 kg ha^{-1}, available K from 327 to 345 kg ha^{-1}, available sulfur from 23 to 29 mg kg^{-1}, exchangeable Ca contents from 1301 to 1943 mg kg^{-1}, and exchangeable Mg from 261 to 312 mg kg^{-1}, respectively (UHF 2006).

Table 2.15 Carbon sequestration; emission and mitigation (t $ha^{-1}yr^{-1}$) in peach-based agroforestry system involving medicinal and aromatic plants

Agroforestry components	Rate of carbon sequestration		Rate of carbon emission		Rate of carbon mitigation	
	1st yr	2nd yr	1st yr	2nd yr	1st yr	2nd yr
Peach (Sole) 1111 plants ha^{-1} at 4.5 m x 2.5 m spacing	1.71	1.91	0.39	0.43	1.32	1.48
Ocimum sanctum (sole)	1.08	1.92	0.81	0.66	0.27	0.26
Withania somnifera (sole)	0.61	0.60	0.34	0.36	0.26	0.25
Andrographis peniculata (sole)	0.81	0.97	0.54	0.68	0.27	0.28
Peach + *O. sanctum*	1.28	1.13	0.95	0.82	0.33	0.30
Peach + *W. somnifera*	0.75	0.87	0.46	0.58	0.28	0.29
Peach + *A. peniculata*	1.09	1.14	0.78	0.81	0.31	0.32
Mean	1.04	1.05	0.73	0.77	0.31	0.30

Source Tripathi (2012)

These *gradonies* not only conserve the moisture but also check the run off during the rainy season. The trenches are generally filled with soil particles (silt and organic matter) due to which the survival and growth of the trees increased. The similar study which was also conducted by Singh et al. (2008) in Shiwalik foot hills on research farm of the Central Soil and Water Conservation Research Training Institute, Dehradun; *Grewia optiva* was planted in the pits (45 × 45 × 45 cm) at 4 × 4 m spacing with a density of 625 trees ha^{-1} and Hybrid Napier grass (*Pennisetum purpureum*) was planted in inter-spaces. After 10 years, it was concluded that *Grewia optiva* trees planted alone or along with Napier hybrid, the biomass of both the components decreased significantly with time after 5 years (lopping and pollarding) of *Grewia optiva* and biomass from cuttings of Hybrid Napier varied from 256 to 2181 kg ha^{-1}.

In other studies when fodder trees, i.e., *Morus alba, Celtis australis, Grewia optiva,* and *Bauhinia variegata* were lopped at different cutting heights of 0.5, 1.0, 1.5, and 2.0 m; the maximum leaf + branch biomass accumulated in *Morus alba*, i.e., 7.38 t ha^{-1} followed by 2.16, 1.44, and 1.31 t ha^{-1} in *Grewia optiva, Bauhinia variegata*, and *Celtis australis*, respectively in 4 years old plantation at 2 m cutting height (Chand et al. 2008). There was significant variation in leaf N, P, K, Ca, and Mg concentrations irrespective of cutting heights. In both the cases the plantation was done in continuous contour trenches, i.e., *gradonies* on the hill slopes. Hence, it is recommended that in the sloppy areas of the region, the plantations should be carried out in the *gradonies* or continuous contour trenches to enhance the survival, growth, and biomass productivity.

Bamboo-Based Agroforestry System

Economic analysis of the bamboo-based agroforestry systems (Table 2.16) reveals that the returns from agricultural crops are quite higher than from the sole bamboo-based systems.

The returns from the bamboo species are in the order of *Dendrocalamus asper*, *D. hamiltonii*, and *Bambusa balcoa,* respectively. *Dendrocalamus asper* culms are edible and are used for making pickles, candy vegetables, etc., by the food processing industries and are sold at higher prices than other bamboo species of the region. Similarly, the *D. asper* clumps are managed at lower heights as young culms are harvested regularly for pickle and candy making thus shade loving agriculture crops viz. turmeric (*Curcuma domestica*), soybean (*Glycine max*), ginger (*Zingiber officinale*), colocacia (*Colocasia esculenta*), and white yam (*Dioscorea alata*) are grown successfully under its canopy (Fig. 2.7). The other bamboo species are not viable options for the rainfed agroforestry systems till the market price of mature bamboo are established. There is an urgent need to establish bamboo-based cottage industries, low cost poly houses, and low cost activated charcoal or pulp and paper industries. The consumption can be made compulsory by incorporating the bamboo pulp with other hard wood pulp for the construction of tents, floors, and houses.

High Hill Temperate Wet Zone

Economics of Apple-Based Cropping System

One study was conducted on apple (*Malus pumila*) based agroforestry in Kullu district of Himachal Pradesh. It revealed that the average cost of cultivation of apple was INR 3,88,850 $ha^{-1}$and the average net benefit from the orchard by selling fruit was INR 10,45,523 ha^{-1} (Table 2.17). The integration of high value crops such as tomato (*Solanum lycopersicon*), pea (*Pisum sativum*), French bean (*Phaseolus vulgaris*), and mustard (*Brassica juncea*) in the system (Fig. 2.8) not only offered diversification in different growing seasons but also generated surplus income without affecting the fruit yield of the orchard (Table 2.18). However, there was significant loss (INR 57,494 ha^{-1}) in the old traditional system when wheat was integrated as a cereal crop under apple. The average data of two consecutive years revealed that growing of pea (*Pisum sativum*) as a substitute crop for wheat in winter season benefitted the farmers to the extent of

Table 2.16 Economic analysis of bamboo-based agroforestry systems

Sr. No.	Bamboo species	Agricultural crops combination	Return from agricultural crop (INR ha^{-1})	Returns from bamboo (INR ha^{-1})	Total (INR ha^{-1})
1.	*Dendrocalamus hamiltonii*	Tulsi	1,40,050	5,817	1,45,867
		Ginger	74,520	5,817	80,377
		Soybean	26,900	5,817	32,717
		Turmeric	81,000	5,817	86,817
		Aloe vera	1,92,030	5,817	1,97,847
2.	*D. asper*	Tulsi	2,62,800	14,293	2,77,093
		Ginger	3,90,400	14,293	4,04,693
		Soybean	30,650	14,293	44,943
		Turmeric	86,000	14,293	1,00,293
		Aloe vera	90,000	14,293	1,04,293
3.	*Bambusa balcoa*	Tulsi	2,53,300	3,253	2,56,553
		Ginger	89,680	3,253	92,933
		Soybean	14,400	3,253	17,653
		Turmeric	76,000	3,253	79,253
		Aloe vera	19,660	3,253	22,913
4.	Open plot	Tulsi	2,59,550	–	2,59,550
		Ginger	1,57,600	–	1,57,600
		Soybean	19,400	–	19,400
		Turmeric	71,000	–	71,000
		Aloe vera	1,98,620	–	1,98,620

Source UHF (2012) US \$ 1 = Rs. 52.40

Fig. 2.7 Bamboo-based agroforestry system: Tulsi, ginger, soybean, turmeric, and aloe vera as inter-crop

INR 2,84,676 ha^{-1} and INR 5,31,966 ha^{-1} in the *kharif* season from the cultivation of tomato.

Hence, it is recommended that high value vegetable crops, such as beans, peas, tomato, cauliflower, cabbage, *broccoli* as off-season vegetable can be integrated in the temperate orchards to increase the farm income.

Agri-silviculture in Kashmir Valley

Experiments were conducted on hill slopes by planting *Ulmus wallichiana* as a tree crop in alleys across the slope at a distance of 1, 1.5, and 2 m row-to-row and were pruned every year at 3 m height for fodder and fuel wood. In between the rows peas and beans were planted in *rabi* and *kharif* seasons, respectively. After 10 years, it was found that there was an increase in average cumulative yield of both *rabi* and *kharif* crops when compared with the average cumulative yield of crops obtained in the control without any plantation. Increase in yield of pea and beans was to the tune of 314.34 and 454.67 kg ha^{-1} over the control, i.e., outside alleys, respectively. (Table 2.19).

Besides the maximum yield of 950 kg ha^{-1} of peas and 1165 kg ha^{-1} of beans were recorded in an alley width of 2 m. In addition to yield of peas and beans, yield from *Ulmus wallichiana* trees in the form of fodder and fuel wood yielded

Table 2.17 Economics of apple-based cropping system in northwestern regions

Years	No. of trees ha^{-1}	Average fruit yield (kg $tree^{-1}$)	Fruit yield (t ha^{-1})	Selling price (Rs kg^{-1})	Gross return (INR ha^{-1})	Total expenses (INR ha^{-1})	Net benefit (INR ha^{-1})	Benefit: cost ratio
1st	278	194.06	54.0	22	11,86,870	3,88,851	7,98,019	3.05
2nd	278	189.06	52.6	32	16,81,877	3,88,851	12,93,026	4.32
Average					14,34,374	3,88,851	10,45,523	3.69

Source Anjulo (2009) 1 US $ = 46.80 INR, Dec., 2009

Table 2.18 Average yield, expenses, and return from annual crops in apple-based cropping system

Crop	Average yield (t ha^{-1})	Selling price (INR kg^{-1})	Gross return (INR ha^{-1})	Total expenses (Rs ha^{-1})	Net benefit (Rs ha^{-1})	Benefit cost ratio
Wheat	2.1 (4.0)	10	21,000 (39,580)	97,074	−26,074 (−57,494)	0.22 (0.41)
Pea	8.8 (11.8)	50	4,40,000 (5,89,400)	15,5,324	+2,84,676 (4,34,076)	2.83 (3.79)
Tomato	73.6 (6.2)	10	7,36,000 (6,24,000)	2,04,034	+5,31,966 (4,19,966)	3.61 (3.06)
French beans	4.5 (5.2)	50	2,25,000 (25,75,00)	99,940	+1,25,060 (1,57,560)	2.25 (2.58)

1 US $ = 46.80, INR Value in parenthesis represent the control, i.e., sole crops with orchards
Source Anjulo (2009)

Fig. 2.8 Cultivation of apple (*Malus pumila*) with mustard (*Brassica juncea*) in Himachal Pradesh

additional benefits. Maximum fodder yield (8.43 kg $tree^{-1}yr^{-1}$) was obtained in closely spaced alleys and maximum fuel wood of 14.22 t ha^{-1} was obtained where alley width was maintained at 1.5 m (Table 2.20).

Thus, agri-silviculture model so devised can help in stabilizing the degraded environment and at the same time helps the farmer for increasing yield and security of food, fuel wood, and fodder.

Table 2.19 Average production of agricultural crops over a period of 10 years

Crop	Yield outside alley (kg ha^{-1})	Yield within alley (kg ha^{-1})	Increase in yield over control (kg ha^{-1})	Percentage increase over control
Rabi (Peas)	554	868.34	314.34	56.77
Kharif (Beans)	566	1014.67	454.67	81.19

Table 2.20 Average yield of crop, fodder, and fuel wood under different alleys

Alley width (Row-to-row)	Crop yield (kg ha^{-1})		Fodder yield (kg $tree^{-1} yr^{-1}$)	Fuel wood (t ha^{-1})
	Rabi	Kharif		
2 m	949.92	1164.67	5.31	10.30
1.5 m	845.84	1080.76	5.40	14.42
1 m	819.26	798.60	8.43	13.09

Source Mughal et al. (2003)

Fruit Trees-Based Pastoral Model in Kashmir Valley

About 2.5 lakh ha area is under apple (*Malus pumila*), almond (*Prunus amygdalus*), cherry (*P. avium*), and other stone fruits in the Kashmir valley. Due to increase in the cattle population, there is an acute shortage of fodder and accordingly growing or cultivating of grasses in orchards is essential to feed them. A scientific fruit trees—based pastoral system was developed in the valley and accordingly evaluated for temperate legumes and grasses. The four grasses, i.e., *Festuca pratense* (fescue), *Dactylis glomerata* (orchard grass), *Trifolium repense* (white clover), and *Trifolium pratense* (clover) were compared with natural undergrowth under fully grown almond orchard. The natural undergrowth of herbaceous vegetation in the orchard was identified as: *Plantago major, Plantago lanceolata, Poa bulbosa* (Poa grass), *Trifolium repense,* and *Indigofera articulata.* A 3 years data showed that all introduced grasses and legumes have higher yield than natural undergrowth grasses (Table 2.21).

The soil analysis revealed that there was negligible change in soil pH and EC; whereas organic matter (OM), and available N increased by continuous cropping and depleted the available P_2O_5 and K_2O by 6.6 and 10.1 %, respectively. Grasses as under storey crops are usually better than crops because the forage grows taller under shade and therefore, associate with trees without loss of yield and there is no root competition. The grasses are also good soil binder.

High Hills Dry Temperate Zone

Aromatic and Medicinal Plants

Cold desert areas of J & K and Himachal are suitable to grow aromatic and medicinal plants along with woody perennials, i.e., salix, poplar, seabuckthorn, etc., which are maintained by the farmers to meet their fuel and fodder requirement. *Salvia sclarea* (Clary sage), an aromatic herb, which produces linalool and linalyl acetate as a main constituent of aromatic oil used in perfumery and generally imported from the France by Indian industries. *Salvia* is a summer crop planted in the first week of May and harvested in the month of September and October for oil extraction. The experiments were carried out at Kashmir valley and Ladakh cold desert area in the open as well as in the polyhouse. The essential oil percentage was found to be 0.2–0.3 % in the open field and 0.5–0.7 % in the polyhouse plantations. Qualitative estimation have shown that the oil from open field conditions contained 26.62 % Linalool and 27.66 % linalyl acetate, whereas under polyhouse conditions it showed improvement and rose to 28.38

Table 2.21 Green forage yield (tons ha^{-1}) of pasture and legumes

Name of grass		Years			Average Yield (t ha^{-1})	% increase over natural vegetation
		1990	1991	1992		
(i)	Natural grass/ vegetation	12.89	17.54	13.50	14.64	–
(ii)	*Dactylis glomerata*	20.95	22.96	22.19	22.03	56
(iii)	*Festuca pratense*	20.47	42.48	21.25	28.07	91
(iv)	*Trifolium repense*	22.80	34.56	17.51	24.96	70
(v)	*Trifolium pratense*	23.00	31.92	18.33	24.58	62
	CD at 5 %	7.89	9.56	4.46	–	–

Source Makaya and Gangoo (1995)

Table 2.22 *Salvia sclarea* (Clarysage) oil: comparison of Linalool and Linalyl acetate in oil from Kashmir and Ladakh

Source		Linalool %	Linalyl acetate (%)
(i)	Kashmir	18.32	32.70
(ii)	Ladakh		
(a)	Open field	26.62	37.66
(b)	Polyhouse	28.38	50.65

and 50.65 %, respectively (Table 2.22). The quantity of essential oil, and linalool and linalyl acetate extracted in polyhouse condition was higher than in open field (Table 2.23).

This indicates that besides vegetables there are good scope of cultivating medicinal and aromatic plants in cold desert area along with indigenous woody perennials like *Hippophae* and *Salix* which are maintained for fuel, fodder, and timber, etc. Further, there is tremendous scope of developing livestock based silvopastoral systems particularly involving small ruminants.

In the Lahaul valley farmers maintain tree species on the boundaries of the cultivated fields in sparse situation or with low density. *Hippophae rhamnoides, Juglans regia, Populus nigra, Prunus armeniaca, Prunus communis,* and *Salix* sp. were noted among the important agroforestry species in the cold dessert of the

Table 2.23 Quality evaluation of oil *Salvia sclarea* from open field and polyhouse conditions in Ladakh

Habitat		Herbage (t ha^{-1})	Different yield parameters			
			Essential oil yield			
			(kg ha^{-1})	Oil (%)	Linalool (%)	Linalyl acetate (%)
(i)	Polyhouse (Stakna, Ladakh site)	12.50	8.70	0.50–0.70	28.38	50.65
(ii)	Open field (Thiksey, Ladakh site)	10.00	3.00	0.20-0.30	26.62	37.66
Down stream products						
Products			Uses			
(i)	Clarysage concrete		High value perfumery product			
(ii)	Clarysage absolute		Excellent modifier yield 3–4 times higher than essential oil			
(iii)	Selareol		Used as best fixative, high value in odorous chemical			

Source Kaul et al. (2006)

Table 2.24 Fuel wood and fodder production (kg) from indigenous agroforestry system of Lahaul valley of Himachal Pradesh

Production	Study area					
	Khoksar[1]		Jahlma[2]		Hinsa[2]	
	kg per tree	kg per ha	kg per tree	kg per ha	kg per tree	kg per ha
Hippophae sp.[3] fuel wood	–	–	31	59	25	10
Salix sp. fuel wood	110	319	213	260	240	144
Fodder	15	44	20.5	25	21.5	13

[1] Harvested after 4 years, [2] Harvested after 3 years, and [3] Harvested after 5 years
Source Kuniyal et al. (2001)

Lahaul valley. *Salix* species (willow) and sea-buck-thorn are the major woody perennial components in the traditional agroforestry system along with annual/perennial medicinal and aromatic plants or arable crops like barley. The woody perennials are maintained on bunds of farmers' fields for fuel, fodder, food, and timber purposes. The fuel wood and fodder production of both the species in different areas of Lahaul valley of Himachal Pradesh have been recorded (Table 2.24).

The fuel wood production of *Salix* varies from 144 to 319 kg ha^{-1} and fodder production varies from 13 to 44 kg ha^{-1}. However, in case of *Hippophae*, the fuel wood production from the two sides varies from 10 to 59 kg ha^{-1} (Table 2.24). The study sites are in the cold arid region having altitude range from 2400 to 6400 m amsl. The soils are sandy and erodible (Kuniyal et al. 2001).

Conclusions

Agroforestry in Northwestern Himalayan regions is a composite, diversified, and sustainable land use system. It provides unique opportunity for integration of different components of the farming systems to optimize the ecosystem functioning and better management of land, water, and biological resources. The traditional agroforestry systems and practices consist of growing trees deliberately with various crops and livestock for multiple benefits, viz., fuelwood, fiber, food, fruits, etc., and are time tested and well adopted in different situations. These systems developed over the years have been found suitable for conservation of natural resources, viz., soil, water, and vegetation. During past couple of decades enough research inputs have been added but still many of these systems need to be further improved with suitable technological interventions considering the local population need, so that the socioeconomic status of the farming communities is uplifted. Fruit trees-based cropping systems having medicinal and aromatic plant species as one of the components have been proved quite remunerative and sustainable systems. As live stocks are very important for these regions, hence improvement in pastures through introduction of high yielding grasses and leguminous forages along with fruit trees also need special attention. In recent times due to rise in average temperature due to climate change the apple belt has shifted toward higher altitudes increasing the total area under apple. This phenomenon needs more research inputs in the region.

References

Anjulo A (2009). Component interactions and their influence on the production of apple based agroforestry system in wet temperate zone of Himachal Pradesh. Ph.D. Thesis, Dr. Y S Parmar UHF, Nauni, Solan (H.P.) India

Bhatt V, Purohit VK, Negi V (2010) Multipurpose tree species of Western Himalaya with an agroforestry perspective for rural needs. J Am Sci 6(1):73–80

Census of India (2006) Population projections for India and states 2001–26 (Revised 2006). Office of the Registrar General and Census Commissioner, Ministry of Home Affairs, Government of India, New Delhi

Census of India (2011) Office of the Registrar General and Census Commissioner, Ministry of Home Affairs. Government of India, New Delhi

Chand K, Mishra VK, Verma KS, Bhardwaj DR (2008) Response of cuttings heights on biomass productivity and plant nutrient. Indian J For 31(2):243–250

Chauhan VK, Dhiman RC (2003) Yield and quality of sugarcane under poplar (*Populus deltoides*) based rainfed agroforestry. Indian J Agric Sci 73(6):343–344

Chauhan VK, Mishra VK, Khosla PK (1997) Growth and productivity evaluation of agri-horti-silviculture system. J Tree Sci 16(2):69–74

Dalvi MK, Ghosh RC (1982) Tree planting and environmental conservation. Forest Research Institute and Colleges, Dehradun. Extension series no. 6. pp 36–52

FAI (2010) Fertilizer statistics 2009–2010. The Fertilizer Association of India, New Delhi

FSI (2011) Indian state of forest report 2010–2011. Forest Survey of India, Ministry of Environment and Forests, Government of India, Dehradun

Ghosh SP (1981) Agroclimatic zone specific research. Indian Council of Agricultural Research, New Delhi, p 539

Grewal SS (1993). Agroforestry systems for soil and water conservation in Shiwalik. In: Agroforestry in 2000 AD for semi-arid and arid tropics. National Research Center for Agroforestry, Jhansi, pp 82–85

HPFS (2011) Himachal Pradesh forest statistics of 2010. Forest Department, Himachal Pradesh, p 16

ICFRE (2010) Annual report 2009–2010. Indian Council of Forestry Research and Education, Dehradun, p 212

Kaul MK, Bakshi SK, Qazi GN (2006) Low-tech agrotechnology in hilly areas: an attempt to convert R&D leads into technology. In: Joshi AP, Agarwal SK, Verma R (eds) Mountain technology agenda: status gaps and possibilities. Bishan Singh Mahendarpal Singh, Dehradun, pp 225–244

Kumar S, Sharma JC, Sharma IP (2002) Water retention characteristics and erodibility indices of some soils under different land uses in North-Western Himalayas. Indian J Soil Conserv 30:29–35

Kuniyal CP, Vishwakarma SCR, Kuniyal JC, Singh GS (2001) Seabuckthorn (*Hippophae* L.)—a promising plant for land restoration in the cold desert Himalayas. In: Singh V, Khosla PK (eds) Proceedings of international workshop on seabuckthorn, 18–21 Feb 2001, New Delhi, pp 1–6

Makaya AS, Gangoo SA (1995) Forage yield of pasture grasses and legumes in Kashmir valley. Forage Res 21(3):152–154

Mhaiskar PR (2012) Vegetative propagation *Pittosporum floribundum* Wight and Arn. through cuttings under mid hill conditions of Himachal Pradesh. M.Sc. Thesis, Dr. Y S Parmar UHF, Nauni, Solan, HP, India

Ministry of Agriculture (2009) Landuse statistics, Government of India, 2008–2009

Mughal AH, Khan MA (2007) An overview of agroforestry in Kashmir valley. In: Puri S, Panwar P (eds) Agroforestry systems and practices. New India Publishing Agency, Pitam Pura, pp 43–53

Mughal AH, Qaisar KN, Khan PA (2003). Agroforestry for conservation of degraded ecosystem and food production in Kashmir hill. In: Rethy P, Darbal PP, Binay Singh, Sood KK (eds) Forest conservation and management. IBD, Dehradun, pp 262–265

Narain P, Singh RK, Sindhwal NS, Joshie P (1998) Agroforestry for soil and water conservation in the western Himalayan Valley Region of India. 1. Runoff, soil and nutrient losses. Agrofor Syst 39:175–189

Ramakrishna YS, Rao GGSN, Kesava Rao AVR, Vijaykumar P (2000) Weather resource management. In: Singh GB, Yadav JSP (eds) Natural resource management for agricultural production in India. Print Asia, New York

RBI (2010) Handbook of statistics on Indian economy 2009–10. Department of Statistics and Information Management, Reserve Bank of India, Mumbai

Saleem M, Gupta LM (2007) Agroforestry for sustainable development of agriculture in North Western Himalayas-with particular reference to Jammu region. In: Puri S, Panwar P (eds) Agroforestry systems and practices. New India Publishing Agency, Pitam Pura, pp 55–65

Sharma JC, Prasad J, Bhandari AR (2002) Effect of watershed characteristics on soil erosion in southern Himachal Pradesh using remote sensing and GIS techniques. In: Dhyani SK, Tripathi KP, Singh R, Raizada A, Sharma AK, Mishra AS, Shrimali SS, Dhyani BL, Sharma AR, Khosla OPS (eds) Resource conservation and watershed management—technology options and future strategies. IASWC, Dehradun, pp 157–163

Sharma K, Thakur S, Sharma R, Kashyap SD (2008) Production and economics of kinnow cultivation with wheat and gobhi sarson in Himachal Pradesh. Indian J Soil Conserv 36(2):114–117

Singh C, Raizada A, Vishwanathan MK, Mohan SC (2008) Evaluation of management practices for a *Grewia optiva*—hybrid napier based silvo-pastoral system for rehabilitating old riverbed lands in the North-Western Himalaya. Int J Ecol Environ Sci 34(4):319–327

State Profile Himachal Pradesh (2010) Ministry of water resources. Central Gorund Water Board Northern Himalayan Region, Dharmshala, Bulletin 14, p 6

Stobdan T, Angchuk D, Singh SB (2008) Sebuckthorn: an emerging storehouse for researchers in India. Curr Sci 94(10):1236–1237

Stone (1992) The state of the world's mountains: a global report. Mountain agenda. Zed Books Ltd., London, p 391

Tewari S, Kaushal R, Purohit R (2007) Agroforestry in Uttaranchal. In: Puri S, Panwar P (eds) Agroforestry systems and practices. New India Publishing Agency, Pitam Pura, pp 95–125

Thakur PS, Dutt V, Thakur A, Raina R (2010) Poplar based agroforestry system: intercropping of medicinal herbs for better production and diversification. Indian J Agrofor 12(1):77–83

Tripathi P (2012) Effect of organic manures on yield and biomass production of medicinal and aromatic plants under peach based agroforestry system. Ph.D. Thesis, Dr. Y S Parmar UHF, Solan

UHF (1987) Annual research report (1986–87). Annual Research Reports submitted to All India Coordinated Research Project on Agroforestry, College of Forestry, Dr Y S Parmar University of Horticulture and Forestry, Solan, p 75

UHF (1988) Annual research report (1987–88) Annual research reports submitted to All India Coordinated Research Project on Agroforestry, College of Forestry, Dr Y S Parmar University of Horticulture and Forestry, Solan, p 56

UHF (2006) Annual research report (2005–06). Annual research reports submitted to All India Coordinated Research Project on Agroforestry, College of Forestry, Dr Y S Parmar University of Horticulture and Forestry, Solan, p 65

UHF (2010) Annual research report (2009–10). Annual research reports submitted to All India Coordinated Research Project on Agroforestry, College of Forestry, Dr Y S Parmar University of Horticulture and Forestry, Solan, p 62

UHF (2012) Annual research report (2011–12) Annual research reports submitted to All India Coordinated Research Project on Agroforestry, College of Forestry, Dr Y S Parmar University of Horticulture and Forestry, Solan, p 48

Verma KS, Thakur NS (2010) Economic analysis of Ashwagandha (*Withania somnifera* L. Dunal) based agroforestry land use system in mid hill Western Himalayas. Indian J Agrofor 12(1):62–70

Verma KS, Bhardwaj DR, Chand K (2007) Agroforestry in Himachal Pradesh. In: Puri S, Panwar P (eds) Agroforestry systems and practices. New India Publishing Agency, Pitam Pura, pp 67–93

Verma KS, Thakur NS, Rana RC (2010) Effect of tree crop combinations and nitrogen levels on herbage yield of sacred basil (*Ocimum sanctum* L.) grown in agrihortisilvipastoral system in mid hill Himalayas. Indian J Agrofor 12(1):71–76

Verma LR (1998) Forestry and agroforestry management practices. In: indigenous technology knowledge for Watershed management in upper North West Himalayas of India. (GCT/RAS/161/NET). In: PWMTA program, Kathmandu, pp 46–60

VPKAS (2011) Vision 2030 Vivekananda Parvatiya Krishi Anusandhan Sansthan Almora, Uttrakhand, p 32

From Shifting Cultivation to Integrating Farming: Experience of Agroforestry Development in the Northeastern Himalayan Region

3

A. K. Singh, A. Arunachalam, S. V. Ngachan, K. P. Mohapatra, and J. C. Dagar

Abstract

Northeast India comprising of eight states is known for biodiversity richness, which is intricately linked to the socio-culture of the indigenous people. In this region, livelihood is mostly dependent on traditional farming practices such as shifting cultivation, homegarden, taungya systems, etc. The crop combinations are based on the household requirements and have the basis of subsistence. Nonetheless, the cultural practices differed between different tribes in the region. As such, tree farming has been a traditional practice and agroforestry is also being practiced in different forms and formats. With the interventions of ICAR, several models of agroforestry right from horticulture-to-fish-based tree farming have been in practice. In the region, a new model of Intensive Integrated Farming System is being tested for its benefit-cost. As the hilly region receives high rainfall, the role of trees on the terrains receives much importance and as so is the influence of agroforestry practices on soil and water resources. Much research needs to be done on crop combinations with good input management for better socio-economic returns to the tribal farmers.

Introduction

Seven hill states that include Arunachal Pradesh, Manipur, Meghalaya, Mizoram, Nagaland, Sikkim, and Tripura are collectively referred to as North Eastern Hill (NEH) region, located between 21°58′ and 29°30′ N latitude and 88°58′ and 97°30′ E longitude spread over 1,83,746 km^2 area (Wasteland Atlas of India 2010), bestowed with several traditional agroforestry systems

A. K. Singh · A. Arunachalam (✉) · J. C. Dagar
Division of Natural Resource Management, Indian Council of Agricultural Research, Krishi Anusandhan Bhawan –II, New Delhi, 110012, India
e-mail: arun70@gmail.com

S. V. Ngachan · K. P. Mohapatra
I.C.A.R. Research Complex for NEH Region, Umiam, Meghalaya 793103, India

J. C. Dagar et al. (eds.), *Agroforestry Systems in India: Livelihood Security & Ecosystem Services*, Advances in Agroforestry 10,
DOI: 10.1007/978-81-322-1662-9_3, © Springer India 2014

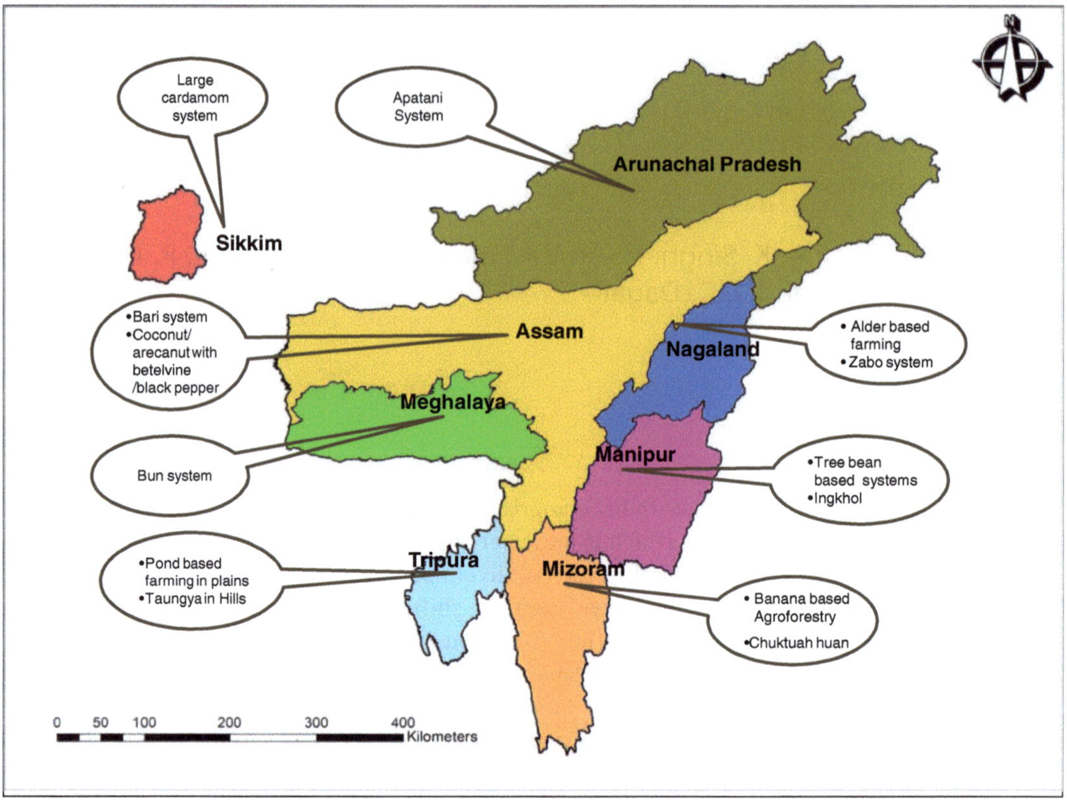

Fig. 3.1 Popular indigenous agroforestry systems in Northeast India

(Fig. 3.1). The region has unique weather and climatic conditions because of its typical geographical location, physiography, highlands in the northern part and their syntaxial bend, presence of alternating pressure cells of North West and Bay of Bengal and presence of tropical maritime air masses (Barthakur 2004).

Climate of the region varies from tropical to alpine type with very high range of variation in precipitation. The altitude ranges from 15 m to more than 5000 m above mean sea level. The temperature varies from as low as 0 °C in the Himalayan ranges (Arunachal Pradesh and Sikkim) to 35 °C in some parts of Tripura. The region receives highest average annual rainfall in the country, with Cherrapunji Plateau receiving over 11000 mm. This extreme variability in the precipitation, temperature, physiography, altitude, etc., has influenced the landuse, land cover, and species composition of the natural flora and fauna of the region. It is inhabited by more than 200 indigenous tribes having diverse lifestyle, cultural ethos, and multiple ethnicities resulting in multidimensional pressure on the natural resources. This unique and diverse climate has probably resulted in rich biological diversity and the region has the highest per capita availability of natural resources in the country. The region may be described as a cultural and genetic paradise and granary of mega biodiversity in terms of flora and fauna. It is one of the 12 mega centers of diversity and shares boundary with three out of the 34 internationally acknowledged hotspots of the world (www.biodiversityhotspots.org). Champion and Seth (1968) classified the forest types of the region as comprised of tropical moist evergreen forest to subalpine and alpine forests. About 80 % of the population lives in rural areas and majority of them depend on agriculture and allied sectors for their livelihood. Six distinct agro-climatic zones in different altitudinal range are there in the NEH (Table 3.1).

Table 3.1 Area under different agro-climatic zones of the NEH region

S. No.	Agro-climatic zone	Altitudinal range (m)	Approximate area (km^2) in the NEH region
1	Alpine zone	>3,500	47,068
2	Temperate subalpine zone	1,500–3,500	33,564
3	Subtropical hill zone	1,000–1,500	29,021
4	Subtropical plain zone	400–1,000	812
5	Mild tropical hill zone	200–800	26,349
6	Mild tropical plain zone	0–200	29,333

Despite having very high per capita availability of natural resources, diversified edapho-climatic conditions, and abundance of biological diversity, the region is still facing food grain deficit. The demand and supply gap in the poultry, meat, milk, etc., is also not positive and the region is importing these food items from other parts of the country. Because of undulating topography, the area under agriculture in the NEH region is very less which varies from 2.04 % of the total geographical area in Arunachal Pradesh to 22.90 % in Tripura. The productivity is low and soil degradation is high. Using Universal Soil Loss Equation (USLE), the annual average soil loss of 51 t ha^{-1} $year^{-1}$ was estimated for Dikrong river catchment in Arunachal Pradesh (Dabral 2002). Mandal and Sharda (2011a, b) reported that about 30 % of the total geographical area of the Northeastern Himalayas (including the NEH, Assam, and hills of West Bengal) are under the category of severe erosion having the potential erosion rate of 40–80 t ha^{-1}. For this region, estimation of acceptable rate of soil erosion (*T*-value) which is defined as the maximum permissible amount of soil erosion at which the quality of soil as a medium for plant growth can be maintained revealed that about 58.94 % of the total geographical area of this region requires various kinds of erosion management. Dhalai (North) of west Tripura hills is the major area of concern that can withstand maximum permissible soil loss of only 5.0 t ha^{-1} $year^{-1}$ (Mandal and Sharda 2011b). The changing climatic scenario is predicting further impact on the agricultural growth in the region. The Indian subcontinent is likely to experience a warming of over 3–5 °C leading to significant changes in precipitation pattern, flood, and drought situation (Ravindranath et al. 2011). The impact is supposed to be more pronounced in the northeastern Himalayan region. According to the recent report by the Ministry of Forest and Environment, Government of India, the yields of rice and maize in NER (including Assam) may decline up to 35 and 40 %, respectively, in the predicted climatic scenario of 2030 (Kumar 2011). Assessment of agriculture vulnerability in the projected future climate scenario (2021–2050) indicated that the districts of Tirap, West Siang, and Changlang will be the most vulnerable districts among all the NEH states. However, the vulnerability of West Sikkim, North Sikkim, East Sikkim, and Imphal will decrease from high to moderate levels (Ravindranath et al. 2011). Therefore, it is imperative to go for agricultural practices that are favorable to conservation of natural resources and resilient to the climatic aberrations. Agroforestry systems have the potential to address multifarious problems and deliver multiple benefits. It is effective in soil and water conservation through provision of permanent cover, improvement of soil health, enhancement of nutrient and moisture use efficiency, and moderation of microclimate. Agroforestry systems also have potential to limit carbon emissions and sequester carbon (Rao et al. 2007). The International Panel on Climate Change (IPCC) Third Assessment Report on Climate Change (IPCC 2001) states that "Agroforestry can both sequester carbon and produce a range of economic, environmental, and socioeconomic benefits. For example, trees in agroforestry farms improve soil fertility through control of erosion, maintenance of soil organic matter and physical properties, increased N, extraction of nutrients from deep soil horizons and promotion of more closed nutrient cycling."

The tribal communities of the northeast region practice varieties of agricultural practices ranging from different types of shifting cultivation systems, fallow system, home gardens, and sedentary agriculture such as wet rice cultivation. They have a natural inclination of keeping trees in their farming practices. The land use usually comprises a mixture of forest trees, animals, and field crops. These components are interdependent and strongly influence the livelihood of people of the region (Sundriyal et al. 1994; Mishra and Ramakrishnan 1982). The farming practices have evolved over centuries from the experiences of the people to meet the food, fuel, fodder, fiber, timber, and other such necessities.

Resource-Based Traditional Landuse Systems of the NEH Region

Shifting Cultivation

Shifting cultivation or *Jhum* is the most primitive and popular farming practiced across the entire NEH region which is essentially an agroforestry system organized both in space and time (Ramakrishnan 1992). The natural forests are removed for cultivation of crops (Fig. 3.2). Jhum is a way of life that is deeply entrenched in the artifacts, sociofacts, and mantifacts of the tribal way of life in the northeast (Murtem et al. 2008; Sailo 2011).

It locally known as *Rep Syrti* in Khasi and *Lo* in Mizo, has been practiced over 9000 years and said to have been originated in the Neolithic era dated by the archeologists to 7000 B. C. (Maithani 2005). At present, with increase in population pressure on land resources, the *Jhum* cycle is getting reduced very fast and reached at 2–4 years at present. This makes the system unstable and lead to severe land degradation as a result of soil erosion and associated factors such as reduction in soil organic matter, nutrients, etc. Total area under shifting cultivation is highest in Nagaland followed by Mizoram and Arunachal Pradesh. In terms of percentage of the total geographical area, Nagaland (17.06 %) and Mizoram (12.42 %) are the most severely affected by *jhum* cultivation. However, there is decline in area under shifting cultivation in most of the north eastern hill states except Nagaland (Table 3.2). It has declined from 1.35 million ha in 2003 to 0.85 million ha in 2005 (excluding the state of Assam).

The region is still rich in the forest resources in the era of rapid urbanization and economic development. The forest cover has not changed very significantly as compared to rest of the country (Table 3.3).

Important Attributes of Shifting Cultivation

One of the important common features of the shifting cultivation is growing of mixed crops after partial or complete removal of vegetation. In the northeastern India, many annual and perennial crops with diverse growth habits are being grown (Table 3.4) having sparsely distributed trees on hilly lands (Toky and Ramakrishnan 1981a). At times cash crops such as potato (*Solanum tuberosum*), rice (*Oryza sativa*),

Fig. 3.2 (*Left*) Forests have been cleared for shifting cultivation. (*Right*) Abandoned Jhum area in north-east India

Table 3.2 Area (ha) under shifting cultivation in different states of the NEH region

States	2005			2003				
	Current *Jhum*	Abandoned *Jhum*	Total	Current *Jhum*	Abandoned *Jhum*	Total	TGA (km^2)	Change (%)
Arunachal Pradesh	102,507	50,639	153,146	111,691	49,622	161,313	83,743	−5.06
Manipur	75,210	10,010	85,220	111,954	369,714	481,668	22,327	−82.31
Meghalaya	29,187	15,712	44,899	62,721	11,662	74,383	22,429	−39.64
Mizoram	102,853	158,903	261,756	114,695	287,046	401,741	21,081	−34.84
Nagaland	123,909	158,865	282,774	111,660	80,130	191,790	16,579	47.44
Tripura	8,928	16,483	25,411	28,489	11,037	39,526	10,486	−35.71
Sikkim	0	0	0	0	0	0	7,096	0
Total	442,594	410,612	853,206	54,121	809,211	1,350,421		−36.82

Source Wasteland Atlas of India 2010 (http://doir.nic.in/wasteland_atlas.htm) *TGA*, Total geographical area

Table 3.3 Change in forest cover in Northeast India

State	Total geographical area (km^2)	Forest cover in 2007, km^2 (%)	Forest cover in 2009, km^2 (%)	Change (km^2)
Assam	78,438	27,692 (35.30)	27,673 (35.28)	−19
Arunachal Pradesh	83,743	67,484[a] (80.58)	67,410 (80.50)	−74
Manipur	22,327	17,280 (77.40)	17,090 (76.54)	−190
Meghalaya	22,429	17,321 (77.23)	17,275 (77.02)	−46
Mizoram	2,1081	19,240 (91.27)	19,117 (90.68)	−123
Nagaland	16,579	13,464 (81.21)	13,318 (80.33)	−146
Sikkim	7,096	3,357 (47.31)	3,359 (47.34)	+2
Tripura	10,491	8,073 (76.95)	7,977 (76.04)	−96
Total North East	262,184	173,911 (66.28)	173,219 (66.07)	−692 (0.26 %)

Source Forest Survey of India (2009 , 2011)

[a] There was an interpretational change and upward revision of forest cover in 2009 in the state of Arunachal Pradesh by 131 km^2

Table 3.4 Mixed crops and their yield under different '*Jhum*' cycles of shifting cultivation in northeastern region of India

Types of crops/ products	Total yield (t ha^{-1}) in 30 years	Yield (kg ha^{-1} $year^{-1}$) under 10 years "Jhum"	Yield (kg ha^{-1} $year^{-1}$) under 5 years "Jhum"
Grain/seed yield	2,585	1,339	130
Leafy and fruit vegetables	262	381	584
Tuber and vegetables	609	381	870
Cocoon (silk)	4	–	–
Pupae (without cocoon)	0.2	–	–

Source (Toky and Ramakrishnan 1981a)

Table 3.5 Changes in rates of accumulation of biomass, and primary productivity of successional communities during fallow periods of shifting cultivation in northeastern regions

Parameters	Successional age of fallow period (years)				
	1	5	10	15	20
Standing biomass (t ha^{-1})	5	23	57	10	150
Accumulation in boles and branches (t ha^{-1} $year^{-1}$)	4	4	7	9	8
Net primary production (t ha^{-1} $year^{-1}$)	5	8	14	17	18
Biomass accumulation quotient (standing biomass/net primary productivity)	0.9	2.5	4.3	5.9	8.3

Source (Toky and Ramakrishnan 1983)

Table 3.6 Nitrogen, phosphorus, and potassium losses (kg ha^{-1} $year^{-1}$) through runoff under *Jhum* cycles at lower elevation

Jhum cycle (years)	Runoff losses			Infiltration losses		
	NO_3–N	PO_4–P	K	NO_3–N	PO_4–P	K
5	5.3	0.9	51.0	9.2	0.1	13.7
10	4.2	1.3	91.2	10.7	0.1	21.2
30	3.7	1.1	64.7	9.8	0.1	15.1

Source Toky and Ramakrishnan (1981b)

maize (*Zea mays*), and ginger (*Zingiber officinale*) are grown in monoculture or mixed culture along with *Pinus kesiya*. Another important attribute of the system is secondary succession of vegetation during the fallow period. Toky and Ramakrishnan (1983) have identified various patterns in Northeastern India. They reported that standing biomass increased from 23 t ha^{-1} in one 5-year fallow to 150 t ha^{-1} in 20-year fallow. However, the rate of accumulation of biomass increased up to 15 years (when it was 9 t ha^{-1} $year^{-1}$) and it declined during succession up to 20 years (Table 3.5).

After the cropping phase of the system while burning the slash, the contents like carbon, nitrogen, and sulfur are lost due to volatilization. After the burn and before the onset of the first rains, sizeable quantities of nutrients are lost through blown-off ash due to strong winds during dry periods. The sediment loss during the cropping period was reported to be 30 t ha^{-1} $year^{-1}$ in a *jhum* cycle of 5 years (Table 3.6) and with the shortening of the cycle, the losses tend to be high (Toky and Ramakrishnan 1981b).

The soil loss from hill slopes (60–79 %) under first year, second year, and abandoned *jhum* was estimated to be 147, 170, and 30 t ha^{-1} $year^{-1}$, respectively (Singh and Singh 1981). During first few years of clearing, carbon and nitrogen levels decrease rapidly. According to one estimate annual loss of top soil was found to be 88.3 million tons and of N, P, and K 10669, 0.372, and 6051 thousand tones, respectively due to shifting cultivation in the region (Sharma 1998). Consequently, the total production from this cultivation is fruitfully low. The paddy yield in Khasi hills (Meghalaya), Garo hills (Meghalaya), Khonsa (Arunachal Pradesh), and Siang (Arunachal Pradesh) are reported to be 128, 504, 408, and 832 kg ha^{-1}, respectively.

Jhum in the region is a complex system with wide variation that depends upon the ecological variation in the area and cultural diversity among various tribal clans. However, there are some commonalities in the basic cropping practice. Usually all the essential crops such as paddy (*Oryza sativa*), maize (*Zea mays*), tapioca (*Manihot esculenta*), colocasia (*Colocassia esculenta*), millets (Species of *Panicum, Eleusine, Pennisetum*), sweet potato (*Ipomoea batatas*), etc., are grown on the same piece of land as mixed crop. *Jhum* in its most traditional form is

Table 3.7 Litter yield and N added through Alder

Alder population (trees ha^{-1})	Litter dry matter (kg per tree)	Litter yield (t ha^{-1})	N added (kg ha^{-1})
60	56.3	3.37	48.3
101	45.3	5.48	74.5
142	58.1	8.25	110.5
166	52.2	8.66	113.5
280	37.5	10.50	142.8
625	21.7	13.56	184.8

Source Sharma and Singh (1994)

not a very unsustainable landuse practice particularly when the *Jhum* cycle is more than 20 years. The soils get enough time to rejuvenate and restore their health and productive capacity. The Angamis tribe from Nagaland used to practice Alder (*Alnus nepalensis*) -based sustainable *Jhum* system that was developed in Khonoma village in Nagaland. It provides about 57 food crops to supplement the staple crop rice. The root nodules of the Alder plants improve soil fertility by fixing atmospheric nitrogen into the soil through *Frankia*. The fallen leaves act as mulches and add humus to the topsoil. The wood is used as fuel wood, for charcoal burning and in construction works. Alder saplings collected from nursery or wild forest are planted in a *Jhum* field located in hills above 1000 m. In the first year in *Jhum* plots, alder trees are pollarded at a height of 2 m from the ground before or after the slash and burn operation (Rathore et al. 2010).

Mixed cropping is repeated in the second year. The field is then left fallow for 2–4 years to allow the Alder trees to grow for pollarding and cropping in the subsequent cycle. Young trees with bole circumference of about 50–80 cm are pollarded for the first time, usually at the age of 7–10 years. Cyclical/subsequent pollarding is performed after 4–6 years. During this operation, the pollarded stumps that coppice profusely are allowed to grow till the harvest of the first year's crop. On the second year, 4–5 selected shoots are retained and the rest is removed. These shoots are allowed to grow till the next *jhum* cycle and the same process is repeated (Mishra and Sharma 2001; Pulamte 2008). Thus with the incorporation of Alder trees in their *jhum* lands, the fertility of the field is increased. It was estimated that the N_2 fixed from Alder plantation (Table 3.7) varied between 48.3 (60 trees ha^{-1}) to 184.8 kg ha^{-1} (625 plants ha^{-1}). Besides fixing atmospheric N_2, the litter added to the soil provided P, K, Ca, and other nutrient through the addition of biomass (Sharma and Singh 1994).

The Konyak tribes in Nagaland also have sound ecosystem knowledge which they use in their shifting cultivation practices. In a study in the Nganchin village of Mon district of Nagaland (Bhan 2009), it was observed that the tree-stand-density nowhere in the Naga system matches the Konyak system where at times about 3,000 small saplings could be observed in 1 ha of land. They gradually reduce the density during the fallow period. They manage the seedlings and saplings of *Macaranga denticulata* on the *Jhum* field and do not uproot them unless the density is too high for cropping. The species grows in poor site conditions and has prolific regenerative activity. Konyaks and the other tribes of Nagaland also keep the trees such as *Trema orientalis*, *Sapium bacatum*, *Grewia spp*, *Quercus spp*, *Schima wallichii*, and *Alnus nepalensis* in the Jhum fields. In the Konyak Jhum field about 42 species could be seen; rice and colocasia being the dominant ones. They have a sound knowledge of mixing rice and colocasia by which the sloping land is covered under vegetation for a greater part of the year, i.e., from April to December. Mixed cropping of rice and colocasia is also practiced by Garo and Khasi tribes of Meghalaya and they cultivate colocasia as a supplementary crop. But, Konyaks grow both the crops as their main crop to meet their food requirements. The common rice varieties cultivated by the Naga tribes are Tangyu, Yamsam, Phuha (Brown rice), Yam, Wungshu, Seshu, Tangyu seshu, Tatak, and Tanyak (black rice). As much as thirteen types of colocasia are grown by the Konyaks. Some of the important ones are Isee, Maywu, Mukshung, Yangshing, Tungmi, Nyakha, Tung, Yakpe, Ngaktung, Tunglu, Tungyey, Tungshu, and Tungkhan. However, with grater urbanization and changing food habit, the area under certain crops such as Job's tears (Coix lacryma-jobi) and some varieties of rice like Yam and Phuwal. The Konyaks have a good sense of fallow management and aware that the leaves and twigs falling from the trees restores the fertility of the *Jhum* land. They count the number of leaf falls and believe that after seven times 'leaf fall' the land becomes mature enough to cultivate. That is why they keep the fallow period as 7 years and deliberately keep the seedlings of tree species for establishment during the resting phase. They religiously protect the Jhum lands from fire during the fallow period. If some accidental fire occurs, the fallow period is extended. This shows the great sense of ecosystem among the Konyak tribes (Bhan 2009).

In the Khasi hills of Meghalaya, shifting cultivation is known as "*Rep Syrti*". Shifting cultivation or *Rep Syrti* practices are of two types—*jhumming* and Bun cultivation. *Jhumming* involves cutting and burning of forest vegetation on sloping lands and using the site for 2–3 years for growing rice, maize, millets, beans, cassava, yam, sweet potato, ginger, chilies, sesamum, and vegetables in mixture thereafter moving to a forest site for repeating the same process (Singh and Prasad 1987; Singh and Dhyani 1996). At times, a single crop of rice is grown in the second year of *jhumming*.

In Bun cultivation (Fig. 3.3), twigs and branches of forest tree species such as *Pinus kesiya*, *Schima wallichii*, *Michelia* species at lower elevations, and *Schima khasiana* in higher elevation along with weed biomass (*Artimisia vulgaris*, *Crotolaria mysorensis*, *Eupatorium odoratum*, *E. adenophorum*, *Imperata cylindrica*, *Inula capa*, *Lantana camara*, *Micania macarantha*, *Panicum khasianum*, *Plectranthus coetsa*, *Rubus ellipticus*, *Saccharum spontaneum*, *Pteridium aquilinum*) from the surrounding areas are kept in heaps at regular interval in the entire area. The buns are usually 2–4 m long, 1–2 m wide, and 0.15–0.35 m in height. They are spaced at 1–2 m depending on the soil depth and are covered with a thin layer of soil in order to burn the whole biomass under anaerobic condition and finally the biomass is slowly converted into ash. The activity is usually done during Feb to March (Singh and Dhyani 1996).

Fig. 3.3 *Bun* method of cultivation along the hill slopes in Meghalaya

Zabo System

Zabo is an indigenous system of farming practice (Fig. 3.4) in many places of Nagaland. *Zabo* means impounding of water. The system has its origin in the Kikruma village of Phek district of Nagaland (Dabral 2002; Pulamte 2008). The system, mostly practiced by the Chakhesang tribe of Nagaland, is a combination of forest, agriculture, animal husbandry, and pisciculture. It is followed up to 100 % slope. Hill top is kept for forestry or catchment area and mid-slope is used for construction of silting ponds and water harvesting tank, land down for animal-yard, and finally the terraced rice fields. These ponds are desilted every year and the material containing soil and forest litter is spread in the fields for manuring. The water harvesting tank is rammed and compacted thoroughly at the bottom and the

Fig. 3.4 *Zabo* system of land management practiced in Nagaland

side walls are plastered with mud and chopped rice straw to prevent seepage losses. The water from the main tank is used for irrigation and is passed through the animal yard to carry dung and urine of the animals to rice fields. This way, around 80–100 kg N, 15–25 kg P, and 50–75 kg K per ha, besides organic matter and micronutrients, are added to the soil annually. A trench is dug-out in the rice field to rear fish. The farmers get reasonably good crop of rice without adding any inorganic nutrient, as well as enough fish for the family. Loss of the soil due to erosion is also below the critical limit (Sharma et al. 1994).

Other Traditional Agroforestry Systems

In the NEH region, trees are deliberately integrated with the crop and livestock production system. A number of crops like maize (*Zea mays*), rice (*Oryza sativa*), oat (*Avena sativa*), ginger (*Zingiber officinale*), turmeric (*Curcuma domestica*), cardamom (*Elettaria cardamomum*), Large cardamom (*Amomum subulatum*), pineapple (*Ananas comosus*), coffee (*Coffea arabica*), and vegetables are grown with forest and fruit tree species such as *Pinus kesiya*, *Alnus nepalensis*, *Schima wallichii*, *Pyrus communis*, *Prunus domestica*, *Areca catechu*, etc. The vegetables include cabbage (*Brassica oleracea* var *capitata*), radish (*Raphanus sativus*), pea (*Pisum sativum*), chilies (*Capsicum frutescens, C. acuminatum, C. annuum*), cowpea (*Vigna unguiculata*), soybean (*Glycine max*), rice bean (*V. umbellata*), pumpkin (*Cucurbita maxima, C. pepo*), cucumber (*Cucumis sativus*), okra (*Abelmoschus esculentus*), brinjal (*Solanum melongena*), tomato (*Lycopersicon esculentum*), bottle gourd (*Lageneria siceraria*), bitter gourd (*Momordica charantia*), potato (*Solanum tuberosum*), *Dioscorea* spp, etc., and climbers like black pepper (*Piper nigrum*), betel vine (*P. betel*), etc., are commonly cultivated. The choice of a particular tree species and intercrop depends upon the climatic conditions of the area and economic importance of the species. Some of the traditional agroforestry systems adopted in the various agroclimatic zones are given below in Table 3.8.

Some of these traditional agroforestry systems found in the region have very high productive potential. The most productive and widely adopted practice in the mild-tropical hills and plain zone is cultivation of pineapple and black pepper with areca nut. This system could generate net return of INR 43,000 ha^{-1}. In the temperate and subalpine zone, plum with potato/cole crops generated a net income of INR 19,000 ha^{-1} (Singh and Dhyani 1996; Bhatt et al. 2001).

Large Cardamom Agroforestry System of Sikkim

Large cardamom (*Amomum subulatum*) is an important cash crop of Sikkim that grows under the forest canopy as well as cultivated with retained trees (Fig. 3.5). Its cultivation is ecologically and economically viable traditional agroforestry system in Sikkim. Earlier, the ancient inhabitants of Sikkim, the Lepchas, used to collect large cardamom from the natural forest. With the progress of civilization, the crop got domesticated; people started cultivating the species by deliberately keeping some trees in the farm to provide the required niche to the cardamom crop. The choice of species was based on their needs experience and traditional knowledge. The crop is now cultivated up to 2,000 m above mean sea level, the lower limit being 600–800 m above msl. Many varieties of cardamom such as Ramsay, Sawney, Golsey, Bharlang, Madhysey, Ramla, etc., are cultivated to get desired quality

Table 3.8 (continued)

Agroclimatic zone	Agroforestry system/ practice	Tree components	Inter crops
		Citrus reticulata	Turmeric + ginger + mustard
		Citrus grandis	Maize + turmeric + cauliflower + mustard leaf + potato
	Agri-silvi-pastoral	*Alnus nepalensis, Schima wallichii, Prunus cerasoides, Terminalia myriocarpa, Castonopsis tribuloides, Litsea polyantha, Macaranga denticulata, Ficus sp.,*	Maize, wheat, pulses, buckwheat, oilseeds, beans, finger millet, broom grass
	Homestead	*Sikkim mandarin, citrus,* Ficus, *guava, pear, pomelo, papaya, pomegranate, avocado, banana,* Urtica sp., Artemisia	Vegetables, tomato, passion fruit, gladiolus, tuberose, marigold, orchids, sugarcane, pig, poultry, cattle, goats, ducks, wild edibles-ferns, nettles, fishery, mushroom, apiary
II. Subtropical Hills to subtemperate hills and Plains (900–1,500 m)	Horti-silvicultural	*Alnus nepalensis/Schima wallichii*	Large cardamom (*Amomum subulatum*)
		Schima wallichii	Pineapple
		Schima wallichii	Ginger/turmeric
		Alnus nepalensis, Schima wallichii, Macaranga pustulata, Albizia sp., Machilus edulis, Saussurea nepalensis, Terminalia myriocarpa, Juglans regia	Large cardamom
	Multi-tier horticultural	Khasi mandarin (Citrus reticulata)	Pineapple/beans/radish/ginger/turmeric/cole crops, etc
III. Subtem perate-Temperate (1,500–2,700 m)	Horti-pastoral system	Pears with vegetables/beans/broom grass	Pears with cabbage, cauliflower, beans, or broom grass
	Horti-silvi-pastoral	Apple, *Juglans regia, Alnus nepalensis, Prunus nepalensis, Quercus* sp., *Betula alnoides, Acer* sp., *Hippophae salicifolia*	Maize, millets large cardamom, potato (table and seed), peas, cabbage, cauliflower, beans, radish
	Horti-silvicultural	Pine with field/vegetable crops	Pine trees with pea, radish, potato, sweet potato, cabbage, turnip, cauliflower, mustard, or maize

(continued)

Table 3.8 (continued)

Agroclimatic zone	Agroforestry system/practice	Tree components	Inter crops
		Plums	Plums, vegetables, pea, radish, cabbage, or cauliflower
	Livestock-based mixed farming	*Betula utilis*, *Acer* sp., *Rubus* sp., *Viburnum erubescens*, *Berberis* sp., *Urtica* sp., *Artemisia* sp.	Goats, pig, sheep, poultry, nomadic herds of yak (dzo's). Grasses like *Imperata cylindrica*, *Arundinella* sp., *Avena* sp., *Eleusine* sp., *Setaria* sp.
	Multi-tier horticultural	Apple with field/vegetable crops	Apple (*Malus pumila*) + potato
Subalpine to Alpine (2,700–4,000 m)	Horti-pastoral-transhumance	*Quercus* sp., *Acer* sp., *Betula utilis*, *Sorbus* sp., *Carex* sp., *Trisetum* sp.	Radish, peas, potato, beans, maize, cabbage, cauliflower, *Brassica juncea* var. *ragosa*, yaks (dzo's) sheep, goats, mules. Grasses *Eragrotis* sp., *Aralia* sp., *Allium* sp., *Iris* sp.
	Livestock-based mixed farming (beyond timberline)-transhumance	*Poa* sp., *Agrostis* sp., *Carex* sp., *Gentiana* sp., *Rumex* sp., *Phlomis rotata*, *Urtica dioca*	Potato, cabbage, peas, *Brassica juncea* var. *ragosa*, yaks (dzos), sheep, and mules

For scientific names of crops and vegetables see the text before the table
Source Chauhan and Dhyani (1990, 1991), Avasti (2006), Bhatt et al. (2006)

and yield. About 30 tree species are used to provide shade to the cardamom plantation and *Alnus nepalensis* is the most commonly used tree species (Singh et al. 1982). The other commonly used woody perennials are *Schima wallichii*, *Engelhardtia acerifolia*, *Eurya acuminata*, *Leucosceptrum canum*, *Maesa chisia*, *Symplocos theaefolia*, *Ficus nemoralis*, *Ficus hookeri*, *Nyssa javanica*, *Osbeckia paniculata*, *Viburnum corylifolium*, *Litsea polyantha*, etc. (Singh et al. 2005). This system supports highly diverse tree species and the diversity index of the cardamom agroforestry system is more than the other agroforestry systems practiced in the region (Sharma and Sharma 1997). The system produces 4.5–5.5 t ha^{-1} $year^{-1}$ woody biomass which is enough to meet the fuel wood requirement (800–1000 kg for curing cardamom produced from one ha of mixed forest) for cardamom processing and other domestic requirement (Sharma et al. 2000).

Majority of the cardamom plantations have Himalayan Alder (*Alnus nepalensis*) as shade tree as the combination of Alder and cardamom is sympatric and proved to be economically viable and ecologically sustainable. *Alnus nepalensis* is a *Frankia* mediated nitrogen fixing tree species and the net primary productivity is more than the other mixed forest species. The practice of using *Alnus nepalensis* as shade tree (Fig. 3.6) has been adopted by the indigenous communities to maintain soil fertility and to sustain productivity (Sharma et al. 1994). Sharma et al. (2000) reported that the yield of finished cardamom under Alder (454 kg ha^{-1} $year^{-1}$) is almost double than the cardamom produced under natural forest canopy (205 kg ha^{-1} $year^{-1}$). The Alder-based cardamom agroforestry system has the potential to generate net income of INR 80,000–90,000 ($2191.6 USD) per ha per annum (Sharma et al. 2009).

Banana-Based Agroforestry in Mizoram

Banana (*Musa* x *paradisiaca*) is a preferred crop in most parts of Mizoram. In the *Khumtung* and *Baktwang* village, people have a belief that

Fig. 3.5 Large cardamom-based agroforestry systems of Sikkim (Northeast India)

Fig. 3.6 Alder (*Alnus nepalensis*)-based agroforestry system in Nagaland (Northeast India

planting banana would protect them against all the natural calamities and misfortune besides giving them good economic prosperity. Along with banana other annual crops, leafy vegetables like Lai Pata (variety of *Brassica juncea*) are grown. Nowadays, people also grow the fruit crops like lemon (*Citrus limon*) and orange (*Citrus sinensis*).

Home Gardens/Homesteads

Home gardens also known as homestead, are usually small plots of land near the house/surrounding the house on which a mixture of annual and perennials are grown together with/without animals and are mostly managed by the family members primarily for their own use and occasionally for commercial purposes. Home gardens are the ancient agroforestry systems with a lot of complexities in structure and multiplicity in function. The composition of the home gardens are governed by the edapho-climatic conditions of the area, socio-economic conditions and cultural orientation, and personal and situation-specific needs of people. Therefore, home gardens usually have an inbuilt strength of meeting the basic needs of people such as spices, condiments, fruits, vegetables, timber, fuel, fiber, etc. Size of these gardens in the northeast India varies from 0.02 to 2.0 ha with a mean from 0.2 to 0.3 ha (Rastogi et al. 1998; Das and Das 2005; Sahoo et al. 2010).

Significance of Home Gardens

Conservation of Biodiversity

They are normally multilayer in structure having a close nutrient cycling system and often act as a repository and testing site for uncommon species, varieties, and land races of plants. Therefore, home gardens are important in situ conservation sites and in accordance with the convention of Biological Diversity Act, Article 7, 8, and 10 (c), inventorization of such areas can help in identification and conservation of biological diversity (Das and Das 2005). For examples, certain very hot chili varieties (varieties of *Capsicum frutescens*) with high capsicin content were selected and cultivated in Lotha and Konyak Naga home gardens. Many wild, rare tree species like *Aquilaria malaccensis* and *Vatika lanceaifolia* are conserved in home gardens because of their commercial value. Konyak Nagas successfully domesticated *Aquilaria agallocha*. Other rare species conserved in the home gardens are *Licuala peltata*, *Streblus asper*, *Meyna laxiflora*, etc. (Das and Das 2005). Multipurpose forest trees are cultivated in home garden of Kara (Rastogi et al. 1998).

Nutritional Security to the Family

Home gardens contain a number of species that fulfills the local dietary requirements, vitamins, and mineral supplements. Food crops dominate most of the home gardens. Sahoo et al. (2010) reported 107 species from different home gardens of Mamit district of Mizoram out of which 32 % were vegetables, 15 % fruits, 13 % medicinal plants, and 6 % spices. Nyishi tribal people in the foothills of Arunachal Pradesh grow on an average 15–20 species in their home gardens. A detail survey found that about 80 species are of ethno botanical importance and are grown by the Nyishi tribe in their home gardens and traditional agroforestry systems (Tangjang and Arunachalam 2009; Deb et al. 2009). In Konyak home gardens in Nagaland, about 157 plant products are used in local diets (Rastogi et al. 1998).

Strengthens Household Economy

Home gardens provide additional benefit and income for people. This is mostly achieved through saving money which otherwise would have been spent to buy the goods from the market. It is reported that in Indonesia and Nicaragua, home gardens contribute about 21.1 and 35 % of the total income, respectively. In southwest Bangladesh (Rahman et al. 2006) and Northeastern Bangladesh (Rahman et al. 2005), the income derived from home gardens is 15.9 and 11.8 % of the total household income, respectively. In Meghalaya, it was 7 % of the annual gross income per household (Tynsong and Tiwari 2010).

Angami Home Gardens

Angami Nagas are well acquainted with the environment around them and have learned to use and manage their limited land and water resources through experience and experimentations over the generations. They have developed excellent system of integrating Alder into their crop production system as they are aware about the beneficial affects of the species thorough their long experience. Angami territory is restricted to the southern Kohima district. In Angami territory, a large amount of land is available for wet or terraced rice cultivation. It is a permanent agricultural system. However, because of intensive cultivation practices and low quality soils, the Angamis sometimes prepare new terraced fields and abandon the old ones. If these abandoned fields are close to the village, they are often used as home gardens.

The tradition of home garden is well developed and many are so extensive that there are areas for cultivation of maize and millets. Also large portion of Angami home gardens are occupied by commercial crops such as oranges and potatoes.

Three different types of homegardens are maintained by the Angami Nagas. They are (a) a small kitchen garden around the house with small plots for chili and vegetables with one or two fruit trees; (b) *tejeye*—larger traditional home gardens away from the house situated on terraced fields on the upper level and cultivated like home gardens but with planned commercial objectives, and (c) *mejeye*—community home gardens on community owned terrace fields cultivated like home gardens. Management practices are different for these three types of home garden systems. In *tejeye,* the main emphasis is on commercially important species such as potato and fruit trees, and intensive water management is observed. *Tejeye* is in fact the extension of terrace fields but is different in terms of plant composition. These home gardens have more perennials than are found in kitchen gardens. Decision making and selection of species are driven by market forces. Greater biodiversity is maintained in the kitchen gardens.

Many indigenous practices have been observed in the *tejeye*, for example, the traditional practices of manuring with *pidi* (*Solanum* leaves) to get rid of pests and insects. Alder trees increase soil fertility and improve the growth of fruit trees. Therefore, a few alder trees are maintained in these gardens. Trees such as *Melia* and *Cedrella* are also planted to increase soil fertility and control insect pests, while *Prunus* spp. are used as a wind break on terrace bunds.

Home Gardens of the *Konyak* Nagas of Nagaland

Konyak Nagas, considered the oldest tribe of Nagaland, are in fact expert agriculturists who, even today live in harmony with nature. Konyak(s) have successfully maintained a sustainable supply of natural resources for many years through various land management systems. These traditional systems of land management have been developed through indigenous knowledge blended with innovative skills. One of these traditional systems of land management is the home gardens maintained around the houses. A study conducted in three villages Liangnyu, Mon, and Tanhai from Mon district of Nagaland highlight the following details of home gardens.

Thirty-two home gardens were surveyed from three villages, of which 10 home gardens were from Liangnyu, 11 home gardens from Mon, and 11 from Tanhai (covering around 10–20 % of the total households). Tanhai had the maximum number of species (122) and the largest average size of home gardens. 87 species have so far been recorded from Mon and 45 species from Liangnyu. Konyak home gardens exhibited a wide diversity in size, shape, location, and composition. The management, organization, and spatial patterns show the maximum use of available land using different combinations of trees, shrubs, and herbaceous plant species in a year. Generally, selection of plants depends on daily necessities. Many of these species have multiple uses. The owners show a tendency toward cultivating mainly edible fruits and vegetables. Other categories include plants yielding timber, fuelwood, fumigators, masticators, species, beverages, construction materials, basketry material, medicines, poisons, and ornaments.

Ingkhol Home Gardens: The Home Gardens of *Meteis* of Manipur

The home gardens in Manipur is a traditional conservation area. The gardens locally known as *Ingkhol* are variable in size and shape. The home gardens vary from hill to valley, reflecting the variations of soil topography, climatic conditions, and cultural practices of the people. It appears that the regimes of moisture and nutrients must be varying in different areas, the penetration of light into gaps creating a heterogeneous landscape and promoting patchiness. In the valley, home gardens are quite diverse and complex. In hilly areas in the village that are permanent, large home gardens contain species for food, fiber, fuel, timber, medicines, and species of sociocultural importance. The hill home gardens have the following features:

(a) Cultivation of wild species such as oaks, cedars, toonas, and red wood.
(b) Economically important plants like *Ericas* and bamboos.
(c) Varieties of chili, tomato, and cabbage along with other local vegetables.
(d) Varied floristic patterns, representing trees form valley and hills.

These home gardens are a mosaic and include the following components:

(a) Perennial, multipurpose trees of horticultural importance, timber or medicines.
(b) Various microhabitats such as water resources with aquaculture of fish, mat grass, *Euryale ferox*, a traditional delicacy, fishery, poultry, cowshed, rabbit house, etc.
(c) Cash crop section for annuals.
(d) Frontal hut, *Shangoy* for socioreligious functions.
(e) Secret basil plant in the center of the court yard (*Shumang*)
(f) Separate areas for raising ornamentals, bananas, spices, vegetables.
(g) Fencing with bamboo species.

Home gardens are fenced by different bamboo species usually in the front and back, whereas the perennial species may demarcate the sides of it. The space for the traditional hut where the socioreligious functions are held is the characteristic feature of the traditional system in which the walls are made from thickly set reeds plastered with mud and the roofs are covered with thatch grass. This is the place where cattle may often be put for want of space. Fragrant flowers like *Thevesia, Nyctanthus, Jasminum, Lilics,* etc., and multipurpose banana plantations surround the deity temple located in front of the house. The back side of the residential unit is grassland with many grasses used for thatching, binding, and other purposes. Perennial wild or cultivated multipurpose trees characteristic of the region demarcate the ultimate boundary. There are five different canopy layers viz., emergent layer (15 m or more tall), main canopy (10–15 m tall), under story (5–10 m tall), shrubs (1–5 m), and herbs (less than 1 m) layer in the traditional home gardens maintained by Meiteis in the plains of the lower altitudes. The emergent layer mainly composed of *Artocarpus heterophyllus*, *Bambusa balcoa,* and *Acacia catechu.* The main caopy layer was dominated by *Mangifera indica*, *Parkia timoriana*, *Tectona grandis*, and *Toona ciliata*. The under story consisted of *Citrus grandis*, *Musa balbisiana*, and *Toona ciliata* and the shrub layer was dominated by *Melastoma malabathricum*, and *Adhatoda vasica*. The herb layer composed of vegetables and tree saplings such as *Clerodendron indicum*, *Corchorus capsularis*, *Areca catechu*, and *Citrus grandis*. The pond dykes were used for planting of *Neptunia prostrata* and *Ipomoea aquatica*. Vegetables formed the predominant category of plants followed by fruits and medicinal plants. Most of the households usually have *Parkia timoriana* and other legumes which they exchange among themselves. Plants such as *Eupatorium birmanicum*, *Ocimum santum,* and *Toona ciliata* are planted by the Meiteis for religious purpose. The plants such as *Pogostemon purpurascens* and *Ageratum conyzoides* are used for haircare. Most of the home gardens have a separate zone for spices like *Allium odorum, Eryngium foetidum*, *Houttuynia cordata,* etc. This practice of growing traditional spice crops in the *Ingkhol* by the Meiteis is a '*living heirloom*' that needs more in depth analysis.

Home Gardens of War Khasi Tribe in Meghalaya

War Khasi people, tribal community in the south Meghala, have a long tradition of forest conservation. Tynsong and Tiwari (2010) in a study in five villages in the south Meghalaya observed that the size of the home gardens varied from 200 to 3,500 m^2. They use seeds, seedlings, and vegetative propagules to regenerate home garden plants. Soil fertility of the home gardens is maintained by the addition of leaf litter, feces of reared animals, and kitchen waste manure. The home gardens usually have four strata. About 197 plant species (70 trees, 41 shrubs, 50 herbs, 23 climbers, and 13 epiphytes) were reported to be used by the tribe. They place the plants in distinct groves such as arecanut grove, banana grove, fruit grove, vegetable garden, orchid grove, etc. Arecanut zones are the most important ones that are common to most of the home gardens. The average income from the home gardens is about 7 % of the total household income.

Chuktuah huan, the Home Gardens of Mizoram

Home gardens, locally known as *Chuktuah huan*, are the age old common practice in Mizoram. The home gardens normally have 3–4 strata/tiers having very high species diversity. The top layer or the dominant canopy consists of perennial tall trees like *Parkia roxburghii*, Jack fruit, Bamboo, etc. The second layer is mostly occupied by papaya, banana, tapioca, and yams. The first layer or the lowest strata consists of vegetables, tuber crop, pineapple, and grasses. The vegetables are also cultivated whereas the woody perennials and MPTs are planted near the periphery and boundary of the backyard. They also rear pigs, cows, chicken, and ducks for meat and egg purposes.

Research and Developments in the Field of Agroforestry in the NEH Region

Agroforestry is a complex landuse system which is practiced by people to derive multiple benefits from a piece of land. The Research and Development (R&D) aims to optimize these production systems with respect to economic viability and utilization of the natural resources like soil, water, and the biological diversity. It is a continuously evolving process and highly dynamic in nature that changes with time and space.

Multipurpose Tree Species for Agroforestry Systems in Northeast

Trees provide a range of significant products and services to rural and urban people. For any agroforestry programs, it is necessary to make an assessment of preferred and useful multipurpose tree species. The choice of species depends on the soil and climatic conditions of the locality and the nature of basic needs of people. Some of the MPTs suitable for agroforestry systems of the northeast region are listed in Table 3.9.

Fruit Trees Based Cropping Systems

NEH region has ample potential for fruit trees based agroforestry systems. For the development of suitable agorforestry system for the region, an experiment was conducted at ICAR Research Complex for NEH Regopn, Umiam where various fruit trees were grown with different combinations of agricultural crops. The results of different tree crop combinations have been summarized (Table 3.10). Khashi Mandarin (*Citrus reticulata*) was planted initially at a tree density of 800 trees ha^{-1}. However, it was observed that 400 trees ha^{-1} produced maximum yield. Average yield of Mandarin was 12.8 kg per tree after 7 years of plantation which increased up to 57.3 kg per tree after 12 years of plantation. In the inter row spaces of fruit trees, groundnut (*Arachis hypogaea*), soybean (*Glycine max*), turmeric (*Curcuma domestica*), ginger (*Zingiber officinale*), and local taro (*Colocacia esculenta*) were cultivated. The average productivity of these crops were 1.6, 1.5, 15.4, 10.0, and 17.0 t ha^{-1}, respectively.

With the increase in age of the tree crop and gradual closure of the over storey canopy, yield of groundnut and soybean started to decline. This opened scope to grow crops like ginger and turmeric and replace these oilseeds and pulse crops in the mandarin-based system. Among the field crops, ginger was the most remunerative. However, major share of income in these systems was generated by Khasi mandarin, which was about 80 % of the total income irrespective of the associated field crops.

In the guava (*Psidium guajava (*cv *Allahabad Safeda)*-based system was planted in association with five field crops namely groundnut, chili, soybean, turmeric, and ginger. Fruit yield of guava increased up to eighth year after which it started to decline because of fruit borer infestation. After 2 years of plantation, fruit yield was about 5.6 t ha^{-1} year^{-1}. Average yield of groundnut, chili, soybean, turmeric, and ginger were 1.8, 1.0, 0.7, 6.4, and 4.8 t ha^{-1}, respectively. Majority of the income was from ginger crop as the market price of guava was low as compared to mandarin.

In Assam, lemon (*Citrus limon*)-based cropping system was relatively more profitable than the guava-based system. Maximum fruit yield was attained after 7 years of plantation (34.8 kg per tree). Different intercrops such as ginger, turmeric, soybean, and radish were cultivated in the inter row spaces of these tree crops. The net return was maximum in Assam lemon + ginger (INR 35,000) followed by Assam lemon + radish crop (INR 14,120).

MPTs-Based Agroforestry Systems

NEH region is characterized by presence of vast tract of forest areas. Therefore, livelihood of the population is also strongly influenced by the forest wealth especially the tree resources. Farmers deliberately keep some of the multipurpose trees (MPTs) in their fields to meet their multifarious requirements. Therefore, some of the MPTs based agroforestry systems were also evaluated at the ICAR Research Complex at Umiam for their suitability for the region. Indigenous trees of the region like *Alnus nepalensis*, *Gmelina arborea*, *Michelia oblonga*, *Parkia roxburghii*, *Prunus cerasoides*, and *Symingtonia populnea* were planted at a density of 416 trees ha^{-1}.

Table 3.9 Important multipurpose tree species suitable for farming in different agroforestry systems

Plant species	Altitudinal distribution (m asl)	Important uses
MPTs		
Aesculus assamica	500–900	Fuel, small timber, fast growing
Albizia chinensis	700–1500	Fuel, fodder, timber, N_2 fixing species fast growing
A. lebbek	350–800	Fuel, fodder, timber, N_2 fixing species fast growing
A. procera	400–700	Fuel, fodder, timber, N_2 fixing species fast growing
Alnus nepalensis	700–2,500	Fuel, fodder, timber, N_2 fixing species fast growing
Altingia excelsa	750–2,100	Fuel, fodder, timber, fast growing
Anogeissus acuminata	800–1,300	Fuel, fodder, timber, slow growing
Anthocephalus chinensis	400–900	Fuel, fodder, timber, fast growing
Artocarpus chaplasha	500–1,500	Fuel, fodder, timber, fast growing
Bauhinia variegata	650–1,500	Fuel, fodder, ornamental, fast growing
Castanopsis indica	650–1,950	Fuel, timber, slow growing
Chukrasia velutina	200–1,400	Fuel, timber, fast growing
Cordia dichotoma	550–1,500	Fuel, fodder, fast growing
Debragesia salicifolia	700–1,800	Fodder
Duabanga grandiflora	150–800	Fuel, fodder, timber, fast growing
Exbucklandia populnea	900–2,400	Fuel, fodder, small timber
Ficus altissima	700–1,200	Fodder, fuel, fast growing
F. curtipes	600–1,000	Fodder, fuel
F. cyrtophylla	500–900	Fodder, fuel
F. elmeri	600–1,200	Fodder, fuel
F. gibbosa	1,000–1,800	Fuel and fodder
F. glomerata	700–1,800	Fuel, fodder, figs edible, fast growing
F. hirta	650–1,350	Fodder, fast growing
F. hispida	600–1,300	Fuel, fodder, figs edible, fast growing
F. hookeri	700–1,500	Fuel, fodder, fast growing
F. oligodon	600–1,400	Fodder, figs edible
Fraxinus floribunda	1,500–2,700	Light timber, ornamental, fast growing
Gmelina arborea	350–1,200	Fuel, fodder, timber, fast growing
Kydia calycina	500–1,200	Fuel, fodder, timber, slow growing
Largerstroemia speciosa	250–1,950	Fuel, fodder, timber
Litsea polyantha	400–900	Fuel, fodder, small timber, fast growing
Livistonia jenkinsiana	500–1,100	Leaves for roof making, new apicals as broom, fruit pulp edible
Macaranga denticulata	800–1,350	Fuel, fast growing
Mesua ferrea	350–1,700	Fuel, fodder, timber, avenue tree
Michelia champaca	200–900	Fuel, timber, avenue, fast growing

(continued)

Table 3.9 (continued)

Plant species	Altitudinal distribution (m asl)	Important uses
M. doltsopa	1,500–2,400	Fuel, timber
Moringa oleifera	100–800	Fodder, tender pods edible, fast growing
Morus alba	250–1,200	Fuel, fodder, sericulture, light timber, fast growing
Parkia roxburghii	500–1,500	Tender pods edible, light timber, fast growing
Pinus kesia	800–2,500	Fuel, timber, slow growing
Prunus cerasoides	800–1,500	Fuel fodder, fast growing
Prunus nepaulensis	1,800–3,000	Fuel, fodder
Quercus griffithii	1,500–2,650	Fodder, fuel, timber, slow growing
Q. serrata	800–2,500	Fodder, timber, slow growing
Q. semiserrata	600–1,500	Fodder, timber, slow growing
Salix tetrasperma	1,200–2,100	Fuel, fodder, small timber, fast growing
Sapindus mukorossi	150–1,500	Fuel, fodder, timber, ornamental, soap nuts as detergents, fast growing
Schima wallichii	350–1,700	Fuel, fodder, timber, fast growing
Shorea assamica	150–800	Fuel, timber
Sorbus cuspidata	120–450	Fuel, timber
Tectona grandis	100–650	Timber, slow growing
Terminalia bellirica	500–1,500	Fuel, timber, fruit medicinal, slow growing
T. chebula	450–1,500	Fuel, timber, fruit medicinal, slow growing
T. myriocarpa	300–1,100	Timber, slow growing
Toona ciliata	150–1,050	Timber, slow growing
Bamboos and canes		
Bambusa balcoa	10–1,200	Young shoot edible, fodder, culm used for various household activities, fast growing
B. bambos	150–350	Live fencing, pole, agricultural implements
B. polymorpha	15–110	Young shoot semi edible, house roofing, partition wall
B. jaintiana	1,100–1,900	Live fencing, fishing rod, winnowing tray, etc.
B. khasiana	1,100–1,900	Live fencing, fishing rod, winnowing tray, etc.
B. muliplex	800–2,100	Young shoot semi edible, live fencing, cane industry
B. nana	100–450	Construction purposes and cane industry
B. nutans	650–1,700	Young shoot edible, fodder, construction purposes
B. tulda	10–850	Young shot semi edible, fodder, construction purposes
B. vulgaris	700–1,700	Ornamental, construction purposes
B. wamin	550–1,000	Ornamental, construction purposes
Chimonobambusa callosa	800–1,300	Young shoot edible, construction purposes
C. hookeriana	500–2,100	Young shoot edible, construction purposes
Dendrocalamus giganteus	35–1,350	Young shoot edible, construction purposes
D. hookerii	1,100–1,900	Young shoot edible, fodder, construction purposes
D. longispathus	600–1,400	Young shoot edible, fodder, construction purposes
D. membranaceus	1,350–1,950	Young shoot edible, fodder, construction purposes
D. sikkimnensis	550–1,150	Young shoot edible, fodder, construction purposes

(continued)

Table 3.9 (continued)

Plant species	Altitudinal distribution (m asl)	Important uses
Gigantochloa rostrata	1,050–2,000	Young shoot edible, construction purposes
Melocanna baccifera	10–1,200	Young shoot edible, construction purposes
Phyllostachys bambusoides	850–1,850	Young shoot edible, cane industry
Schizostachyum dullooa	15–250	Young shoot edible, live fencing, cane industry
Teinostachyum wightii	550–1,100	Young shoot edible, live fencing, cane industry
Calamus spp	200–700	Furniture and various other household activities
Daemonorops spp	150–600	Furniture and various other household activities
Hedgerow species		
Cajanus cajan	150–950	Fuel, fodder, seeds edible, N_2 fixing, fast growing
Crotalaria pallida	150–960	Fuel, N_2 fixing, fast growing
Desmodium rensonii	400–1,000	Fuel, fodder, N_2 fixing, fast growing
Flemingia macrophylla	200-950	Fodder, N_2 fixing, fast growing
Indigofera tinctoria	250–1,200	Fuel, fodder, N_2 fixing, fast growing
Leucaena leucocephala	200–950	Fodder, fuel, N_2 fixing, fast growing
Mllettia ovalifolia	500–1,200	Fodder, N_2 fixing, fast growing
Tephrosia candida	300–1,000	Fuel, fodder, N_2 fixing, fast growing
Thysanolaena maxima	500–1,500	Fodder, Spikes as broom, fast growing

Compiled from various sources

After 12 years of growth, volume production was assessed for each species besides fuel and foliage yields. Volume production varied among different species and it was highest (2.07 m^3 per tree) for *Parkia roxburghii*, and lowest (0.43 m^3 per tree) for *Symingtonia populnea.* Though monetary input for each species was not considerably different, the output was highest (INR 1,854 per tree) for *Parkia roxburghii* followed by *Gmelina arborea* (INR 1,625 per tree) and *Michelia oblonga* (INR 1,157 per tree). After 12 years, on an average, farmers could get benefit of INR 361,000 from one hectare cultivation of these tree species only.

With these tree species, field crops soybean (*Glycine max* cv *Alankar*) and linseed (*Linum usitatissimum*) were intercropped up to fifth year. The net return was INR 1,625 ha^{-1} for soybean. Linseed was not economical as the net return was only INR 389 ha^{-1}. After that, pineapple (*Ananas comosus* cv *Kew*) was introduced with a density of 32,625 plants ha^{-1}. This is also most remunerative fruit crop of NEH region. The average net profit through intercropping of pineapple was INR 18,805 ha^{-1}, irrespective of tree species. After 5 years of plantation, crop composition was changed, and partial shade loving ginger (*Zingiber officinale* cv *Nadia*), turmeric (*Curcuma domestica* cv *RCT-1*), and local taro (*Colocasia esculenta*) were intercropped with MPTs. The net return was highest for ginger as compared to turmeric and taro, irrespective of tree species. Thus, based on 12 years research findings, ginger was found to be the most profitable intercrop, followed by pineapple.

Table 3.10 Performance of fruit tree-based cropping systems in the NEH region (No. of trees in each case was 400 ha^{-1})

Tree crop	Field crop	Variety of field crop	Net return (INR ha^{-1})
Khasi Mandarin	Groundnut	JL-24	4,541
	Soybean	Alankar	19,625
	Turmeric	RCT-1	30,375
	Ginger	Nadia	33,416
	Taro	Local	18,583
Guava	Groundnut	JL-24	3,000
	Soybean	Alankar	916
	Turmeric	RCT-1	2,750
	Ginger	Nadia	15,791
	Chilies	Local	1,125
Assam Lemon	Soybean	Alankar	2,583
	Turmeric	RCT-1	1,916
	Ginger	Nadia	36,625
	Radish	Japanese White	2,583

Prices are based on the market price of late 1990s
Source Annual Reports of ICAR Regional Research Complex of NEH

Three-Tier Agroforestry System in the NEH Region

In one experiment Alder (*Alnus nepalensis*-promising nitrogen fixing tree species) was planted as a tree crop during 1987 and tea (*Camellia sinensis*) was planted in 1993 as second storey crop at a density of 12,350 plants ha^{-1}. The investment for Alder and tea was INR 11,398 and 36,035 ha^{-1}, respectively. Besides tea, large cardamom (*Amomum subulatum*), turmeric, ginger, taro, and black pepper (*Piper nigrum*) were intercropped. Alder produced 850 kg ha^{-1} biomass of pruned material and 2.4 t ha^{-1} biomass of foliage. Green bud production of tea ranged from 4.4 to 6.4 t ha^{-1} for a period of 5 years with an average production of 5.9 t ha^{-1}. Productivity of large cardamom was 640 kg ha^{-1}. Ginger, turmeric, and taro produced 7.9, 16.5, and 17.2 t ha^{-1}, respectively. Black pepper was found to be sensitive to frost injury. Therefore, no significant yield could be obtained from this crop. Among various crops, net benefit was maximum (INR 33,111 ha^{-1}) through large cardamom, followed by tea and ginger. On an average, the multistoried agroforestry system could generate a net annual return of INR 12,884 ha^{-1}.

Silvopastoral Systems

Som (*Machilus bombycina*) tree is suitable for raring of *Munga* silkworm in Assam and its wood is used for tea-boxes. The tree attained average height of 6.75 m, 10.30 cm dbh and 0.046 m^3 volume in 5 years of plantation. Maize (*Zea mays*, cv. *Vijay* Composite) and broom grass (*Thysanolaena maxima*) were intercropped with it. Broom grass was cultivated on the terrace risers, covering total area of 480 m^2. Average grain production of maize was 1.2 t ha^{-1} in association with this tree crop as compared to 1.35 t ha^{-1} in control plots. Broom grass produced 6.3 t ha^{-1} of flower (most remunerative part of it), 8.6 t ha^{-1} of green fodder, and 3.6 t ha^{-1} of dry fuel wood. This system generated net return of INR 23,444 ha^{-1}.

Bhatt et al. (2010) evaluated seven MPTs such as *Acacia auriculaeformis*, *Alnus nepalensis*, *Bauhinia purpurea*, *Exbucklandia populnea*, *Ficus hookeri*, *Michelia champaca*, and *Michelia oblonga* with broom grass in the understorey in the mid hill conditions of Meghalaya. After 10 years, highest standing volume was recorded in *Acacia auriculaeformis* (220 m^3 ha^{-1}) followed by *Exbucklandia populnea* (120 m^3 ha^{-1}) and *Alnus nepalensis* (114 m^3 ha^{-1}). Yield of broom grass cultivated in the tree interspaces varied from 3.8 t ha^{-1} dry biomass (under *Bauhinia purpurea*) to 2.4 t ha^{-1} (under *Alnus nepalensis*).

Nonarable hilly areas with high slopes (>45 %) and low soil depth (<0.6 m) can be managed under suitable tree and grass combinations under livestock-based silvo-pastoral system. In an experiment at ICAR Research Complex for NEH Region, Meghalaya, 13.54 t ha^{-1} (dry matter) forage yield was obtained from combination of stylo (*Stylosanthes*) legume and *Setaria* grass with Alder. A combination of stylo and guinea grass (*Panicum maximum*) with Alder could produce 11.30 DM t ha^{-1}. In addition, Alder could provide 1.32 t ha^{-1} of fuel wood from the pruned branches.

Fish-Based Agroforestry System

The composite unit of aquaculture was consisted of paddy, vegetables, large cardamom, and fish culture besides bean cultivation on bund area of pond. It was revealed that among various components, fish culture generated maximum monetary returns (INR 36,000 ha^{-1}), followed by radish (INR 33,850 ha^{-1}), and cured large cardamom (INR 29,000 ha^{-1}) and brinjal (INR 25,500 ha^{-1}) cultivation, respectively. Average income from aquaculture-based system was INR 16,976 ha^{-1}.

Sericulture-Based Agroforestry System

Seven mulberry (*Morus alba, M. serrata, M. laevigata*) varieties, seven silkworm breeds including a bivoltine breed (NB-18) were studied for their yield and rearing performance. The results obtained are presented in the Table 3.11.

Intensive Integrated Farming System

Intensive integrated farming system (IIFS) is based on the concept that there is no waste in the system and the later is only a misplaced resource which can become a valuable material for another product (Edward et al. 1986). It is a more refined and holistic approach of land use system through practices in which a number of production components are integrated with the primary objective of developing a self-sustainable system. In IIFS all the components of agriculture like crop, fish, forestry, and horticulture are integrated in a complementary way (Bhatt and Bujarbaruah 2005). Fish fingerlings were introduced in each ponds @ 6000 fingerlings ha^{-1} with species composition of catla (*Catla catla*)—20 %, rohu (*Labeo rohita*)—10 %, mrigal (*Cirrhinus mrigala*)—20 %, silver carp (*Hypopthalmichthys molitrix*)—20 %, grass carp (*Ctenopharyungodon idella*)—20 %, and gonius (*Labeo gonius*)—20 %. Duck (Indian Runner and Khaki Campbell), pig (Large Black), layer birds (White Leghorn), goat (Black Bengal), and cow (*Holstein*) were reared and integrated with fishery. One pond was kept as control to compare the fish growth without integration of livestock/poultry/ducks. Vermicompost, liquid manure, and mushroom cultivation were started in IIFS. The five subsystems of IIFS were developed as detailed in Table 3.12.

The monetary input and output has also been calculated for each subsystem (Table 3.13). The total output/input ratio was highest (1.76) in crop–fish–dairy–MPTs–fruit trees–hedge rows–vermiculture–liquid manure–broom grass followed by Broiler chicken–crop–fish–duck–horticulture–nitrogen fixing hedge row (1.58). The monetary output/input could further increase if family labor is engaged for adopting IIFS (For detailed report, refer to Bhatt and Bujarbaruah 2005).

Water Conservation and Utilization

Out of 10 ha experimental site, 3.31 ha area was marshy where cultivation of crop was not possible. To rehabilitate such land, seven earthen water harvesting structures were created. Average cost involved for establishing these small water harvesting structures of 0.10–0.15 h, was INR 43,200 per pond. The average capacity of

Table 3.11 Yield of mulberry and silkworm cocoon in sericulture based agroforestry system

Mulberry variety	Plant height (m)	Yield (t ha^{-1} $year^{-1}$)			Net returns from cocoon (INR ha^{-1})
		Leaf	Cocoon	Fuelwood	
TR-4	1.70	19.1	0.81	6.4	33,449
TR-10	1.69	16.6	0.70	6.3	27,125
BC-259	1.44	15.2	0.65	5.7	23,627
S-1635	1.51	18.2	0.77	6.1	31,085
C-7635	1.52	16.5	0.70	5.6	26,865
Kanva-2	1.43	14.1	0.60	5.7	21,715
Local	1.28	9.1	0.39	4.1	8,215

Dhyani et al. (1996)

water retention ranged from 1000 to 1800 m^3 and average cost of one cubic meter water harvesting was estimated to be INR 32.36. It indicated that one liter of water could be harvested/conserved at price of INR 0.03 in first year itself which includes the cost of excavation, ramming, slope stabilization, plantation cost of planting Congo and guinea grass, spillway making, etc. Second year onward there was no cost involved except the maintenance cost whereas water could be harvested regularly. The details of water used for various purposes have been shown in Table 3.14.

Effect of Agroforestry on Soil and Water Resources

Effect on Soil Physico-Chemical Properties

Tree species ameliorate soil by adding both above and below ground biomass into the soil system. However, variations do exist in the inherent capacity of different tree species in rehabilitating degraded lands. Five different trees species suitable for agroforestry systems were studied at ICAR Research Complex for NEH Region at Umiam, Meghalaya by Saha et al. (2007). Soil samples were collected from 0–15 cm and 15–30 cm soil depth under five multipurpose tree species such as Khasi pine (*Pinus kesiya*), Alder (*Alnus nepalensis*), Tree bean (*Parkia roxburghii*), Champak (*Michelia oblonga*), and Gambhar (*Gmelina arborea*). A control plot in the form of natural fallow was also maintained near these tree-based land use systems for the purpose of comparison. Effect of tree species on bulk density (BD), organic carbon (OC), and porosity of the soil was significant. All the tree species lowered BD, and increased OC and porosity as compared to the natural fallow (Table 3.15).

The water stable aggregates (>0.25 mm) increased significantly under the different multipurpose tree species. Water stable aggregates were highest for the soils under *Pinus kesiya* (82.4 %) followed by *Michelia oblonga* (78.5 %) and *Alnus nepalensis* (77.6 %). Soil erodibility decreased with the tree species to the extent of 23.1–43.6 % as compared to control. Therefore, these species were instrumental in decreasing erodibility of soils of the NEH region. Protection of soils directly against erosive forces of raindrop and surface run off by improving physical and hydrological parameters of soil have been reported in many studies in India (Grewal and Abrol 1986; Deb et al. 2005; Jha and Mohapatra 2009).

Effect on Soil Hydrological Properties

Tree species improved moisture retention capacity of soil as compared to the control (Table 3.16). At -0.03 M Pa suction, soil moisture under different tree species was 21–36 % more than that of the control. Similar was also the trend in available water under the different tree-based systems.

Table 3.12 Description of IIFS models

Farming system	Land use component with area (ha)	Area (ha)	Description
Broiler chicken–crop–fish–duck–horticulture–nitrogen fixing hedge row	Pond—0.15 Pond dyke—0.03 Duck shed—0.016 Broiler shed—0.006 Field crop—0.75	1.06	In upland area, finger millet (0.18 ha), maize (0.30 ha), and rice bean (0.12 ha) followed by ginger and turmeric were cultivated during *kharif*. In lowland area, paddy (0.65 ha) and mustard 0.30 ha were cultivated. During *rabi* season potato, tomato, cabbage, knol khol, and radish were cultivated. Nitrogen fixing shrubs were planted on contour bunds, fodder grasses, and fruit trees were raised on pond dykes and farm boundaries. Ducks were reared (72 nos.) on pond dykes. Composite fish culture was practiced and 900 fingerlings were stocked
Crop-fish-poultry-multipurpose trees	Pond—0.12 Pond dyke—0.04 Poultry shed—0.01 Field crop—0.80	0.97	In upland area, paddy (0.45 ha) and rice bean (0.05 ha) during *kharif* and buckwheat (0.50 ha) in *rabi* season was cultivated. In lowland area paddy (0.30 ha) in *kharif* and potato (0.25 ha) and french bean (0.05 ha) were cultivated. Fodder grasses and fruit trees were raised on pond dyke and farm boundaries. Layer birds (52 nos) were raised on pond dykes. Composite fish culture was practiced and 720 fingerlings were stocked
Crop–fish–goat–MPTs–hedge rows	Pond—0.10 Pond dyke—0.035 Goat shed—0.008 Field crop—0.80 Hedge row—0.10	1.04	In upland area, paddy (0.30 ha), ginger (0.30 ha), turmeric (0.20 ha) during *kharif* and mustard (0.30), tomato (0.40 ha), and radish (0.10 ha) during *rabi* season were grown. Fodder grasses, MPTs and fruit trees were cultivated on pond dike and farm boundary. Goats (6 nos) were reared on pond dyke. Composite fish culture was practiced and 600 fingerlings were stocked
Crop–fish–pig–bamboo–MPTs–fruit trees–hedge rows	Pond—0.12 Pond dyke—0.035 Pig shed—0.001 Field crop—0.80 Hedge row—0.09 .	1.05	In upland area, paddy (0.30 ha), colocasia (0.10 ha) and maize (0.40 ha) during *kharif* and brinjal (0.10 ha), radish (0.05 ha), potato (0.30 ha), and buckwheat (0.15 ha) during *rabi* season were cultivated. MPTs and fruit trees were raised on pond dykes and farm boundaries. Edible bamboo species were also cultivated on farm boundary. Hedge rows of different species were planted on contour bunds. Vermicompost was prepared in two units each of 12′ × 6′ x 2′ size. Pigs (2 nos) on pond dykes. Composite fish culture was practiced and 720 fingerlings were stocked

(continued)

Table 3.12 (continued)

Farming system	Land use component with area (ha)	Area (ha)	Description
Crop–fish–dairy–MPTs–fruit trees–hedge rows–vermiculture–liquid manure–broom grass	Pond–0.12 Pond dyke—0.06 Dairy shed—0.016 Field crop—0.80 Hedge row—0.17	1.17	In upland area paddy (0.60 ha) was cultivated. Broom grass (0.10 ha) and job's tear (0.10 ha) were cultivated along the water channels. MPTs and fruit trees with fodder grasses were raised on pond dyke and farm boundary. Cattle (2 milch cows and 2 calves) was reared. Oyster mushroom was cultivated in 8 m × 3 m × 2.5 m size unit. Liquid manure was prepared in 3 units 3′ × 3′ × 2.5′ capacity. Vermicomposting was done in 6 units of 1 m × 1 m × 0.75 m. Composite fish culture was practiced in the six ponds. Composite fish culture was practiced and 720 fingerlings were stocked
Upland crops, and fish farming without integration (control)	Pond—0.10 Pond dyke—0.05 Crop area—0.80	0.95	In upland area, paddy (0.40 ha) and maize (0.40 ha) during *kharif* season and buck wheat (0.20 ha) and frenchbean (0.30 ha) were grown. Fruit trees were grown on pond dyke. Composite fish culture was practiced and 600 fingerlings were stocked

Table 3.13 Monetary output/input pattern (INR year^{-1}) of IIFS

Farming system	Total input	Total output	Output/input ratio (including labour component)	Output/input ratio (excluding labour component)
Broiler chicken–crop–fish–duck–horticulture–nitrogen fixing hedge row	1,05,722	1,67,331	1.58	2.24
Crop–fish–poultry–multipurpose trees	60,137	90,625	1.51	2.12
Crop–fish–goat–MPTs–hedge row	59,442	91,880	1.55	2.40
Crop–fish–pig–bamboo–MPTs–fruit trees–hedge rows	77,273	1,09,887	1.42	1.86
Crop–fish–dairy–MPTs–fruit trees–hedge rows–vermiculture–liquid manure–broom	1,70,120	2,98,735	1.76	2.48
Upland crops, and fish farming without integration (control)	31,773	34,894	1.09	1.50

Table 3.14 Water harvesting and utilization pattern in IIFS

IIFS	Water harvested in pond (m^3)	Water utilization (m^3)
Broiler chicken–crop–fish–duck–horticulture–nitrogen fixing hedge row	1,000	Fishery—924
		Vegetables—70
		Fruit trees—5.3
Crop–fish–poultry–multipurpose trees	1,800	Fishery—1675
		Vegetables—83
		Duckery—37
		MPTs—4.5
Crop–fish–goat–MPTs–hedge row	1,200	Fishery—1003
		Vegetables—67.5
		Poultry—126
		MPTs—3.2
Crop–fish–pig–bamboo–MPTs–fruit trees–hedge rows	1,300	Fishery—1170
		Vegetables—89.5
		Goat—36.0
		MPTs—4.5
Crop–fish–dairy–MPTs–fruit trees–hedge rows–vermiculture–liquid manure–broom	1,320	Fishery—1123
		Vegetables—76.5
		Pig—54.0
		Fruit trees—3.4
		Vermiculture—63.1

Table 3.15 Effect of various multipurpose trees on soil physical properties

Tree species	Organic C (g kg^{-1})	Bulk density (mg m^{-3})	Total porosity (%)	Micro aggregates (<0.25 mm)	Dispersion ratio	Erosion ratio	Erosion index
Pinus kesiya	3.54	1.04	54.3	17.6	0.21	0.20	0.11
	±0.33	±0.12	±6.22	±5.68	±0.09	±0.03	±0.01
Alnus nepalensis	3.22	1.09	55.6	22.4	0.23	0.23	0.12
	±0.47	±0.09	±5.87	±3.30	±0.05	±0.01	±0.02
Parkia roxburghii	2.31	1.23	52.2	28.8	0.26	0.30	0.14
	±0.61	±0.20	±3.20	±8.22	±0.11	±0.04	±0.01
Michelia oblonga	3.36	1.05	55.5	21.5	0.23	0.22	0.11
	±0.96	±0.32	±4.58	±7.45	±0.03	±0.03	±0.03
Gmelina arborea	2.86	1.14	52.4	38.0	0.25	0.24	0.12
	±1.24	±0.09	±6.04	±8.69	±0.04	±0.02	±0.02
Control (no tree)	1.56	1.32	48.7	44.2	0.35	0.39	0.15
	±0.92	±0.11	±8.09	±6.02	±0.06	±0.03	±0.03
LSD ($P < 0.05$)	0.39	0.15	5.06	3.05	0.06	0.05	0.03

Source Saha et al. (2007)

Table 3.16 Effect of various multipurpose trees on soil water retention characteristics

Tree species	Available water ($m^3\ m^{-3}$)	Infiltration rate ($mm\ h^{-1}$)	Hydraulic conductivity ($mm\ h^{-1}$)	Profile moisture storage (cm/60 cm)	
				In dry season	In rainy season
Pinus kesiya	0.220 ± 0.03	8.04 ± 1.28	5.44 ± 2.02	20.45 ± 3.22	24.60 ± 1.04
Alnus nepalensis	0.201 ± 0.02	7.28 ± 0.95	4.82 ± 1.46	19.44 ± 2.50	22.68 ± 0.98
Parkia roxburghii	0.192 ± 0.01	4.85 ± 0.56	3.23 ± 2.11	13.85 ± 3.61	18.52 ± 0.62
Michelia oblonga	0.210 ± 0.02	6.10 ± 1.23	4.84 ± 1.54	18.54 ± 2.37	21.66 ± 1.10
Gmelina arboria	0.183 ± 0.01	5.36 ± 0.82	3.50 ± 1.65	14.60 ± 2.11	19.41 ± 0.24
Control (No tree)	0.151 ± 0.02	3.84 ± 1.46	2.12 ± 2.35	11.45 ± 2.05	15.34 ± 0.72
LSD ($P < 0.05$)	0.11	1.06	0.18	2.17	2.30

Values for soil parameters are the means of three replications under two soil depths (0–15 and 15–30 cm) and two seasons across the year. Infiltration of water in the soil was also influenced by the tree vegetation. Infiltration rate under *Pinus kesiya* was almost twice that of the control (3.84 mm hr^{-1}).

Conclusions

In NEH Region agroforestry is a composite, diversified, and sustainable production system. It provides unique opportunity for integration of different components of the farming systems. This helps to optimize the ecosystem functioning and better management of land, water, and biological resources. In northeast hill region trees are deliberately grown with various crops and livestock under traditional production systems. Some of the systems developed for the NEH region have positive impact on the soil and water resources. These systems need to be further improved with suitable technological interventions to carter the needs of the local populace and help in improving the socioeconomic conditions of the farming communities. In NEH Region, agroforestry has gone a long way from traditional shifting cultivation to multienterprise farming systems where many components (based on the needs of the farmers) are integrated for getting optimum income providing the livelihood security.

References

Avasti RK (2006) Non timber forest products: options for broad based agroforestry systems in Sikkim Himalaya. In: Bhatt BP, Bujarbaruah KM (eds) Agroforestry in Northeast India: opportunities and challenges. ICAR Research Complex for NEH Region, Umiam, pp 213–235

Barthakur M (2004) Weather and climate. In: Singh VP, Sharma N, Shakhar S, Ojha P (eds) Brahmputra Basin water resources. Kluwer Academic Publisher, Netharlands, pp 17–30

Bhan S (2009) A case study on shifting cultivation practices in Mon district of Nagaland. J Soil Water Conserv 8(2):8–13

Bhatt BP, Bujarbaruah KM (2005) Intensive integrated farming system: a sustainable approach of landuse in eastern Himalayas. Technical Bulletin No. 46. ICAR Research Complex for NEH region, Umiam, p 43

Bhatt BP, Sachan MS, Singh K (2006) Production potential of traditional agroforestry systems of Meghalaya: a case study. In: Bhatt BP, Bujarbaruah KM (eds) Agroforestry in Northeast India: opportunities and challenges. ICAR Research Complex for NEH Region, Umiam, pp 337–349

Bhatt BP, Singh R, Mishra LK, Tomar JMS, Singh M, Chauhan DS, Dhyani SK, Singh KA, Dhiman KR, Datta M (2001) Agroforestry practices and research:

an overview. In: Verma ND, Bhatt BP (eds) Steps towards modernization of agriculture in NEH region. ICAR Research Complex for NEH region, Umiam, pp 365–392

Bhatt BP, Singha LB, Satapathy KK, Sharma YP, Bujarbaruah KM (2010) Rehabilitation of shifting cultivation areas through agroforestry: a case study in Eastern Himalayas. India J Trop For Sci 22(1):13–20

Champion HG, Seth SK (1968) A revised survey of the forest types of India. Government of India Press, Delhi

Chauhan DS, Dhyani SK (1990) Traditional agroforestry practices in Northeast Himalayan regions of India. India J Dry Land Agric Res Devel 4(2):73–81

Chauhan DS, Dhyani SK (1991) Existing agroforestry practices of Meghalaya. J Hill Res 4(1):19–23

Dabral PP (2002) Indigenous techniques of soil and water conservation in Northeastern region of India, Vol III: Technology and method of soil and water conservation. In: Proceedings of 12th ISCO conference, 26–31 May 2002, Beijing, China. Organized by Ministry of Water Resources, People's Republic of China, pp 90–96

Das T, Das AK (2005) Inventorying plant biodiversity in homegardens: a case study in Barak Valley, Assam, Northeast India. Curr Sci 89(1):155–163

Deb S, Arunachalam A, Das AK (2009) Indigenous knowledge of Nyishi tribes on traditional agroforestry systems. India J Tradit Knowl 8(1):41–46

Deb S, Tangjang S, Arunachalam A, Arunachalam K (2005) Role of litter, fine roots and microbial biomass in soil C and N budget in traditionally managed agroforestry systems. In: Bhatt BP, Bujarbaruah KM (eds) Agroforestry in Northeast India: opportunities and challenges. ICAR Research Complex for NEH region, Umiam, pp 491–506

Dhyani SK, Chauhan DS, Kumar D, Kushwaha V, Lepcha ST (1996) Sericulture based agroforestry system for hilly area of Northeast India. Agrofor Syst 34:247–258

Edwards P, Kaewpaitoon K, McCoy EW, Chantachaeng C (1986) Pilot scale crop/livestock/fish integrated farm. AIT Research Report, 184, Bangkok, Thailand, p 131

Forest Survey of India (2005, 2009, 2011) State of Forest Reports (2005, 2009 and 2011). Ministry of Environments and Forests, Government of India, Dehradun

Grewal SS, Aborl IP (1986) Agroforestry on alkali soils: effect of some management practices on soil initial growth, biomass accumulation and chemical composition of selected tree species. Agrofor Syst 4:221–232

IPCC (2001) Climate Change: the scientific basis. Contribution of the Working Group 1 to the Third Assessment Report of the IPCC. Cambridge University Press, Cambridge

Jha P, Mohapatra KP (2009) Leaf litter, fine root production and turnover in four major tree species of the semi-arid region of India. Plant Soil. doi: 10-1007/s11104-009-0027-9

Kumar M (2011) Northeast India: soil and water management imperatives for food security in a changing climate. Curr Sci 101(9):1119

Maithani BP (2005) Shifting cultivation in the Northeast India: policy issues and options. Mittal Publications, New Delhi, p 179

Mandal D, Sharda VN (2011a) Assessment of permissible soil loss in India employing a quantitative biophysical model. Curr Sci 100(3):383–390

Mandal D, Sharda VN (2011b) Appraisal of soil erosion risks in the eastern Himalayan region of India for soil conservation planning. Land Degrad Manag. doi: 10.1002/ldr.1139

Mishra AK, Sharma UC (2001) Traditional wisdom in range management for resource and environment conservation in northeastern region of India. ENVIS Bulletin, vol 9, issue 1. Himalayan Ecology

Mishra BK, Ramakrishnan PS (1982) Energy flow through a village ecosystem with slash and burn agriculture in Northeastern India. Agicultural Syst 9:57–72

Murtem G, Sinha GN, Dopum J (2008) Jhumias view on shifting cultivation in Arunachal Pradesh. Bull Arunachal Forest Res 24(1&2):35–40

Pulamte L (2008) Indigenous agricultural systems of Northeast India. Report of India Science and Technology. National Institute of Science and Technology Development Studies (NISTADS), CSIR, p 346

Rahman MM, Furukawa Y, Kawata I, Rahman MM, Alam M (2005) Homestead forest resources and their role in household economy: a case study in the villages of Gazipur Upazila of Central Bangaladesh. Small Scale Forest Econ Manag Policy 4(3):359–376

Rahman MM, Furukawa Y, Kawata I, Rahman MM, Alam M (2006) Role of homestead forests in the household economy and factors affecting forest production: a case study in southwest Bangladesh. J Forest Res 11:89–97

Ramakrishnan PS (1992) Shifting agriculture and sustainable development: an interdisciplinary study from Northeastern India. UNESCO-MAB series, Paris. Parthenon Publications, Cornforth, p 424

Rao KPC, Verchot LV, Laarman J (2007) Adaptation to climate change through sustainable management and development of agroforestry systems. SAT eJournal 4(1):1–30 (An open access journal published by ICRISAT. ejournal.icrisat.org)

Rastogi A, Godbole A, Shengii P (1998) Applied ethnobotany in natural resource management- Traditional home gardens. Highlights of the training workshop held at Kohima, Nagaland, India. 18–23 June 1997. International Centre for Integrated Mountain Development, Kathmandu, Nepal

Rathore SS, Karunakaran K, Prakash B (2010) Alder based farming system, a traditional farming system in Nagaland for amelioration of Jhum land. India J Tradit Knowl 9(4):677–680

Ravindranath NH, Rao S, Sharma N, Nair M, Gopalakrishnan R, Rao AS, Malaviya S, Tiwari R, Sagadevan A, Munsi M, Krishna N, Bala G (2011) Climate

change vulnerability profiles for Northeast India. Curr Sci 101(3):384–394
Saha R, Tomar JMS, Gosh PK (2007) Evaluation and selection of multipurpose tree for comparing soil hydro-physical behaviour under hilly ecosystem of north east India. Agrofor Syst 69:239–247
Sahoo UK, Rocky P, Vanlalhriatpuia K, Upadhyaya K (2010) Structural diversity and functional dynamism of traditional homegardens of Northeast India. The Bioscan 1:159–171
Sailo A (2011) Alley cropping: an alternative to slash and burn in the slope lands of the Mizo hills. J Nat Res Life Sci Educ 40:78–81
Sharma UC (1998) Water management by the tribes of northeastern region: the traditional wisdom and future strategies. Fertilizer News, 43:43-49
Sharma E, Sharma R, Singh KK, Sharma G (2000) A boon for mountain populations: large cardamom farming in the Sikkim Himalayas. Mt Res Devel 20(2):108–111
Sharma G, Sharma R, Sharma E (2009) Traditional knowledge system in large cardamom farming: biophysical and management diversity in Indian mountainous regions. India J Tradit Knowl 8(1):17–22
Sharma HR, Sharma E (1997) Mountain agricultural transformation process and sustainability in the Sikkim Himalayas, India. Discussion paper series no. MFS97/2, International Centre for Integrated Mountain Development (ICIMOD), Nepal
Sharma R, Sharma E, Purohit AN (1994a) Dry matter production and nutrient cycling in agroforestry systems of cardamom grown under *Alnus* and natural forest. Agrofor Syst 27:293–306
Sharma UC, Singh KA (1994) Agriculture with Alder—a potential indigenous farming system of Nagaland. In: Proceedings of the national seminar on developments in soil science, 28 Nov–1 Dec 1994, New Delhi, India, pp 570–571
Sharma UC, Prasad RN, Sonowal (1994) An indigenous technique of soil and water conservation in north eastern region—the Zabo system of farming. In: Bhushan LS, Abrol IP, Ram Mohan Rao MS (eds) Proceedings of the 8th ISCO conference. Oxford and IBH Publication Co. Pvt. Ltd., New Delhi, pp 969–975
Singh A, Singh MD (1981) Soil erosion hazards in north eastern hill region. Research Bulletin No. 10, ICAR Research Complex for NEH Region, Umiam, Meghalaya
Singh BP, Dhyani SK (1996) An appraisal of indigenous landuse systems in Meghalaya for developing sustainable rainfed hill agriculture. In: Kohli RK, Arya KS (eds) In: Proceedings of IUFRO-DNAES international meet on resource inventory techniques to support agroforestry and environemntal activities. DNAES, Chandigarh. 1–3 Oct 1996, pp 61–72
Singh BP, Prasad RN (1987) Alternatives to shifting cultivation; Technologies for better crops. 42. Indian Council of Agricultural Research, New Delhi, India, p 18
Singh KA, Rai R, Bhutia DT (1982) Large cardamom (*Ammomum subulatum*) plantation—an age old agroforestry system in Eastern Himalayas. Agrofor Syst 1:241–257
Singh KK, Krishna AP, Chetri S (2005) Large cardamom Agroforestry based sustenance farming areas of Khangchendranga Biosphere Reserve, Sikkim, Himalaya. In: Bhatt BP, Bujarbaruah KM (eds) Agroforestry in Northeast India: opportunities and challenges. ICAR Research Complex for NEH Region, Umiam, pp 71–78
Sundriyal RC, Rai SC, Sharma E, Rai YK (1994) Hill Agroforestry systems in South Sikkim, India. Agrofor Syst 26:215–235
Tangjang S, Arunachalam A (2009) Role of traditional home garden systems in Northeast India. India J Tradit Knowl 8(1):47–50
Toky OP, Ramakrishnan PS (1981a) Runoff and infiltration losses related to shifting agriculture in Northeastern India. Environ Conserv 8:313–321
Toky OP, Ramakrishnan PS (1981b) Cropping and yields in agricultural systems of the northeastern hill region of India. Agro-Ecosystems 7:11–25
Toky OP, Ramakrishnan PS (1983) A comparative study of the energy budget of the hill agro-ecosystems with emphasis on the slash and burn system (*jhum*) at lower elevations of north-eastern India. Agric Syst 9:143–154
Tynsong H, Tiwari BK (2010) Plant diversity in homegardens and their significance in the livelihoods of War Khasi community of Meghalaya, Northeast India. J Biodivers 1(1):1–11
Wasteland Atlas of India (2010) Report of Ministry of Rural Development. Department of Land Resources, Government of India, p 140

Agroforestry Inroads from the Traditional Two-Crop Systems in Heartlands of the Indo-Gangetic Plains

4

P. S. Pathak, Jagdish Chander Dagar, Rajesh Kaushal, and O. P. Chaturvedi

Abstract

The Indo-Gangetic Plains formed by the sedimentation from Himalayas are one of the most fertile regions of the world. It covers an area of 15.3 % of the country and provides shelter to about 33 % human and 35 % livestock population of the country. The region has been heavily exploited for agriculture by the indiscriminate use of high yielding varieties, fertilizers, pesticides, chemicals, etc., which has resulted in soil degradation and unsustainable use of natural resources. Declining productivity, soil quality, water table; waterlogging; and salinity have now become a major problems in this region which needs immediate attention. Agroforestry in this region has high potential to address the issues of degradation and sustainability. Recognizing the importance of agroforestry, organized research was initiated in Indo-Gangetic plains in early 1970s through industry participation in plantations of *Populus deltoides*. The research, however, gained momentum with the allocation of a seven coordinating center of All India Coordinated Research Project on Agroforestry (AICRP AF) by Indian Council of Agriculture Research (ICAR) in 1983. Simultaneously, other organizations like ICFRE, WIMCO, Pragati Biotechnologies, ITC, Star paper mill, etc., and some NGOs also started research in agroforestry. Joint efforts of research in last 3 decades have resulted in development of different agroforestry models. Management and cultural practices of trees and annual crops have also been standardized. Quality planting material of many tree

P. S. Pathak (✉)
Ex-ADG (Agroforestry) ICAR,
New Delhi, 110012, India
e-mail: pathakprem@gmail.com

J. C. Dagar
Krishi Anusandhan Bhawan II, ICAR,
Pusa, New Delhi 110012, India

R. Kaushal · O. P. Chaturvedi
Central Soil and Water Conservation Research and Training Institute, Dehradun, Uttarakhand 248195, India

J. C. Dagar et al. (eds.), *Agroforestry Systems in India: Livelihood Security & Ecosystem Services*, Advances in Agroforestry 10,
DOI: 10.1007/978-81-322-1662-9_4, © Springer India 2014

species has been developed which has resulted in productivity enhancement and large scale adoption of agroforestry by the farmers. Agroforestry has also played a major role in the rehabilitation of lands degraded by salinization, ravines, gullies, and other water and wind erosion hazards in this region. The impact of agroforestry in this region is well recognized. Poplar and Eucalyptus based agroforestry, and agrihorticulture systems are some of the successful models which have been adopted by farmers on large scale. The present chapter is an attempt to review the progress made in agroforestry R&D during the last 3 decades. Some of key issues which are required for the development of agroforestry have also been discussed.

Introduction

The Indo-Gangetic Plains (IGP) also called as The North Indian Rivers Plains are large and fertile plain areas encompassing most of the northern and eastern India, the most populous parts of Pakistan, parts of southern Nepal, and virtually all of the Bangladesh. The Indo-Gangetic plains are formed by the sedimentation from Himalayas carried by the Indus and Ganga rivers and their tributaries. The plains support one of the most populous areas on Earth, being home to nearly one billion people (or around 1/7th of the world's population) on 700,000 km^2. In India, Indo-Gangetic plains cover about 15.3 % of the country and provide about 50 % of the total food grains to feed 40 % of the population and shelter about 35 % of livestock population. The region contributes 51.9 % to the national food grain output and about 72 % of the area is under cultivation. The western region (subregions 1, 2, and 3) is a food surplus region characterized by high investment, high productivity, heavy use of fertilizers, and groundwater for irrigation, and an influx of labor (Fig. 4.1). In contrast, the eastern region (subregions 4 and 5) is a food deficit region characterized by low productivity, low inputs of fertilizer and water, risk of flooding, poor infrastructure, and labor emigration.

Ecology and General Features

The Indo-Gangetic Plains represent eight agro-ecological regions (AERs) and 14 agro-ecological subregions. It covers the states of West Bengal, Bihar, Uttar Pradesh, Haryana, and Punjab along the Himalayan ranges across the northern India. These plains are formed through sedimentation from Himalayas carried by rivers Indus and Ganga and their tributaries. The Indo-Gangetic plains extend from the delta of the river Indus to the delta of rivers Ganga and Brahmaputra. The Indo-Gangetic plains are bound on the north by the abruptly rising Himalayas, which feed its numerous rivers and are the source of the fertile alluvium deposited across the region by the two river systems. The southern edge of the plain is marked by the Vindhya and Satpura Range, and the Chota Nagpur Plateau. The altitude ranges from 150 to 600 m. The western section, comprising Punjab, Haryana, Chandigarh, Delhi, and western Uttar Pradesh, is slightly higher in elevation than the eastern section of the plain. The area is classified in 4 meso-level regions (Singh 1971) viz., trans- upper-, middle-, and lower- Gangetic plains. The area, rivers, and climate of different regions are given in Table 4.1.

Climate

Planning Commission of India conceptualized 15 agro-climatic zones, of which this area represents zone III to VI. Climate of the region varies from hot-humid monsoonal to semi-arid. The region shows a gentle gradient in rainfall and temperature regimes. In lower Gangetic plain (LGP) region, the mean rainfall ranges from 1200 to 1700 mm. It shows 3 months receiving more than 300 mm rainfall (June, July,

Fig. 4.1 Indo-Gangetic plains (*Source* www.gecafs.org) (*1* and *2* trans-Gangetic plain, *3* upper Gangetic plain, *4* middle Gangetic plain, and *5* lower Gangetic plain)

Table 4.1 Area, rivers, and climate of Indo-Gangetic plains

Regions	Area	Rivers	Climate	Temperature	Rainfall
Lower Gangetic plains	About 78 % area of West Bengal Jharkhand (6.94 M ha)	Ganga, Hugli, Damodar, Rupanarayan, Bhagirathi	Hot and humid monsoonal	Mean minimum temperature 21 °C in south to 2.2 °C in north, summer temperature from 16 to 32.7 °C	High rainfall (1200–1700 mm) very high relative humidity
Middle Gangetic plains	This zone covers eastern districts of U.P. and the northern part of Bihar (17.03 M ha)	Ganga, Gomati, Ghagra, Great Gandak, Little Gandak, Kosi, Mahananda	Humid to sub-humid monsoonal	Mean minimum temperature ranges from 8.9 °C in January to 27.2 °C in June and the mean maximum temperature from 23.3 °C (Jan) to 37.6 °C (June)	Moderate to high 1000–1500 mm except in the western area (<1000 mm)
Upper Gangetic plains	Central and Western half of UP (13.87 M ha)	Ganga, Yamuna, Ramganga, Gomati, Ghagra	Sub-humid (four well marked seasons hot summer, wet summer, pre-winter transition, and winter	Maximum mean temperature <40 °C in the north and >45 °C in the south. minimum temperature in January can reach 2 °C with average low of 20 °C	500 to 1400 mm from west to east. (The northern parts receive more than 1200 mm). Winter rain common
Trans-Gangetic plains	States of Punjab, Haryana, Delhi, Chandigarh, Ganganagar district of Rajasthan (12.5 M ha)	Vyas, Ravi, Satluj, Yamuna, Ghaggar	Semi-arid and sub-humid in different districts with three distinct seasons of winter, summer, and rainy	Rises up to 43 °C in summer months, heat waves, average annual temperature ranges from 23.1 to 25.1 °C	190–1150

Source Pathak et al. (2000)

and August) while September receives >250 mm rain. It has 5 months dry (January, February, March, November, and December), but all the 12 months do receive some rainfall. Number of rainy days are 81. The mean maximum and minimum temperature are 36.4 and 11.7 °C. July, August, and September record the minimum difference in the maximum and minimum temperature.

The middle Gangetic plain (MGP) has 6 months dry in east and 8 months toward west. The monthly rainfall exceeds 400 and 300 mm in August and September, respectively. June and July months receive >100 mm rain. Rainfall in March and December are negligible. Mean maximum and minimum temperature are 35.8 and 8.9 °C. In the western part monthly rainfall does not exceed 300 mm in any month but it exceeds 250 mm in July and August and 200 mm in September. June receives >100 mm rainfall. Total rainfall is 289 mm less than eastern part with only 49 rainy days. So between the two zones of the region, rainfall quantum, pattern, and number of rainy days show that western part is much drier compared to eastern. April and November months show negligible rainfall. The mean maximum and minimum temperature are 39 and 7.5 °C indicating that the zone is warmer and cooler than the eastern zone. The variation in maximum and minimum temperature is also wider than the eastern zone of this region.

The upper Gangetic plain (UGP) is represented by Central and Western parts of Uttar Pradesh. It receives 80–120 cm annual rainfall which descends from north–east to south–west. It has 8 months dry; the total rainfall is 103 mm less than western MGP. The wet months are July to October while April and December receive negligible rainfall. The mean maximum and minimum temperature are 40.6 and 7.9 °C indicating that the place is warmer than MGP. The variation in maximum and minimum temperature is wider. The western *Tarai* region records >400 mm rainfall in the July, >350 mm in the August, and 250 mm in the September. The wet-months are from June to September. The total rainfall is 568 mm more than the eastern part with 62 rainy days. October and November months receive negligible rainfall. The mean maximum and minimum temperatures are 37.6 and 6.3 °C, respectively showing that the place is cooler than the previous part. Temperature variation between maximum and the minimum is very narrow during June to September and very wider during April to May when the peak high temperature is recorded. In general, relative humidity varies from 30-35 % during summer to 85–89 % during peak rain season.

In trans-Gangetic plain (TGP), Haryana has semi-arid climate in southwest and a semi-arid climate in remaining parts. The mean rainfall ranges from 190 mm on Rajasthan border to 1150 mm in areas adjoining Punjab and Himachal Pradesh. It has 7–9 months dry in Haryana where monthly rainfall does not exceed 150 mm. It is more than 100 mm in July and August. The total rainfall is 421 mm lesser than UGP with only 27 rainy days. Wet months are July to September. All the months receive some rain with November and April receiving negligible rainfall. The maximum and minimum temperatures are 40.8 and 5.1 °C. The variation in maximum and the minimum temperature is lowest in July and August while it is widest in December. Summers are characterized by gusty winds, dust storms, and thunder storms, whereas during winter the temperature occasionally drops below freezing point with severe frost in month of December and January. The western part receives 200 mm rainfall in July >100 mm rainfall in August and >150 mm in September. The total rainfall is 259 mm more than Hisar with 69 rainy days. The wet-months are July to October and January, November, and December receive negligible rainfall. The variation in the maximum and minimum temperature is narrow in July and maximum in April with mean maximum and minimum of 39.7 and 6.0 °C, respectively. The place is comparatively lesser warm.

Biotic Features

The two major components of biotic factor viz., human and livestock population in the region are considered for their population density and also

Table 4.2 Population in states of Indo-Gangetic plains

State	Area (M ha)	Population (M)	Population density (no/km ^{2})	Population growth rate	Rural (%)	Urban (%)
a: Human (year 2011)						
Bihar	9.41	103.80	1102	25.01	86.70	11.30
Haryana	4.42	25.35	573	19.90	65.21	34.79
Punjab	5.03	27.70	550	13.73	62.51	37.49
Uttar Pradesh	24.09	199.58	828	20.09	77.72	22.28
West Bengal	8.87	91.34	1029	13.93	68.11	31.89
Total of India	328.73	1210.19	382	17.64	68.84	31.16

b: Livestock and poultry (year 2007): figure in thousands

State	Area (Mha)	Cattle	Buffaloes	Sheep	Goats	Total livestock	Total poultry
Bihar	9.41	12559	6690	218	10167	30342	11420
Haryana	4.42	1552	5953	601	538	8859	28785
Punjab	5.03	1777	5062	208	290	7408	10685
Uttar Pradesh	24.09	18883	23812	1188	14793	60272	8754
West Bengal	8.87	19188	764	1577	15069	37419	86210
Total of India	328.73	199075	105343	71558	140537	529698	648830

Source a: http://www.census2011.co.in
Source b: Ministry of Agriculture (2007):18th livestock census, department of animal husbandry, dairying, and fisheries, Government of India

the growth (Table 4.2). It is observed that the human population is highest in Uttar Pradesh followed by Bihar while it is lowest in Haryana. The population density is highest in Bihar followed by West Bengal and the lowest in Punjab. The livestock population is highest in Uttar Pradesh followed by West Bengal while it is lowest in Punjab. Cattle and buffaloes population are highest in West Bengal and Uttar Pradesh, respectively, while sheep and goats are highest in West Bengal. Total poultry population is highest in West Bengal followed by Haryana.

Land Use and Prevalent Cropping Systems

Land use

The Indo-Gangetic Plains are the densely populated area in the country. The increasing population is making the man to land ratio highly unfavorable as a result very high proportion of land is used for cultivation (Table 4.3). Net area sown is highest in Uttar Pradesh. The percentage of the reported area under cultivation varied from 59 % in West Bengal to 82 % in Punjab, as compared to 42 % in the country. These are creating long-term threats to the sustainability of agricultural systems in the rural livelihoods.

Cropping Systems

Overall the region has a mixed economy with good growth of agriculture, horticulture, and animal husbandry. The major cropping systems are rice (*Oryza sativa*) and wheat (*Triticum aestivum*) with local and occasional variations occurring in different subregions of the area under rainfed lands. Similarly, the choice of cropping systems and number of tree species on farm also change. Tree number per hectare ranged from 15 to 33 in the rainfed areas

Table 4.3 Land use statistics (2009–10) in the Indo-Gangetic plains (thousand ha)

Particulars		Bihar	Haryana	Punjab	U.P.	West Bengal
Geographical area		9416	4421	5036	24093	8875
Reporting area		9360	4371	5033	24170	8684
Forests		622	40	295	1662	1174
Not available for cultivation	Area under non-agricultural use	1690	470	503	2801	1799
	Barren and uncultivated land	432	104	25	494	22
	Total	2121	574	528	3295	1820
Other uncultivated land excluding fallow land	Permanent pastures and other grazing lands	16	28	4	65	6
	Land under miscellaneous tree crops and grooves (not included in net sown area)	243	12	5	360	55
	Culturable wasteland	45	29	3	431	31
	total	304	70	12	856	92
Fallow land	Fallow land other than current fallow	122	5	4	537	20
	Current fallows	858	133	37	1232	323
	Total	980	138	40	1769	342
Net area sown		5332	3550	4158	16589	5256
Area sown more than once		2159	2801	3717	8175	4274
Total cropped area		7491	6351	7875	24764	9530

Source Ministry of Agriculture (2011a, b): directorate of economics and statistics
Totals may differ because of rounding up of the figures

compared to 10–500 in the irrigated lands. High tree density under irrigation is being practiced in trans-Gangetic plains where poplar (*Populus deltoides*) trees are planted by the farmers on short rotation as agrisilviculture system and also on farm boundary. In most of the rainfed croplands the trees are either planted on the boundary or natural growth is maintained in the field as scattered trees. The area under rainfed croplands decreased toward trans-Gangetic plains. Use of fruit trees on croplands is more prevalent in the lower and middle Gangetic plains compared to the rest of the area. *Acacia nilotica, Azadirachta indica, Dalbergia sissoo*, and *Mangifera indica* are some of the common tree species throughout the region.

Fruit trees and vegetables form the component of agrihorticulture or agrisilviculture in the farming systems. In IGP, fruits and vegetables production ranged from 7.7 (Haryana) to 19.7 t ha^{-1} (Punjab) for fruits and from 13.4 (Haryana) to 21.3 t ha^{-1} (U.P.) plains for vegetables (Table 4.4). The mean production in the country is 11.7 t ha^{-1} for fruits and 17.3 t ha^{-1} for vegetables. In Punjab, fruit production efficiency is comparatively very high (19.7 t ha^{-1}) in comparison to other states and India (11.7 t ha^{-1}).

Natural Resources

Vegetation

The region has four different types of forest covers as described by Champion and Seth (1968). In the lower Gangetic plains along Ganga-Brahmaputra delta littoral and swamp forests predominate with species of mangroves and *Syzygium cuminii*. In the rest of the area of lower Gangetic plains, tropical dry deciduous forests predominate. The trees have less than 25 m height. Prominent species include *Tectona grandis, Shorea robusta, Terminalia tomentosa, Acacia catechu, Boswellia serrata*, etc. These forests with species variations are also predominant in the middle and upper Gangetic plains, where most of these forests are degraded. Along

Table 4.4 Fruits and vegetables production in Indo-Gangetic plains (area in 000 ha, production in 000 t)

	Fruits		Vegetables		Mean production (t ha^{-1})	
State	Area	Production	Area	Production	Fruits	Vegetable
Bihar	296.4	3911.8	845.0	14630.2	13.2	17.3
Haryana	46.3	356.6	346.4	4649.3	7.7	13.4
Punjab	69.8	1373.2	174.1	3585.8	19.7	20.6
Uttar Pradesh	324.8	5368.4	829.4	17679.4	16.5	21.3
West Bengal	211.6	2952.8	1349.7	26725.5	14.0	19.8

Source Kumar et al. (2011)

the foot hills of Himalayas in Haryana, Uttar Pradesh, and part of Bihar, tropical moist deciduous forests are predominant due to high moisture and moderate temperature. Dominant tree species are deciduous with lower storey showing evergreen habit. *Terminalia, Madhuca indica, Acacia catechu,* and *Shorea robusta* predominate along with many species of secondary importance. The degraded forests have been now afforested with species of *Eucalyptus, Anthocephalus cadamba, Morus alba, Shorea robusta, Populus deltoides, Dalbergia sissoo,* etc. In the trans-Gangetic plains tropical thorn forests have species of *Prosopis cineraria, Acacia nilotica, A. leucophloea, A. catechu, Prosopis juliflora, Ziziphus sp., Mimosa sp., Aegle marmelos*, etc. In this region, most of the characteristic woody species have changed due to plantation of fast growing species. Looking to the national average, the region is a deficit zone so far as forestry is concerned.

Recorded forest area, forest cover and tree cover of Indo-Gangetic plains have been given in Table 4.5. Area-wise Uttar Pradesh has the largest forest cover followed by West Bengal. In terms of percentage of forest cover with respect to geographical area, West Bengal with 14.62 % has highest followed by Bihar (7.27 %). Trees outside forests (tree cover) in Punjab (3.37 %), Haryana (3.15 %), Uttar Pradesh (3.06 %), Bihar (2.52 %), West Bengal (2.63 %), have gained prominence even under high biotic pressure. Punjab followed by Bihar and Haryana States showed positive changes in the forest cover. In Bihar, enhanced plantation activity outside forest areas in recent times contributed to increase in forest cover while, in Punjab, growth of young plantation carried out under externally funded projects and agroforestry activities in tree outside forests areas have been responsible for increase in forest cover (FSI 2011). The potential production of timber and

Table 4.5 Recorded forest area, forest cover, and tree cover in Indo-Gangetic plains (area in km^2)

State	Geographical area (GA)	Recorded forest area			Total forest area	Forest cover	Tree cover	Change in forest cover as compared to 2009
		Reserved forest	Protected forest	Unclassified				
Bihar	94163	693	5779	1	6473 (6.87)	6845 (7.27)	2369 (2.52)	+41
Haryana	44212	249	1158	152	1559 (3.53)	1608 (3.64)	1395 (3.15)	+14
Punjab	50362	44	1137	1903	3084 (6.12)	1764 (3.50)	1699 (3.37)	+100
Uttar Pradesh	240928	11660	1420	3503	16583 (6.88)	14338 (5.95)	7382 (3.06)	−3
West Bengal	88752	7054	3772	1053	11879 (13.38)	12995 (14.64)	2335 (2.63)	+1

Figures in parentheses are % of geographical area
Source FSI (2011)

Table 4.6 Potential production of timber and fuelwood from TOF in the states of Indo-Gangetic plains

State	Growing stock (million cum)	Annual production of timber (million cum)	Annual availability of fuelwood (000 tons)
Bihar	47.20	1.21	945
Haryana	15.27	1.98	825
Punjab	19.31	2.65	920
Uttar Pradesh	81.68	5.08	2253
West Bengal	45.69	1.44	529
Total	209.15	12.36	5472

Source FSI (2011)

fuelwood from trees outside forest (TOF) in the region was found to be 209 million cum growing stock, 12.4 million cum annual production of timber, and 5472 thousand tons annual availability of fuelwood (Table 4.6).

Soils

The rivers such as the Indus, the Ganga, the Brahmaputra, and their tributaries bring enormous load of silt from the mountains and deposit them on the beds due to which soils are of of recent origin with good depth and fertility. The soils of lower Gangetic plains are derived from material transported through Himalayan- rivers and their tributaries and are categorized as Inceptisols and Entisols. The soils are lighter in texture in north and west but heavier in south and east. The soil pH is acidic or neutral with medium organic matter and available phosphorus and potash. The coastal subregion soils are generally saline with sporadic patches of saline alkaline and degraded alkaline soils. In some parts of the coastal region, acid sulfate soils are also present. In some parts, the soils are laterite with red and gravelly patches. The coarse texture and well-drained nature in the uplands makes it highly susceptible to erosion.

In middle Gangetic plains, the soils of northwestern Bihar are developed on alluvium of Gardak, Ghagra, and their tributaries. These soils are calcareous having high level of $CaCO_3$ (3–8 %). In part of eastern Uttar Pradesh, soils are sandy loam to loam and calcareous clay to deep alluvium. In part of Bihar, soils are heavy textured sandy loam to clayey and in the northeast these are sandy to silty loam, with pH 5.6 to 7.6 (alkali soil with high pH available in patches particularly in Bihar). In general the soils are good in fertility with occasional patches of problematic soils with high amount of calcium carbonate and salinity problems (described in Chap. 9).

The alluvium soils of upper Gangetic plain are flat to gently sloping generally well drained and deep. These soils are developed from the alluvium deposited by Ganges, Yamuna, and their tributaries. Gangetic alluvial soils are neutral to moderately alkaline usually coarse textured and calcareous at lower depths, while soils from Yamuna alluvium are dark to very dark gray, relatively fine textured, calcareous and possess shrink swell properties. These are the most fertile lands of the country. The region has been divided in four distinct physiographic units namely, upper *tarai* and *bhabhar* belts, eastern alluvial, western plains, and middle alluvial plains. In the *tarai* and *bhabhar* belts there is a complex association of soils of Inseptisol and Alfisol.

Eastern alluvial plain has coarse and fine loamy type of Udic Haplustalfs, Typic/Udic Ustochrepts, and Ustifluvents. The soils in western part are coarse textured. Accumulation and redistribution of salts in profile are a major problem. In the southeastern area Aravali soils exist.

In trans-Gangetic plains, the soils of Punjab are developed on alluvium of Satluj and Beas river system, whereas there of Haryana mainly from the alluvium deposited by Yamuna and its tributaries. The soils of foot hills of Shivalik subregion are reddish brown to olive brown and have ochric epipedon underlain by cemtic or clay enriched horizon, pH almost neutral with sandy loam to clay loam texture. In aridic areas, the major soil groups are Camborthids and Caciorthids associated with Torripsamments and Torrifluvents. The ustic areas in central Punjab are represented by Ustochrepts and Haplustalfs soil groups. Associated sandy soils are Ustipsamments. The major soil formations of north-eastern Punjab (udic area) are Eutrochrepts, Udifluvents, and Hapludolls. In Haryana, Seirozen soils are found in the areas receiving less than 500 mm annual rainfall, while the arid brown soils (Natrustalfs, Ustochrepts, and Haplaquepts) occur in the areas receiving 500–750 mm rainfall. These soils are calcareous with $CaCO_3$ pan at 0.75 to 1.5 m depth. In large part of area, soils have deteriorated with varying degree of sodicity and salinity, whereas in scarce rainfall areas, soils are deserted calcareous with the nodules in the subsurface horizon (Yadav et al. 2000). Problematic salt-affected soils have been dealt in Chap. 9.

Water resources

(i) ***Rainfall:*** Average annual rainfall in Indo-Gangetic plains ranges from 539 to 1203 mm (Table 4.7) which clearly reveals extreme diversity in rainfall. Major share of rainfall is received through southwest monsoon during June–September.

(ii) ***Surface and groundwater resources:*** The IGP developed mainly by the alluvium of the Indus, Yamuna, Ganga, Ramganga, Ghagra, Rapti, Gandak, Bhagirathi, Silai, Damodar, Ajay, and Kosi rivers. The surface water resources of IGP are given in Table 4.8. Maximum length of river and canal is in Uttar Pradesh followed by Punjab.

The groundwater resource estimate in different states of IGP is given in Table 4.9. Although the rainfall variations are large across the region but available groundwater resource for irrigation is largest in Uttar Pradesh followed by Bihar and the lowest in Haryana. The level of groundwater development is maximum in Punjab (141.51 %) followed by Haryana (80.21 %) and the lowest in Bihar (34.25 %). This provides a basis for considering the choice of tree species and agroforestry model for adoption.

Agroforestry in Indo-Gangetic Plains

Due to heavy demographic and biotic pressure, natural resources in the IGP are in constant threat. Further, indiscriminate use of high yielding varieties, fertilizers, pesticides, chemicals, etc., in the region has resulted in large scale land degradation which has raised the concerned on long-term sustainability of natural resources. Environmental threats (floods, secondary salinization, pollution, fast decreasing water table, etc.), declining size of the land holdings, demographic pressure, and other socioeconomic determinants are some of the major challenges which need to be addressed. Agroforestry has a great potential to meet some of these challenges in the region. Agroforestry has been an age old practice in the region (Pathak and Dagar 2001). However, with changing time, new agroforestry technologies have been developed which are being adopted by farmers on large scale. Some of the traditional and modern agroforestry practices in the region are:

Table 4.7 Annual and seasonal rainfall in different meteorological divisions of Indo-Gangetic plains

Meteorological division	Annual rainfall (mm)	Seasonal rainfall (June-Sept) (mm)	Percentage of annual rainfall
Bihar	1203	1023	85
Haryana	539	450	83
Punjab	602	461	76
Uttar Pradesh (east plains)	1008	889	88
Uttar Pradesh (west plains)	831	723	87

Source Handbook of Agriculture (1986)

Table 4.8 Surface water resources in Indo-Gangetic plains

State	Length of rivers and canals (km)	Reservoirs	Other water bodies (million ha)			Total
			Tanks/ponds	Wells, lakes, and derelict water bodies	Brackish water	
Bihar	2000	0.06	0.64	0.04	–	1.60
Haryana	5,000	Negligible	0.09	0.01	0.30	0.20
Punjab	15,270	Negligible	0.07	–	–	0.07
Uttar Pradesh	28,500	1.380	1.61	0.13	–	4.45
West Bengal	2,526	0.17	2.76	0.04	2.10	5.45

Source Ministry of Agriculture (2012): Department of animal husbandry and dairying

Table 4.9 Groundwater resource (BCM yr^{-1}) in Indo-Gangetic plains

State	Total replenishable water	Utilizable water for irrigation	Net draft	Level of groundwater development (%)
Bihar	29.19	27.42	9.39	34.25
Haryana	9.31	8.63	9.10	80.21
Punjab	23.78	21.44	30.34	141.51
Uttar Pradesh	76.35	70.18	45.36	64.63
West Bengal	30.36	27.46	10.84	39.47

Note Estimates for 2004; BCM = Billion Cubic Meters
Source Central Groundwater Board (2006); Ministry of Water Resources (2007)

Traditional Agroforestry Practices

Scattered Trees on Farm

The practice of growing agricultural crops under scattered trees on farm lands is being practiced for centuries in this region. These practices have also been discussed in earlier chapter for rainfed situations but here trees are retained in irrigated fields. The density of the trees varies from 10 to 50 trees ha^{-1}. Some of the practices are very extensive and highly developed. Trees are mainly used for shade, fodder, fuelwood, fruits, vegetables, and medicinal uses and have not much commercial value. In Punjab, farmers retain few local tree species on their agriculture land for meeting the daily needs of fuelwood, fodder, timber, etc. The tree species include: *dek* (*Melia azedarach*), *kikar* (*Acacia nilotica*), *dhak* (*Butea monosperma*), mango (*Mangifera indica*), *shisham* (*Dalbergia sissoo*), *siris* (*Albizia lebbeck*), *jand* (*Prosopis cineraria*), etc., and local bushes for their domestic use.

In submontane foot hill zone (Kandi Area) of Punjab, *kikar, dhak, shisham,* mango, *khair* (*Acacia catechu*), *phulai* (*Acacia modesta*), *ber* (*Ziziphus mauritiana*), *toot* (*Morus alba*), *neem* (*Azadirachta indica*), *dhau/chhal* (*Anogeissus laitfolia*), *kachnar* (*Bauhinia variegata*), *rajain* (*Holoptelea integrifolia*), *bel* (*Aegle marmelos*), *aonla* (*Emblica officinalis*), etc., are commonly found on farms. Though the traditional agroforestry has undergone various changes depending upon the locality and socioeconomic conditions of the farmers, but the native tree species like *kikar, jand, shisham, toot, neem, dek*, mango, *khair, jamun* (*Syzygium cuminii*), *beul* (*Grewia optiva*), etc., are still grown as scattered trees or as boundary plantations in different zones of Punjab because of their multiple uses. Under irrigated lands, the number of *kikar* trees retained on field bunds vary from 5 to 10 ha^{-1}. Under terraced agricultural areas, having no irrigation facilities, about 35–50 *kikar* trees are generally seen on the field bunds of each hectare. But in case of coarse textured

Table 4.10 Tree density and phytobiomass calculated in different TOF types in Southern Haryana

TOF type	Number of trees per ha	Phytobiomass (tons per ha)	Phytobiomass range (tons per ha)
Linear	171–1556	40.71	8.3–91.3
Road	478–557	41.59	8.0–14.3
Rail	852–1440	11.15	12.7–92.0
Canal		45.21	
Block	447–1200	18.24	6.8–64.9
Scattered	170–416	7.15	1.3–25.4
Urban	132–336	9.53	2.0–13.5
Rural	64–164	6.79	3.3–12.4
Agroforestry		6.33	

Source Singh and Chand (2012)

undulating lands with uncertain prospects of getting sustainable crops, 50–100 trees per ha are retained on the fields (Sarlach et al. 2007)

In southern parts of Haryana particulars in Bhiwani and Mohindergarh districts, farmers retain 5–10 trees ha^{-1} of *Prosopis cineraria* (Khejri). The density however, is continuously decreasing due to cutting of the trees for use in drought year and introduction of mechanical farming (Nandal and Bangarwa 2007). Recently, Singh and Chand (2012) using synthesis approach of remote sensing reported that trees outside forest (TOF) phytobiomass in southern Haryana varied from 1.3 tons per ha in the scattered trees to 91.5 tons per ha in the dense linear area along the canals. The density of trees varied from 64 to 164 in agroforestry to 852–1440 along canal (Table 4.10).

The total phytobiomass of linear stands (in 1387 ha area), block (1254 ha), and scattered (2502 ha) stands was observed to be 208, 81.5, and 77.6 tons per ha, respectively and total carbon was observed in respective stands to be 98.8, 38.7, and 36.8 tons per ha. Presence of *Prosopis cineraria* has been found to enhance the growth of adjacent crops (Arya et al. 1991). Puri and Bangarwa (1992) in the semi-arid regions of Haryana found no significant effect of *Azadirachta indica* and *Prosopis cineraria* on the yield, of wheat. The yield, however, decreased significantly in the vicinity of *Dalbergia sissoo* and *Acacia nilotica*

Farmers in sub-humid *tarai* region of Uttar Pradesh and Uttarakhand prefer *Dalbergia sissoo, Syzygium cuminii*, and *Trewia nudiflora.* In Bihar, *Dalbergia sissoo, Litchi chinensis,* and *Mangifera indica* are frequently retained on fields (Tewari et al. 2007).

Home Gardens

Home gardens are traditionally found in W. Bengal, Eastern Uttar Pradesh, and Bihar where multiple crops are present in a multitier canopy configuration. The leaf canopies of the components are arranged in different vertical layers with the tallest components having foliage tolerant to strong light and high evaporation demand and the shorter components having foliage requiring or tolerating shade and high humidity. In West Bengal, home gardens comprise of 3–4 tiers of coconut (*Cocos nucifera*), areca nut (*Areca catechu*), banana (*Musa paradisiaca*); vegetables and flowers in both villages and cities. Arecanut (*Areca catechu*) is grown in backyard of every house, but not a single tree in the agriculture fields (Panwar 2007).

In eastern Uttar Pradesh, majority of the small and marginal farmers develop agroforestry around their homes. *Artocarpus heterophyllus, Azadirachta indica, Dendrocalamus strictus, Psidium guajava, Musa paradisiaca*, and *Citrus* spp are some of the multipurpose trees grown by the farmers (Rana et al. 2007). In Bihar, the average size of the home garden varies between 0.05 and 0.42 ha where *Dendrocalamus strictus, Dalbergia sissoo, Mangifera indica, Litchi chinensis, Musa paradisiaca*, and *Carica papaya* are the common species. A good number of vegetables and spices are also found in these home gardens (Chaturvedi et al. 2001, Chaturvedi and Das 2007).

Trees on Farm Boundaries

In Haryana and Punjab, *Eucalyptus* and *Populus deltoides* are commonly grown along the field boundaries or bunds. In Yamunanagar district in Haryana, 18.8 % of the cultivated area is under boundary plantation of poplar (*Populus deltoides*) and only 4.8 % under block plantation of poplar. Under this practice, trees are planted on field boundaries or on along irrigation channels either on one side or on both sides of the channel (Pathak et al. 2006). Farmers of Haryana have cultivated cloned *Eucalyptus* as bio-drainage plantation on farm boundaries to check the waterlogging in Canal Command areas and to increase farm productivity (Jeet Ram et al. 2011; see in Chap. 9 for details). Farmers also plant *Dalbergia sissoo* on the field boundary at a spacing of 2–4 m. Up to first 3 years the competition with field crops has been found to be negligible. Planting these trees in north–south direction minimizes the losses to crops. Sale of quality timber from this species compensates for crop losses grown in association. In Bihar, *Dalbergia sissoo,* D. latifolia and *Wendlandia exserta* are most common boundary plantations.

Silvopastoral Systems

In Kandi areas of Punjab, fodder trees like *beul* (*Grewia optiva*), *bahera* (*Terminalia bellirica*), *kachnar* (*Bauhinia variegata*) *dhak* (*Butea monosperma*), *dhau/chhal* (*Anogeissus latifolia*), *khirak* (*Celtis australis*) for feeding cattle and *jhar ber* (*Ziziphus nummularia*), *ber* (*Ziziphus mauritiana*) are used for feeding goats. These trees are raised on the field bunds and their branches are lopped during winter to get nutritious leaf fodder (Sarlach et al. 2007). In Bihar, scattered tree species like *Acacia nilotica, Acacia catechu , Ziziphus mauritiana, Dalbergia sissoo,* and *Mangifera indica* are often found growing on grazing lands with grasses like *Dichanthium annulatum, Chrysopogon fulvus, Heteropogon contortus,* and *Cenchrus ciliaris* on flood prone and salt-affected areas. Lopped prunings from trees are often used for fuel and fodder. The tree also provides shelter to the grazing animals during summer and rainy seasons. The controlled grazing helps in improvement of the productivity of these grazing lands (Chaturvedi and Das 2007).

Social forestry and tree plantation drive

On the recommendation of National Commission on Agriculture (NCA 1976), the Social Forestry programmes were initiated by the Government of India for developing trees outside the conventional forest area with participation of local communities (Tewari 1992). Under the social forestry, large scale plantations of *Eucalyptus tereticornis* (Mysore hybrid) and *E. citriodora* were undertaken in West Bengal to provide raw materials to 14 paper mills for filling the gap between total demand (204.3 thousand tons per year) and supply (42 thousand tons per year) from forest department (Thapliyal 1986). The Government Distillation plant at Beliatore (Bankura District) used *E. citriodora* for the oil and local people used lops of *Eucalyptus* tree as fire wood.

In Bihar, planting of *E. tereticornis* was started on a large scale during 1962 in patches for restocking blanks in degraded forests. This tree was grown in block plantation in the existing blanks in the forests and as boundary plantation in wheat fields, on road/rail sides and also on private lands. It performed nicely on deep red loam soil having >1 m soil depth and gained popularity among rural and urban areas. The yield of *Eucalyptus* hybrid was 32.5 to 66 t ha^{-1} at 2×2 m spacing in Singhbhum district. In Uttar Pradesh, *Eucalyptus* plantation was started in 1961–1962. Plantation of *Populus deltoides* in the Uttar Pradesh (*Terai* region) was fairly successful at Mala and Bareily (28–29°N) but outstanding in the area between 29 and 30°N. Planting of various species of multipurpose trees in multiple rows along the Sharda canal by the farmers was a success story (Pathak and Roy 1993).

In Haryana, large scale planting of *E. tereticornis* began in 1962 in wastelands for pulpwood. This tree was planted as strip plantation on road and rail track sides and canal banks. During 1986 the annual targets of 45,000 ha were fixed. The yield varied from 60–90 m^3 ha^{-1} at 3 × 3 m spacing and 40 % of its standing volume was converted into timber (solid volume), another 40 % into pulpwood, and remaining 20 % into fuelwood (Pillai 1986). In Haryana, multiple row planting (strip planting) of various trees was also accepted by the local people along with western Yamuna canal (Pathak and Roy 1993). An area of 0.28 m ha currently water logged was proposed for bio-drainage where the water table was within 3–5 m below ground level and showing rising trend (Diwan 1996). Jeet Ram et al. (2011) have proved that bio-drainage is an eco-friendly tool to combat waterlogging and increase farm productivity (also see in Chap. 9). Based on the success story the Ministry of Water Resources has approved an implementation project for planting *Eucalyptus* trees on 19 million ha waterlogged area in Haryana (Personal communication with Jeet Ram).

In Punjab, planting of *E. tereticornis* started in the early 1960s and after the reorganization of the states in 1966 most of the existing forests went to Himachal Pradesh hence more attention was given on Social Forestry programmes. *Eucalyptus* was planted mainly on the farmers' land as it provided higher financial returns at times even higher than that of annual crops on poor or average quality agricultural lands. The area under *Eucalyptus* plantation was 0.71 m ha during 1984–1985 (Dogra and Sandhu 1986). In Punjab an area of 0.19 m ha was proposed under bio-drainage (Diwan 1996). In wastelands, *E. tereticornis* at 1.5 m spacing and *Populus/Azadirachta indica* at 3.0 m spacing were proposed while in the agricultural fields on the north south field boundaries *Eucalyptus* and *A. indica or Ziziphus mauritiana* were planned (Diwan 1996).

The afforestation activity under 20 Point Programme covered 18.1 m ha area in the country after 1985. A perusal of Table 4.11, shows larger emphasis in the region was on the tree seedling distribution and covering the area in different states of Indo-Gangetic plains during the 8th Plan. Maximum number of seedlings planted and the area was in Uttar Pradesh followed by West Bengal while the minimum was in Punjab. The choice of species varied from place to place but the prominent multipurpose tree species included *Eucalyptus tereticornis, Albizia lebbeck, A, procera, Leucaena leucocephala, Prosopis juliflora, Cassia siamea, Terminalia arjuna, Populus deltoides,* etc.

Agroforestry Research

Organized agroforestry research was initiated in IGP in early 1970s through industry participation in plantations of *Populus deltoides*. The research however, gained momentum with the allocation of a coordinating center of All India Coordinated Research Project on Agroforestry (AICRP AF) in 1983 at BCKV, Jhargram; RAU, Pusa; NDUAT, Faizabad; CSUAT, Kanpur; GBPUAT, Pantnagar; CCSAU, Hissar and PAU, Ludhiana. Under AICRP Agroforestry, initially there was three core projects namely Diagnostic survey and appraisal of existing agroforestry practices, collection and evaluation of promising MPTs, and management of agroforestry systems. The outcome of these three core projects are discussed as under:

Diagnosis and Design (D&D)

During the early phase of the AICRP on Agroforestry, D&D survey was conducted by different centers (BCKV, Jhargram; RAU, Pusa; NDUAT, Faizabad; CSUAT, Kanpur; GBPUAT, Pantnagar; CCSAU, Hissar and PAU, Ludhiana). The survey generated important information and identified important agroforestry practices in different parts of Indo-Gangetic plains (Table 4.12).

The results of the study revealed that there is a good scope of inclusion of fruit trees in agroforestry in West Bengal and Bihar while in Uttar Pradesh, Haryana, and Punjab (Fig. 4.2). A ideal blend of fruit trees—vegetable crops/forage

Table 4.11 Progress of afforestation programme under social forestry in the Indo-Gangetic plains (seedlings in million, area in 000 ha)

State	Eighth plan (1992–97)		1997–98	
	Seedling distribution for planting on private lands	Area (public lands including forest lands)	Seedling distribution for planting on private lands	Area (public lands including forest lands)
West Bengal	395.754 (7.09)[a]	189.4 (3.74)	23.8 (2.21)	18.2 (1.88)
Bihar	64.212 (1.15)	86.1 (1.7)	11.0 (1.07)	14.2 (1.47)
Uttar Pradesh	1381.896 (24.77)	433.0 (8.55)	197.8 (19.13)	88.0 (9.11)
Haryana	71.263 (1.28)	156.7 (3.09)	3.4 (0.32)	17.9 (1.85)
Punjab	34.121 (0.61)	70.1 (1.38)	6.6 (0.64)	5.0 (0.52)
Total	1947.246 (34.9)	935.3 (18.46)	24.6 (23.37)	143.3 (14.84)
Total of India	5579	5067	1033.6	965.6

[a] Values in parenthesis are percentage of total of India

legumes can make the system more sustainable. It was observed that some tree species such as *Mangifera indica* and *Psidium guajava* are being used throughout the plains. Some other timber species are also present in the major part of plains except areas of some centers. For example, *Acacia nilotica* (except West Bengal, and Bihar), *Eucalyptus* and *Azadirachta indica* is being used by the farmers of Bihar, Uttar Pradesh, and West Bengal. For fodder, *Leuceana leucocephala* was grown in Bihar and Uttar Pradesh (probably at Research stations then and later it has become popular all around) while in dry region of Haryana and Punjab, xerophytic species (*Prosopis cineraria* and *Acacia nilotica*) are in use. Other *Acacia* species in use are *Acacia nilotica* (Haryana and Punjab), *Acacia catechu* (Bihar), *Acacia tortilis* (Haryana). In some areas, some valuable timber species are also being used as fuelwood such as *Dalbergia sissoo* in Bihar, *Eucalyptus* in West Bengal and Punjab, *Terminalia arjuna* in West Bengal, *Mangifera indica* in Bihar, etc. This situation shows the need to initiate awareness among the farmers toward the economic value of timber wood and also for adopting fast growing fuelwood trees such *as L. leucocephala, A. nilotica, Casuarina equisetifolia*, etc.

Fruit trees such as *Mangifera indica, Psidium guajava*, and *Aegle marmelos* are grown in the whole Indo-Gangetic plains. *Litchi chinensis, Artocarpus heterophyllus*, and *Emblica officinalis* are common in Bihar while *Vitis vinifera, Pyrus communis,* and *Prunus persica* are grown in Punjab. Many Citrus orchards have been raised in Punjab as commercial plantations. Farmers also have specific requirements for example in Punjab, *tarai* areas of Uttar Pradesh and Uttarakhand, *Populus* is grown for plywood; in Bihar, *Madhuca indica, Azadirachta indica* for oil, *Morus alba* and *Terminalia arjuna* for sericulture, and *Cassia fistula, Bahunia variegata*, etc., for apiculture; and in Uttar Pradesh, *Cassia fistula, Polyalthia longifolia* and *Cassia siamea* for ornamental purposes. Generally, the main tree species being used by the farmers are different in the areas are based on the needs of the local stakeholders.

Collection and Evaluation of Germplasm

After D&D exercise germplasm of suitable multipurpose trees was collected and evaluated by different centers of the AICRP on agroforestry and based on the need of local farmers the priority species were identified and listed (Table 4.13).

Initially, each AICRP on agroforestry center was allocated two species for germplasm collection and provenance trials. Organizations like WIMCO, Forest nursery, Lalkuan, were also involved in the development of clonal planting

Table 4.12 Major tree species used by farmers in Indo-Gangetic plains

States	Tree species	Fodder	Firewood	Fruits	Others
West Bengal	*Gmelina arborea, Terminalia arjuna, Acacia. auriculaeformis, Adina cordifolia, Dalbergia sissoo, Mangifera indica, Holoptelia integrifolia, Azadirachta indica*	*T. arjuna, Acacia nilotica, Z. mauritiana, Ficus, Tectona grandis*	*Acacia auriculaeformis, Cassia siamea, Eucalyptus, Terminalia arjuna, Acacia nilotica*	*Mangifera indica, Psidium guajava, Syzygium cuminii, Aegle marmelos*	Many trees/ shrubs in live fences, boundaries of fields
Bihar	*Acacia lenticularis, Azadirachta indica, Wendlandia exserta* (Banket) *Gmelina arborea, Terminalia arjuna, T. grandis, A. lebbeck, D. sissoo*	*D. sissoo, Moringa oleifera, Dendrocalamus strictus, Sesbania sesban, L. leucocephala, Butea monosperma, A. nilotica, Gliricidia sepium.*	*Pongamia pinnata, Cassia fistula, Sesbania grandiflora, Wendlandia exserta, Pithecellobium dulce, Acacia catechu, Trewia nudiflora, Bombax ceiba, Mangifera indica, D. sissoo, Bauhinia variegata*	*Litchi chinensis, Mangifera indica, Artocarpus heterophyllus, Aegle marmelos, Psidium guajava, Cocos nucifera, Citrus aurantifolia, Syzygium cuminii, Carica papaya, Musa paradisiaca, Emblica officinalis*	Oil: *Madhuca indica, Azadirachta indica, Pongamia pinnata, Moringa oleifera.* Sericulture: *Morus alba, Terminalia arjuna.* Apiculture: *Cassia fistula, Bauhinia variegata, Bombax ceiba, A. lebbeck, Azadirachta indica, Mangifera indica, Litchi chinensis.*
Eastern U.P	*D. sissoo, A. nilotica, Eucalyptus* hybrid, *Azadirachta indica,*	*Azadirachta indica, L. leucocephala, P. juliflora*	*Madhuca latifolia, Leucaena leucocephala*	*M. indica, P. guajava, Emblica officinalis, A. marmelos, Z. mauritiana*	Live fences, Bamboo as pulpwood
Tarai region of U.P and Uttarakhand.	*D. sissoo, A. lebbeck, A. nilotica*			*Mangifera indica, Morus alba, Emblica officinalis, Aegle marmelos, Prunus persica, Citrus aurantifolia, Pyrus communis, Psidium guajava, Musa paradisiaca*	Ornamentals: *Cassia fistula, Polyalthia longifolia, Cassia siamea.*

(continued)

Table 4.12 (continued)

States	Tree species	Fodder	Firewood	Fruits	Others
Haryana	*D. sissoo, A. nilotica, Eucalyptus tereticornis, Populus deltoides*	*P. cineraria, A. nilotica, Ailanthus excelsa*	*P. juliflora, A. nilotica, P. cineraria, A. tortilis*	*Z. mauritiana, Citrus aurantifolia, Cordia rothii, Psidium guajava, Emblica officinalis*	Plywood *Populus deltoides, Ailanthus exselsa*
Punjab	*D. sissoo, Eucalyptus tereticornis, A. nilotica, Toona ciliata, Madhuca indica*	*G. optiva, A. lebbeck, Butea monosperma, Morus alba, L. leucocephala, Bauhinia variegata, A. nilotica, A. catechu*	*A. nilotica, Eucalyptus* hybrid, *A. lebbeck, D. sissoo, G. optiva, Mangifera indica, Melia azedarach, Acacia nilotica*	*Mangifera indica, Psidium guajava, Vitis vinifera, Syzygium cuminii, Citrus* spp, *Morus alba, Pyrus communis, Prunus persica*	Plywood: *Populus deltoides*

Source Pathak et al. (2000)

Fig. 4.2 *Left* Litchi (*Litchi chinensis*)—wheat; and *right* Mango (*Mangifera indica*) turmeric based cropping systems in Bihar

Table 4.13 MPTs based on priority in the Indo-Gangetic plains

Agro-climatic zones	1	2	3	4	5
Lower Gangetic plains region	Eucalyptus hybrid	*Acacia auriculaeformis*	*Gmelina arborea*	*Acacia nilotica*	*Azadirachta indica*
Middle Gangetic plains region	*Populus deltoides*	*Anthocephalus cadamba*	*Eucalyptus* hybrid	*Dalbergia sissoo*	*Acacia nilotica*
Upper Gangetic plains region	*Populus deltoides*	*Eucalyptus* hybrid	*Dalbergia sissoo*	*Anthocephalus cadamba*	*Leucaena leucocephala*
Trans-Gangetic plains region	*Populus deltoides*	*Eucalyptus* hybrid	*Dalbergia sissoo*	*Melia azadirach*	*Leucaena leucocephala*

Source Planning Commission (2001)

material. However, with passage of time tree improvement work was restricted to few species because of lack of funding and retirement of the staff engaged in tree improvement research. In past 3–4 decades, suitable clones of Poplar (*Populus deltoides*) have been developed or introduced. These included G3, G48, S7C4, S7C8, S7C15, S7C20, D121, Pant Poplar-5 from

Fig. 4.3 Breeding work in *Populus deltoides* at GBPUAT, Pantnagar

GBPUAT, Pantnagar; Kranti, Uday, Bahar, W series from WIMCO; L & PL series from Forest nursery Lalkuan and PAU). Clones from *Eucalyptus* (7, 10, 413, 405, Hybrid clone 2070, K series from KFRIM, etc.), *Acacia nilotica* (KSH-9, 33, 89, 104, PT-4, PT-14, Provenances from Banaskantha, Akola, Adilabad, Shivpuri, Hastinapur), and *Dalbergia sissoo* (PS-2, PS-6, PS-84, PS-283, PS-280, 9, Provenances from Raipur, Mauranipur) were also introduced. However, now the tree improvement research in agroforestry is restricted to few organizations (Fig. 4.3).

Agroforestry Systems and Management Research

Location specific agroforestry models are highly productive and conserve natural resources efficiently. Some of the agroforestry systems developed and recommended (Table 4.14) during the last 3 decades under AICRP Agroforestry in IGP (Pathak et al. 2000). The technology extension project on agroforestry was also conducted by the AICRP centers with funds from Department of Wasteland Development, Ministry of Environment and Forests. It strengthened the transfer of agroforestry technologies in the farmers' fields.

Agrisilvicultural system with *Populus deltoides* has a great potential in the irrigated areas. *Madhuca indica* and *Syzygium cuminii*, *Leucaena leucocephala*, *Eucalyptus*, *Acacia auriculaeformis*, *Terminalia arjuna* are also some important tree species in agrisilvicultural systems in West Bengal. For the rainfed areas of Haryana, *Ailanthus excelsa* and *Prosopis cineraria* and *Ziziphus mauritiana* are important while for rice and wheat systems of central plains of the Punjab, boundary plantation of *A. nilotica* and *Eucalyptus* have been recommended. For these systems agricultural crops are wheat (*Triticum aestivum)* for Bihar, *tarai* region of Uttar Pradesh, and flood prone areas of Punjab; rice and wheat rotation (in irrigated central plains of Punjab and Haryana), oat (*Avena sativa)* for Bihar and Punjab; and sesamum (*Sesamum indicum*), *toria (Brassica napus*), French bean *(Phaseolus vulgaris)*, coriander (*Coriandrum sativum*), etc. and *Litchi chinensis* based cropping system for alkali soils for Bihar. Alley cropping of *Leucaena leucocephala*, or *Sesbania sesban* with maize (*Zea mays*), wheat (*Triticum aestivum*) turmeric (*Curcuma domestica*), and coriander has also been proposed for Bihar region. *Emblica officinalis, Psidium guajava*, and *Ziziphus mauritiana* based agrihorticultural systems have been recommended for Haryana. For West Bengal, *Annona squamosa* and *Punica granatum* have been suggested. For eastern and central Uttar Pradesh, forest and fruit trees based cropping systems involving *Casuarina equisetifolia, Emblica officinalis*, *Zyziphus mauritiana,* and *Psidium guajava*); and silvopastoral systems with *L. leucocephala* + napier grass and *Albizia amara* + grass + *Stylosanthes*: silviculture

Table 4.14 Recommended agroforestry systems for Indo-Gangetic plains

State	System	Agroforestry intervention proposed
West Bengal	Farm boundary and homesteads	*Annona squamosa*, *Emblica officinalis*, *Ziziphus mauritiana*, *Punica*
		granatum, *Madhuca latifolia*, *Syzygium cuminii*, can be grown with agricultural crops for supplementing food
		For fodder *L. leucocephala;* fuelwood *Eucalyptus*, *Acacia auriculaeformis,* and *T. arjuna* can be combined as woody components of agroforestry system
Eastern Uttar Pradesh	Farm boundary; fruit and forest trees based cropping systems; orchards; homesteads	In degraded land *Casuarina equisetifolia* can be grown in combination with fruit species such as *Emblica officinalis, Psidium guajava*, and *Ziziphus mauritiana*
Bihar	Farm boundary, shelter belt, agrisilviculture, silvopastoral, forest and fruit trees based cropping systems, homesteads	1. In agrisilviculture, *Populus* (10–12 months old saplings) are planted at 5 × 4 m spacing from mid January to mid February. The most suitable clone for north-Bihar is G3. Wheat, sesamum, vegetables, pulses, seed spices, oats, etc. are grown in the interspaces for 5–6 years
		2. *L. leucocephala* is the most suitable MPT for alley cropping followed by *Sesbania sesban* at an interval of 6 m with a plant to plant spacing of 50 cm. Maize, wheat, turmeric, coriander, kidney bean can successfully be grown in the alleys for 5–6 years in *L. leucocephala* and 3–4 years with *Sesbania sesban*
Central Uttar Pradesh	Agrisilviculture (scattered and boundary plantation)	1. *Casuarina* can be grown for profit on salt-affected soils
		2. In flood prone tracts *Terminalia arjuna*, *A. nilotica*, and *Syzygium cuminii* are the most ideal
		3. Silvopastoral with *L. leucocephala* + napier grass; and *Albizia amara* + grass + stylo is highly beneficical
		4. Forest and fruit trees based cropping systems are more profitable than pure agriculture crops
		5. Agrisilviculture with fisheries for low lying areas and ponds is quite remunerative
Tarai region of Uttar Pradesh and Uttarakhand and western Uttar Pradesh	Agrisilviculture (random trees on farm, block plantations) and fruit trees based cropping systems	1. Under the canopy of *Populus* (clone G 3) planted at 5 × 4 m spacing, wheat, turmeric, potato, coriander, and radish for seeds are successful
		2. Wheat yield was relatively higher under *Melia azedarach, D. sissoo, A. lebbek,* and *Trewia nudiflora* planted at 5 m x 4 m spacing in comparison to other tree species

(continued)

Table 4.14 (continued)

State	System	Agroforestry intervention proposed
Haryana	Agrisilviculture, farm boundary agrihorticulture	1. For irrigated areas: *Populus* based agrisivicultural system
		2. For rainfed areas: *Ailanthus excelsa* and *P. cineraria* based agrisilviculture and *Ziziphus mauritiana* and *Emblica officinalis* based agrihorticultural systems
Punjab	Agrisilviculture, farm boundary	1. For the flood prone area: *Populus* based agrisilviculture as block plantation (5 × 4 m) with sugarcane as intercrop for the 1st and 2nd years and later on wheat, oats, and barseem in the Rabi season, but in Kharif season inter-cultivation is difficult due to close canopy
		2. Central irrigated plains with rice–wheat rotation system: Boundary plantations of *Eucalyptus* and *A. nilotica*-based agroforestry intervention

Source Pathak et al. (2000) (For scientific names of crops see the text)

(*Casuarina*) on salt-affected soils; and agri-silvi-fisheries in low lying areas have been recommended.

Agroforestry Adoption by Farmers

Joint efforts of research in last 3 decades have resulted in development of different agroforestry models. Management and cultural practices of trees and annual crops have also been standardized. Quality planting material of many tree species have been developed which has resulted in productivity enhancement and large scale adoption of agroforestry by the farmers. Agroforestry has also played a major role in the rehabilitation of lands degraded by salinization, ravines, gullies, and other water and wind erosion hazards in this region. Agricultural Universities and Research Institutes in the region have initiated agroforestry demonstration and training programs with the assistance of different agencies. The impact of agroforestry in IGP is well recognized. Based on the findings of the research, farmers are adopting the technology. In states like Haryana and Punjab, nearly 90 % of the timber produced annually is generated outside the forests through sustainably managed plantations.

Although agroforestry has been a way of life in this region, its adoption by people has not been systematically studied. In 1970–1975, only 7 % of the farmers were planting *Eucalyptus* and fetched good prices at the harvest resulting in increase in adoption of *Eucalyptus* (46 % farmers in 1980–1985). But when these plantations become harvestable (in late 1980s to early 1990s) the prices crashed due to glut in the market of E*ucalyptus* timber and low consumption by paper/pulp mills. This resulted into sharp decline in the percentage of E*ucalyptus* growing farmers (5 %) in 1990–1991. This situation diverted the farmers' attention to *Populus* plantation which increased from 3 % (in 1970–75) to 48 % (in 1990–95).

In a detailed study on the adoption potential of agroforestry in Punjab, it was observed that the farm income was the most effective factor favoring the adoption of agroforestry followed by absenteeism and boundary plantation in Punjab (Dhanda 1997). The study also revealed that the major constraints expressed by farmers to adoption of agroforestry were marketing (46 %), crop damage (30 %), and long rotation (21 %). Other constraints are apprehensions (19 %), bad experience (11 %), and other

(35 %) such as small land holdings, labor, technical skills, etc. This survey indicated good adoption of agroforestry which could be further increased by providing good planting material and technical guidance, improving marketing infrastructure, assuring support prices, and also the insurance cover.

The results of agroforestry diagnostic and design survey in *Tarai* areas of Uttar Pradesh and Uttarakhand by GBPUAT, Pantnagar revealed that small land holdings, nonavailability of good quality seedlings of trees and laborers, adverse effects on crop yields, and ignorance of farmers about scientific methods of tree management were the major constraints (Status Report on Agroforestry Research at Pantnagar 1995). A general observation of the *Eucalyptus* plantation in the area shows vast variation in the growth form of trees. In many block plantations hardly 30 % crop has attained desirable growth and size resulting in poor yield and loss to these farmers which was attributed to poor quality of seedlings.

Socioeconomic Characteristics of Tree Farmers

In the sociological analysis of tree plantation and agroforestry, Saxena and Srivastava (1995) found that block plantation was associated with supervision constraint and higher value for the hired labor and non-crop income while boundary planting had no labor implication. They also found that in Punjab, Haryana, and western Uttar Pradesh trees were required neither for ecological function nor to satisfy subsistence needs. Farmers took to tree farming because of the push and pull factors of intensive agriculture, tenancy rules, low cost of the output, and efficiency of supervision. *Eucalyptus* was thus a promising alternative. They further concluded that tree growing was an outcome of agrarian capitalism, i.e., production for market, but with reduced labor. This was proved by their uprooting the same when market for *Eucalyptus* wood collapsed in northern India. In another study of farmlands in Uttar Pradesh, Saxena (1995) observed that earlier trees were raised neither for fuel nor for land fertility since the production systems were geared to crops and livestock production. Due to spread of irrigation, tree density declined and further tree planting was undertaken for additional income and not to produce fuelwood for home consumption. Wood production was more evident in advanced villages of western Uttar Pradesh, where farmers had large holdings, diversified income sources, and tradition for marketing cash crops and not in subsistence villages of eastern Uttar Pradesh where holdings were smaller. *Eucalyptus* planting became attractive due to reduced costs, improving profitability, minimizing danger of encroachment, and cut down on the supervision time in western Uttar Pradesh compared to the eastern part. Further, it was reported that tree planting was pursued more aggressively by well off farmers belonging to upper caste. Some small farmers with non-crop incomes also took up to this activity. However, many surveys show the emphasis on tree planting increases with supportive and remunerative markets in the wood catchments by paper-pulp and ply board industries. In all such innovations, farmers are growing prosperous (Pathak et al. 2006). In recent years, farmers are interested in cloned *Eucalyptus* in Canal Command areas where water table has come at surface declining the crop yields (Jeet Ram et al. 2011). Further, many fruit orchards are coming up in many IGP areas.

Success Stories

Poplar-Based Agroforestry

Poplar (*Populus deltoides*) is widely grown in Punjab, Haryana, Uttarakhand, and Uttar Pradesh. Nearly 15 million poplars plant are raised under agroforestry plantations in these states every year covering roughly 30 thousand ha area. Similar area is harvested yielding about 0.36 million m^3 wood of poplars yearly supporting the safety match industry and a large number of veneer and plywood units. The tree due to its deciduous nature is compatible with various agriculture crops (Fig. 4.4).

Fig. 4.4 Poplar-wheat based agrisilviculture system

Spacing of 5 × 5 m is appropriate for crop yield and tree productivity but usually spacing of 5 × 4 m is preferred to accommodate more number of trees by farmers for better overall economics with insignificant loss to crops and trees (Gandhi and Dhiman 2010). All *rabi* and *kharif* crops can be grown during the first 3 years except paddy. However, inter-cultivation of sugarcane (*Saccharum officinarum*) is preferred for first 2 years as it is more profitable. Third year onwards cultivation of wheat (*Triticum aestivum*), cabbage (*Brassica oleracea* var *capitata*), chilly (*Capsicum acuminatum, C. annuum*), tomato (*Lycopersicon esculentum*), barley (*Hordeum vulgare*), coriander (*Coriandrum sativum*), turmeric (*Curcuma domestica*), ginger (*Zingiber officinale*), strawberry (*Fragaria chiloensis*), oats (*Avena sativa*), berseem (*Trifolium alexandrinum*), mustard (*Brassica juncea. B.oleracea, B.nigra*), etc., can be raised throughout the rotation. Poplar can give yield of 150–200 m^3 ha^{-1} in block plantation and 12–20 m^3 ha^{-1} in boundary plantation at 8 years rotation under good care. It can give gross income of INR 0.6–0.8 million ha^{-1} under block plantation and productivity of 20 m^3 ha^{-1} yr^{-1} while, income from intercrop is additional. Better tree growth is recorded at wider spacing as compared to 5 × 4 m recommended spacing (Lal 2004). The above ground biomass production in poplar in Indo-Gangetic plains was found to vary from 7.4 t ha^{-1} (1-year) to 105.4 t ha^{-1} (10-years) in block plantation and 7.9 to 49.5 t ha^{-1} in boundary plantation (Table 4.15).

In block plantation system, however, a decreasing trend in yield of crops is observed with the increase in age of poplar (Venkatesh et al. 2009). The crop yield in first year is considered equivalent to the open mono crop condition. The yield reduction ranged between 42 % (wheat) and 45 % (paddy) in final rotation period at the 7th year (Table 4.16).

Dhiman (2012) identified intercrops grown in poplar-based agroforestry and reported that in 98 % of the poplar block plantations farmers

Table 4.15 Comparision of above ground biomass at different densities and age of *Populus deltoides* in Indo-Gangetic plains

Age	Density	Above ground biomass (t ha^{-1})	Reference
9–10	500 (Pl)	105.4	Singh (1989)
5–8	400 (Pl)	67.4–134.3	Lodhiyal et al. (1995)
1–4	666 (Pl)	7.4–89.3	Lodhiyal and Lodhiyal (1997)
8	–	202.59	Singh and Lodhiyal (2009)
3–7	500 (AS)	63–128.58	Rizvi et al. (2011)
	200 (BP)	16.39–49.51	
9	500 (AS)	48.7–51.46	Yadava (2010)
	130 (BP)	12.83	
8	500 (AS)	50.1	Kanime et al. 2012)
	70 (BP)	7.87	

(*Pl* plantation, *AS* agrisilviculture, and *BP* boundary plantation)

Table 4.16 Poplar growth and crop yield reduction along age series in block plantation (spacing 5 × 4 m)

Age (yrs)	Height (m)	GBH (cm)	Light reduction (%)	Crop yield (t ha^{-1})		
				Sugarcane/wheat	Wheat	Paddy
1	6.0	18	21.3	59.1[a] (0 %)	4.3 (0 %)	4.7 (0 %)
2	8.0	30	32.3	39.4[a] (33 %)	3.9 (7 %)	4.3 (8 %)
3	12.5	35	37.9	3.7 (12 %)	3.7 (12 %)	4.2 (14 %)
4	15.0	46	46.7	3.4 (20 %)	3.4 (20 %)	3.7 (23 %)
5	22.8	69	50.8	3.0 (30 %)	3.0 (30 %)	3.1 (34 %)
6	26.3	79	53.5	2.4 (40 %)	2.4 (40 %)	2.7 (42 %)
7	26.8	84	53.5	2.5 (42 %)	2.5 (42 %)	2.6 (45 %)

(Values in parenthesis indicate % yield reduction over first year) [a] Yield of sugarcane, since for the initial 2 years sugarcane only is grown and from third year onwards wheat is grown. *Source* Venkatesh et al. 2009

grow intercrops and only a few absentee land owners or casual growers avoid intercrops. On an average, reduction in grain yield was (var PBW 343) 20.10 % under 1 year old poplar plantation which increased to 54 % under 4-year plantation (Chauhan et al. 2010). In the boundary plantation at 2 m spacing around the field, the trees attained height of 27.3 m with GBH of 99.3 cm in 7 years and tree growth was superior than the block plantation. With the aging of trees, reduction in crop yield ranged from 6–7.5 % in 4 years and almost double in 7th year.

Intercropping of agriculture crops with poplar have been found enhancing tree growth and productivity (Chauhan et al. 2012). As diversification option, Chandra (2001) recorded 27.5 t ha^{-1} and 15 t ha^{-1} of lettuce (*Pisonia grandis*) and beet roots (*Beta vulgaris*), respectively under Poplar. Gandhi and Joshi (2002) reported an earning of INR 0.4 million from the inter-cultivation of strawberry (*Fragaria chiloensis*) per annum. Dhiman and Gandhi (2010) recorded comparable yield of garlic (*Allium sativum*) under poplar than open conditions.

Eucalyptus-Based Agroforestry

Eucalyptus is grown in Punjab, Haryana, Uttarakhand, and Uttar Pradesh to cater the pulpwood, plywood and timber industries. *Eucalyptus* is mainly grown on the farm bunds as boundary or a long canal banks and road sides. It causes loss to the associated crops yet it is planted due to higher economic value and grown at spacing of 3 × 2 m or 4 × 2 m. Cotton (*Gossypium arboretum/hirsutum*), chillies (*Capsicum* spp), blackgram (*Vigna mungo*), greengram (*Vigna radiata*), rice (*Oryza sativa*), and groundnut (*Arachis hypogaea*) are cultivated as intercrops during the first year of plantation cycle. Intercrops are also taken up during the second and third years of plantation (Fig. 4.5), but the yield reductions are observed up to 70 %. *Eucalyptus* trees are harvested after 5th year for pulpwood and after 8th year for timber/plywood.

A minimum of three coppices are taken up. Intercrops are also taken up during the establishment as well as during second and third coppice. Average productivity of 20 to 25 m^3 ha^{-1} yr^{-1}has been achieved under unirrigated conditions. However, productivity standards have been redefined by many farmers by achieving unprecedented record yield of 50 m^3 ha^{-1} $yr^{-1}$from cloned plantations as compared to 6 to 10 m^3 ha^{-1} yr^{-1} from normal seedlings. Because of better soils, adequate irrigation facilities and more progressive farmers in Punjab and Haryana, the productivity of clonal eucalyptus plantations is to establish new records and farmers can expect net returns of INR 1 million ha^{-1} at 10 years rotation from irrigated clonal eucalyptus plantations assuming yield of 200 tons (Lal 2004). The performance

Fig. 4.5 Eucalyptus-wheat agrisilviculture system

of wheat, potato (*Solanum tuberosum*), mustard (*Brassica juncea. B.oleracea, B.nigra*), lentil (*Lens esculenta*), and berseem (*Trifolium alexandrinum*), was compared under 2 years old eucalypts planted at 6 × 2 m spacing (Nandal and Bangarwa 2007). The results revealed maximum yield reduction in mustard (82.4 %) followed by potato (80.8 %), wheat (63.2 %), berseem (52.1 %), and lentil (16.3 %).

Field studies in Haryana have established that in the upper soil surface (0–60 cm) water content before each irrigation in wheat sown under eucalypts was two to three times less than the control. Surprisingly, the water content beyond 80 cm soil depth was higher under eucalypts than open field. Similarly, Malik and Sharma (1990) reported that in 3 and half years old *Eucalyptus tereticornis* bund plantation was found to extract five times more water than mustard from 0–15 cm soil profile. Jeet Ram et al. (2007) studied effects of two 18 years old eucalypts hybrid plantations on the shallow groundwater table at Dhob-Bhali research plot located in Rohtak district of Haryana and reported deeper groundwater table underneath the eucalypts plantation. The average draw down of groundwater table underneath the plantation was 1.01 m below the average natural groundwater table level (4.15 m).

Jeet Ram et al. (2011) also observed drawdown of 85 cm water from 3 years old plantation raised on bunds in agriculture fields. The above ground biomass production in *Eucalyptus* as reported by different workers is given in Table 4.17.

Rehabilitation of Degraded Lands

In Indo-Gangetic plains, agroforestry played a major role in the rehabilitation of lands degraded by salinization, ravines, gullies, and other water and wind erosion hazards. By using the agroforestry technologies developed at Central Soil Salinity Research Institute, Karnal, the state forest departments, non-governmental agencies (NGOs), National Wastelands Development Board (NWDB), and other developmental agencies have rehabilitated more than 1 million hectares (out of a total of 6.75 million hectares) of salt-affected soils, particularly the village community lands, government lands, absentee lands, areas along road sides, canals, and railway tracks (Tomar 1997, Dagar 2012). The status and rehabilitation of wastelands including mined, ravine, and degraded watersheds are dealt in detail in Chap. 8 while rehabilitation of salt-affected and waterlogged areas through agroforestry has been dealt in Chap. 9.

Rehabilitation of Salt-Affected and Waterlogged Soils

In Uttar Pradesh Indian Farming Forestry Development Cooperative Limited promotes farm forestry on degraded lands through village level cooperatives and integrated farming

Table 4.17 Comparisons of above ground biomass for Eucalyptus in Indo-Gangetic plains

Age (years)	Density	Above ground biomass (t ha^{-1})	References
5–8	2000	54.1–101.8	Bargali et al. (1992)
	192	27.7–43.3	Yadava (2010), Yadava (2011)
10	120	21.2	Kanime et al. (2012)

system approach. They operate in three states of the Indo-Gangetic plains. Their major activities are in Uttar Pradesh where four districts have been selected where area under wastelands specially the saline sodic soils is more. They have so far planted 3.8 million trees. Species used are *Dalbergia sissoo* (20 %), *Leucaena leucocephala* (11.8 %), *Prosopis juliflora* (26.6 %), *Eucalyptus* (10.4 %), and remaining 31.3 % include *Tectona grandis, Albizia procera, Terminalia arjuna, Acacia nilotica, Syzygium cuminii,* etc., and about 0.6 million trees are ready for harvest. These trees cover 2611 ha area where 789 ha is treated for soil conservation planting 2.63 million trees. *Prosopis juliflora* (Mesquite) was found to improve the physical and chemical properties of degraded sodic soils in Haryana. It reduced pH, electrical conductivity and exchangeable Na levels and increasing infiltration capacity, OC, total N, available P and exchangeable Ca, Mg, and K levels (Bhojvaid et al. 1996, Singh 1999).

More details are given in Chap. 9 in this volume.

Agroforestry for Biofuel and Bioenergy Production

The tree borne oil-seeds (TBOs) species like Karanj (*Pongamia pinnata*), Neem (*Azadirechta indica*), Mahua (*Madhuca indica*), Paradise tree (*Simarouba glauca*), and Jojoba (*Simmondsia chinensis*) offer enormous potential on degraded lands in Indo-Gangetic plains. Research on *Jatropha curcas* and *Pongamia pinnata* revealed that these species can be intercropped with annual crops such as cowpea (*Vigna unguiculata*), sesame (*Sesamum indicum*), sunflower (*Helianthus annuus*), French bean (*Phaseolus vulgaris*), black gram (*Vigna mungo*), green gram (*V. radiata*), and groundnut (*Arachis hypogaea*), etc. In well-maintained plantation *Jatropha* gave about 2 kg of seed per plant. If planted in hedges, the reported productivity of *Jatropha* was from 0.8–1.0 kg of seed per meter length of live fence which was equivalent to seed production between 2.3 and 5.0 t ha^{-1}, depending upon whether the soils conditions for plantations and between 2.5 and 3.5 t ha^{-1} yr^{-1} for hedges (Planning Commission 2004). However, the economics and viability of the *Jatropha* plantation and biofuel production are still at initial stage and require long-term studies. (Some examples are also mentioned in Chap. 2).

Agroforestry for Carbon Sequestration and Mitigating Climate Change

In Indo-Gangetic plains, climate change could represent an additional stress on ecological and socioeconomic systems that are already facing tremendous pressures due to green revolution. Fast growing trees in Indo-Gangetic plains make important contributions toward atmospheric CO_2 assimilation and can play a significant role in the mitigation of atmospheric accumulation of greenhouse gases. Agroforestry has great potential for eco-restoration of degraded lands/forests and mitigating the accumulation of C as CO_2 in the atmosphere by maintaining and increasing the land-based carbon sinks. In Indo-Gangetic plains, an area of about 312 thousand ha is planted with *Populus deltoides* out of which 60 % is under block plantation and 40 % as bund plantation (ICFRE 2012). Utilizing the above data, it is estimated that at national level, planting of poplar may lead to sequestration of more than 542 thousand t C yr^{-1} when wood products are not accounted for and about 1 million t C yr^{-1} inclusive of wood products (Gera 2012).

In Yamunanagar and Saharanpur districts of Haryana and Uttar Pradesh, respectively, contribution of poplar plantations to carbon storage is estimated to be 27–32 t ha^{-1} in boundary systems and 66–83 t ha^{-1} in agrisilvicultural system at a rotation period of 7 years (Rizvi et al. 2011). Biomass, carbon storage and carbon dioxide mitigation potential of plantations of *Populus deltoides, Eucalyptus tereticornis, Dalbergia sissoo, Mangifera indica, Litchi chinensis,* and *Prunus salicina* were assessed in *tarai* region of Uttarakhand. The maximum total biomass (94.8 t ha^{-1}) was observed in *D. sissoo* plantation, followed by *P. deltoides* when grown in block plantation (63.0 t ha^{-1}). Carbon stocks ranged from 4.51 t ha^{-1} in *P. deltoides* boundary plantation to 43.39 t ha^{-1}in *D. sissoo* plantation (Kanime et al. 2012). Chauhan et al. (2010) reported that total biomass carbon storage after 7 years was equivalent to 62.48 t ha^{-1} (8.92 t ha^{-1}yr^{-1}). Pal et al. (2009) reported that there is great clonal variation in carbon sequestration in poplar. WIMCO-22 clone was the best in terms of sequestering carbon stock; while WIMCO-42 was the poorest one in this respect. Dhiman (2009) estimated that only 1.04 million tones C out of 2.5 million tones C from poplar production system in India is locked in wood-based products for different durations and the remaining is released back in the form of fuel and only a marginal fraction of 0.3 million t C is added to soil through the leaf litter every year. The carbon sequestration potential of different tree species in Indo-Gangetic plains as estimated by recent studies is presented in Table 4.18. The variation in the sequestration potential may be attributed to the mean annual increment which varies with site, density, plantation as well as quality of planting stock (Gera 2012).

Agroforestry for Income and Employment Generation

Agroforestry can enhance the income of farmers by generating more employment opportunities thereby providing livelihood security. In addition to income and profitability, numerous paper and plywood industries like ITC, WIMCO Ltd., Green Ply Ltd, Shirdi Industries Ltd, and Star paper mills are promoting tree cultivation and are providing huge employment opportunities. Many of the agroforestry centers in the country have worked out economics of the system and advocated most remunerative combinations (Patil et al. 1979; Pathak et al. 2000). Attempts have been at accounting only the cost of inputs and outputs in material terms. But the costs of goods and services produced by the agroforestry system in its temporal and spatial dimensions have been left out.

Mathur and Sharma (1983) observed that growing *Populus* up to 8 years with crops ensured high economic returns (NPV 11046 and benefit/cost ratio of 3.22 at 12 % interest). Mathur and Sharma (1984) reported the net present value of *Eucalyptus* growing on field bunds as INR 34371 at 10 years rotation and INR 22556 at 8 years rotation at 10 % rate of interest. Financial analysis of clonal eucalypts plantation planted in April, 1999 and harvested in March, 2005 on 1.4 hectare of a farmer in Kurukshetra district of Haryana revealed a net income of INR 53357 ha^{-1} year^{-1} (Forest News 2005). Dwivedi et al. (2007) reported that returns per hectare from poplar-based agroforestry is 2–5 times more than the traditional crop rotation. Rani et al. (2011) recorded cost: benefit ratio as high as 5.51 with annual flower (*Petunia hybrida*) seed production under poplar. Kinnow (*Citrus* sp) based horti-agricultural system has been found to be highly remunerative in Haryana. In general, average gross income of INR 125 thousand to INR 250 thousand per ha is obtained from Kinnow plantation after 6–7 years of plantation depending upon tree density and management. Some of the progressive farmers even earn more than INR 100 thousand ha^{-1} (Nandal and Bangarwa 2007). The B:C ratio of some of the recommended agroforestry systems in Indo-Gangetic plains is given in Table 4.19 (Planning Commission 2001).

In Bihar, litchi (*Litchi chinensis*)-based cropping system has been found to generate employment opportunities of 130–140 man days ha^{-1}. Chaturvedi and Khan (2009) reported that silvo-pastoral system on an average cycle of 10 years can generate employment of 120 man days ha^{-1} per year.

Research and Policy Issues

A lot of work has been done in agroforestry in last 3–4 decades which have helped in better understanding of agroforestry in the region. The coordinated effort has lead to development and adoption of new technologies which has clearly demonstrated significant impact on production and protection aspect. However,

Table 4.18 Carbon sequestration potential of different agroforestry systems in Indo-Gangetic plains

Species	Nature of plantation	Carbon sequestration potential (t C ha^{-1} yr^{-1})	References
E. hybrid	Boundary	0.34–0.88	Yadava (2010); Yadava (2011)
E. tereticornis	Boundary	0.84	Kanime et al. (2012)
P. deltoides	Block	12.02	Singh and Lodhiyal (2009)
	Block	2.01–2.54	Gera et al. (2006); Gera et al. (2011a, b)
	Boundary	1.33–1.42	
	Block	1.98	Hooda et al. (2007)
	Block	9.42–11.87	Rizvi et al. (2011)
	Boundary	3.86–4.56	
	Block	2.06	Yadava (2010)
	Boundary	0.52	
	Block	2.75	Kanime et al.(2012)
	Boundary	0.43	
D. sissoo	Block	1.04	Yadava (2011)
	Block	2.73	Kanime et al.(2012)
Litchi chinensis	Block	0.94	
Mangifera indica	Block	1.43	

Table 4.19 Agroforestry systems with respective B: C ratios in Indo-Gangetic plains

Zone	Tree	Number of trees ha^{-1}	Crops	Years	B:C ratio at 15 % discount factor
Lower Gangetic(rainfed)	*Acacia auriculaeformis*	60	Jute	15	1.17
Middle Gangetic(rainfed)	*Dalbergia sissoo*	280	Sesamum	20	1.15
Middle Gangetic (non-arable land)	Bamboo	250	Marvel grass (*Dichanthium annulatum*)	10	1.76
Upper Gangetic (irrigated)	*Populus deltoides*	500	Wheat	7	3.31
Trans-Gangetic (rainfed)	*Acacia nilotica*	400	Bajra, guar	20	2.61
Saline soils	*Prosopis juliflora*	600	MPTs	10	1.83
Biodrainage	*Populus deltoides*	500	MPTs	7	4.17

Source Planning Commission (2001)

still many issues need to be resolved for which more coordinated, focused, determined efforts are required from all the research institutions, SAUs, forest departments, panchayats, NGOs, and farmers. Future areas of agroforestry need to include the following issues:

- Diversification of agriculture through agroforestry and other alternate land use systems.
- Develop agroforestry models for small land holdings for optimizing multiple output production, resource conservation, carbon sequestration, and biodiversity conservation.
- Screening of shade tolerant varieties of annual crops.
- Development of soil working, agronomic practices, and management techniques to reduce below ground competition.
- Refinement of existing technologies to make it more useful to the stakeholders.
- Development of quality planting material for productivity enhancement.
- Necessary arrangement for registration of nurseries and certification of genetically improved clones.

- Advancing carbon sequestration through agroforestry to enhance livelihoods while mitigating climate change.
- Quantification and economic evaluation of intangible benefits of agroforestry.
- Capacity building of the stakeholders through trainings on nursery production, tree growing, soil and water conservation techniques, biofuel production, value addition of agroforestry products.
- Adaptive research trials on farmers' field, establishment of pilot scale demonstration in collaboration with different line departments, media support for extending agroforestry technologies.
- The current import policy to allow duty free import of timber and pulp and ban on export of farm-grown timber needs to be revised in view of tough competition to the tree growers.
- Forest acts, rules, and regulations in respect of tree felling, transit permit, etc., need to be amended so that it can be made farmers friendly.
- The markets for the agroforestry produce are not stable in the region and are neither supported by any government due to which farmer do not get good returns from trees.
- Provision of support price for farm-grown timber and marketing of agroforestry products through Mandi Parishad Act and the Forest Corporations of the States is required to check exploitation of farmers in the region.
- Need to develop mechanism for collection and dissemination of information regarding prices, import, export, and domestic trade of wood products.
- Institutional finance through NABARD and other banks needs to be provided for agroforestry plantations and nurseries.
- Provision of insurance cover at low premium.

Conclusions

The IGP in the country is spread over 50.3 m ha area in the country. It occupies major parts of West Bengal, Bihar, Uttar Pradesh, Haryana, and Punjab. The region is large and fertile and contributes 51.9 % to the national food grain output showing the presence of good infrastructural support for high crop productivity. Great diversity in soil, water, and climate resources occur in this region. Overall the region has a mixed economy with good growth of agriculture, horticulture, and animal husbandry. The region is densely populated due to which it has been exploited indiscriminately by intensive cultivation which has raised the concern on long-term sustainability of natural resources particularly water. Declining productivity, soil quality, water table; waterlogging; salinity have now become serious problems in the region. Agroforestry has a great potential to meet some of these challenges. Agroforestry in this region is in-built in the farming system. Social Forestry programmes were initiated in this region in early 1970s after the recommendation of National Commission on Agriculture which diverted the attention of policy makers for scientific research in agroforestry. Organized agroforestry research in this region started with participation of industries in social forestry programmes. The research further gained momentum with the allocation of a different coordinating center of All India Coordinated Research Project on Agroforestry (AICRP AF) by ICAR in 1983. A lot of work has been done in agroforestry in last 3–4 decades. Proven agroforestry technologies have been developed by many of research organizations in the region particularly for rehabilitation of degraded environments. The adoption of agroforestry in the region has been very high. Poplar and Eucalyptus based agroforestry models have been very successful in the region. Numerous paper and plywood industries are now promoting tree cultivation and are providing huge employment opportunities. In states like Haryana and Punjab, nearly 90 % of the timber produced annually is generated outside the forests through sustainably managed plantations. Attention needs to be given toward development of market oriented infrastructure for establishment of fruit-based industries so that fruit-based agroforestry systems get priority as farmers are ready to accept these enterprises.

In this region, agroforestry also played a major role in rehabilitation of lands degraded by salinization, ravines, gullies, and other water and wind erosion hazards. More than 1 million hectares of salt-affected soils, particularly the village community lands, government lands, absentee lands, areas along road sides, canals, and railway tracks have been rehabilitated by agroforestry technology. Despite of commendable progress in agroforestry, issues like quality planting material, screening of shade tolerant annual crops, refinement of existing technology, strict laws and procedures, import and export policies, marketing of agroforestry produce need immediate attention.

References

Arya S, Toky OP, Bisht RP Tomer R (1991) *Khejri*. Intensive Agriculture, p 248

Bargali SS, Singh SP, Singh RP (1992) Structure and function of an age series of eucalyptus plantations in central Himalaya. I Dry matter dynamics. Ann Bot 69:405–411

Bhojvaid PP, Timmer VR, Singh G (1996) Reclaiming sodic soils for wheat production by *Prosopis juliflora* afforestation in India. Agrofor Syst 34:139–150

Central Ground Water Board (2006) Dynamic Groundwater Resources of India. Ministry of Water Resources, Government of India, Faridabad

Champion HG, Seth SK (1968) A revised survey of forest types of India. Manager of publication, Government of India, New Delhi

Chandra JP (2001) Scope of Poplar cultivation. Indian Forester 127(1):51–60

Chaturvedi OP, Das DK (2007) Agroforestry systems and practices prevailing in Bihar. In: Puri Sunil, Panwar Pankaj (eds) Agroforestry systems and practices. New India Publishing Agency, New Delhi, pp 277–304

Chaturvedi OP, Das DK, Jha AN (2001) Household fuel situation and agroforestry in north Bihar plains-A diagnostic study. Range Manag Agrofor 22(1):94–105

Chaturvedi OP, Khan IK (2009) Agroforestry research perspectives and Potential. In: Chaturvedi OP, Venkatesh A, Dwivedi RP, Singh R, Dhyani SK (eds.) Agroforestry: natural resource sustainability, livelihood and climate moderation. Satish Serial Publishing House, Delhi, India, p 27–42

Chauhan SK, Sharma R, Dhillon WS (2012) Status of intercropping in poplar based agroforstry in India. ENVIS Forest Bull 12(1):49–67

Chauhan SK, Sharma SC, Beri V, Ritu, Yadav S, Gupta N (2010) Accounting poplar and wheat productivity for carbon sequestration agri-silvicultural system. Indian Forest 136(9):1174–1182

Dagar JC (2012) Utilization of degraded lands/habitats and poor quality water for livelihood security and mitigating climate change. Indian J Agrofor 14(1):1–16

Dhanda RS (1997) Final report research project on 'study on the adoption potential and prospects of agroforestry in Punjab'. Punjab Agricultural University, Ludhiana, p 63

Dhiman RC (2009) Carbon footprint of planted poplar in India 2009. ENVIS Forest Bull 9(2):70–81

Dhiman RC (2012) Diagnosis of intercrops in poplar based agroforestry in India. Indian Forest 138(7):600–609

Dhiman RC, Gandhi JN (2010) Varietal differences in onion and garlic yield when grown with and without mulch in 3 years old poplar plantation. Indian J Agrofor 12(1):40–45

Diwan PL (1996) A view point of waterlogging and salinity in parts of Punjab and Haryana. In: Joshi LK, Dinkar VS (eds.) Waterlogged saline and alkali lands-prevention and reclamations. Ministry of Water Resource, Government of India, New Delhi, pp 93–102

Dogra AS, Sandhu SS (1986) Role of *Eucalyptus* in the economy of Punjab. In: Shrama JK, Nair CTS, Kedarnath S, Kondas S (eds) Eucalyptus in India, past, present and future. Kerala Forest Research Institute, Peechi, pp 261–266

Dwivedi RP, Kareemulla K, Singh R, Rizvi RH, Chauhan J (2007) Socio-economic analysis of agroforestry system in Western Uttar Pradesh. Indian Res J Extension Educ 7:18–22

Forest News (2005) Plantations of clonal eucalypts-success story. Forest News 2:5–6

FSI (2011) India State of Forest Report. Forest Survey of India, Dehardun

Gandhi JN, Dhiman RC (2010) Performance of wheat intercropped under different tree spacings of poplar (*Populus deltoides* Marsh) plantations. Indian J Agrofor 12(2):28–31

Gandhi JN, Joshi BC (2002) Poplar based agroforestry and planting stock availability. In: Sanjeev K, Chauhan R, Gill SS (eds.) Tree crop interface. Department of Forestry and Natural Resources, PAU, Ludhiana, pp 157–161

Gera M (2012) Poplar culture for speedy carbon sequestration in India: a case study from tarai region of Uttarakhand. Forest Bull 12(1):75–83

Gera M, Gera M, Bisht NS (2011a) Carbon sequestration potential of selected plantation interventions in terai region of Uttarakhand. Indian Forest 137(3):273–289

Gera M, Mohan G, Bisht NS, Gera N (2006) Carbon sequestration potential under agroforestry in Rupnagar district of Punjab. Indian Forester 132(5):543–555

Gera M, Mohan G, Bisht NS, Gera Neelu (2011b) Carbon sequestration potential of agroforestry under CDM in Punjab state of India. Indian J Forest 34(1):1–10

Handbook of Agriculture (1986) Indian Council of Agriculture Research. New Delhi, India

Hooda N, Gera M, Andrasko K, Sathaya J, Gupta MK, Vasistha HB, Chandran M, Rassaily SS (2007) Community and farm forestry climate mitigation projects: case studies from Uttaranchal, India. Mitig Adapt Strat Glob Change 12(6):1099–1130

ICFRE (2012) Country report on poplars and willows. Indian Council of Forestry Research & Education, Dehradun

Jeet R, Garg VK, Toky OP, Minhas PS, Tomar OS, Dagar JC, Kamra SK (2007) Bio-drainage potential of *Eucalyptus tereticoornis* for reclamation of shallow water table areas in North-West India. Agrofor Syst 69:147–165

Jeet R, Dagar JC, Lal K, Singh G, Toky OP, Tanwar RS, Dar SR, Kumar M (2011) Biodrainage to combat water logging, increase farm productivity and sequester carbon in canal command area of North-West India. Curr Sci 100(11):1673–1680

Kanime N, Kaushal R, Tewari SK, Raverkar KP, Chaturvedi S, Chaturvedi OP (2012) Biomass production and carbon sequestration in different tree based systems of Western Himalayan Tarai Region. Forest Trees and Livelihood (in press)

Kuldeep Singh, Pritam Chand (2012) Above-bround tree outside forest (TOF) phytomass and carbon estimation in the semi-arid region of southern Haryana: a synthesis approach of remote sensing and field data. J Earth Syst Sci 121(6):1469–1482

Kumar B, Mistry NC, Singh B, Gandhi CP (2011) Indian horticulture database. National Horticulture Board. Ministry of Agriculture, Government of India

Lal P (2004) Integerated development of farm forestry plantation and wood based industries. Indian Forester 130:71–78

Lodhiyal LS, Lodhiyal N (1997) Variation in biomass and net primary productivity in short rotation high density central Himalayan poplar plantations. For Ecol Manage 98:167–179

Lodhiyal LS, Singh RP, Singh SP (1995) Structure and function of an age series of poplar plantation in central Himalaya.I Dry matter dynamics. Ann Bot 76:191–199

Malik RS, Sharma SK (1990) Moisture extraction and crop yield as a function of distance from a row of *Eucalyptus tereticornis*. Agrofor Syst 12:187–195

Mathur NK, Sharma AK (1984) *Eucalyptus* in reclamation of saline and alkaline soils in India. Indian Forester 110:7–15

Mathur RS, Sharma AK (1983) Poplars in India. Indian Forester 109:591–631

Ministry of Agriculture (2007) 18th Livestock census, department of animal husbandry, dairying and fisheries. Ministry of Agriculture, Government of India, New Delhi

Ministry of Agriculture (2011a) Annual report, Directorate of economics and statistics, department of agriculture and cooperation. Ministry of Agriculture, Government of India, New Delhi

Ministry of Agriculture (2011b), Agricultural statistics at a glance, directorate of economics and statistics, ministry of agriculture. Government of India, New Delhi (http://eands.dacnet.nic.in)

Ministry of Agriculture (2012) Annual report 2011–12. Department of animal husbandry, dairying and fisheries. Ministry of Agriculture, Government of India, New Delhi

Ministry of Water Resources (2007) Annual report 2006–07, Ministry of Water Resources, Government of India, New Delhi

Nandal DPS, Bangarwa KS (2007) Agroforestry systems and practices in Haryana. In: Puri Sunil, Panwar Pankaj (eds) Agroforestry systems and practices. New India Publishing Agency, New Delhi, pp 149–162

NCA (1976) National Commission on Agriculture. Report on forestry Vol. 8 and 9. Ministry of Agriculture and Irrigation, Govt. of India, New Delhi

Pal R, Melkania U, Dhiman RC (2009) Inter-clonal variation in carbon pool of *Populus deltoides* Bartr. Indian Forester 135(9):1209–1216

Panwar P (2007) Agroforestry systems and practices in West Bengal. In: Puri Sunil, Panwar Pankaj (eds) Agroforestry systems and practices. New India Publishing Agency, New Delhi, pp 319–332

Pathak PS, Dagar JC (2001) Traditional agroforestry systems: an over view. Landmarks in Botany in India CAS in Botany, BHU, Varanasi

Pathak PS, Pateria HM, Solanki KR (2000) Agroforestry systems in India: a diagnosis and design approach. NRCAF and ICAR, Jhansi, p 166

Pathak PS, Roy MM (1993) Agroforestry in the Indo-Gangetic plain. In: Bentley WR, Khosla PK, Seckler K (eds.) Agroforestry in South Asia: problems and applied research perspectives. Winrock International, U.S.A. and Oxford and IBH Publishing Co. Pvt. Ltd., New Delhi, pp 299–314

Pathak PS, Solanki KR, Rai P, Handa AK, Pateria HM (2006) Agroforestry In: hanbook of agriculture. Indian Council of Agriculture Research, New Delhi

Patil BD, Deb Roy R, Pathak PS (1979) Agroforestry: problem analysis. Potentialities and projected research and development with special reference to Indo-Gangetic plains. In: Proceedings of agroforestry seminar, Imphal, May 16–18, pp 40–68

Pillai VNK (1986) Management of *Eucalyptus* plantations in Haryana. In : Shrama JK, Nair CTS, Kedarnath S, Kondas S (eds.) Eucalyptus in India, past, present and future. Kerala Forest Research Institute, Peechi, pp 54–59

Planning Commission (2001) Report of the task force on greening India for livelihood security and sustainable development. Planning Commission, New Delhi, Government of India, New Delhi

Planning Commission (2004) Report of the committee on development of biofuels. Planning Commission, New Delhi, Government of India, pp 200–222

Puri S, Bangarwa KS (1992) Effects of trees on yield of irrigated wheat crop in semi-arid regions. Agrofor Syst 20:229–241

Rana BS, Saxena AK, Parihar KS (2007) Status and scope of agroforestry development in eastern part of Uttar Pradesh. In: Puri Sunil, Panwar Pankaj (eds)

Agroforestry systems and practices. New India Publishing Agency, New Delhi, pp 191–212

Rani S, Chauhan SK, Kumar R, Dhatt KK (2011) Bioeconomic appraisal of flowering annuals for seed production under poplar (*Populus deltoides*) based agroforestry system. Trop Agric Res 22:125–133

Rizvi RH, Dhyani SK, Yadav RS, Singh R (2011) Biomass production and carbon stock of popular agroforestry systems in Yamunanagar and Saharanpur districts of Northwestern India. Curr Sci 100:736–742

Sarlach HS, Singh Baljit, Chauhan SK (2007) Promising agroforestry practices in Punjab. In: Sunil P, Pankaj P (eds.) Agroforestry systems and practices. New India Publishing Agency, New Delhi, pp 127–148

Saxena NC (1995) Trees on farm lands in Uttar Pradesh: socio-economic charecteristics of tree planters. In: Saxena NC, Ballabh V (eds) Farm forstry in South Asia. Sage Publications, New Delhi, pp 210–244

Saxena NC, Srivastava R (1995) Tree growing as an option option for supervision-constrained farmers in North–West India. In: Saxena NC, Ballabh V (eds) Farm forstry in South Asia. Sage Publications, New Delhi, pp 260–273

Singh B (1989) A study of silvicultural practices, biomass productivity and nutrient cycling (NPK) in poplar plantation of sub-himalayan forest of U.P. state, India. Ph.D. Thesis, Kumaun University, Nainital, India

Singh G (1999) Management of alkaline and saline soils through agroforestry. In: Solanki KR, Bisaria AK, Handa AK (eds.) Sustainable rehabilitation of degraded lands through agroforestry. National Research Centre for Agroforestry, Jhansi, pp 52–61

Singh P, Lodhiyal LS (2009) Biomass and carbon allocation in 8-year-old poplar (*Populus deltoides* Marsh) plantation in tarai agroforestry systems of central Himalaya India. New York Sci J 2(6):49–53

Singh RL (1971) India - a regional geography. National Geographical Society of India, Varanasi

Status Report on Agroforestry Research at Pantnagar (1995) AICRP on agroforestry. G.B. Pant University of Agriculture and Technology, Pantnagar, p 39

Tewari DN (1992) Monograph on *Eucalyptus*. Surya Publications Dehradun, p 361

Tewari S, Kaushal R, Purohit R (2007) Agroforestry in Uttaranchal. In: Puri Sunil, Panwar Pankaj (eds) Agroforestry systems and practices. New India Publishing Agency, New Delhi, pp 95–126

Thapliyal GB (1986) performance of *Eucalyptus* plantations in the state of West Bengal with special reference to the laterite tract. In: Sharma JK, Nair CTS, Kedarnath S, Kondas S (eds.) Eucalyptus in India-present, past and future. Kerala Forest Research Institute, Peechi pp 47–53

Tomar OS (1997) Technologies of afforestation of salt-affected soils. Int Tree Crops J 9:131–158

Venkatesh A, Tewari RK, Rajendra S, Kareemulla K, Kumar A, Chaturvedi OP, Dhyani SK (2009) Biometrics and economics of poplar based agroforestry systems in northern India. In: Chaturvedi OP, Venkatesh A, Yadav RS, Badre A, Dwivedi RP, Ramesh S, Dhyani SK (eds.) Agroforestry, natural resource sustainability, livelihood and climate moderation. Satish Serial Publishing House, Delhi pp 661–674

Yadav RL, Singh SR, Prasad K, Dwivedi BS, Batta RK, Singh AK, Patil NG, Chaudhary SK (2000) Management of irrigated agroecosystem. In: Yadav JSP, Singh GB (eds.) Natural resource management for agricultural production in India. In: Proceedings of international conference on managing natural resources for sustainable agricultural production in the 21st Century, New Delhi, India, 14–18 Feb 2000, pp 775–870

Yadava AK (2010). Carbon sequestration: underexploited environmental benefits of Tarai agroforestry systems. Report Opin 2(11): 35–41

Yadava AK (2011) Potential of agroforestry systm in carbon sequestration for mitigating climate changesvin Tarai region of central Himalaya. Nat Sci 9(6):72–80

Agroforestry as a Strategy for Livelihood Security in the Rainfed Areas: Experience and Expectations

5

G. R. Korwar, J. V. N. S. Prasad, G. Rajeshwara Rao, G. Venkatesh, G. Pratibha, and B. Venkateswarlu

Abstract

Agroforestry in rainfed areas increases livelihood security through simultaneous production of food, fodder, and firewood, and an increase in total productivity per unit area of land. Management of trees in conjunction with crops in rainfed areas minimizes the risk associated with stress period through diversified components and through efficient utilization of limited natural resources. The major objective of agroforestry in rainfed areas is gainful utilization of off-season precipitation, income stabilization, soil and water conservation, insurance against weather aberrations and mitigation of climate change. In India, agroforestry practices in rainfed agriculture have been used to manage scattered trees on farm lands, trees on farm bunds, wood lots as block plantations, trees on range lands, and vegetative live hedges for ecological, social, and economic functions. To enhance rural livelihood security among the dryland farmers, several improved agroforestry systems, commercial plantations and biofuels and bioenergy systems came into being for adoption. Agroforestry plantation-based success stories reveal livelihood security of small, marginal, and landless farmers. Steps to promote basic and promotional agroforestry research in dryland agriculture and appropriate policy responses with extension outreach may potentially deliver better results in rainfed agriculture. Rainfed agroforestry for livelihood security reflects the positive way in utilization of rainfed area resources.

Introduction

Dryland or rainfed areas are the areas of meager or undependable rainfall in which the average precipitation is deficient in relation to water requirements (Srinivas et al. 1999). Rainfed agriculture is, by definition, agriculture dependent upon the vagaries of weather, especially rainfall (Ramakrishna et al. 1999). Rainfed

G. R. Korwar (✉) · J. V. N. S. Prasad · G. R. Rao · G.Venkatesh · G. Pratibha · B. Venkateswarlu
Central Research Institute for Dryland Agriculture (CRIDA), Santoshnagar, Hyderabad, Andhra Pradesh 500059, India
e-mail: grkorwar@rediffmail.com

J. C. Dagar et al. (eds.), *Agroforestry Systems in India: Livelihood Security & Ecosystem Services*, Advances in Agroforestry 10,
DOI: 10.1007/978-81-322-1662-9_5,

agriculture is challenged with several biophysical and socioeconomic problems. Population explosion has forced marginal and submarginal lands into cultivation, posing a serious threat to natural ecosystem. Dryland are fragile and less productive, resulting in low and uneconomic yields. They are characterized by high soil and climatic variability, land degradation, and resource poor farmers. Agroforestry is a land use practice, which integrates trees with agricultural production systems enabling diversification for increased social, economic, and environmental benefits. Farmers' objective of introducing woody perennials in their farming systems is not only to cover the risk of crop failures but also to meet the regular demand of fuel, food, fruit, fodder, and small timber.

Deforestation and overgrazing are the two principal factors responsible for soil degradation. Livestock, a vital component in the dryland farming systems, plays an important role in imparting stability. Farmers' preference in the agroforestry systems is for fruit trees compared to fodder and fuel wood species. Agri-horticultural systems are economically more viable compared to arable farming. Ready and assured market and procurement at farm gate attract farmers into pulpwood production. Apart from fruit, fodder, pulpwood, and timber trees, alternate choices of high-value export-oriented crops like medicinal and aromatic crops can also be integrated in agroforestry systems. The major objective of agroforestry in rainfed areas is gainful utilization of off-season precipitation, income stabilization, soil and water conservation, insurance against weather aberrations, and mitigation of climate change. Agroforestry plantation-based success stories reveal livelihood security of small, marginal, and landless farmers.

Ecology and General Features

Rainfed agriculture is largely practiced in arid, dry semi-arid, wet semi-arid, and dry sub-humid regions in the country (Singh 1999). With about 68 % of rural population (Kumar et al. 2009), these regions are also home to 81 % of rural poor (Rao et al. 2005). In India, 70 % of the cultivated area is rainfed; based on the percent rainfed area, districts in India were categorized into 3 classes, i.e., <35 % rainfed area, 35–70 %, and >70 % rainfed area (Fig. 5.1). Except for few districts in coastal areas of AP, Tamil Nadu, and IGP, rest of the districts are having more than 35 % area as rainfed. Most of the districts of Central India, parts of AP and Karnataka, are having more than 70 % area under rainfed condition and also come under dry and moist semi-arid climate. Parts of Orissa, Chhattisgarh, and West Bengal having more than 70 % rainfed areas come under moist sub-humid climate (NRAA 2012). Coarse cereals (85 %), pulses (83 %), oilseeds (70 %), and cotton (65 %) are the predominant rainfed crops grown in India (CRIDA 2007). Normally, it is considered that soils in rainfed regions are not only thirsty but also hungry (Venkateswarlu 1986). Rainfed agricultural scenario is influenced by both biophysical and socioeconomic factors and their interaction.

Biophysical Parameters

Rainfall

India on an annual basis receives about 4×10^3 km^3 (400 M ha m) of precipitation (FAI 1994) out of the 5×10^5 km^3 precipitation received globally (Lal 1994). India's share, thus, is about one percent of the global precipitation. Based on the amount of rainfall an area receives, Central Research Institute for Dryland Agriculture, Hyderabad estimated that (Kanwar 1999) 15 million ha of rainfed cropped area lies in the arid region and receives <500 mm rainfall, another 15 million ha is in 500–700 mm rainfall zone, and 42 million ha is in 750−1150 mm rainfall zone. About 20 million ha lies in 1150 mm rainfall belt. The rainwater availability in different monsoon periods indicates that major contribution comes from the southwest monsoon (74 %) as compared to northeast monsoon (10 %). Of late, there is an evidence of erratic

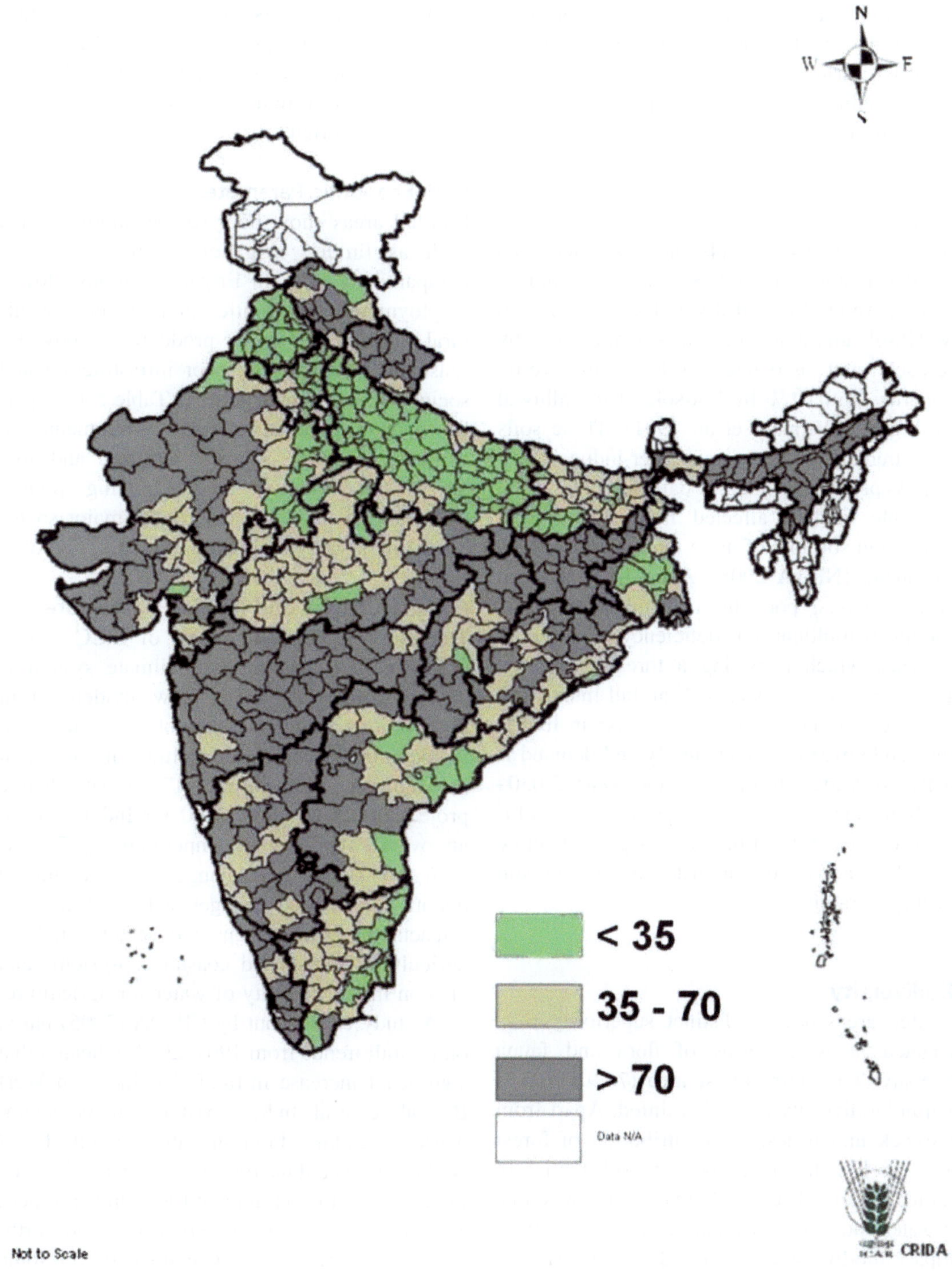

Fig. 5.1 Rainfed areas as percent of net sown area in India (Adapted from NRAA, 2012)

behavior of rainfall like increased frequency of high-intensity rains, reduced number of rainy days, and shift in rainfall. Most future scenarios indicate that there may not be a significant change in the total rainfall in the country in the years to come but the distribution could become more erratic with high-intensity storms followed by long dry spells. The challenge is to harvest rainwater during excess rainy events and reuse the same efficiently during dry spells to improve yield and income.

Soils

Vertisols, Alfisols, Entisols and associated soils are the major soil orders found in rainfed regions. About 30 % of dryland area is covered by Alfisols and associated soils, while 35 % by Vertisols and associated soils having vertic properties and 10 % by Entisols of the alluvial soil regions (Virmani et al. 1991). These soils are extremely prone to rainwater-induced erosion. As per the harmonized database, 73.3 m ha of arable land is affected by water erosion resulting in soil loss of more than 10 tons per ha per annum (NRAA 2008). Apart from soil and nutrient losses, poor nutrient management is leading to multinutrient deficiency of essential nutrients, which is posing a threat to rainfed agriculture. Due to increased rainfall intensities, land degradation is likely to increase in future. The gap between nutrient supply and demand is likely to widen further by the year 2,030–2,050 m tones of NPK. Incorporation of multipurpose trees (MPTs) biomass as green manure is to be encouraged in order to sustain soil fertility in rainfed areas.

Biodiversity

Rainfed areas have a distinct superiority over irrigated areas in terms of flora and fauna diversity. The entire forest area (76.9 m ha) is rich in biodiversity and it is rainfed. Apart from livestock and fishes, the contribution of forest areas to livelihoods of poor is substantial in rainfed regions. The forest area is a major source of water and fodder. Climate change is likely to impact biodiversity in rainfed regions significantly. This impact will be complex with differential responses between interacting species, such as, crops, pests, and pathogens. Certain genotypes may be preferred over others. This may lead to loss of useful gene pool, and hence *ex situ* conservation in identified hot spots becomes important.

SocioEconomic Parameters

Rainfed areas show poor socioeconomic status such as limited irrigation facility—15 % as compared to 48 % in irrigated regions—lower employment opportunities and higher agricultural labor force, lower productivity, poverty, seasonal out-migration, poor infrastructure and social developmental indices (Table 5.1). Apart from above, population pressure, fragmentation of land holdings, tenancy farming, and low investment capacity, credit, pricing policy, marketing, and wage rates are the major socio-economic issues that impact rainfed agriculture.

Climate Change and Rainfed Agriculture

The fourth assessment report of IPCC (2007) observed that "warming of climate system is now unequivocal, as is now evident from observations of increases in global air and ocean temperatures, widespread melting of snow and ice, and rising global sea level". Climate change projections made up to 2100 for India indicate an overall increase in temperature by 2–4 °C with no substantial change in precipitation quantity. All these changes will have adverse impacts on climate sensitive sectors such as agriculture, forest, and coastal ecosystems and also on the availability of water for agriculture.

A study carried out by CRIDA (2005) based on rainfall trends from 1901–2004 indicated that significant increase in rainfall is likely in West Bengal, central India, coastal regions, south-western Andhra Pradesh, and central Tamil Nadu. Significant decreasing trend was observed in central part of Jammu and Kashmir, northern MP, central and western part of UP, and northern and central part of Chhattisgarh. In some areas, both the rainfall and number of rainy days are decreasing which is a cause of concern. It is

Table 5.1 Relative characteristics of rainfed *vis-à-vis* irrigated regions

Parameter	Rainfed regions	Irrigated regions
Poverty ratio (%)	37	33
Proportion of agricultural labor (%)	30	28
Land productivity (INR[#]/ha)	5716	8017
Proportion of irrigated area (%)	15	48
Per capita consumption (kg/year) of:		
Cereals	240	459
Pulses	20	12
Total food grains	260	471
Cooperative credit (INR[#]/ha)	816	1038
Bank Credit (INR[#]/ha)	1050	1650
Infrastructure development index	0.30	0.40
Social development index	0.43	0.44
Number of predominant crops	>34	1[a], 2[b]

Note [a] Rice–Rice in southern India and [b] Rice–Wheat or Cotton–Wheat in northern India
[#] 1 US $ = 54.2931 INR, January 2013
Source 1. Source of Estimates of 17th to 48th rounds: NSSO 1992 Report No. 407
2. GoI (2006)–NSSO 2006, p. 18

to be noted here that the negative deviations in the monsoons are accompanied by a fall in food grain production in India.

Importance of Agroforestry in Rainfed Areas

As we know that through agroforestry, we may optimize land productivity by harnessing positive interactions between tree-crop-livestock on a land area. It is a holistic concept that involves various organisms sharing habitat and its abiotic and biotic components having ecological and economic interactions in their spatial and temporal dimensions. Conservation of natural resources and optimization of productivity are vital in the functioning of the system. With small land holdings it is a livelihood strategy, while having larger holdings it may take a commercial dimension. Though forestry activities are mainly with government, rural people have been practicing tree planting in their farms and homesteads, to meet household requirements for fuel, fodder, poles, timber, fruits, and nontimber forest produce. The objective of agroforestry is to take advantage of the complimentary relationships between trees, crops, and livestock in such a way that the productivity, stability, and sustainability of the total system exceed most of the single cases. Currently, the area under agroforestry (including farm forestry) in rainfed areas is over 6 million hectares and nearly 10 million hectares are covered with rubber, cashew, coconut, mango, and other species. As is the case in other areas, agroforestry in rainfed areas also promotes sustainable agriculture enriching the soil by fixing nitrogen, and improving drainage, efficient nutrient cycling, and opportunity for vertical expansion to optimize land productivity and diversity in output to meet domestic needs and improving economy of farmers. The detailed benefits have been mentioned in Chap. 11 of this publication.

Natural Vegetation

The popular dominant species of semi-arid areas are species of *Acacia, Adina, Aegle, Ailanthus, Albizia, Anogeissus, Asparagus, Azadirachta, Bambusa, Bauhinia, Boswellia, Buchanania, Butea, Capparis, Carissa, Cassia, Cordia, Dalbergia, Dendrocalamus, Diospyros, Emblica, Erythroxylon, Euphorbia, Grewia, Hardwickia, Indigofera, Cydia, Lagerstroemia, Madhuca, Melia, Mesua, Pongamia, Prosopis,*

Pterospermum, Salvadora, Shorea, Sterculia, Syzygium, Tectona, Tephrosia, Terminalia and *Ziziphus.* Subdominant species in these forests are grasses like species of *Andropogon, Cenchrus, Chrysopogon, Cymbopogon, Dichanthium, Heteropogon, Panicum,* and *Sehima* etc.

Agroforestry Systems/Practices

Though dry land agroforestry research is a new concept, the philosophy of dry land agroforestry is an age-old practice and has been rural way of life since time immemorial. The practice of sole agriculture in dry land tract is not sustainable due to very high rainfall variability. Hence, integration of multipurpose tree species in dry land systems accrues enhanced overall productivity per unit area on sustainable basis. Because most of the trees are drought resistant and provide fuel wood, food, fiber, fodder, and other products even if the crop fails. Woody components integrated in dry land ecosystem can take advantage of the invaluable services of climate moderation provided by adjacent natural or restored ecosystems. Most often, this perennial integration is specific to topographic and edaphic factors. Traditional agroforestry practices benefit biodiversity through *in situ* conservation of tree species on farms, reduction of pressure on remaining forests, and the provision of suitable habitat for plant and animal species on farmland. Agroforestry systems/practices of rainfed areas may be classified as traditional and advanced agroforestry systems.

Traditional Agroforestry Systems of the Region

Different traditional dry land agroforestry systems followed by farmers may include:

- Scattered trees on farm lands/Parkland systems
- Trees on farm boundaries/boundary plantations
- Farm wood lots/block plantations
- Trees on range lands
- Vegetative live hedges/live fences.

Scattered Trees on Farm Lands/ Parkland Systems

Trees are often intentionally planted or allowed to persist as scattered from natural regeneration in crop fields. Mature trees on farm lands frequently follow a random or irregular sequence pattern, such species do not receive or require much canopy management operations during the cropping seasons, and they provide a range of products and services on farmlands. According to Nair and Dagar (1991), although the farmers and land owners integrate a wide variety of woody perennials in their crop fields and livestock enterprises, most of these practices are location specific. For example, growing of *Acacia nilotica* in Indo-Gangetic plains and scattered trees of *Borassus flabellifer* in peninsular and coastal regions (Pathak et al. 2000). Intercropping under scattered trees is the simplest and most popular form of agroforestry under smallholder management (Nair 1993). Jambulingam and Fernandes (1986) have documented the cultivation of *Acacia nilotica* trees on rice bunds (raised risers) in Tanjavur and Thiruchirapalli districts of Tamil Nadu. A study on the influence of three agroforestry tree species, viz., anola (*Emblica officinalis*), tamarind (*Tamarindus indicus*), and *Acacia senegal* on the growth and yield of castor (*Ricinus communis*) and green gram (*Vigna radiata*) under rainfed conditions revealed that the yields of arable intercrops were significantly influenced by the trees (Korwar et al. 2006). Production performance of chick pea in terms of relative grain yield and straw yield was superior in association with *Dalbergia sissoo* as compared to *Albizia lebbeck, Azadirachta indica* and *Acacia nilotica* (Gill 2009). Farmers in Koppal and Bijapur district of northern Karnataka practiced scattered planting of woody perennials in their crop lands as most prominent traditional practice next to boundary planting (Table 5.2). Though these compete with crops, farmers are ready to accept them, as these trees have other economic advantages. Canopy management and root pruning are important to reduce the competition of the trees with the crops in the parkland system of agroforestry (Korwar 1999).

Trees on Farm Bunds/Boundary Plantations

Trees in dry land tracts of agricultural crop fields are often planted on field risers (bunds) and/or borders or farm boundaries, to demarcate the field boundaries and also to serve as windbreaks/shelterbelts/live fences. In this system, the space of the bund is gainfully utilized. If a tree species has an erect growth habit, the associated crop growth is not adversely affected by shading (Pathak et al. 2000; Korwar 1999). In Bihar, *Dalbergia sissoo* and *Wendlandia exserta* are the most common boundary plantations. *Casuarina equisetifolia* is extensively planted on field bunds in coastal areas of Orissa (Pathak et al. 2000). Paddy field risers are usually planted with tree species such as *Borassus flabellifer*, *Casuarina equisetifolia, Eucalyptus tereticornis, Gliricidia sepium,* etc. in many parts of the country (Kumar 1997).

In a survey of traditional agroforestry practices in Bagalkot, Gulbarga, Koppal, and Raichur districts in northern dry tract of Karnataka, it was found that nearly 88.4 % farmers followed bund planting as the most prominent practice under rainfed situation. The other potential traditional agroforestry practice was boundary planting in all the five districts with 76.8 % of farmers adopting this practice in Gulbarga district (Table 5.2). The common tree species grown as boundary plantations in dry land systems are *Tectona grandis*, *Leucaena leucocephala* (pollarded for fodder), *Borassus flabellifer*, *Cocos nucifera*, *Acacia nilotica var cupressiformis*, *Dalbergia sissoo* and *Prosopis juliflora* (Pathak et al. 2000, Korwar 1999). In Maharashtra state, of the total 143.21 lakh farmers, 85 % were dry land farmers and their main occupation is seasonal agriculture integrated with perennial woody components along boundary and bunds of the crop fields (Table 5.3).

Woodlots as Block Plantations

Recently, the development of wood lots has been on the rise due to shortage of fuel wood, timber, and demand for poles or pulpwood in industry. This system cannot be evaluated as agroforestry system unless the site resources is fully utilized and rotated with arable crops completely, especially in the early stages of stand development. In semi-arid regions of India, the block plantations of *Leucaena leucocephala*, *Casuarina equisetifolia,* and species of *Eucalyptus* are very common due to multiple uses and demand for pulp wood and poles from paper mills, and other industries in the surrounding vicinity. The woodlots are being raised mostly on large farms to avoid the risks of crop failures due to increase in labor costs and labor management, vagaries in monsoon, and non-availability of other inputs.

Table 5.2 Percentage of dry land farmers following traditional agroforestry practices in northern Karnataka

Agroforestry practices (plantation)	Dry land regions of northern Karnataka					
	Bijapur	Bagalkot	Gulbarga	Koppal	Raichur	Average
Bund	92.5	95.4	90.2	81.2	82.7	88.4
Boundary	52.4	45.6	76.8	18.7	72.5	53.2
Scattered	25.2	21.2	12.8	37.4	18.4	23
Block	5.8	6.4	4.2	12.6	4.2	6.6
Avenue	2.6	3.5	3.8	14.2	5.4	5.9

Source Devaranavadgi et al. (2010)

Table 5.3 List of woody species integrated in traditional dry land agroforestry systems of Maharashtra

Traditional agroforestry system in dry land areas of Maharashtra	Multipurpose woody species
Traditional farm forestry	*Tectona grandis*, *Gmelina arborea*, *Terminalia arjuna*, *Albizia odoratissima*, *Acacia nilotica*, *Azadirachta indica*, *Butea monosperma*, *Mangifera indica*, *Ziziphus mauritiana*, *Terminalia bellerica*, *Syzigium cuminii* and *Annona squamosa*
Farm boundary plantations	*Pongamia pinnata*, *Acacia nilotica*, *Dalbergia sissoo*, *Azadirachta indica*, *Leucaena leucocephala*, *Eucalyptus* spp, *Gliricidia maculata*, *Acacia catechu*, *Albizia lebbeck*, *Albizia procera*, *Tectona grandis*, *Gmelina arborea*, *Dendrocalamus strictus*, *Bambusa arundinacea* and *Bambusa vulgaris*
Block plantations/Farm wood lots	*Tectona grandis*, *Dalbergia sissoo*, *Eucalyptus* spp, *Acacia mangium*, *Casuarina equisetifolia*, *Leucaena leucocephala* and *Dendrocalamus strictus*
Natural silvopasture	*Hardwickia binata*, *Albizia lebbeck*, *Ziziphus mauritiana*, *Acacia nilotica*, *Mangifera indica*, *Erythrina indica*, *Emblica officinalis* and *Annona squamosa*
Live hedges	*Prosopis juliflora*, *Ziziphus oenoplia*, *Caesalpinia sepiaria*, *Dodonaea viscosa*, *Lawsonia inermis*, *Lantana camara*, *Acacia senegal*, *Vitex negundo*, *Ipomoea carnia*, *Jatropha gossypifolia*, *Jatropha curcas* and *Bambusa* spp

Source Ilorkar et al. (2011)

Trees on Range Lands (Silvopastoral System)

Silvopasture denotes land use system in which grasses and legumes are integrated with woody perennials on the same unit of land management to improve the productivity per unit area, apart from increasing the period of fodder availability. These systems are more common in drier areas, and contribute a substantial amount of leaf fodder as top feeds for the animals during lean periods of fodder availability, through lopping/pruning of trees. In the semi-arid regions, silvopastoral systems involving native tree species (e.g., *Albizia procera*, *Albizia lebbeck*, *Acacia* spp., *Azadirachta indica*, *Dalbergia sissoo*, *Morus alba* and *Pongamia pinnata*) have been practiced for many years (Singh and Roy 1993). Rai et al. (1999, 2001) have reported increase in land productivity through silvopastoral systems in the shallow red gravely soils under semi-arid condition at Jhansi in central India, using trees such as *Acacia nilotica* var. *cupressiformis*, *Albizia lebbeck*, *Albizia procera* and *Hardwickia binata*. Encouraging results have been reported on the performance of grass (*Cenchrus*) and legume (*Stylosanthes*) grown under various tree species such as *Albizia amara*, *Albizia lebbeck*, *Acacia tortilis*, *Dalbergia sissoo* and *Leucaena leucocephala* (Rao and Osman 1994; Pathak 2002).

Marginal lands which have poor productivity for grain crops can be put under silvopastoral system. The tree species for this system should be chosen depending upon the farmers' needs and marketability. This system assures 10t $ha^{-1}yr^{-1}$ biomass production (as against <1 t^{-1} from natural stands) at 10 year rotation in dry zones besides assuring soil conservation, healthy environment, and employment generation (Pathak et al. 1995). Trees such as *Borassus flabellifer*, *Casuarina equisetifolia*, *Eucalyptus tereticornis*, *Albizia lebbeck*, *Cassia siamea*, *Tamarindus indica*, *Acacia nilotica*, *Acacia leucophloea* and many others co-exist in the complex systems prevailing in the semi-arid regions of peninsular India (Pathak et al. 2000).

Banni, the India's largest single stretch of grassland located on the northern border of Bhuj Taluka of Kachchh district in the Gujarat State supports the growth of perennial and palatable grasses of high productivity. The grasses include *Sporobolus marginatus, Sporobolus pallidus, Sporobolus helvolus, Dichanthium annulatum, Chloris barbata, Cenchrus ciliaris, Cenchrus setigerus, Desmostachya bipinnata* etc., and sparsely distributed colonies of the *Cynodon dactylon, Dactyloctenium sindicum* and *Eleusine compressa*. Other vegetation includes sedges *Eleocharis atropurpurea, Cyperus rotundus,* and *Cyperus alopecuroides*. The pioneer colonizing species for highly saline soils are *Aeluropus lagopoides* and *Cressa cretica*. The shrub and tree strata are mainly composed of *Prosopis cineraria. Acacia nilotica, Acacia leucophloea, Acacia senegal, Salvadora persica, Salvadora oleoides, Capparis decidua, Tamarix* spp. and *Prosopis juliflora*. Among the tree species, *Acacia nilotica* was once distributed all over Banni but today, palatable grasses (as mentioned above) and tree species like species of *Acacia* and *Salvadora* though present; their abundance has decreased very significantly due to the massive invasion and dominance of *Prosopis juliflora*. Free livestock grazing (Figs. 5.2 and 5.3) are usually allowed for local inhabitants from adjoining grassland habitats, as per traditional rules and regulation governed by the local bodies (Panchayat) of each village (Singh et al. 1998; Oswin 2004; Joshi et al. 2009).

The Kangayam region in Tamil Nadu state in peninsular India has a range of technological and social practices applied for over a century, which have made the traditional grassland a sustainable production system (Fig. 5.4). Eight species of perennial grasses, 6 annual grasses, 9 legumes, and 16 forbs have been reported in this grazing land (Kumar et al. 2011). *Acacia leucophloea* is the predominant tree species with occasional presence of *Acacia nilotica, Acacia planiformis, Albizia amara, Azadirachta indica,* and palmyra (*Borassus flabellifer*) trees.

Vegetative Live Hedges/Fences

Vegetative live hedge/fence is an age-old and traditional agroforestry practice in semi-arid tropics to protect the crops and garden lands. It may be defined as "a way of establishing a boundary by planting a line of trees and/or shrubs the latter usually from large stem cuttings or stumps, at relatively close spacing and by fixing wires to them". Live fences can be extremely diverse, low risk systems that provide farmers with numerous benefits. Besides their primary purpose of controlling human and animal movement, live fences may provide fuel wood, fodder, and food, act as wind-break, or enrich the soil, depending on the species used. Other auxiliary benefits like shade, seed bank, mulch, medicine, nutrient trap, soil and water conservation, and wildlife habitat make the use of living

Fig. 5.2 Maldharis, a pastoral community, grazing their livestock in the Banni grasslands dominated by woody component *Prosopis juliflora*

Fig. 5.3 Semi-arid Banni grassland in monsoons with new flush for grazing, marked with trees and Conical Hill (Keero Dungar) at the back ground Source: http://www.indianaturewatch.net/displayimage

Fig. 5.4 Mecheri sheep grazing in the traditional silvopasture (*Cenchrus ciliaris* + *Acacia leucophloea*) with live fence (*Balsamodendron berryi)* at Kangayam region of Tamil Nadu, Southern India (Adapted from Kumar et al. 2011)

fences a more widespread practice (Rocheleau et al. 1988). The live fencing practices may be single species or multiple species and depending on the species it may be used around field crops, backyards, homesteads, and orchards/plantations, depending on the landscapes being protected. Long back, Buchanan (1807) reported that many of the hedges in Coimbatore district were of Mullukiluvai (*Balsamodendron berryi syn. Commiphora berryi*), which made good fences. It formed a good fence over the greater part of Coimbatore and cattle trespass was rare, cattle and crops were protected, large quantities of fuel supplied, and protection was given to growing trees. The fields are protected using this species now also (Fig. 5.5).

Fig. 5.5 Traditional silvopasture protected with live hedge of *Balsamodendron berryi* at Coimbatore district, Tamil Nadu, Southern India (Adapted from Kumar et al. 2011)

Other common hedge species are *Gliricidia sepium*, *Agave sisalana*, *Jatropha curcas*, *Jatropha gossypifolia*, *Lantana camara*, *Cactus* spp, *Opuntia* spp, *Prosopis juliflora*, *Balanites aegyptiaca*, *Pithecellobium dulce*, *Lawsonia inermis*, *Carissa carandas*, and *Parkinsonia aculeata*. Regular pruning and canopy management is very important to check them from competing with the crops and also check the undesirable seed production, especially the weedy species.

Improved Agroforestry Systems and Their Contribution to Livelihood Security

Dry land regions are characterized by erratic rainfall and high temperature where the prospects of successful arable cropping are limited. Apart from the climatic variability, shallow soils and poor water retention cause lower yields. Integration of perennials into the arable systems not only provides additional income to the farmers but also reduces the risk during droughts. As described elsewhere the tree-based systems have multifarious benefits in comparison to sole cropping under rainfed situations. Some of the important improved systems/practices are discussed below.

Forest Trees and Arable Crops-Based Systems

In this system, intercropping of annual food crops and woody perennials in a land use system enhances total productivity and also ensures sustainability of land. Involving nitrogen fixing tree species, the system offers immense possibilities of supplementing the nitrogen requirements of crops grown in association, besides providing rich organic matter and improving the soil structure preventing land degradation. Most of the nitrogen fixing trees commonly planted include species of *Acacia*, *Albizia*, *Hardwickia*, *Gliricidia*, *Leucaena*, *Pongamia*, and *Sesbania* (Solanki and Ramnewaj 1999). Several tree species have been tried in different parts for their potential for integration into arable systems. Some experiments were conducted on leguminous trees (e.g. *Faidherbia* and *Acacia*) at Central Research Institute for Dryland Agriculture (CRIDA), Hyderabad; at Indian Grassland and Fodder Research Institute (IGFRI), Jhansi; National Research Centre for Agroforestry (NRCAF) Jhansi and on *Gmelina arborea* at Raipur. Results of the experiments conducted under rainfed conditions at CRIDA on yield of intercrops with two densities of trees (156 and 625 trees ha^{-1}) of *Faidherbia albida* from 5 to12 years of tree age are presented in Table 5.4.

Table 5.4 Crop yields in *Faidherbia albida* agroforestry system at various ages and densities under rainfed conditions at CRIDA, Hyderabad

Tree age (years)	Tree density (per ha)	Crop(s)[a]	% increase (+) / reduction (−)
5	156 and 625	Sorghum, groundnut	+57 and +32 % higher sorghum and groundnut yield (156 trees ha^{-1}) in comparison to higher tree density of 625 trees ha^{-1}
6	156	Sorghum, groundnut	Sorghum −6 %, Groundnut +14 % over sole crop control
6	625	Sorghum, groundnut	Sorghum −46 % and groundnut −24 % over sole crop
8–9	625	Cowpea	−9 % in grain yield and −10 % in dry matter production in comparison to control
8–9	625	Sorghum	−12 % in grain yield and −13 % in dry matter production
10	156	Sorghum, cowpea	Sorghum −11 %, cowpea −8 % over sole crop control
11–12	625	Green gram, black gram, cowpea	Green gram −26–29 %, cowpea −59–64 %, and black gram −91–96 % grain yield in comparison to control
11–12	156	Green gram, black gram	Green gram +12–20 %, black gram + 24–27 % grain yield in comparison to control

[a] For scientific names see text *Source* Korwar and Pratibha (1999)

At 5 years of tree age, sorghum (*Sorghum bicolor*) and groundnut (*Arachis hypogaea*) yields were higher in lower tree density by 32–57 % in comparison to higher tree density. Tree population did not affect the dry matter production in both the groundnut and sorghum crops at 5 years of tree age. At 6 years of tree growth, though, groundnut yields improved when compared to the lower tree density, there was a marginal reduction in yields of sorghum in comparison to the sole crop. The crop yields were significantly affected under higher tree densities of 625 trees ha^{-1} during the sixth year. Seed and dry fodder yields of sorghum were reduced by 12 and by 13 %, respectively, when intercropped with 8 to 9-year-old *Faidherbia albida* at a density of 625 trees ha^{-1} in comparison with the sole crop. However, the extent of reduction in *Faidherbia albida* is lowest in comparison to other N-fixing tree species such as *Acacia ferruginea* and *Albizia lebbeck*. At 10 years of age, the yield reduction in sorghum and cowpea (*Vigna unguiculata* cv. C 152) was to the extent of 8–11 %, respectively, in comparison to the sole crops. However, in 11 to 12-year-old trees, the seed yields of green gram (*Vigna radiata* cv. ML 267) and black gram (*Vigna mungo* cv. T 9) were higher under the lower tree density (156 plants ha^{-1}, spacing 8 m x 8 m) than when monocropped or grown under the higher tree density (625 plants ha^{-1}), whereas cowpea yields were lower under the trees at both densities than when cultivated as monocrop (Korwar and Pratibha 1999).

With wider tree spacing, the negative effect of trees was reduced. The extent of yield reduction was also influenced by the amount of rainfall received during the cropping period apart from the tree age. For example, the extent of yield reduction was unusually high during 1991–1992, which coincided with lower rainfall. During the years of low rainfall, the yield reduction is more with the higher tree density. Under no input conditions, viz., under no nitrogen application situation, the *Faidherbia albida* + sorghum and *Faidherbia albida* + cowpea system recorded higher grain yields in comparison to sole crop controls and the extent of yield increase was 54 and 32 % in sorghum and cowpea, respectively (CRIDA 1996).

Experiments conducted at Jhansi on several trees species with three spacings (2 m × 4 m,

2 m × 6 m and 2 m × 10 m) with four cropping sequences showed that the intercrop yields were highest during the first year tree growth and gradually declined with the advancement of tree growth (Gill 1995). The tree biomass obtained by pruning ranged from 0.6 to 2.8 tons ha^{-1} from the third year of tree growth and tree pruning is one of the ways to reduce the tree–crop competition in these systems particularly under rainfed conditions. Though yield reduction was observed in crops, the biomass from trees contributed to the stability of the system. Similarly, the yield reduction in arable crops was observed up to 15 % in a 6-year-old *Hardwickia*-based system developed in Jhansi and the reduction was higher at higher tree density. In *Albizia*-based system at Hyderabad in alfisols, yield reduction in intercrop was observed up to 20 % due to tree competition which could well be compensated by value from pollarded portions. The benefit–cost ratio (2.68) was marginally high with pure arable crop as compared to tree pollarding (2.40) after considering the value of tree prunings (Joseph et al. 1999). In the last few years, systematic experiments are being conducted at several locations of All India Coordinated Research Project on Agroforestry (AICRPAF) as a result of which several recommendations have emerged (Pathak and Solanki 2002) for different agroclimatic regions of the country for rainfed conditions (Table 5.5). Various arable crops included coarse cereals (Species of *Echinochloa, Sorghum* and *Pennisetum*); pulses such as green gram (*Vigna radiata)*, black gram (*V. mungo)*, pigeon pea (*Cajanus cajan)*, cowpea (V. *unguiculata*, gram (*Cicer arientinum),* lentil (*Lens culinaris*) soya bean, (*Glycine max*); oil seeds such as sesame (*Sesamum indicum*), mustard (*Brassica oleracea, B. juncea*); paddy (*Oryza sativa*), tomato (*Lycopersicum esculentum*), chillies (*Capsicum annuum*), curry leaf (*Murraya koenigii*) and groundnut (*Arachis hypogaea*).

Though there are successful examples of tree-based cropping systems viz. *Faidherbia albida* system much popular in Africa and *Prosopis cineraria* system in the semi-arid parts of India; large-scale integration of trees into arable systems is not happening in India particularly under rainfed conditions. Trees like *Azadirachta indica*, *Acacia nilotica* and *Gmelina arborea* have economic uses and are acceptable to farmers. These systems compete with crops after 6–7 years of tree age. Canopy management, root pruning, and reducing the tree density are some of the important techniques to reduce the tree–crop competition. With some more research inputs, these practices may easily be made adaptable by the farmers.

Alley/Hedge Row Cropping

Alley cropping also known as hedge row intercropping is an agroforestry practice in which food (arable) crops are grown in alleys formed by hedge rows of trees or shrubs (Kang and Wilson 1981). The hedge rows are cut back at crop planting and kept pruned during the cropping season to prevent shading and to reduce competition with food crops. The hedge rows are allowed to grow when there are no crops and normally pruned during the season and the pruned material is either used as mulch, fodder, or source of nitrogen for crops. Trees such as species of *Leucaena*, *Gliricidia*, *Sesbania, Cajanus,* and *Cassia* have been widely tested in alley cropping system. Short duration rainy crops such as pearl millet and sorghum were found to be compatible with *Leucaena leucocephala* and *Gliricidia sepium* than long duration species such as castor (*Ricinus communis*) and pigeon pea (*Cajanus cajan*). Wider alleys and low cutting heights were found to give higher intercrop yields in leucaena system in semi- arid condition (Korwar 1992; Rao et al. 1991). Alley cropping of *Leucaena* with sorghum (*Sorghum bicolor*) and cowpea (*Vigna unguiculata*) was studied by Osman et al. (1998) at Hyderabad to quantify the effect of shoot pruning, fertilization, and root barriers around *Leucaena leucocephala* trees on crop production under rainfed conditions. Crops grown with pruned trees attained higher dry matter and leaf area index than did those with unpruned trees.

Table 5.5 Recommended forest tree-based agricultural practices for different regions of the country for rainfed conditions

Agroclimatic region	Suggested trees	Crops
Eastern plateau and hill region (Maharashtra, Uttar Pradesh, Orissa and West Bengal)	*Acacia nilotica* *Borassus flabellifer*	Coarse cereals, pulses, oil seeds, soybean, lentil, mustard
	Tectona grandis	
	Gmelina arborea	
Central plateau and hill region (Madhya Pradesh, Rajasthan and Uttar Pradesh)	*Azadirachta indica*	Ground nut, sorghum, pigeon pea, gram, lentil, mustard, soyabean, paddy, sesamum
	Acacia nilotica	
	Hardwickia binata	
	Dalbergia sissoo	
Western plateau and hill region (Maharashtra, Madhya Pradesh and Rajasthan)	*Azadirachta indica*	Groundnut, sorghum, pigeon pea, gram, lentil, mustard, soyabean, paddy, sesamum
	Acacia nilotica	
	Hardwickia binata	
	Dalbergia sissoo	
	Leucaena leucocephala	
Southern plateau and hill region (Andhra Pradesh, Karnataka and Tamil Nadu)	*Tamarindus indica*	Tomato, chili, curry leaf
	Ailanthus excelsa	Cowpea, sesamum, sorghum, pearl millet
	Albizia lebbeck	Cowpea, sesamum, sorghum, pearl millet
East coast plains and hill region (Orissa, Andhra Pradesh, Tamil Nadu and Pondicherry)	*Acacia nilotica*	Paddy/pulse, ground nut, sesamum
	Gmelina arborea	
	Dalbergia sissoo	
	Anogeissus acuminata	
	Prosopis juliflora	

[a] For scientific names of crops see the text

Sorghum or cowpea intercropped with trees responded to fertilizer application more strongly than did their respective pure crops, suggesting an increased need for fertilizer application in this system over that currently used for pure crops. Impact of root barriers was small on either crop. Irrespective of root barriers, a high response to tree pruning suggested above ground competition for light-dominated tree-crop interactions in this system.

Alley cropping of *Leucaena leucocephala* with pearl millet (*Pennisetum typhoides*) and pigeon pea (*Cajanus cajan*) reveals that reduction of yields with the increase of alley widths (10, 15, and 20 m) might be due to dense shade of the tree (Patil et al. 1999). Mittal and Singh (1983) reported that hedge row plantation of *Leucaena leucocephala* with maize, black gram, and cluster bean reduced the yields of maize (*Zea mays*),black gram (*Vigna mungo*), and cluster bean (*Cyamopsis tetragonoloba*) by 38, 34, and 29 %, respectively. This reduction in yield was compensated by relatively higher fodder and fuel production of *Leucaena.* Maximum returns were obtained with intercropping compared to growing of pure trees or crops. Alley cropping systems are effective in reducing the runoff and soil loss (Rao et al. 1991) and production of green fodder/green leaf manure (Korwar and Radder 1994). However, this system has not become popular among the farming community as some of the studies reported yield losses due to competition for moisture (Rao et al. 1991), and farmers did not see the direct economic benefit and stray cattle menace during the off season is a major problem in some of the areas (Korwar 1999). In recent years, due to international market of cluster bean (*Guar*) gum, the crop has become a boon for low rainfall areas.

Fruit Tree-Based Cropping Systems Involving Grasses and Arable Crops

Agricultural crops are normally grown in the interspaces of fruit trees planted at a spacing of 5–7 m apart. The fruit trees are managed for 30–35 years and they give regular income and the intercrops, mostly annual crops, provide seasonal returns. Fruit crops such as mango (*Mangifera indica*), lime (*Citrus* sp), lemon (*C. limon*), guava (*Psidium guajava*), pomegranate (*Punica granatum*), goose berry (*Emblica officinalis*), jamun (*Syzygium cuminii*), wood apple (*Feronia limonia*), and tamarind (*Tamarindus indica*) can be grown in areas where the rainfall is more than 600 mm. Tree species like custard apple, ber (*Ziziphus* spp), Phalsa (*Grewia subinaequalis*), Karonda (*Carissa carandas*), Lasura (*Cordia* spp.), and Bel (*Aegle marmelos*) can be grown in areas where the rainfall is less than 500 mm under dryland situations. On marginal and degraded lands ber, custard apple (*Annona squamosa*), and jamun (*Syzygium cumini*) have shown good promise. Short-duration millets, pulses, or oil seeds can be profitable intercrops under rainfed conditions. Studies conducted at CRIDA showed that fruit-based agricultural systems gave highest benefit–cost ratio in comparison to other alternate land use systems (Table 5.6) under rainfed conditions.

However, depending on the requirements, perennials such as grasses and legume stylo (*Stylosanthes hamata*) can also be grown in the interspaces of fruit trees. Examples from some individual fruit-based systems have been mentioned below:

In recent years, there has been an increasing trend toward the adoption of fruit tree-based systems particularly the mango in view of the growing demand for fruits, higher returns, support from developmental programmes like National Horticulture Mission (NHM), and microirrigation. Mango requires about 5 years for attaining commercial yields. Wide range of intercrops can be grown in the interspaces during the first 5 years (Reddy and Sudha 1988). Sheep and goat rearing has become an attractive option for many farmers. On farm experiments conducted have shown that introducing pasture component has substantially improved the profitability of mango-based systems even in low rainfall regions (CRIDA 2005).

Experiments conducted with four varieties of goose berry (*Emblica officinalis*) namely *Chakaya, Kanchan, Krishna,* and NA-7 at two spacings of 10 m × 6 m and 5 m × 6 m at Jhansi have shown that about 25–100 kg fruits per plant could be produced after 6 years (Ram Newaj and Gill 1995). Prasad et al. (1997) have reported that highest average girth of 38.4 cm in association with grass *Chrysopogon fulvus* and 32.4 cm with Napier grass (*Panicum maximum*) when compared to 37.7 cm of pure stand of goose berry. Fruit yield and number of fruits per tree were also minimum in association with Napier where as maximum when grown with *Chrysopogon fulvus*. Experiments have shown that in *Emblica officinalis* fruit yields were significantly higher (5,277 kg ha^{-1}) in sole trees than intercropped ones (3,478 kg ha^{-1}) due to competition for moisture. During low rainfall years, the arable crops like Ragi (*Eleusine*

Table 5.6 Benefit–cost ratios of different alternate land use systems (CRIDA 1993–2010)

System	Period (years)	Benefit: cost ratio
Arable farming	1	1.34
Forest tree + sorghum + pigeon pea	10	1.65
Fruit trees + arable crops	30	5.53
Forest trees (with castor intercrop)	10	1.99
Forest trees + grasses	10	2.45

coracana), common millet (*Panicum miliaceum*), and species of *Echinochloa* and *Setaria* failed where as some fruit yield in goose berry was obtained which shows the resilience of the tree-based systems. Among the nutrient management practices, significantly highest yields (5,396 kg ha^{-1}) were obtained with a combination of farm yard manure (FYM) and inorganic fertilizers (IF); and the lowest (3,170 kg ha^{-1}) was with vermin-compost (VC) plus inorganic fertilizers (IF) combination. However, VC, FYM, and IF were on par with FYM + IF. The fruit quality in terms of vitamin C was highest with FYM + inorganic fertilizers (IF) followed by vermin-compost (VC) and FYM and the lowest was with VC + IF. Reducing sugars were highest with VC and lowest with VC + IF.

Experiment conducted on Alfisols in Hyderabad has shown that the fruit yields of *ber* (*Ziziphus mauritiana*) in association with different arable crops are comparable, but the association of pigeon pea caused considerable reduction in fruit yield. The mean gross monetary returns were higher from intercropping of *ber* with sunflower followed by intercropping with castor but far less than with pigeon pea. They found that under rainfed conditions *ber* fruit tree-based cropping system on poor and marginal lands is remunerative and there was maximum return (Table 5.7) with sunflower (*Helianthus annuus*) and castor (*Ricinus communis*).

Patil et al. (1999) revealed that the fruit trees like *ber (Ziziphus mauritiana)* and goose berry (*Emblica officinalis*) had no adverse effect on yield of arable crops. The advantage of growing pearl millet (*Pennisetum typhoides*) + pigeon pea (*Cajanus cajan*) along with *ber* was higher (INR 8305 ha^{-1}) when planted at 10 m alley width and highest monetary returns obtained when *ber* planted at 10 m alley width followed by 20 and 30 m. Under rainfed condition on a light soil, the cropping system with *ber,* planted in 10 m alley width was more remunerative. Pearl millet and pigeon pea were found suitable companion crops with *ber* at different alley widths. Studies conducted at IGFRI have shown that growing grass with *ber* has no effect on fruit yield of *ber* (Sharma 1996) and fruit production increased with stylo (*Stylosanthes hamata*) in comparison to control and in association with grass.

In a study at CRIDA (1996), 95 % survival of guava (*Psidium guajava*) was recorded after 6 years of planting by controlling the buffalo grass (*Paspalum conjugatum*) in the basins (2 m in diameter) and pot watering of plants during the dry period. Yield reduction of stylo (*Stylosanthes* sp) was less under widely spaced plants (8 m × 5 m) compared to closer spacing (5 m × 5 m), indicating the inevitability of wider spacing for fruit trees when grown with stylo. The fruit yield also increased significantly with closer spacing and also when forage crops were not

Table 5.7 Gross monetary returns from *ber*-based cropping system

Treatment	Gross monetary returns (INR ha^{-1})				Returns/ INR invested
	Fruit	Fuel wood	Grain from crops	Total	
Ber + Sunflower	16185	460	2874	19519	0.93
Ber + Castor	15990	460	1926	18376	0.78
Ber + Pigeon pea	13845	472	2158	16475	0.68
Sunflower	–	–	2982	2982	0.86
Castor	–	–	2268	2268	0.26
Pigeon pea	–	–	2245	2245	0.73
Ber	17745	496	–	18241	1.15
Sem±	–	–	–	401	–
CD (0.05)	–	–	–	867	–

Source (CRIDA 1993,) Please see text for scientific names of crops

planted in association (Table 5.8). Recommendations on the suitable fruit trees along with several arable crops for various regions of the country are available for rainfed conditions (Table 5.9).

Agroforestry Systems Based on Commercial Plantations

In India, with the rapid increase in population and due to the emphasis on paper as environment friendly packaging material lead to a rapid increase in paper consumption for various uses such as news print, printing paper, value added paper etc. The rate of annual production is about 4.1 % where as the growth in the consumption is at an average rate of 5.1 % a year. In India, the annual per capita consumption of paper and paperboard is only 4.5 kg compared to 42 kg in China and 55 kg worldwide.

Due to the inability of forests to meet the raw material requirements, industry to a great extent depends on the imported pulp. The imports were to the tune of 768,000 tons costing about US $ 435 m during 2002–2004 (FAOSTAT 2006). Of the annual pulp production of 1.9 million tons using 6.8 million tons of wood, nearly 20 % of wood is procured from forests through government sources and remaining 80 % is from private land holders through agroforestry systems (Kulkarni 2008).

Many fast growing plants have been screened for their suitability for pulp and paper making. These includes species of bamboo, *Populus* spp, *Casuarina equisetifolia, Ailanthus excelsa, Eucalyptus globulus, Eucalyptus grandis, Eucalyptus tereticornis, Sesbania aculeata, Moringa oleifera, Prosopis juliflora, Leucaena leucocephala, Grevillea robusta, Pinus taeda, Pinus radiata, Sesbania aegyptiaca, Acacia auriculaeformis, Gmelina arborea,* and *Hibiscus sabdariffa.* Among these trees, *Eucalyptus, Leucaena, Casuarina,* and *Bamboo* are grown widely for the pulpwood under rainfed conditions where as poplar is generally grown under irrigated conditions of North India.

Eucalyptus based systems

About 170 species, varieties, and provenances of eucalyptus were tried in India (Bhatia 1984) the favored has been the *Eucalyptus* hybrid, a form of *Eucalyptus tereticornis* known as Mysore gum. The important characteristics of contributing to its popularity under Indian conditions are its fast growth, capability of tolerating weed competition during the early stages, coppices well, is fire hardy, browse resistant, and has the ability to adapt to a wide range of edaphoclimatic conditions (Kushalappa 1985). The average yield from the plantations is around 5 tons $ha^{-1}yr^{-1}$. Often trees are spaced at 1.5 m × 1.5 m and in case of bund plantations close planting of 1 m × 1 m is adopted. A rotation of 7–9 years has been recommended for the pulpwood, where as for commercial wood, a rotation of 13–14 years is recommended. The annual increment of eucalyptus peaks in the fifth year and then drops (Chaturvedi 1989). The clonal eucalyptus plantations raised for pulpwood in Andhra Pradesh are generally harvested at 4 year intervals and intercropping in these systems is generally not feasible due to close spacing of 3 m × 2 m.

Table 5.8 Fresh yields of forage and fruit in guava-based horti-pastoral system

Spacing (m)	Forage yield (t ha^{-1})		Fruit yield (kg $plant^{-1}$)
	Stylo legume	Buffalo grass	
5 m × 5 m	5.22	2.45	95.4
8 m × 5 m	6.56	2.14	99.7
Control	8.76	2.55	–
SE	0.29	0.55	3.09
CD (0.05)	0.88	NS	NS

Table 5.9 Recommended different fruit tree-based cropping systems for different rainfed regions

Regions	Fruit trees	Crops
Eastern plateau and hills region (Maharashtra, Uttar Pradesh, Orissa and West Bengal)	*Annona squamosa*	Coarse cereals, pulses, oil seeds, soybean
	Tamarindus indica	
	Emblica officinalis	
	Psidium guajava	
	Achrus zapota	
	Sapindus trifoliatus	
	Citrus spp	
	Artocarpus heterophyllus	
	Mangifera indica	
Central and western plateau and hills regions (Maharashtra, Madhya Pradesh, Rajasthan and Uttar Pradesh)	*Ziziphus mauritiana*	Ground nut, sorghum, pigeon pea, gram, lentil, mustard, soybean, paddy, sesamum
	Anona squamosa	
	Emblica officinalis	
	Citrus spp	
	Punica granatum	
East coast plains and hill region (Orissa, Andhra Pradesh, Tamil Nadu and Pondicherry)	*Annona squamosa*	Pulses, oil seeds, black gram, green gram, pigeon pea, paddy
	Mangifera indica	
	Anacardium occidantale	
	Cocos nucifera	

Source Pathak and Solanki (2002) For scientific names of arable crops mentioned in last column see text

Agroforestry experiments conducted by CRIDA in collaboration with ITC Paperboards & Speciality Papers Division (PSPD), Sarapaka have shown that altering the tree geometry by increased tree row spacing (or alley width), effectively improved the intercrop yield during the entire 4-year period, especially with the triple row arrangement, which produced 73 and 66 % of the sole cowpea yields in 2003 and 2004, respectively (Prasad et al. 2010a). Cowpea (*Vigna unguiculata*) yields in other wider-row arrangements varied from 50 to 62 % of the sole crop yields in the third year and 39 to 59 % of the sole crop yields in the fourth year of tree plantation. The fodder yields of grasses such as Congo signal (*Brachiaria mutica*) and guinea grass (*Panicum maximum*) vars. Mukuni and Riversdale also decreased in the presence of trees as compared to sole grasses (Table 5.10).

In an experiment at Hosakote (Karnataka), application of N (320 kg), P (30 kg), and K (100 kg) applied in three split doses at 9, 18, and 27 months after planting increased the dry biomass of stem by 23 % and also increased the

Table 5.10 Crop yields in sole and *Eucalyptus* agroforestry systems during 4 years (2001–2005) of study in Andhra Pradesh, India

Treatment (spacing)	Post rainy season Cowpea (kg ha^{-1})				Rainy and post rainy season 2005 Dry fodder yield (t ha^{-1}) of grasses		
	(*Rabi*) 2001	(*Rabi*) 2002	2003	2004	Makuni	Riversdale	Congo signal
10 m ×1.5 m (Triple rows)	1097	726	525	401	1.65	1.67	1.46
1 m × 1 m (Paired rows)	945	645	456	340	1.95	2.01	1.58
7 m × 1.5 m (Paired rows)	869	505	411	237	1.58	1.33	1.25
6 m × 1 m	908	443	358	277	1.69	1.40	1.65
3 m × 2 m (Farmer's practice)	893	309	114	144	0.90	0.91	0.74
Control (crop without tree)	975	734	716	594	2.09	2.21	2.12
LSD (p = 0.05)	NS	132	110.8	62	Tree geometry −0.62 Intercrops −0.18		

branch weight significantly. In *Eucalyptus camaldulensis*, application of fertilizers resulted in increased height and volume index from 1 to 3 years of tree age, but the level of response declined with the increase in tree age in sodic soils with water logging (Singh et al. 2005). Among the oil cakes tried, neem (*Azadirachta indica* tree) cake applied at 1.0 kg per pit significantly improved tree height and collar diameter over mustard (*Brassica oleracea*) cake and mahua (*Madhuca indica* tree) cake in sodic soils (Gupta et al. 2006).

Leucaena Leucocephala-Based Systems

The utility of *Leucaena* for the fodder is widely reported and its use for wood is a recent phenomenon. The wood is generally described as strong, light in weight, easy to work, and able to give attractive finish (Rao 1984). *Leucaena leucocephala* has a porous wood structure, longer fiber than that of other hardwoods, high holocellulose and α-cellulose, and low lignin content with xylan-type hemicellulose, making it a suitable raw material in pulp and paper industry (Malik et al. 2004). In India, interest in leucaena was revived during mid 1970s with the introduction of giant arboreal varieties from Hawaii and the Philippines, which were evaluated over various climatic regions and soil types. Kushalappa (1980) observed that varieties K8 and K28, raised in the drier regions of Karnataka having a rainfall of about 500 mm, had very poor growth. Qureshi and Desai (1981) made similar observations for Maharashtra region. The growth of Hawaiian Giant and other introduced leucaena cultivars were not satisfactory in Punjab and drier parts of Tamil Nadu, respectively (Dutt and Jamwal 1988). The poor growth of the introduced leucaena cultivars in India could be due to different edaphic and climatic conditions which are not similar to that in Mexico and Central America, and the lack of effective *Rhizobium* strains and their failure to establish effective symbiotic association with VAM fungi (Dutt and Jamwal 1988). However, K8 variety could produce a biomass yield of 18t ha^{-1} yr^{-1} at 6 years when planted at spacing of 1 m x 1 m and providing protective irrigation (Table 5.11).

Maslekar (1984) reported 24.78 and 15.33 t ha^{-1} yield, respectively, for irrigated and rainfed conditions of a 2.5-year-old plantation in Maharashtra. However, some of the varieties like K636 have yielded higher biomass productivity in Australia (Mullen and Gutteridge 2002) and in India (Prasad et al. 2011b). *Leucaena* varieties are commercially propagated through

Table 5.11 Fresh and dry biomass (t ha^{-1}) of different parts of *Leucaena leucocephala* planted in different spacings at harvest (51 months) at Bhadrachalam in Andhra Pradesh, India

Treatments (spacing)	Marketable bole biomass	Dry biomass				
		Bole	Branch	Leaf	Bark	Total
1 m × 1 m	144.9	81.0	4.9	2.2	4.3	92.4
1.3 m × 1.3 m	124.6	70.6	3.4	1.9	3.2	79.1
3 m × 0.75 m	136.6	78.1	3.8	1.5	3.7	87.1
3 m × 1 m	113.9	65.9	3.0	1.2	2.9	73.0
5 m × 0.8 m	117.4	68.6	3.7	2.1	2.7	77.1
3 m × 2 m	94.8	52.1	2.6	1.4	2.4	58.5
SED	10.7	6.3	0.9	0.5	0.6	7.1
LSD (p = 0.05)	23.8	14.1	2.1	NS	1.4	15.7

seeds or through seedlings grown from seeds. The seeds of *Leucaena* are to be treated either with hot water or acid for breaking the dormancy before sowing. In *Leucaena*, spacing found to have noticeable effect on diameter growth parameters (Prasad et al. 2011a). Trees responded to wider spacing and lower competition with greater individual diameter growth. The tree density of 1 m × 1 m recorded significantly lower height and diameter at breast height (DBH), whereas trees in 3 m × 2 m density recorded higher tree height and DBH growth. Trees in stand density of 1 m × 1 m partitioned 88 % of the total biomass into the stem whereas the stand density of 3 m × 2 m partitioned only 82 % into stem in small-sized trees (2.5–5.0 cm). *Leucaena* trees in the narrow spacing and higher stand density partitioned slightly more biomass in boles and less in branches and foliage than trees in lower stand densities (3 m × 2 m).

Much of the literature on *Leucaena* and arable cropping systems is on alley cropping systems (Rao et al. 1991). In pulpwood systems, trees are closely planted at about 0.7 m × 0.7 m which restricts the intercropping. Experiments conducted by CRIDA have shown that increasing the row width permits intercropping during the initial 2 years.

Further, increasing the tree row spacing beyond 3 m was found to increase the intercrop yield till second year but not in the third year due to the expansion of leucaena tree canopy resulting in complete shade to the underneath crop (Table 5. 12). Hence, increasing the inter-row distance beyond 3 m is also not suitable for enhancing the intercropping yield (Prasad et al. 2010b).

In soils where leucaena is grown for the first time, *Rhizobium* inoculation is helpful. Application of *Rhizobium* strains resulted in increasing the dry matter accumulation, shoot length, and leaf number significantly (Datta and Das 1997). Hegde et al. (1988) reported a biomass productivity of 161 t ha^{-1} in 6 years where as the dry wood yield is 107 t ha^{-1} with the K6 variety with four irrigations per year. Agarwal et al. (1985) reported a mean wood biomass of 66.83 t ha^{-1}yr^{-1} under irrigated conditions and 14.13 t ha^{-1}yr^{-1}in rainfed conditions. In on-farm trials conducted in Andhra Pradesh, the fresh biomass of *Leucaena leucocephala* (var. K-636) was obtained to be 149 t ha^{-1} whereas the dry biomass was 92 t ha^{-1} in a 4-year-period (Prasad et al. 2011a). Biomass contributed by different tree parts to the total was in the order of: bole > branch > bark > leaf. Mishra et al. (1995) reported 23–43 t ha^{-1} yr^{-1} annual dry biomass of *Leucaena leucocephala* in Uttar Pradesh, India.

Experiments conducted by Saikia and Sharma (1994) revealed that plant growth and yield parameters were found to be affected by density, by the fifth year the maximum (160 t ha^{-1}) DM production was in the closest spacing and minimum (45 t ha^{-1}) obtained in the wider spacing. With the increase in population, the biomass yield/unit area also increased.

Table 5.12 Post-rainy season cowpea yields (g m^{-2}) in sole and leucaena agroforestry systems during 4 years (2001–2003) of study in Andhra Pradesh, India

Treatment	Post rainy season (*Rabi*) 2001	Post rainy season (*Rabi*) 2002	Post rainy season (*Rabi*) 2003
13 m × 1 m (Four rows)	98.1	62.6	21.5
10 m × 1 m (Triple rows)	106.0	60.0	15.6
7 m × 1 m (Paired rows)	96.0	56.0	11.0
5 m × 0.8 m	98.6	31.5	8.0
3 m × 2 m	101.5	23.5	7.6
3 m × 1 m	98.5	23.0	6.9
3 m × 0.75 m	100.2	12.7	8.5
Sole crop (crop without tree)	102.9	85.6	71.6
LSD (p = 0.05)	NS	13.2	3.1

Bamboo-Based Systems

Bamboo is considered one of the world's fastest growing plants. Nearly, half of the bamboo produce is now used in pulp and paper industries. The use of bamboo as industrial raw material is leading to their over exploitation and a steady decline in their natural stands. Several bamboo species were tried and *Dendrocalamus strictus* and *Bambusa bambos* are quite common. The results were quite encouraging in the states of Gujarat and Rajasthan, where the survival percentage ranged from 68 to 100 % for both the species (Srivastava et al. 2008). From a study conducted, on two bamboo species namely *Melocanna baccifera* and *Dendrocalamus longispathus* in Mizoram, it was found that maximum average height, number of leaves, and culm emergence in both the species were observed under the 3 m × 3 m spacing, while minimum height was observed in the 2 m × 2 m spacing (Jha and Lalnunmawia 2003). The growth attributes of *Bambusa tulda* and *Bambusa balcooa* were statistically at par when planted at closer spacing (10 m × 10 m) than wider spacing (12 m × 12 m) except number of internodes (Banerjee et al. 2009).

Interspaces between bamboo rows can be effectively utilized for intercropping with ginger (*Zingiber officinale*), turmeric (*Curcuma longa*), large cardamom (*Amomum subulatum*), and dinanath grass (*Pennisetum pedicellatum*) up to a distance of 11–15 m from the bamboo row where shading can be expected for a part of a day. Rice (*Oryza sativa*), finger millet (*Eleusine coracana*), soyabeans (*Glycine max*), setaria (*Setaria sphacelata*), and stylo (*Stylosanthes*) were suitable crops beyond this distance. Light interception in bamboo varied from 84 to 95 % to a distance of 11 m and was less than 44 % near the bamboo row. Turmeric, ginger and dinanath grass (*Pennisetum pedicellatum*) produced higher yields near bamboo rows (1–2 m), and the yield decreased with greater light transmittance away from the rows (Ali et al. 2006). However, relatively greater yield was observed at 12 m distance from the bamboo rows.

Cultivation of only shade-loving crops such as turmeric, ginger and dinanath grass are recommended under bamboo clumps. Application of urea at 60 g per plant, superphosphate at 40 g per plant, and muriate of potash at 40 g per plant in three split doses at intervals of 3 months recorded higher height growth, average number of leaves, maximum culm emergence in *Melocanna baccifera* and *Dendrocalamus longispathus,* respectively, in 3 m x 3 m spacing (Totey et al. 1989). A complete harvest of *Bambusa bambos* at 6 years age removed 2,341 kg ha^{-1} N, 22 kg ha^{-1} P, 2,653 kg ha^{-1} K, 1,211 kg ha^{-1} Ca, and 1,356 kg ha^{-1} Mg (Shanmughavel and Francis 2001) and any fertilizer application schedule should consider the nutrient removal by bamboo.

A 7-year-old (*Dendrocalamus strictus*) based agroforestry system at Parbhani, Maharashtra, during the year 2000–2001 produced a biomass of 17.87 kg bamboo per tree having a canopy area of 20.11 m^2. The annual biomass production in the bamboo agroforestry system was 7,150 kg ha^{-1} yr^{-1} (Patil et al. 2004). Aboveground biomass of thorny bamboo (*Bambusa bambos)* clumps averaged 2,417 kg per clump with an average accumulation of 241.7 t ha^{-1} in the home gardens of Thrissur, Kerala (Kumar et al. 2005). The productivity of *Bambusa bambos* showed that the total biomass increased with age. It was 2.2t ha^{-1} in the first year and reached to 297.8t ha^{-1} in the sixth year (Shanmughavel and Francis 2001).

Economics of Pulpwood Systems

The financial evaluation of *Eucalyptus* and *Leucaena* systems in comparison to the cowpea crop is presented in Table 5.13.

Cost of cultivation, gross, and net returns were higher with *Eucalyptus* agroforestry system followed by sole eucalyptus system, *Leucaena leucocephala* intercropping system, sole *Leucaena* in the decreasing order. Returns from the tree systems were substantially higher in comparison to the arable cropping system. Returns were higher in *Eucalyptus* agroforestry by 3 times in comparison to cowpea. Financial indicators such as net present value and benefit–cost ratio were higher with tree systems in comparison to annual crops. Though, tree cultivation required additional expenditure which was mainly due to the cost of the planting material, its transportation to the field, pitting and planting, returns more than compensated the costs. Intercropping in the initial years allows better cash flow during the initial years of plantation cycle when the returns from tree are not forthcoming.

These tree species are accepted by farmers and are well integrated into the existing land use systems. The acreage under eucalyptus and leucaena has increased rapidly in the southern states particularly in Andhra Pradesh. It is estimated that Prakasam district of Andhra Pradesh alone produces 0.7 million tones of wood worth of INR 560 million annually from the private holdings of the farmers (Saigal and Kashyap 2002). Tree growing as a commercial short-rotation coppice has increasingly become a profitable land use with the establishment of company–farmer relationships, trading of wood in the open market, assured market, high returns from trees, and supportive policies of the state government during the last decade.

Biofuels and Bioenergy Systems

Among the energy sources, oil and biomass are the widely used sources. India, with 3.2 % consumption of the world oil production, stood sixth among the top consumers in the world. The annual requirement of petroleum products of the country is approximately 124 million tons, whereas the domestic production of crude oil and natural gas is approximately 34 million tons during 2006−2007 leaving a huge gap between

Table 5.13 Benefit–Cost ratio (B: C) and Net Present Value (NPV) of net income at different discount rates for two tree systems in Andhra Pradesh, India

Systems	6 %		12 %		18 %	
	NPV	B:C	NPV	B:C	NPV	B:C
Agroforestry systems						
Eucalyptus + Cowpea	62074	2.1	44237	1.9	31584	1.7
Leucaena + Cowpea	56730	2.3	41228	2.1	30195	1.9
Sole *Eucalyptus*	54808	2.3	36905	2.0	24292	1.7
Sole *Leucaena*	44903	2.3	29838	2.0	19240	1.7
Arable cropping (Cowpea)	24374	1.9	21859	1.9	19750	2.0

demand and supply and is met only by import. The annual imports are to the tune of INR 272,000 crores during 2007–2008 necessitating the need to look for alternate sources. Biodiesel is defined as the fatty acid alkyl esters of vegetable oils, animal fats, or waste oils and is an environmentally friendly alternative to conventional feeds of livestock. In India, biofuels from nonedibles such as *Jatropha* and *Pongamia* are seen as potential sources that can perform under limited resources (Rao et al. 2010).

Jatropha-Based Systems

Jatropha curcas is generally found in all parts of India and is commonly grown as hedge and as live fence around agricultural fields as it is not browsable by animals and can be grown over a wide range of arid and semi-arid climatic conditions with an average annual rainfall between 480 and 2,380 mm and an average temperature range of 11–28 °C. The plant has the ability to tolerate the extreme drought and survive light frost. Flowering normally occurs during the rainy season and fruiting during winter. *Jatropha* can be grown in shallow, gravely, and sandy soils with low fertility (Rao et al. 2010). The oil content varies from 35 to 40 % in the seeds to 50–60 % in the kernel (Ginwal et al. 2005). Seed yields under cultivation can range from 1,500 to 2,000 kilograms per hectare, corresponding to extractable oil yields of 540 to 680 liters per hectare (Rao et al. 2010). The oil is used as an illuminant, lubricant, and in soap industry (Kaushik et al. 2007). *Jatropha* oil cake is a rich source of nitrogen, phosphorous, and potassium and can be used as organic manure. The oil is extensively used in soap industry due to its high saponification value (Divakara et al. 2009). Several experiments were conducted at CRIDA to standardize the production technology of *Jatropha* biofuel trees. Seeds or cuttings of *Jatropha* can be planted in the main field with the onset of monsoon. For vegetative propagation, cutting of 3–4 cm thickness and 15–20 cm length are recommended. Seedlings can be raised in nursery and transplanted when 8–10 weeks old.

Pongamia Pinnata-Based Systems

Pongamia pinnata, commonly known as Karanja is a medium-sized tree with a short crooked trunk and broad crown of spreading or drooping branches. The seeds contain oil used as a substitute for diesel, fatty acid methyl esters (FAMEs) of seed oils have already been found suitable for use as fuel in diesel engines. The oil is also used as a lubricant, water paint binder, pesticide, and in soap and tanning industries. The tree is adaptable to wide agronomic climatic conditions. There are various other uses of *Pongamia* also. It is recommended as a shade tree for pastures and as windbreak for tea. The leaves are said to be a valuable lactagogue (promotes milk production in cattle) fodder, especially in arid regions. It is often intercropped with fodder grasses that grow well in its shade. Dried leaves are used in stored grains to repel insects. The seedcake is used as cattle and poultry feed, and biogas made out of fermented seedcake is used as kitchen stove fuel in rural villages. Furthermore, the waste pulp is used as an organic fertilizer which provides good income to the rural poor.

Pongamia can be propagated through seeds, cuttings, and grafts. Seedlings should be raised in nurseries. Seedlings should be planted in the field with the onset of monsoon when the seedlings attain a height of 60 cm (4–5 months of age). It is necessary to dip the cuttings of 30 cm length and 4–5 cm diameter in 500 ppm IBA solution for 5 s to get higher rooting percentage. *Pongamia* can be successfully propagated through cleft grafting during December–January. 1-year-old rooted cuttings or cleft-grafted plants are planted in the field. Vegetatively propagated plants reduce the variation among population and increase the seed yields. *Pongamia* plants are hardy and can sustain heat and drought. Graft is recommended to reduce the gestation period, height, and harvest cost. In addition, this facilitates more number of plants to be accommodated per hectare. Intercrops like castor, pigeon pea, black gram and green gram have performed well in *Pongamia* plantations at CRIDA, Hyderabad, till 5 years after the establishment of the *Pongamia* plants.

Simarouba glauca-Based Systems

Simarouba glauca commonly known as aceituno, paradise-tree, or bitter wood, is an important tree species grown in the forests of Central and South America. It grows up to 20 m high and has a trunk 50–80 cm in diameter. It produces bright green leaves 20–50 cm in length, small white flowers, and small red fruit. It grows well up to 1000 m above sea level in all types of well-drained soils. Its cultivation depends upon the rainfall distribution and the kind of soil. The tree is able to withstand a temperature range of 40–45 °C. It can be cultivated in all the tropical regions of the world. It has a wide adaptability in terms of altitude as it can be grown from 0 to 1500 m above mean sea level. *Simarouba* can adapt to varying soil conditions, from sandy, lateritic, gravelly to black soils and with a pH ranging from 5.0 to 8.5. It can be grown in degraded soils which are very poor in nutrients and unsuitable for cultivation of other crops. It can grow in soil types with moderately good porosity. *Simarouba* requires 700–1000 mm rainfall for normal growth; the crop can withstand relatively long (6–8 months) dry spells in a year (Rao et al. 2010). In rainfed areas, 2–3-month-old saplings have to be used for transplanting. Direct seeding can be done in lands with minimum limitation for water. About 200 trees can be accommodated in one hectare. It produces fruits similar in size, shape, and color to olives. There are two varieties: one produces greenish white fruit and the other violet to black fruits. The tree starts flowering during December and bears fruits in January and February. The fruits are ready to be harvested in May. Though the tree commences bearing fruit from the fourth year of planting, economic yields (20 kg fruits per tree) can be harvested only from the 10th year of planting. The average yield of fruit from a hectare of a 10-year-old plantation is about 6,000–8,000 kg.

Planting pattern can be suited to alley cropping, boundary planting, bund planting, block plantations, or as avenue trees. Alley cropping with regular crops in interspace should have a spacing of 2 m × 10 m. Planting a cluster of 2–3 seedlings in a pit is done at the onset of monsoon. Trenches and basins should be made to facilitate proper rainwater harvesting. Mulching of organic matter around the plant is done during post-monsoon period. Lateral bud pruning is done till the saplings grow to about 3 m height so that the trees grow tall and straight. This gives space for operations in intercrops in a *Simarouba*-based agroforestry system. The high yielders should be retained after 5–6 years of planting, the low yielders are cut and the wood can be used for timber. After spreading of the canopy, shade-loving crops or fodder crops can be grown as intercrops to get additional income and to prevent weed growth. Pruning of unwanted and criss-cross branches is done in June–July months to get better yield. A spacing of 4 m × 5 m should be followed. Female and male plants should be maintained in the ratio of 10:1 and should be arranged in proper geometry in block plantation for effective pollination.

Silvopastoral Systems

Silvopastures are the land use systems in which trees and/or shrubs are combined with livestock and pasture for forage and fuel wood production on the same unit of land. It is an ideal combination of grasses, legumes, and trees for optimizing land productivity, conserving plants, soils, and nutrients and producing forage, fuel wood, timber etc. on a sustainable basis (Rai et al. 1999). Trees and shrubs in the silvopastoral system often contribute substantial quantities of leaf fodder especially in arid and semi-arid regions as top feeds and contribute toward alleviation of fodder scarcity during the lean periods of fodder availability.

Experiments conducted in red gravelly soils with five tree species (*Albizia lebbeck*, *Acacia tortilis*, *Hardwickia binata*, and *Leucaena leucocephala*) in Jhansi and Datia under three spacings (4 m × 3 m, 4 m × 4 m, and 4 m × 6 m) revealed that trees can be grown up to density of 825 trees per ha for 11–12 years without causing reduction in the under storey biomass (Rai et al. 1999). The leaf fodder yields depend on species, initial age, lopping intensity, and interval as well as agroclimatic conditions. In semi-arid conditions, from a 8–9-years-old silvopastoral system, green and dry fodder production of 9.63 and 5.28 t ha^{-1}, respectively, can be obtained annually through lopping at one-third intensity in case of *Albizia procera* and 6.24 and 2.78 t ha^{-1}, respectively, in case of *Albizia lebbeck* through biannual lopping at two-third intensity (Deb Roy 1988). Similarly, from 8 to 10-years-old *Acacia tortilis–Cenchrus* silvopastoral system, top feeds provide 2.75–3.50 kg per tree on annual lopping. Roy et al. (1987) suggested that annual lopping at 2/3 intensity was more suitable in *Albizia amara* for higher growth and production as compared to biannual lopping. A 5-years-old plant of *Dichrostachys cinerea* provided dry leaf fodder of 2.4 kg per tree per year when biannual looping was done at half intensity. Similarly, top feed production of six fodder tree species of Bundelkhand region was 11.38 kg per tree for *Albizia procera*, 11.20 kg per tree for *Albizia amara*, 4.21 kg per tree for *Albizia lebbeck*, 3.67 kg per tree for *Hardwickia binata*, and 2.76 kg per tree for *Dichrostachys cinerea* and the minimum yield of 0.51 kg per tree was noted for *Acacia tortilis* at the age of 10 years (Roy 1991). In *Sesbania grandiflora* and *Sesbania sesban*, the dry leaf fodder yield of 0.3 kg per tree was reported at 3.5 years of growth when grown at the density of 5,000 per ha (Rai et al. 1983). In an evaluation of 18 tree species under natural grass land in red gravelly soil in semi-arid region, maximum dry leaf fodder of 2.69 and 3.80 kg per tree was obtained when pruned up to 50 and 75 % of the tree height, respectively, from *Leucaena leucocephala* followed by *Albizia procera* and minimum yield was received from *Anogeissus pendula* (Rai et al. 1995). Study conducted at Mettupalayam, Tamil Nadu, Dastagir and Suresh (1991) reported that production of forage and growth of *Acacia leucophloea* planted at 4 m × 4 m spacing increased in mixtures (*Cenchrus ciliaris* + *Cilitoria ternatea*, *Cenchrus ciliaris* + *Desmodium* species or *Stylosanthes spp* and *Cenchrus ciliaris*) as compared to sole cropping.

Studies were conducted at NRCAF, Jhansi on 15-months old established silvopasture consisting of *Albizia amara* and *Leucaena leucocephala* as tree component and *Dichrostachys cinerea* as shrub, *Chrysopogon fulvus* as grass, *Stylosanthes hamata* and *Stylosanthes scabra* as pasture legumes and natural *Sehima-Heteropogon* grassland to know the comparative performance of adult sheep and goats in this system. Results showed that goats and sheep grazed on silvopasture gained in their body weight at the rate of 28.6 and 2.1 g g $h^{-1}day^{-1}$, respectively, in a total grazing period of 241 days (Rai et al. 1994). On the other hand, animals grazed on natural grassland, goats gained in their weight at the rate of 27.4 g $h^{-1}day^{-1}$ during same period even after supplementation of 1.5 kg g $h^{-1}day^{-1}$ green leaf of *Leucaena leucocephala* to the animals. Grazing studies with six male lambs + six male kids (4–6 months age) on 2-years-old silvopastoral system and natural grassland from August 1992 to June, 1994 (690 days) revealed that animals grazing on both the pastures continued to gain in their weight up to 1.8 years of their age without any supplementation of concentrate feeds However, gain in body weight of lambs and kid grazing on silvopasture was 33.0 and 39.4 % higher as compared to natural grassland, respectively. The better performance of sheep and goats on silvopasture was due to the availability of nutritious herbage from tree leaves and pasture legumes.

Different economic models have been prepared for specific site based silvopastoral systems by several workers. Singh and Reddy (1986) compared the economics of different agroforestry systems where they found that a 10-

years-old silvopasture gave B/C (benefit–cost) ratio of 2.56 and 1.68 at 11 and 20 % discounting leading to an IRR (Internal rate of return) of 34 %. Similarly, National Wasteland Development Board (NWDB 1987) reported the economics of silvopastoral system based on different types of areas like plain, undulating and hilly, ravenous and desert at 10 years. The B/C ratio varied from 1.05 to 1.21 at 20–30 % discounting rate and IRR of 10.99–21.21 %. Pathak (1991) reported that establishment cost of silvopasture in central Uttar Pradesh was found to be INR 31,842 per ha and returns of INR 45,240 per ha with 12 years rotation. He observed B/C ratio of 1.42 with an IRR of 22.67 %.

Boundary Plantations/Windbreaks and Shelterbelts

It has been reported that trees grown on field bounds as windbreak or shelterbelt primarily reduce the wind velocity and change microclimate which are reflecting in growth and development of the nearby crop and ultimately in crop yield (Marshall 1967). Windbreaks/shelterbelts concept gained an important place in dryland agriculture as they show impact on various characteristics of climate and soil, which directly or indirectly play a key role in dry land agriculture. The main effect of a tree windbreak is to provide shelter, i.e., a windbreak alters the mean wind speed, wind direction, and turbulence of the airflow. As a result, the surrounding aerial, plant, and soil environments are modified. In one observation, maximum increase in mango productivity was observed in eastern dry zone of Karnataka when *Casuarina equisetifolia* trees were used as windbreaks. In another observation, *Acacia auriculaeformis* exhibited the greatest canopy spread (5.72 m E–W and 4.36 m N–S) followed by *Casuarina equisetifolia* and these exhibited the highest reductions in wind velocity (72 and 62 %, respectively). Three rows of windbreaks increased the growth and yield of mango more effectively than single and double rows of windbreaks (Reddy et al. 2000).

Growing trees on farm boundaries is a common practice in India. Many of the boundary plantations also help as shelterbelts and windbreaks, particularly in fruit orchards. In northern parts of India, particularly in Haryana and Punjab, *Eucalyptus* and *Populus* are commonly grown along field boundaries or bunds of paddy fields; other trees which are found grown as boundary plantations or live hedge include *Acacia nilotica*, *Dalbergia sissoo*, and *Prosopis juliflora*. Farmers of Sikkim grow bamboo (*Dendrocalamus*) all along irrigation channels. In coastal areas of Andhra Pradesh, *Borassus* is the most frequent palm. In Andamans, farmers grow *Gliricidia sepium*, *Jatropha curcas*, *Ficus retusa*, *Ceiba pentandra*, *Vitex trifoliata*, and *Erythrina indica* as live hedges (Dagar 1995). In Bihar, *Dalbergia sissoo* and *Wendlendia exserta* are most common boundary plantations. *Casuarina equisetifolia* is extensively planted on field bunds and along sandy coastal areas in Orissa.

Farm bunds could be resourcefully used for rising nitrogen fixing shrubs and trees to generate nitrogen affluent lopping. *Gliricidia* on farm bunds serves dual purpose of producing green leaf manure rich in N and also helps in conserving soil through reduced soil erosion (Srinivasarao et al. Srinivasa Rao et al. 2011). Growing of *Gliricidia sepium* at a spacing of 75 cm on farm bunds could provide 28–30 kg nitrogen/annum in addition to a valuable organic matter (Wani et al. 2009). A study on adoption pattern of different agroforestry systems in north Bihar revealed that larger farmers (80 %) adopted boundary plantation (Ali and Chaturvedi 2008). In Gulbarga and Raichur districts, bund planting with neem, *Acacia nilotica* and eucalyptus were predominant under rainfed agroecosystems. In Bidar and Raichur districts of Karnataka, the preference was for fruit yielding tree species, whereas in Gulbarga district highest preference was for timber and fruit yielding tree species (Madiwalar et al. 2007).

Live fencing around plantations of fruit plants and croplands is essential in areas where the entry of cattle and wild animals and trespassing is rampant. Apart from their protective role (serving as a barrier), living fences play an important part in environment management, and may provide fuel wood, fodder and food, act as windbreaks and enrich the soil, depending on the species used. Other auxiliary benefits like shade, seed bank, mulch, medicine, nutrient trap, soil and water conservation, and wildlife habitat make the use of living fences a more widespread practice. In northern parts of India, particularly in western Uttar Pradesh, Haryana, and Punjab, *Eucalyptus* and *Populus* species are commonly grown along field boundaries or bunds; other such trees include *Dalbergia sissoo* and *Prosopis juliflora.* Farmers prefer multiple species in several rows and fill gaps by placing dry thorny twigs of species like *Ziziphus mauritiana, Ziziphus nummularia, Acacia nilotica* etc. Further, *Aloe vera, Cactus,* and *Agave sislana* plantations along farm boundary are common in north India (Das and Das 2005). In dry regions, several cactoides and *Agave sisalana* are quite commonly grown on farms bunds.

Shade Trees in Plantations

Over the years, Silver oak (*Grevillea robusta*) has evolved into perhaps the most preferred shade tree species in tea plantations. Its unique leaves effectively filter light and provide enough shade during the dry months. Because of its deep roots, it does not compete for nutrients and moisture with the tea plants. Moreover, it can withstand pollarding, lopping, and desuckering. It recovers and grows fast, is immune to common pests and diseases, and can serve as windbreaks. In rainfed plantations, tea growers always worry about the loss of soil moisture during the dry season. The topography of Western Ghats limits the construction of sprinkler irrigation systems. Hence, tea growers rely on the shade trees to maintain soil moisture. Preliminary studies found that the leaf litter of shade trees helped maintain soil moisture in tea plantations without competing with the tea plants during the dry months. Shade trees also lessen fluctuations in soil temperature which harm the root growth of tea plants, increase relative humidity, and provide high organic matter. This organic matter compensates for the loss of nutrients in tea plantations. Studies also showed that shaded conditions not only favored increased production but also a high chlorophyll content of tea plant leaves because of the reduced leaf temperature. Integrating shade trees in tea plantations is a low-cost but effective carbon sequestration strategy. It is estimated that a 20-year-old Silver oak shade tree can sequester up to 41.8 tons per ha of carbon. The life span of Silver oak is up to 40 years before it can be cut and replanted. With the current market price for Silver oak timber estimated at US$ 265 per m^3, an additional income of US$ 65 950 can be earned per hectare after 40 years.

Success Stories

Pulpwood-Based Industries

In India, the paper industry is primarily rural based with close linkages with farming community. Over the years, it has evolved into an agro-based industry from its earlier character of a forest-based industry. Currently, Indian paper industry is consuming only about 7 million tones or about 3 % of the total wood consumed in India; about 90 % is consumed as fuel wood. Additional raw material requirement by 2012–2013 is anticipated to be about 8 million tones of wood which will be about 6 % of total consumption of the country. This would require afforesting 2 million ha of land to maintain proper ecological balance. As a result of the implementation of the Forest (Conservation) Act, 1980, and following the provisions of the National Forest Policy of 1998, supply of raw material from government forests to wood-based industries has gradually declined. This has forced the industry to look for alternative sources such as imports of timber and pulp, and sourcing timber through the farm forestry sector. Wood-based corporations like ITC Bhadrachalam Paperboards Limited, JK Paper Corporation

Limited, Star Paper Mills of Saharanpur (Uttar Pradesh), and Ballarpur Industries Limited initiated tie-up with farmers by supplying seedlings to these farmers, arranging micro-credit, providing technical backup, and entering into buy-back arrangements. Because of the farmers defaulting on loans and then selling products at open markets at higher price, most of the formal arrangements have now been abandoned. The companies are now selling high-quality clonal seedlings of eucalyptus and poplar as well as seedlings of subabul through local nurseries to the farmers and purchasing raw material from open market. Farm forestry/agroforestry is now a viable land use option in many parts of the country for farmers as it helps to diversify their income from agriculture and at the same time reducing the risk of growing only agricultural crops on their farmlands.

A notable aspect of the development in the farm forestry sector has been the investment by the companies, though limited in R & D activities. For instance, the annual productivity of clonal seedlings developed by ITC Badrachalam range from 20 to 58 m^3 per ha per year, while the productivity of seed raised eucalyptus plantation is only 4–5 m^3 per ha per year (TERI 2005-06).

While all pervading myth that continues to slur the image of the industry is that it perpetually uses forest raw material and consequently denudes natural forests of the country, the truth is that over the last decade, industry led farm/social forestry have brought around 0.25 million hectares under pulp wood plantations, mainly degraded marginal lands of farmers. At the current estimate, wood-based segment of the industry uses 80 % of the total requirement from farm produced wood. In particular, IPMA member mills have been actively promoting agroforestry with private land holders/farmers to meet imminent raw material needs in a sustainable manner thereby, positively impacting the Greening India Mission of the Government. Also, the industry initiative has created major employment in the remote areas in close proximity to the manufacturing facilities thereby, helping rural development.

Promotion of Fruit-Based Agroforestry for Sustainable Livelihood

This success story belongs to Bor Roppada Kaval (B R Kaval) hamlet in Hangodu Village in Hunsur Taluka of Mysore District, Karnataka, India. Most of the families in this region being landless poor, they had to depend on nonagricultural occupations for their livelihood. Some of the poor had illegally settled down since 15–20 years on 1500 ha of deforested land at B.R. Kaval, which were legally handed over to them for cultivation. However, due to unfertile land and harsh climate, the crop yields were very low and most of these families owning less than 2 ha of land earned a mere INRs 2,000–3,000 annually. Realizing the scope for providing sustainable livelihood to resource poor families through promotion of hortiforestry, Bhartiya Agro Industries Foundation selected B R Kaval as one of the project villages for launching an agroforestry project started in 1991 and ended in 1999 and was funded by the Danish International Development Agency (DANIDA), through Council for Advancement of Peoples' Action and Rural Technology (CAPART). The beneficiaries of the projects were 100 families of B R Kaval, who owned less than two hectares of land.

In this project, three groups were formed with the participating families. Under the guidance of an NGO, Bhartiya Agro Industries Foundation (BAIF) staff and the groups met at least once a week and reviewed the work carried out by the members in their respective fields and planned the farming activities to be completed during the following 1–2 weeks. The following interventions were undertaken to obtain the desired results:

- After transplanting mango and cashew, the participants raised food crops such as sorghum and finger millet in the interspaces. They collected all the crop residues and branches pruned from the trees to prepare vermicompost.
- All the members of the participating families were involved in the development of their

orchards. They raised many vegetables during the rainy season. Tuber crops were grown under the shade of the trees. Trees planted on boundaries attained a height of 12–15 m in 5 years.

- The families who were earlier dependent on forests for fuel wood were now able to harvest the branches from their own trees for fuel.
- With plenty of forage and grasses available from the field and breeding services provided by BAIF, the productivity of their livestock improved.
- The group as a whole decided to stop free grazing of animals and adopted stall-feeding. This resulted in easy collection of a substantial quantity of dung.
- Soil and water conservation, along with increased manure application, helped improve crop yields despite losing 20 % of land to tree species. After 5 years of hard work, the plots owned by these farmers had a pleasant look even from a distance. A closer view of the plots revealed a wide diversity with respect to trees and crops.
- Self-Help Groups at B.R. Kaval promoted small enterprises. Under the aegis of their cooperative society named as "Deepika Fruit Processing Committee", the group produced about 500 kg of mango pickle and sold nearly half of this quantity in the local market through linkages established with shops, visitors to the project site, and the BAIF cattle breeding center at B R Kaval. They also loaned INR 75,000 for production, processing, and marketing of mango pickles as well as for providing credit to members of the group for taking up dairy cattle production as an income generating activity. These groups donated benches, a water drum, and stone poles for constructing a boundary wall, to the local primary school.

During the last 5 years of the project, this agroforestry programme has made a substantial change in the lives of these 100 families who were earlier eking out a living on government-owned degraded land for nearly 15 years. These are times of prosperity and happiness at B R Kaval. Not many visitors can believe the rapid pace at which this village has developed. Looking to the success of this project, 60 ha of agroforestry plots have been established by the farmers in the periphery of the core project villages. The villagers are now aware about the importance of trees and the value of multipurpose tree species. They are taking keen interest in identifying the sources for seeds and purchase of seedlings. These are sown by them on their degraded lands or in the backyard of their houses. Seedlings of fuel wood species are available at a cost of half a rupee per seedling and seedlings of teak and silver oak cost INR 2 to INR 3. They have also taken up cultivation of turmeric, ginger and tapioca under partial shade. Most of the trees planted on the bunds in the year 1992 achieved a height of nearly 20 m. They are no longer dependent on the forests for their supply of firewood. Having adopted improved technologies for better crop yield through NADEP composting and vermicomposting, the use of chemical fertilizers has declined. The water table has increased in open wells and bore wells and the village has brought prosperity adopting agroforestry as way of life.

Ram Lamb Production in Orchards (NATP-CRIDA, Hyderabad)

This success story is based on National Agricultural Technology Project (NATP) funded to CRIDA at village level in Andhra Pradesh where in integrated farming system, fruit and forest trees, livestock and crops were integrated components. Income from ram lamb production under horti-pastoral systems was found remunerative in all the on-farm experiments in Ranga Reddy (Village-Timmareddygudem, Shabad Mandal) and Mahbubnagar (Village-Amarchinta, Atmakur Mandal) Districts of Andhra Pradesh. Net gain from the horti-pastoral systems ranged from INR 40 to 70,000 through ram lamb production. Further, higher income was observed with supplementary grazing on established pasture or supplementation of *Leucaena leucocephala* foliage. The present study reveals that the mango and sweet orange orchards could be developed as horti-pastoral systems with suitable understory grass species and boundary plantation of *Leucaena leucocephala* for higher biomass production and income generation. These systems can be efficiently integrated for ram lamb production in rainfed areas. Season plays a lot in availability of nutrients and subsequently the growth of the ram lambs under horti-pastoral systems in rainfed areas; hence the lambs could be introduced preferably in the mid-rainy season (September month) for maximum weight gain. It is a unique example of sustaining a poor family through small ruminants.

Watershed Management Through Agroforestry Brings Back Smile to Dhani Ram of Ubarao Village in Madhya Pradesh

Dhani Ram Kushwaha alias Dhanua, a farmer of village Ubaro, Tehsil Niwari, Distt. Tikamgarh (MP) owns about 2 ha cultivated land adjacent to forest area which is far away from his village. His holding is sloppy with typical red soil having poor nutrient status. This is the only source of livelihood for Dhani Ram. Due to age-old practices and continuous drought, income of INR 14,000 per annum from the land could barely meet his expenses. Under the guidance of NRCAF scientists, he opted for the demonstration of agroforestry systems in his field and installation of two gabion structure in nallah passing through his field. The Centre demonstrated guava (*Psidium guajava*) and citrus-based agroforestry system on his less than 2 ha crop land. Sixty-eight guava plants (cv. Allahabad Safeda) and 42 citrus plants (Kagzi) were planted at a spacing of 6 m × 8 m. Expenditure on pitting, FYM application, weeding, and watering were met by farmer himself. During *kharif* and *rabi,* demonstrations for ground nut and wheat were arranged which included truthful labeled seed of groundnut var. Kaushal and wheat var.WH-147 and balanced doze of fertilizer in terms of N, P, and K.

During the year 2007, check dam near Dhani Ram's field overflew twice in monsoon resulting in 2 m water recharge in well. His income doubled in the very first year from crops. Farmer was supplied with seedlings of Lasoda (*Cordia sp.*) for plantation along roadside, Karonda (*Carissa carandas*) as live fence along one side of field boundary, and teak on boundary bordering forest. He was given goose berry (*Emblica officinalis*) plants (cv Krishna) for planting in crop land on other side of *nallah (water channel).* He sowed sweet chili in an area of 0.5 ha on the other side of *nallah.* He contributed 40 % for field bunding across the slope. Good rains, plenty of water in well, improved seed of wheat accompanied with across the slope field bunding resulted in bumper wheat yield to the tune of 3.5 tons per ha. After *rabi* harvest, he sowed vegetables like brinjal (*Solanum melongena*), bitter gourd (*Momordica dioica*), chili (*Capsicum annum*), bottle gourd (*Lagenaria siceraria*), etc. in about 0.2 ha area for regular cash income during summer. In 2009, due to good antecedent moisture condition, the check dam over flew and again he harvested bumper *kharif* and *rabi* crops. Guava (*Psidium gaujava*) plants flowered and yielded about 200 kg fruits which added to his regular income. A small patch of about 50 m^2 area is transplanted with napier grass (*Panicum maximum*) for feeding to his bullocks. During *rabi* 2009, he had 21 crops/vegetables/spices/fruit trees in combination in the same field. The vegetables, fruits, crops, and grasses are grown on the same land almost simultaneously. He never leaves field without crop and maintains the fertility of soil through organic manure. His family is getting year round employment and earning regular income. For the last 2 years, he is earning around 40–50 thousand/annum from his land. The smile brought back to Dhani Ram.

Joint Forest Management (Agroforestry Approach) in Shivalik Region of Haryana (COMPETE 2009)

The area under the Shivaliks, which was once covered by dense forests with a variety of flora and fauna, reached its worst form of degradation in the early 1970s. Reckless felling of trees, frequent forest fires, and increasing biotic pressure destroyed the vegetation in the area. Large tract of lands was cleared for agriculture. The problem of grazing was so serious that in heavily grazed areas, 4–6 cm of topsoil used to disappear after just one heavy shower. On the other hand, because of the poor economic conditions of the people, forest laws, and traditional methods of forest regeneration proved ineffective. Against this background, an intervention was designed in the foothills of Shivalik with three criteria, namely ecological viability, economic feasibility, and social desirability (social and political acceptability).

The program site is located in the Himalayan foothills (Shivaliks) of northern India covering about 3,000 square km of north and northeastern Haryana. The tract is hilly with rugged and undulating topography. The slopes are gentle to very steep. The seasonal torrents, which originate from the hills and get wider as they enter the plains, are a peculiar feature of the drainage system of the area. The area falls under two territorial forest divisions, namely Morni Pinjore and Yamunanagar, on the forest administration map. The economy of the area is primarily dependent on agriculture and livestock. However, agricultural productivity in the area was beholden to the whims of nature in the absence of irrigation. Livestock, the other main source of livelihood, consisted of mainly unproductive stocks because fodder was scarce. The *Bhanjdas* (the basket-making community) and *Banjaras* (those who make ropes from a grass locally known as *bhabbar*) are directly dependent on availability of such NTFP (non-timber forest products) as bamboo and "*bhabbar*" (*Eulaliopsis binata*) grass.

The Energy and Resource Institute (TERI) began a JFM (joint forest management) support program in the Haryana Shivaliks in July 1990 in collaboration with the Haryana Forest Department (HFD), with financial support from the Ford Foundation. The program envisaged peoples' participation in the management of forest resources of the state jointly with HFD. TERI has been providing all the necessary backup support in developing and implementing the program. Village-level resource management institutions, popularly known as HRMS (hill resource management societies), were formed in 55 villages in the two forest divisions, which worked in close collaboration with the local forest department officials. The other strategies that have been undertaken to elicit people's participation in the management and protection of degraded forests are as follows:

A series of small check dams in the upper hilly catchment and earthen water-harvesting dams at suitable sites have been constructed for soil conservation and water harvesting. As a result, cultivation that was earlier restricted to wheat and maize has now diversified to include rice, sorghum, pearl millet, groundnut, and vegetables. The productivity of wheat, the staple *rabi* crop, rose from 0.9 t ha^{-1} to 45–50 t ha^{-1} in response to irrigation. Similarly, the productivity rose to 5.5 tons for maize and 0.2 tons for rice. Plots of forests were leased to HRMS for harvesting of *bhabbar* (*Eulaliopsis binata*) grass and other fodder grasses to contribute to the general economic improvement of the local communities. The internal rate of return worked out to approximately 80 % when *bhabbar* grass was leased out to the community; when sold to the contractor, the rate was as low as 17 %. After the beginning of the JFMP in the area, the monthly quota of bamboos (felling permit of bamboos) was increased from 50 to 100 per family, at INR 7 per 100 bamboo poles and the felling season was extended from 6 to 9 months. The key achievements are summarized below:

- The maximum average yield of *bhabbar* grass was 850 kg ha^{-1} under 6 years of community protection compared to the yield of 300–360 kg ha^{-1} in the unprotected areas.
- The total number of trees per ha increased from a minimum of 700 in unprotected forest areas to a maximum of 3,960 in case of 10 years of protection.
- The number of weedy shrubs per ha is maximum in unprotected forest areas about 14,000 whereas in areas protected for 10 years, it is as low as 3,000 and so.
- Water-harvesting structures help villagers to increase their earning from farming through diversifying the agricultural activities.

- Supply of bamboos to *Bhanjda* community at concessional rates provides employment and a source of income.
- Leasing out of forest areas to HRMS for extraction of *bhabbar* contributed to the development of village infrastructure and also economic improvement of the local communities, especially *Banjaras*.
- Leasing out of forest areas to HRMS for extraction of fodder grasses helped the pastoral community to restock their livestock with more productive breed.

Improved Agroforestry Practices Led to Prosperity in Munger District

It is the success story of Narayan Das popularly known as 'Narayanji' of village of Itwa, Dharhara Block, in Munger district. About 7 years ago, while growing pearl millet in Kharif and Chick pea in Rabi, like his father and others in the village, it downed on him that this mundane form of cultivation would lead him nowhere. It was time to shift gears. Not wasting time, he adopted improved agroforestry practice and planted trees in this farm bunds. Today, there is no looking back for him as he cashes in on the roaring success of his wise decision.

His first step was to level the land and install a pump for irrigation from the nearby Nallah. Availability of water meant intensive agriculture with three cropping season a year. It also increased the range of crops that he could grow onion, potato, tomato, and several other vegetables. Inspired by training at Krishi Vigyan Kendra (KVK), Munger and later on with Agricultural Technology Management Agency (ATMA) officials, Narayanji planted 1,000 seedlings of *Gmelina arborea* on the bunds of his 5 acre farm in 1996. Another 1,000 seedling of *Gmelina arborea* were added the following year. Seedlings were planted in a single row with about 2 m between the plants. Because of irrigation facilities, about 95 % of seedlings survived.

To minimize shading of the adjacent crop field, lower branches of the trees were pruned at the onset of the rainy season. As these trees did not affect the adjacent crop fields, neighbors did not object to having these trees on his field bunds. Narayanji started harvesting *Gmelina* for poles in 2001. So far, he has sold about 1,000 poles. A pole from a 40-months-old *Gmelina* tree which can be used for making farm implements fetches INR 120. Poles from trees older than 50 months fetches about Rs 150 per tree, also used for house construction. Farmers from nearby villages and traders from the Tehsil come to his farm to buy the poles. The branches and broken trunks are sold as firewood at the rate of just INR 2 per kg, while the foliage is fed to cows and goats in the household. His total income from *Gmelina* so far is about INR 130,000 and he has another 800 trees left in the field.

Sharing his success story, Bhoge, who owns 5.6 ha of land in Nilaj village in Bhandara district, said he started agroforestry long way back since 1963. Apart from two crops a year, he also planted saplings of Teak, Nilgiri, and Chivan, which is helping him to earn INR 40,000 extra annually. Bhoge has started motivating other farmers in his hometown and also visits nearby villages to spread advantages of agroforestry. Moreover, both state and central governments provide subsidy and have launched various schemes for agroforestry and guidance, he explained. He, however, pointed out that for agroforestry farmers have to keep patience as it takes time to bear results. For his good work, Bhoge also bagged many prestigious awards including Shetinishstha Puraskar (in 1992), Vasantrao Naik Samiti Puraskar (Pusad), Vanrai, and Krishi Bhushan to name a few.

Cooperative of Local Communities for Profitable Agroforestry (Agroforestry Federation)

It is a success story of a pioneer cooperative of tree farmers with production of *Eucalyptus* wood as a viable economic and environmental alternative to other agricultural crops under seasonal drought conditions. Efficient production methods, effective use of the limited water available, and a wide market end use acceptance and economic returns are cited in its favor. The formation of local community cooperatives offering technical and marketing expertise and downstream processing facilities which promote "on farm" returns to the producer are important activities. The credible marketing system developed is stated as a major factor promoting the operation. It is now increasingly realized that cooperative agroforestry can be used as an effective medium to fight the battle against environmental degradation as well as for alleviation of poverty in rural areas by increasing farm income. Following are some important observations:

- The cooperative formula has been working successfully in Maharashtra for a long time particularly in the field of agriculture and the cooperative culture is well established in the farmers' psyche. It cannot succeed where people are individualistic in their approach.
- The farmers' cooperatives have been promoted in Maharashtra by experienced and capable leaders with a socialistic bent of mind within a democratic framework. The leadership fulfills the managerial and entrepreneurial needs of the cooperative organization, even though the individual members of the cooperatives—being poor and illiterate—lack the managerial capability. The cooperatives can take care of such inadequacies. The economic viability of the cooperative agroforestry venture depends upon an efficient and credible marketing system which can ensure remunerative prices to the grower members.
- The choice of species is governed by the ability to grow well on a particular site, fast rate of growth, low gestation period, and above all ready marketability of the produce preferably backed up by an industry link up. The choice of species must be left to the farmers and their cooperatives. In this context, *Eucalyptus* spp. still have great potential along with other species like teak and bamboo.
- A pragmatic and liberal approach by the financial institutions as regards choice of species and spacing with an eye on economic viability, can give a boost to cooperative agroforestry. The financial institutions are also generally more willing to support private farmers through the cooperatives.
- Agroforestry is an effective means to enhance the area under tree growth and thus relieve pressure on Government reserved forests, by meeting the demand for firewood, poles, and timber from agroforestry crops raised on private lands. Incentives in the form of simplified procedures for harvesting and relaxing of forest rules applicable to harvesting and transport of forestry produce can give a boost to agroforestry cooperatives. The Government of Maharashtra has been very responsive and prompt in removing the bottlenecks notified by the cooperatives. Such responses cannot be expected by individual farmers.
- The Govt. of India's National Wastelands Development Board has initiated a novel scheme of providing incentives to agroforestry cooperatives for the technology extension programme. As it is truly said "seeing is believing", demonstration farms practicing agroforestry go a long way in convincing and motivating the farmers to accept agroforestry as an alternative land use system and in establishing the economic viability of such a system. The farmers, being illiterate, are slow to change. But, once they see the things happen, they pick up the concept very fast.
- An interesting and indeed very significant lesson which has been learned from the experience of growing *Eucalyptus* by the Nashik District Eucalyptus Growers' Society is that, by using superior quality planting stocks, which will ensure a uniform girth and height growth, the income from agroforestry produce can easily be enhanced three to four times compared to that obtained by using planting stocks of seed origin. The Agroforestry Federation has entered this high tech field by undertaking field trials of tissue culture plants of elite origin with the help of National Chemical Laboratory, Pune. The Federation will be collaborating with similar agencies to ensure a supply of genetically

superior planting stocks of eucalyptus to the grower members. This will give a further boost to cooperative agroforestry and ensure its success in the ecological rehabilitation of barren wastelands and thus usher in an era of all round socioeconomic development on a self-sustainable basis.

Research Gaps and Way Forward

i. Research, policy, and practice should direct toward maintenance, upgradation, and refinement of proven traditional and improved dryland agroforestry models to strengthen basic research for on-farm value addition and innovation.
ii. Many aspects of functional dynamics of multiple-use species are less known in dry land areas. Research on context-specific silvicultural and farming systems models on long-term basis with management implications of below ground and above ground systems involving multiple-use species are required.
iii. Dynamics of tree-intercrop-soil nutrient–microbe relationships and their role in source-sink relations toward soil amelioration, carbon sequestration, and reduction in greenhouse gas emission needs to be explored as a continuous cycle for its climate change mitigation and adaptation potential.
iv. More research efforts need to be mounted towards identification of viable agroforestry models in dry lands for balancing ground water level without affecting economic crop productivity to the land holders.
v. Intensive research efforts are required for participatory domestication of unexploited and under-exploited multipurpose trees, bushes, and grasses that yield commercially high-value medicinal, aromatic, industrial, and esthetic products in the dry land systems. It will provide wide options for livelihood improvement through improved biodiversity.
vi. Promotional research on production and supply of genetically improved agroforestry planting material using latest plant breeding and genetic tools for improved productivity, value addition and quick returns is needed.
vii. Agroforestry extension and communication networks are choked. This calls for evolving appropriate policy and development of intervention packages to popularize the potential of agroforestry. Strategy to transfer technology and effective feedback through well-coordinated efforts by all agricultural universities, ICAR institutes, open universities, extension departments, and nongovernmental organizations should be given priority for further research.
viii. Emphasis for national and international collaborative research linkage programmes for effective dissemination of knowledge pertaining to identical social and economic scenarios.
ix. Marketing issues and value addition to products of agroforestry
x. More emphasis on tangible benefits (viz., cash returns) from agroforestry systems over the intangible ones.

Conclusions

In a vast country like India, with a high unemployment rate, scarcity of land and capital, deforestation, and a growing population, agroforestry systems are ideal land use options to use the available natural resources in the fragile rainfed ecosystem for generating assured income with minimum risk. Harnessing agroforestry enterprise in rainfed areas has lot of scopes to enhance livelihood security. In view of the growing importance of agroforestry for rainfed agriculture, there is an urgent need for further research to continuously refine and improve the system. To meet both present and future demands, infrastructure and policies need to be supportive for the development and adoption of traditional and improved agroforestry systems based on synergy with nature. There is also a potential to utilize agroforestry in rainfed areas to solve problems like global warming (through increased carbon sequestration), and for biodiversity conservation, which will require that greater policy support for agroforestry in rainfed areas.

References

Agarwal A, Dureja SN, Garg SC (1985) Some observations on growth characteristics and pulping qualities of *Leucaena leucocephala* in Nepanagar. J Trop For 1:285–301

Ali MS, Chaturvedi OP (2008) Pattern and constraints of adoption of agroforestry systems in north Bihar. J Trop For 24(1/2):10–15

Ali MS, Das DK, Chakraborty AK, Sattar A (2006) Performance of intercrops under bamboo-based agroforestry system in Bihar. J Agro Meteorol 8(2):266–268

Banerjee H, Dhara PK, Mazumdar D (2009) Bamboo (*Bambusa* spp) based agroforestry systems under rainfed upland ecosystem. J Crop Weed 5:288–292

Bhatia CL (1984) *Eucalyptus* in India—Its status and research needs. Indian Forester 110:91–96

Buchanan F (1807) A Journey from Madras through the Countries of Mysore, Canara and Malabar, vol II. Blumar & Co, London

Chaturvedi AN (1989) Silvicultural requirements of Eucalyptus for small farms. In: Withington D, MacDicken KG, Sastry CB, Adams NR (eds) Multipurpose tree species for small farm use, International Institute for Agricultural Development. Winrock, USA, p 282

COMPETE (2009) Best practices and failures from Asia and Latin America (Report on India). Second Periodic Activity Report (01.01.2008–31.12.2008)

CRIDA (1993, 1996) Annual Reports. Central Research Institute for Dryland Agriculture (CRIDA), Hyderabad, India

CRIDA (2005) National Agricultural Technology Project- Rainfed Agro-ecosystem production system: Completion Report 1994–2004. Agroecosystem Directorate, CRIDA, Hyderabad, p 202

CRIDA (2007) Perspective Plan, Vision 2025. Central Research Institute for Dryland Agriculture, Hyderabad, Andhra Pradesh

CRIDA (1993-2010) Annual Reports. Central Research Institute for Dryland Agriculture (CRIDA), Hyderabad, India

Dagar JC (1995) Agroforestry systems for Andaman and Nicobar Islands. Int Tree Crops J 8:107–128

Das T, Das AK (2005) Inventorying plant diversity in home gardens: A case study of Barak Valley, Assam, North-East India. Curr Sci 89:155–163

Dastagir G, Suresh KK (1991) Forage production under silvipastoral system. J Trop For 7:206–209

Datta A, Das PK (1997) Evaluation of nodulation and its impact on biomass in some agroforestry tree species. Range Manag Agrofor 18(1):91–97

Deb R (1988) The growth and aerial biomass production of *Albizia procera* under silvipastoral system. In: Sing P and Pathak (eds) Rangelands: Resources and Management. IGFRI Jhansi, pp 338–342

Devaranavadgi SB, Wali SY, Patil SB, Jambagi MB, Kambrekar DN (2010) Survey of traditional agroforestry systems practiced in northern dry tract of Karnataka. Karnataka J Agric Sci 23(2):277–281

Divakara BN, Upadhyaya HD, Wani SP, Gowda CL (2009) Biology and Genetic improvement of *Jatropha curcas* L: A review. Appl Energy 86(11):2273–2282

Dutt AK, Jamwal U (1988) *Leucaena* growth in India. Indian J For 11:190–198

FAI (1994) Fertilizer Statistics (1993-94) III-21 to IV-86. Fertilizer Association of India, India

FAOSTAT (2006) http://faostat.fao.org

Gill AS (1995) Agri-silvi-cultural/Agri-silvi-horticultural studies. Research Highlights (1989–1994), NRCAF, Jhansi

Gill AS (2009) Performance of trees in the semiarid subtropics for production of chick pea in rain fed conditions. Indian Forester 135(6):831–837

Ginwal HS, Phartyal SS, Rawat PS, Srivastava RL (2005) Seed source variation in morphology, germination and seedling growth of *Jatropha curcas* Linn., in Central India. Silvae Genet 54:76–80

GoI (2006) Report of the Working Group on Animal Husbandry and Dairying, 11th Five Year Plan (2007–2012), Planning Commission, Government of India, New Delhi, India, p 232

Gupta MK, Sharma SD, Jha MN, Pandey Rajiv (2006) Response of organic and chemical fertilizers to the establishment and growth of eucalyptus in sodic land of Uttar Pradesh, India. Indian Forester 132:726–736

Hegde NG, Withington D, MacDicken KG, Sastry CB, Adams NR (1988) Multipurpose trees for small farmers in India. In: Mishra PJ, Dash SK, Sahu D (eds) Establishment pattern of live hedge species in North-Eastern Ghat of Orissa. Indian J of Soil Conserv 28:86–87

Ilorkar VM, Suroshe SB, Jiotode DJ (2011) Agroforestry interventions across different agro-climatic zones in Maharashtra, India. Indian J Forestry 34(1):105–109

IPCC (2007) Summary for Policymakers In: Solomon S, Qin D, Manning M, Chen Z, Marquis M, Averyt KB, Tignor M, Miller HL (eds) Climate Change 2007. The Physical Science Basis. Contribution of Working Group I to the Fourth Assessment Report of the Inter-Governmental Panel on Climate Change Cambridge University Press, Cambridge, USA

Jambulingam R, Fernandes ECM (1986) Multipurpose trees and shrubs in Tamil Nadu State (India). Agrofor Syst 4:17–32

Jha LK, Lalnunmawia F (2003) Agroforestry with bamboo and ginger to rehabilitate degraded areas in North East India. J Bamboo Rattan 2(2):103–109

Joseph B, Rao LGG, Sreemannarayana B (1999) Effect of pruning on yield of sunflower in *Albizzia lebbeck* based agri-silvi system. Indian J Agrofor 1(2):129–133

Joshi PN, Kumar V, Koladiya M, Patel YS, Karthik T (2009) Local perceptions of grassland change and priorities for conservation of natural resources of Banni, Gujarat, India. Front Biol 4(4):549–556

Kang BT, Wilson GF (1981) The development of alley cropping as a promising agroforestry technology. In: Agroforestry. A Decade of Development. pp 227–243

Kanwar JS (1999) Need for future outlook and mandate for dryland agriculture in India. In: Singh HP, Ramakrishna YS, Sharma KL, Venkateswarlu B (eds) Fifty Years of Dryland Agriculture Research in India. Central Research Institute for Dryland Agriculture, India, pp 1–19

Kaushik N, Kumar K, Kumar S, Roy S (2007) Genetic variability and divergence studies in seed traits and oil content of Jatropha (*Jatropha curcas L.)* accessions. Biomass Bioenergy 31:497–502

Korwar GR (1992) Influence of cutting height of *Leucaena* hedgerows on alley cropped sorghum and pearl millet. Indian J Dryland Agric Res Dev 7:57–60

Korwar GR (1999) Alternate land use systems: trees and bushes. In: Singh HP, Ramakrishna YS, Sharma KL, Venkateswarlu B (eds) Fifty Years of Dryland Agriculture Research in India. Central Research Institute for Dryland Agriculture, India, pp 507–512

Korwar GR, Pratibha G (1999) Performance of short duration pulses with African winter-thorn (*Faidherbia albida*) in semi-arid regions. Indian J Agric Sci 69(8):560–562

Korwar GR, Pratibha G, Ravi V, Palani Kumar D (2006) Performance of castor (*Ricinus communis*) and green gram (*Vigna radiata*) in agroforestry systems in semi-arid tropics. Indian J Agron 51(2):112–115

Korwar GR, Radder GD (1994) Influence of root pruning and cutting interval of Leucaena hedgerows on performance of alley cropped rabi sorghum. Agrofor Syst 25:95–109

Kulkarni HD (2008) Private farmer and private industry partnerships for industrial wood production: a case study. Int For Rev 10:147–155

Kumar A, Natarajan S, Biradar NB, Trivedi BK (2011) Evolution of sedentary pastoralism in south India: case study of the Kangayam grassland. Pastoralism Res Policy Pract 1:7

Kumar BM (1997) Bamboos in the Home gardens of Kerala - a shrinking resource base. J Non Timber For Prod 4:156–159

Kumar BM, Rajesh G, Sudheesh KG (2005) Aboveground biomass production and nutrient uptake of thorny bamboo [*Bambusa bambos* (L.) Voss] in the homegardens of Thrissur Kerala. J Trop Agric 43:51–56

Kumar P, Joshi PK, Birthal PS (2009) Demand projections for food grains in India. Agric Econ Res Rev 22(2):237–243

Kushalappa KA (1980) Status of subabul in Karnataka. In: Proceedings of seminar on Ipil–Ipil, held in Gandhinagar, December 29–30, 1980. pp 84–85

Kushalappa KA (1985) Productivity and Nutrient Recycling in Mysore Gum Plantations near Bangalore. Ph.D. Thesis, Mysore University, p 178

Lal R (1994) Water management in various crop production systems related to soil tillage. Soil Tillage Res 30:169–185

Madiwalar SL, Channabasappa KS, Deshmukh RC (2007) Survey and documentation of agroforestry models in Bidar, Gulbarga and Raichur districts. Karnataka J Agric Sci 20(1):93–95

Malik RS, Dutt D, Tyagi CH, Jindal AK, Lakharia LK (2004) Morphological, anatomical and chemical characteristics of *Leucaena leucocephala* and its impact on pulp and paper making properties. J Sci Ind Res 63:125

Marshall JK (1967) The effect of shelter on the productivity of grasslands and field crops. Field Crop Abstr 20:1–14

Maslekar AR (1984) Biomass production in rainfed and irrigated subabul plantations. Indian Forester 110:749–753

Mishra PN, Tewari SK, Dheer S, Katiyar RS, Singh D (1995) Effect of coppicing height on the regeneration and productivity of certain firewood shrubs in alkaline soils of North Indian plains. Biomass Bioenergy 9:459–463

Mittal SP, Singh P (1983) Studies on intercropping of field crops with fodder crops of subabul under rainfed conditions annual report. Central Soil & Water Conservation Research & Training Institute, India

Mullen BF, Gutteridge RC (2002) Wood and biomass production of *Leucaena* in sub-tropical Australia. Agrofor Syst 55:195–205

Nair PKR (1993) An introduction to agroforestry. Kluwer, The Netherlands

Nair PKR, Dagar JC (1991) An approach to developing methodologies for evaluating agroforestry systems in India. Agrofor Syst 16:55–81

NRAA (2008) Harmonization of wastelands/degraded lands datasets of India. Published by National Rainfed Area Authority (NRAA), Ministry of Agriculture, Government of India. NASC Complex, DP Shastri Marg, New Delhi, p 5

NRAA (2012) Prioritization of Rainfed Areas in India, Study Report 4. NRAA, India, p 100

NSSO (1992) Land and Livestock Holding Survey, Report on Operational Land Holdings in India: Salient Features Round 48th, Report No. 407, National Sample Survey Organization, Ministry of Statistics and Programme Implementation. Government of India, New Delhi

NSSO (2006) Level and Pattern of Consumer Expenditure (2004-05). National Sample Survey Organization, Ministry of Statistics and Programme Implementation, Government of India, New Delhi

NWDB (1987) Cost Models for Establishment of Silvipasture Forms in Different Regions. Monograph,

National Wasteland Development Board (NWDB), New Delhi

Osman M, Emminham WH, Sharrow SH (1998) Growth and yield of sorghum or cowpea in an agrisilviculture system in semiarid India. Agrofor Syst 42(1):91–105

Oswin DS (2004) A Review - Wetland ecosystems and coastal habitat diversity in Gujarat. India J Coastal Dev 7(2):49–64

Pathak PS (1991) Economic analysis of silvipastoral systems. Colloquium on Forestry Economics Research in Asia. Faculty of Forestry Kasetsart University, Bangkok, Thailand

Pathak PS (2002) Common pool degraded lands: technological and institutional options. In: Marothia DK (ed) Institutionalizing Common Pool Resources. Concept Publishing Co., India, pp 402–433

Pathak PS, Gupta SK, Singh P (1995) IGFRI approaches: Rehabilitation of degraded lands. Bulletin Indian Grassland and Fodder Research Institute, Jhansi, pp 1–23

Pathak PS, Pateria HM, Solanki KR (2000) Agroforestry systems in India - A diagnosis and design approach. All India Coordinated Research Project on Agroforestry, ICAR, National Research Centre for Agroforestry, Jhansi, India

Pathak PS, Solanki KR (2002) Agroforestry Technologies for Different Agro-climatic Regions of India. Indian Council of Agricultural Research (ICAR), p 41

Patil EN, Bonde RD, Salunkhe CD, Borse RH (1999) Effect of fruit and fodder trees on rainfed arable crops. Indian J Dryland Agric Res Dev 14(2):81–83

Patil VD, Sarnikar PN, Adsul PB, Thengal PD (2004) Profile studies, organic matter build-up and nutritional status of soil under bamboo (*Dendrocalamus strictus*) based agroforestry system. J Soils Crops 14(1):31–35

Prasad JVNS, Korwar GR, Rao KV, Mandal UK, Rama Rao CA, Rao GR, Venkateswarlu B, Rao SN, Kulkarni HD, Rao MR (2010a) Tree row spacing affected agronomic and economic performance of *Eucalyptus* -based agroforestry in Andhra Pradesh, Southern India. Agrof Syst 78(3):253–267

Prasad JVNS, Korwar GR, Rao KV Srinivas K, Rama Rao CA, Srinivasa Rao Ch, Venkateswarlu B, Rao SN, Kulkarni HD (2010b) Effect of modification of tree density and geometry on intercrop yields and economic returns in Leucaena based agroforestry systems for wood production in Andhra Pradesh, Southern India. Exp Agric 46(2):155–172

Prasad JVNS, Korwar GR, Rao KV, Mandal UK, Rao GR, Srinivas I, Venkateswarlu B, Rao SN (2011a) Optimum stand density of *Leucaena leucocephala* for wood production in Andhra Pradesh, Southern India. Biomass Bioenergy 35:227–235

Prasad JVNS, Korwar GR, Rao KV, Srinivas K, Srinivasa Rao Ch, Pedababu B, Venkateswarlu B, Rao SN, Kulkarni HD (2011b) On-farm evaluation of two fast growing tree species for biomass production for industrial use. New Forests 42:51–61

Prasad R, Saha B, Sarma JS, Agnihotri Y (1997) Development of hortipastoral land use system in degraded land. Annual Report. Central Soil & Water Conservation Research & Training Institute, India, p 44

Qureshi LM, Desai BP (1981) State experience papers-Maharashtra. In: Kaul RN, Gogate MG, Mathur NK (eds) *Leucaena leucocephala* in India. In: Proceedings of National Seminar, held in Urulikanchan (Maharashtra), June 26–27, 1981, pp 97–98

Rai MP, Pathak PS, Deb Roy R (1983) Biomass production of *Sesbania grandiflora* in different habitats. Int J Ecol Environ Sci 9:21–27

Rai P, Deb Roy R, Verma NC, Rao GR (1994) Performance of small ruminants on silvipasture and natural grassland under grazing condition. In: Singh P, Pathak PS, Roy MM (eds) Agroforestry Systems for Degraded Lands. Oxford and IBH publishing Co. Pvt. Ltd., New Delhi, pp 773–779

Rai P, Deb Roy R, Rao GR (1995) Evaluation of multipurpose tree species in rangelands under semi-arid condition of Uttar Pradesh. Range Manag Agrofor 16:103–113

Rai P, Solanki KR, Rao GR (1999) Silvipasture research in India - a review. Indian J Agrofor 1:107–120

Rai P, Yadav RS, Solanki KR, Rao GR, Singh R (2001) Growth and pruned production of multipurpose tree species in silvo-pastoral systems on degraded lands in semi-arid region of Uttar Pradesh, India. Forests Trees Livelihoods 11:347–364

Ramakrishna YS, Vijaya Kumar P, Ramana Rao BV (1999) Crop-Weather Relationship Studies in Dryland Agriculture. In: Singh HP, Ramakrishna YS, Sharma KL, Venkateswarlu B (eds) Fifty Years of Dryland Agriculture Research in India. Central Research Institute for Dryland Agriculture, India, pp 211–226

Ram N, Gill AS (1995) Effect of *Leucaena* and intercrop on *anola (Emblica officinalis)* under rainfed conditions. Van Anusnadhan 12(ii):51–55

Rao GR, Prasad YG, Prabhakar M, Srinivas I, Rao KV, Korwar GR, Venkateswarlu B, Shankar AK, Ravindra Chary G, Reddy NN (2010) Biofuel Crops for Dryland Cultivation and Processing Issues. CRIDA, p 38

Rao JV, Osman M (1994) Studies on silvipastoral systems in non-arable dry lands, In: Singh P, Pathak PS, Roy MM (eds) Agroforestry Systems for Degraded Lands. Oxford & IBH, New Delhi, pp 755–760

Rao KPC, Bantilan MCS, Singh K, Subrahmanyam S, Deshingkar P, Rao P. Parthasarathy, Shiferaw B (2005) Overcoming Poverty in Rural India: Focus on Rainfed Semi-Arid Tropics. Patancheru, Andhra Pradesh, India: International Crops Research Institute for the Semi-Arid Tropics, Hyderabad, p 96

Rao MR, Ong CK, Pathak P, Sharma MM (1991) Productivity of annual cropping and agroforestry systems on a shallow Alfisol in Semi-arid India. Agrofor Syst 15:51–63

Rao YV (1984) *Leucaena* plantations- A farming experience. Leucaena Res Rep 5:48–49

Reddy VS, Sankaranarayana V, Janakiraman N, Gowda TNV, Venkataramana P, Sivappa V (2000) Effect of wind breaks for improving productivity of crops in arable lands in eastern dry zone of Karnataka. Crop Res (Hisar) 20(3):415–419

Reddy YVR, Sudha M (1988) Economics of different land use systems in dryland farming. Annual Report, CRIDA, Hyderabad. pp 62–63

Rocheleau D, Weber F, Field-Juma A (1988) Agroforestry in Dryland Africa. ICRAF, Kenya, p 18

Roy MM (1991) Some tropical fodder trees for sustained fodder and firewood availability during lean periods. J Trop For 7:196–205

Roy MM, Deb Roy R, Pathak PS (1987) *Dichrostachys cinerea*: a potential fodder shrub for waste land plantation. Silver Jubilee Publication No.6. IGFRI, Jhansi

Saigal S, Kashyap D (2002) The second green revolution: Analysis of farm forestry experience in western tarai region of Uttar Pradesh and coastal Andhra Pradesh. Ecotech Services Private Limited, New Delhi 180

Saikia CN, Sharma TC (1994) *Leucaena leucocephala:* its culture and use as a source of high alpha – cellulose pulp. Adv For Res India XI:205–233

Shanmughavel P, Francis K (2001) Bioproductivity and nutrient cycling in bamboo and acacia plantation forests. Bioresour Technol 80(1):45–48

Sharma KK (1996) Agroforestry in farming systems development. Indian For 122(7):547–559

Singh HP (1999) Integrated Watershed Management in Drylands –A System Perspective. In: Singh HP, Ramakrishna YS, Sharma KL, Venkateswarlu B (eds) Fifty Years of Dryland Agriculture Research in India. Central Research Institute for Dryland Agriculture, India, pp 259–270

Singh P, Roy MM (1993) Silvopastoral systems for ameliorating productivity of degraded lands in India. Ann For 1:61–73

Singh RP, Reddy YVR (1986) Degradation and rehabilitation of drylands some technical and economical aspects. Paper presented in Regional Workshop for 9 countries of South and Southeast Asia on Economics of Dry and Degradation and Rehabilitation held from August 25–29, 1986 at New Delhi

Singh SB, Srivastav A, Bhatnagar S (2005) Response of *Eucalyptus camaldulensis* and *Dalbergia sissoo* to fertilizers in sodic soil. Indian For 131:1171–1177

Singh YD, Vijay Kumar V, Patel IL, Sharma YK, Dangar SV, Subba Rao SV (1998) Ecorestoration of Banni Grassland, First Annual Technical Report, Gujarat Ecology Commission, GERI Campus, Race Course Road, Vadodara, India, pp 16–18

Solanki KR, Ramnewaj AK (1999) Agroforestry, an alternate land use system for dryland agriculture. In: Singh HP, Ramakrishna YS, Sharma KL, Venkateswarlu B (eds) Fifty Years of Dryland Agricultural Research in India CRIDA. Hyderabad, pp 463–473

Srinivas K, Vittal KPR, Sharma KL (1999) Resource characterization of dryland: Soils. In: Singh HP, Ramakrishna YS, Sharma KL, Venkateswarlu B (eds) Fifty Years of Dryland Agriculture Research in India. Central Research Institute for Dryland Agriculture, Hyderabad, India, pp 41–55

Srinivasa Rao CH, Venkateswarlu B, Dinesh Babu M, Wani SP, Dixit S, Sahrawat KL, Kundu S (2011) Soil health improvement with *Glyricidia* green leaf manuring in rainfed agriculture: on-farm experiences. CRIDA Hyderabad, p 16

Srivastava RL, Ashok Kumar, Sharma SK, Emmanuel CJS K, Tomar UK (2008) Initial performance of tissue culture raised bamboos in Rajasthan and Gujarat. Indian For 134(7):916–922

TERI (2005-2006) Annual report. The Energy Research Institute, New Delhi

Totey NG, Arun P, Kapoor KS, Kulkarni R, Khatri PK, Chouhan JS, Bhowmik AK, Dahia VK (1989) Effects of nitrogen and phosphorus fertilization on growth and yield of bamboo (*Dendrocalamus strictus*) seedlings. Indian J For 12(2):106–111

Venkateswarlu J (1986) Efficient Resource Management for Drylands. In: 15 years of Dryland Agriculture Research, CRIDA, Hyderabad, India, pp 42–58

Virmani SM, Pathak P, Singh R (1991) Soil related constraints in dryland crop production in Vertisols, Alfisols and Entisols of India. In: Soil Related Constraints in Crop Production. Bulletin 15, Indian Society of Soil Science, pp 80–95

Wani SP, Sridevi TK, Rockstrom J, Ramakrishna YS (2009) Rainfed Agriculture-past trend and future prospectus. In: Wani SP, Rockstrom J, Oweis T (eds) Rainfed Agriculture: Unlocking the potential Comprehensive Assessment of Water Management in Agriculture Series. CAB International, Wallingford. UK, pp 1–35

Livelihood Improvements and Climate Change Adaptations Through Agroforestry in Hot Arid Environments

6

J. C. Tewari, Moola Ram, M. M. Roy, and Jagdish Chander Dagar

Abstract

Of the total area of arid zones of the world, Africa, accounts for 46.1 % followed by Asia (35.5 %). Majority of rest 19.4 % of arid zones are spread over in Australia, North America (Mexico and Southern part of USA), and South America. The hot Indian arid zone is spread in 31.7 million hectare area of which major part is in northwestern India (28.57 m ha) and some in southern India (3.13 m ha). The arid regions of Rajasthan, Gujarat, Punjab, and Haryana together constitute Great Indian Desert, better known as *Thar*. Arid western Rajasthan covers 61 % of total hot arid areas of the country and thus, forms the principal hot arid region of the country. Major part of it occurs between Aravalli ranges on the east and southeast and *Thal* desert of Pakistan (*Thal* desert is simply the western extension of *Thar*, only name has been changed) which is spread up to Suleman Kithara ranges in extreme west. The production and life support system in this part of hot Indian arid zone are constrained by climatic limitations. Sand dunes are dominant land formation of principal hot arid zone. More than 64 % area is sandy and intensities of dune vary from place to place. The human population density of this part of the country is quite high (127 person/km^2) as compared to global average of 6–8 persons per km^2 for arid zones. Large tracts of lands in *Thar* desert region of Rajasthan having widely scattered trees/shrubs of various species in association with crops of food grain and fodder as the best example of traditional agroforestry. The people of the region have evolved agroforestry-based drought protective mechanism through their ingenuity and centuries old experience, which has descended from one

J. C. Tewari (✉) · M. Ram · M. M. Roy
Central Arid Zone Research Institute, Jodhpur, Rajasthan 342003, India
e-mail: drjctewari@gmail.com

J. C. Dagar
Krishi Anusandhan Bhawan II, Indian Council of Agricultural Research, New Delhi, 110012, India

J. C. Dagar et al. (eds.), *Agroforestry Systems in India: Livelihood Security & Ecosystem Services*, Advances in Agroforestry 10,
DOI: 10.1007/978-81-322-1662-9_6, © Springer India 2014

generation to other. Depending upon climatic, edaphic, socioeconomic, and cultural situation, drought hardy woody perennials, which are multipurpose in nature, are integrated in farming systems to develop productive systems in form of traditional agroforestry. This paper describes the structure and production functions of traditional agroforestry systems of the region and their improvement with emphasis on livelihood improvement and climate change adaptation potential of these systems.

Introduction

According to conservative estimates, arid zones and their advances affect around 600 million people directly or indirectly. Desertification or land degradation processes represent a struggle in which man and land are engaged. At the world scale GLASSOD database indicates that 349.6 million ha of land in arid zone, are affected by light to moderate degree of soil degradation and 42.9 million ha by strong to extreme. It must be admitted that man himself or the increase of his population and of his livestock contribute to the expansion of desertification or land degradation processes. And yet, he suffers the worst consequence of such processes by having to be satisfied with left over land. Of the total land area of the world, arid zones cover 18.8 % and are located in almost all the continents of the world. Of the total area of arid zones of the world, Africa, accounts for 46.1 % followed by Asia (35.5 %). Majority of rest 19.4 % of arid zones are spread over in Australia, North America (Mexico and Southern part of USA), and South America (UNEP 1992). Europe has very small area under arid zones (<0.5 %of total arid zone of world).

The arid zones of USA, Arabian Peninsula, some central Asian and middle-east countries, Australia, and other petroleum rich countries meet out their domestic energy need through petroleum products and natural gas, however, contrary to this, in arid regions of India, woody vegetation resources form the main sources of fuel, besides, contributing to requirement of small timber and fencing material (Tewari et al. 1999). The hot Indian arid zone is spread in 31.7 million ha area of which major part is in northwestern India (28.57 m ha) and some in southern India (3.13 m ha). Arid western Rajasthan covers 61 % of total hot arid areas of the country and thus, forms the principal hot arid region of the country. This principal hot arid region is better known as *Thar* desert. The image of desert one has is that of the vegetative less, dense unbroken mantle of sand, and inaccessible terrain condition, however, shattered upon entering the *Thar* desert in most of the parts barring extreme western fringes. Sparsely distributed trees and other woody taxa with underneath growth of arable crops (especially during '*kharif*' season, as agriculture is predominantly rainfed) and/or grasses in long stretches interspersed with distantly distributed village settlement and "*Dhanis*" (a unique settlement pattern characteristic of *Thar* desert, meant for living of agriculturists families during active cropping period away from the villages but, nearby their fields) are actual features of the region (Tewari et al. 1999). Thus, rural folk in this principal hot arid region of the country have been practising arable cropping in association with scattered trees on crop fields since time immemorial. As most trees are drought resistant, these are still able to provide fuel, fodder, and other products when arable crops fail due to drought which is a common feature in the region. This chapter describes distribution, structure, and production functions of traditional and improved agroforestry systems with emphasis on livelihood improvement and climate change adaptations .

General Features and Ecology

Location and Distribution

The arid tropics or hot arid regions of India lie between 24–29 ° N latitude and 70–76 ° E longitude covering an area 31.70 million ha involving seven states viz. Rajasthan, Gujarat, Punjab, Haryana, Andhra Pradesh, Karnataka, and Maharashtra (Fig. 6.1). Area wise break up of hot arid regions in said states in given in Table 6.1. The arid regions of Rajasthan, Gujarat, Punjab, and Haryana together constitute Great Indian Desert, better known as *Thar*. The arid areas of Andhra Pradesh, Karnataka and Maharashtra located in peninsular India and have relatively better rainfall regions and experience less severe extremes of temperatures as compared to arid western Rajasthan, which is general has typical arid to hyper arid environment. In Chap. 5 also role of agroforestry systems for livelihood security in rainfed areas has been described but in this Chapter the information is exclusively for arid hot environments and no repetition has been made.

Physiography

Arid western Rajasthan is spread over in 12 districts of western Rajasthan. Major part of it occurs between Aravalli ranges on the east and southeast and *Thal* desert of Pakistan (*Thal* desert is simply the western extension of *Thar*, only name has been changed) which is spread up to Suleman Kithara ranges in extreme west. The arid areas of peninsular India are mostly sandy plains and small rocky-gravelly hilly terrains are found in a few pockets. In general, five major land forms viz. (i) sand dunes and sheet deposit (ii) comparatively dune free area, commonly known as sandy plains (iii) shallow soil areas (iv) rocky-gravelly terrain, and (v) saline soils. Saline and brackish water areas are recognized in arid western Rajasthan, the principal hot arid region of the country (Sharma and Tewari. 2001).

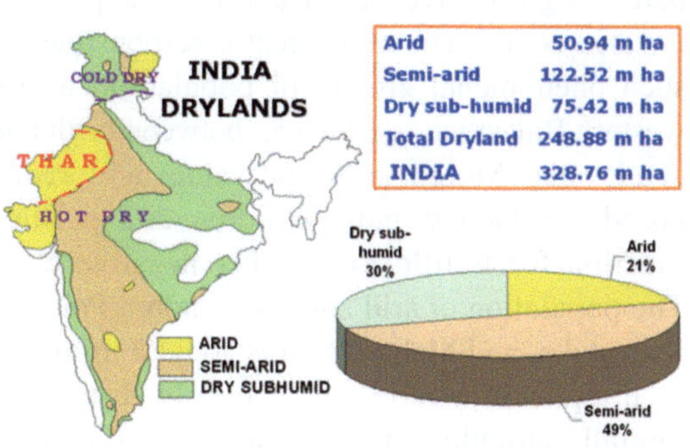

Fig. 6.1 Arid, semi-arid and dry subhumid regions of India

Fig. 6.2 *Prosopis cineraria* - *Tecomella undulata* - *Salvadora* spp. based traditional agroforestry system. Underneath crop is of cumin. Lopped trees are of *T. undulata*

Climate and Soils

The entire area stretching from the Sahara to *Thar* appears to be meteorologically homogenous one and this evidence supports the view that *Thar* desert of arid western Rajasthan is not an isolated desert. The production and life support system in this part of hot Indian arid zone are constrained by climatic limitations such as low annual precipitation (100–300 mm year^{-1}); very high temperature during summer season (mean maximum temperature 41 °C) touching a maximum of 48–50 °C and short (December to

Table 6.1 Distribution of arid area in different states

State (s)	Area (million hectares)	Percent of total
Rajasthan	19.61	61.0
Gujarat	6.22	19.6
Punjab and Haryana	2.73	9.0
Andhra pradesh	2.15	7.0
Karnataka	0.86	3.0
Maharashtra	0.13	0.4
Total	31.70	11.8 % of India

mid-February) cool and dry winters (the mean winter season temperature varies from 10 to 14 °C); high wind speed (30–40 km hr^{-1}); high evapotranspiration; and general low humidity regime (aridity index 0.045–0.19).

The arid areas of peninsular India, which accounts for about 10 % of total hot Indian arid zone receive total annual rainfall between 300 and 350 mm, and experiences mean maximum summer temperature not more than 38 °C. Winters are mild with average temperature ranging between 15 and 19 °C and minimum temperature never go below 6–8 °C.

Sand dunes are dominant land formation of principal hot arid zone. More than 64 % area is sandy and intensities of dune vary from place to place. In general, soils contain 1.8–4.5 % clay, 0.4–1.3 % silt, 63.7–87.3 % fine sand, and 11.3–30.3 % coarse sand. They are poor in organic matter (0.04–0.12 %) and low to medium in phosphorus content (0.05–0.10 %). The nitrogen content is mostly low ranging between 0.20 and 0.07 %. The infiltration rate is very high 7–15 cm h^{-1}. Due to complete absence of any aggregation, the soils are highly erosive. In Kutch area, the soils are moderate to highly saline in nature. Moreover, in Indira Gandhi Nahar Pariyojana (formerly Rajasthan canal) command areas and in many other parts there is also problem of soil salinity from moderate to highly saline in nature. The red soils are common in the arid areas of peninsular India. Such soils are derived mainly from granite rocks. They are highly variable in depth and exhibit a well-developed profile character. The percentage of clay increases markedly from the top soil to subsurface soil. Often murrum layer, which is gravelly disintegrated rock, found beneath the red soils.

The Population and Livelihood Resources

From the analysis of available authentic records the population of arid western Rajasthan, the principal hot region of the country registered an increase of 490 % from 1901 to 1991. The present population of the region is approximately 30.0 million. Decennial variation in population from 1901 to 1991 in *Thar* showed a growth rate of 186 % (between 1901 and 1971) as compared to 132 % for whole country. The population growth rate for the decade 1971–1981 and 1981–1991 was 36.7 and 30.7 %, respectively. The changes that occurred in the density of population in *Thar* between 1971 and 2011 are as under:

Years	Density of population (persons per km^2)
1971	48
1981	69
1991	89
2001	101
2011	127

The population density is quite high as compared to global average of 6–8 persons per km^2 for arid zones. The major factor responsible for such phenomenal growth of population in arid western Rajasthan is wide gap between birth and death rates. Moreover, social values are predisposed for having more children and positive sanction for fertility outnumber negative ones. The population of arid zones of Andhra Pradesh, Karnataka, and Maharashtra is about 3 million.

In spite of erratic and unevenly distributed rainfall, agriculture is mainstay of rural population, and mixed crop-livestock farming, mixed livestock-crop farming and livestock farming form the spectrum of economic activities. The distribution of agricultural land holding by size is fairly large, the average size of an agricultural land holding with a cultivator being 4–6 ha. Next to land livestock constitutes the most important asset of the cultivators. The number of livestock in arid western Rajasthan estimated to be 30 million in year 2000, which mainly includes cattle, sheep, goats, camels, and buffaloes. On an average a household has 1.32 buffaloes, 2.28 cows, 1.48 cow young stock, 0.38 buffaloes young stock, 6.48 sheep, 6.50 goats, and 0.49 camel. The other sources of rural livelihood are caste occupations like carpentry, black smith, oil pressing, pottery making, tanning, leather work, dyeing, gold smithy, etc.

Vegetation

From the floristic studies carried out since 1952 to date, it emerges that principal hot Indian arid regions have 682 species belonging to 352 genera and 87 families of flowering plants (Bhandari 1978; Kumar 1998), of these nine families, 37 genera and 63 species are introduced. The majority of introduced species are woody ones.

Phytogeographically, 37 % of the plant species represent African elements, 20.0 % oriental elements, 14 % species being tropical, 10.3 % cosmopolitan, and 9.3 % Australian. Nearly 9.4 % species are endemic to the regions. Besides, a large number of species show polymorphism, which is more pronounced in herbaceous plants.

Classified broadly under "Tropical Thorn Forest" (Champion and Seth 1968), however, the vegetation of hot Indian arid zone has been categorized into six major formations by Satyanarayan (1964). These formations are: mixed xeromorphic thorn forest, mixed xeromorphic woodlands, mixed xeromorphic riverine thorn forest, lithophytic scrub desert, psammophytic scrub desert, and halophytic scrub desert. This categorization still continues to be widely accepted.

Traditional Agroforestry Systems

Historians are generally of the view that organized settlements in the region began in the fourth century BC, when number of tribes migrated from fertile Indo-Gangetic plains into this environmentally hostile, but otherwise secure tract in response to waves of the invasions from the west. The settlements increased and expanded, and by the sixth or seventh century AD much of the desert region of western Rajasthan was not only settled but also politically organized (Tod 1832). Since then land use pattern of the region has witnessed several changes. The mixed crop-livestock farming, mixed livestock-crop farming, and livestock farming form the spectrum of economic activities in arid regions of India ranging from settled agriculture (in true sense, settled agroforestry) to nomadism (particularly, pastoralism). Tewari (1997) described large tracts of lands in *Thar* desert region of Rajasthan having widely scattered trees/shrubs of various species in association with crops of food grain and fodder as the best example of traditional agroforestry. These systems entail excellent 'multipurpose' species such as *Prosopis cineraria*, *Acacia nilotica*, *Ziziphus* species, etc., providing large number of end-use products. While contributing to food security in difficult time, trees on agricultural fields are important source of income. Boffa (1999) opined that multipurpose tree of traditional agroforestry systems are important for rural food security and income generation, and also for ensuring social and cultural stability to some extent. In addition, these systems actively maintain a rich pool of the forest genetic diversity on farmers' land. However, demographic environmental, social, and economic developments in past 40 years have put tremendous biotic and abiotic pressure on this traditional land use structure and its production attributes.

Distribution and Structure on Qualitative Basis

The people of the region have evolved agroforestry-based drought protective mechanism through their ingenuity and centuries old experience, which has descended from one generation to other. Depending upon climatic, edaphic, socioeconomic, and cultural situation, drought hardy woody perennials, which are multipurpose in nature, are integrated in farming systems to develop productive systems in form of traditional agroforestry.

Many classifications pertaining to distribution and structure of traditional agroforestry systems of the region have been proposed. Based on dominant trees/shrubs Malhotra et al. (1986) identified seven types of agroforestry practices in different regions of western Rajasthan (Table 6.2).

Harsh et al. (1992) described the distribution and structural components of traditional agroforestry systems of arid western Rajasthan on the basis of reconnaissance of administrative districts of western Rajasthan. Only prominent representative components (both woody and herbaceous) of the systems have been enlisted. The scheme of the classification is given in Table 6.3. The crops with trees included pearl millet (*Pennisetum typhoides*), green gram (*Vigna radiata*), moth (*V. acontifolia*), black gram (*V. mungo*) and cluster bean (*Cyamopsis tetragonoloba*) in rainfed condition and wheat (*Triticum aestivum*), rice (*Oryza sativa*), cotton (*Gossypium arboreum/hirsutum*), mustard (*Brassica juncea/nigra*), chillies (*Capsicum annuum*), sorghum (*Sorghum* sp), isabgol (*Plantago ovata*), cumin (*Cuminum cyminum*), sesame (*Sesamum indicum*) and green gram in irrigated condition.

Saxena (1997) defined traditional agroforestry systems of the region on the basis of vegetation complex of different areas. Rainfall zones, habitats and district boundaries were also taken into account to define the traditional agroforestry systems (Table 6.4).

Quantitative Analysis of Structure of Traditional Agroforestry Systems

Structural Features

All the above-mentioned studies defined the traditional agroforestry systems of arid western Rajasthan in simple qualitative terms. Pratap Narain and Tewari (2005) first time described the traditional agroforestry systems of the region on the basis of distribution and densities of tree species. This was perhaps first report, where quantifications were done to describe the systems. In general, rainfall appears to be governing

Table 6.2 Traditional agroforestry systems in arid zone of Rajasthan (Malhotra et al. 1986)

Dominant trees/shrubs with agricultural crops	Zone
Ziziphus nummularia ("Bordi") grown with agricultural crops *Prosopis cineraria* ("Khejri") trees are very few	Jaisalmer-Pugal-Girab (below 150 mm isohyets)
Crops grown in association with *P. cineraria* and *Z. nummularia*	Barmer-Osian-Bikaner belt
Crops mainly grown in association with *P. cineraria* and a limited number of *Z. nummularia* plants are available	Jodhpur-Nagaur-Shekawati belt
Mainly *P. cineraria* dominate in agricultural fields but few *Ailanthus excelsa* trees are also in agricultural fields	Nawalgarh-Sikar
The dominating trees species are *Acacia nilotica* var. *cupressiformis* and *A. nilotica* var. *rediana*. Both the "*Kharif*" (summer monsoon) and "*Rabi*" (winter) crops are grown with these trees	Pali
Prosopis cineraria dominates but *Acacia leucophloea* is also present. Some crops are grown in association with these species	Arid and semi-arid areas bordering Eastern Ajmer
In this tract *Salvadora persica* and *S. oleoides* dominate; crops are grown with these trees	Jalore Districts

Table 6.3 Components of traditional agroforestry systems in various districts of Rajasthan (Harsh et al. 1992)

Main tree/shrub species	Main crops	Prominent grass species	District
Prosopis cineraria, Acacia nilotica sub sp. *indica, Acacia tortilis*	Pearl millet, green gram and cluster bean (rainfed). Wheat, cotton, rice and green gram (irrigated)	*Lasiurus sindicus*	Ganganagar and Hanumangarh
Prosopis cineraria, Ziziphus nummularia, Calligonum polygonoides, Acacia jacquemontii	Green gram, moth bean, cluster bean and pearl millet	*Lasiurus sindicus*	Bikaner
Calligonum polygonoides, Ziziphus nummularia, Prosopis cineraria, Acacia senegal, Capparis decidua	Green gram, pearl millet and cluster bean	*Lasiurus sindicus*	Jaisalmer
Prosopis cineraria, Tecomella undulata, Ziziphus nummularia, Capparis decidua	Pearl millet, green gram and cluster bean	*Lasiurus sindicus, Cenchrus ciliaris*	Barmer
Prosopis cineraria, Ziziphus nummularia, Capparis decidua, Acacia senegal	Pearl millet, green gram and cluster bean (rainfed). Wheat, chillies, mustard and green gram (irrigated)	*Cenchrus ciliaris*	Jodhpur
Prosopis cineraria, Maytenus emarginata, Ziziphus nummularia	Pearl millet, green gram and cluster bean	*Lasiurus sindicus, Cenchrus ciliaris*	Churu, Jhunjhunu and Sikar
Prosopis cineraria, Acacia nilotica	Pearl millet and green gram (rainfed). wheat, green gram and mustard (irrigated)	*Cenchrus ciliaris*	Naguar
Prosopis cineraria, Salvadora persica, Salvadora oleoides, Acacia nilotica, Punica granatum (fruit tree)	Pearl millet, green gram, isabgol, sorghum and cumin	*Cenchrus ciliaris*	Jalore
Acacia nilotica subsp. *indica, Acacia nilotica* var. *cupressiformis, Acacia leucopholea, Acacia catechu, Salvadora* spp	Sorghum, pearl millet, green gram and cluster bean	*Cenchrus ciliaris Cenchrus setigerus*	Pali

factor for evolution of traditional agroforestry systems in the Region.

Upper Transect

On the basis of rainfall, four types of major traditional agroforestry systems have been identified in upper transact of arid western Rajasthan extending from Danta-Ramgarh (transitional zone between arid and semi-arid region) to extreme western fringes of Bikaner/Ganganagar districts (Table 6.5). Over 400 mm rainfall zone predominant system is *Prosopis cineraria—Acacia nilotica* based; between 300 and 400 mm rainfall zone *P. cineraria* based; between 200–300 mm rainfall zone *Ziziphus* spp—*P. cineraria* based; and in less than 200 mm rainfall zone *Ziziphus* spp—*P. cineraria—Salvadora* spp based. The tree/shrub density of system is also clearly governed by rainfall regime. Data indicated with decrease in rainfall, the density of woody component of the system substantially declined. *P. cineraria* is found in all the rainfall zones, however, its density tend to decrease with decreasing rainfall from east to west ($r = 0.87$; $p < 0.0001$).

Lower Transect

The lower transect in arid western Rajasthan encompassed Pali, Jalore, parts of Jodhpur, and Barmer district in rainfall gradient

Table 6.4 Traditional agroforestry systems of arid western Rajasthan (Saxena 1997)

Tree/shrub species association	Annual rainfall (mm)	Habitats	Associated crops/grasses
Calligonum-Haloxylon-Leptadenia	110–150	Sand dunes, interdunes	Pearl millet, cluster bean, *Lasiurus sindicus*
Ziziphus-Capparis	150–200	Rocky, gravelly pediments	Pearl millet, green gram, moth bean, cluster bean/*Cymbopogon jwarancusa, Aristida* spp., *Cenchrus ciliaris*
Calotropis-Calligonum-Clerodendrum	200–250	Sand, gravelly pediments	Pearl millet, green gram, moth bean, sesame, *Cenchrus ciliaris*
Prosopis-Ziziphus-Capparis	250–300	Alluvial plains, soils often-with "Kankar pans" at 80-150 cm soil depth	Pearl millet, cluster bean, green gram, moth bean, sesame, *Cenchrus ciliaris, C. setigerus*
Salvadora-Prosopis-Capparis	250–300	Alluvial plains but soils are moderately saline	Cluster bean, pearl millet, sesame, wheat (irrigated areas) with *Cenchrus setigerus, Sporobolus* spp.
Prosopis-Tecomella	275–325	Sandy plains (rainfed)	Pearl millet, cluster bean, green gram, moth bean/*Cenchrus ciliaris* with *C. setigerus*
Prosopis	300–350	Alluvial plains (rainfed)	Pearl millet, cluster bean, green gram, moth bean/*Cenchrus ciliaris*
Prosopis-Acacia	300–350	Alluvial plains (irrigated)	Sorghum, cumin, pearl millet, mustard, Wheat

Table 6.5 Traditional agroforestry systems of *thar* desert

Rainfall zone (mm)	Agroforestry system	Trees/shrubs (Nos. ha^{-1})	% density of prominent species
>400	*P. cineraria–A. nilotica* based	31.4	80.5
300–400	*P. cineraria* based	14.2	80.0
200–300	*Ziziphus* spp–*P. cineraria* based	91.7	100.0 (91.7 % *Ziziphus* spp)
<200	*Ziziphus* spp–*P. cineraria–Salvadora* spp based	17.2	87.2 (65 % *Ziziphus* spp)

of >500–200 mm from southeastern margin to western part (CAZRI 2008). Four distinct agroforestry systems have been identified in the study region (Table 6.6).

In southeastern margin, where total annual rainfall is >500 mm, the predominant agroforestry system is *Azadirachta indica–Acacia nilotica* var. *cupressiformis* based. This system extends from extreme eastern part of Pali district to western part of Jalore district. In the rainfall zone >400–500 mm, *Prosopis cineraria—Azadirachta indica* based system is prevalent. The density of woody component in the system was found to have only 8.2 individuals per ha with a very low total basal cover (0.6 m^2 ha^{-1}). In rainfall zone of 300-<400 mm, the major traditional agroforestry system was *P. cineraria—Tecomella undulata—Salvadora* spp. based (Fig. 6.2). The total basal cover contributed by system farming species is as high as 92 % there by indicating that maximum biomass was accumulated on system farming species as biomass corresponds to basal cover. In western part of Barmer district (rainfall zone >200–300 mm), the most prevalent system is *P. cineraria—Salvadora oleoides—T. undulata* based.

Table 6.6 Features of woody species in identified traditional agroforestry systems in lower transect of arid western Rajasthan

Rainfall zone	System	Total trees/shrubs density (No. ha^{-1})	Total basal cover (m^2 ha^{-1})	Total canopy cover (% ha^{-1})
>500	*Azadirachta indica-Acacia nilotica* var. *cupressiformis* (12[a])	15.9	1.1	2.3
<500–400	*P. cineraria-A. indica* (8)	8.2	0.6	1.5
<400–300	*P. cineraria-T. undulata-Salvadora* spp (8)	32.1	2.2	4.7
<300–200	*P. cineraria- Salvadora oleoides-T. undulata*	33.8	4.3	4.8

[a] Values in parentheses are total number of tree/shrub species in the system

Production Attributes

Dhir (1982) has shown that incidence of agriculture holding during late medieval times was only one-fifth of today. The process of rapid expansion of agriculture started with increase in population. In 1950s, much of agriculturally usable lands were brought under plough with predominant fallow system. The trend since then in already established agricultural tract has been intensification of farming and breaking new lands in the environmentally marginally suited parts of the region. The main productive functions fulfilled by traditional agroforestry systems of western Rajasthan are production of cereals and other food grains, oil seeds, fuel wood, fodder, wood for agricultural implements, minor timber, minor tree products like fruits, gum, resins, and vegetables, etc. Production attributes of traditional agroforestry system defined by Pratap Narain and Tewari (2005) and CAZRI (2008) are given in following description:

Upper Transect

Arable crops: In upper transect of arid western Rajasthan rainfall seems to play determinant role in crop production. *P. cineraria—A. nilotica* based agroforestry system was found to be highly productive and *Ziziphus* spp – *P. cineraria—Salvadora* spp system was least productive (Table 6.7). It was very interesting that yield of pearl millet, main cereal crop of the region, below the canopy of woody components in any system was not affected at all, rather in *Ziziphus* spp—*P. cineraria* and *Ziziphus* spp—*P. cineraria—Salvadoran* spp systems, it was substantially increased below the tree canopies in comparison of open spaces. However, legumes such as moth bean, cluster bean, and green gram exhibited slightly yield reduction below the tree canopies, 5.0 % (*Ziziphus* spp—*P. cineraria* system) to 19.5 % (*P. cineraria* system); 12.5 % (*Ziziphus* spp—*P. cineraria—Salvadora* spp system) to 18.2 % (*Ziziphus* spp–*P. cineraria* system); and 10.0 % (*P. cineraria* system) to 33.7 % (*P. cineraria—A. nilotica* system), respectively. Yield reduction under tree canopies for sesame ranged from 26.3 % (*P. cineraria—A. nilotica* system) to 46.1 % (*P. cineraria* system).

Fuel wood and leaf fodder: Foliage of trees constitutes nutritious components of animal feed in arid parts of Rajasthan. The leaf fodder of some tree species is as nutritious as leguminous fodder and is comparable with grass production from the pastures. Fuel wood and leaf fodder production potential of *P. cineraria–A. nilotica* system was maximum, while it was minimum for *Ziziphus* spp.- *P. cineraria–Salvadora* spp. system (Table 6.8). Thus, production from woody components of traditional agroforestry systems also follows the rainfall gradient like that of arable crops.

Table 6.7 Crop production (t ha^{-1}) in traditional agroforestry system in arid region of western Rajasthan

AF systems crops	*P. cineraria—A. nilotica* based	*P. cineraria* based	*Ziziphus* spp—*P. cineraria* based	*Ziziphus* spp—other spp based
Pearl millet				
below the canopy	10.5	7.9	2.0	2.0
away from canopy	10.8	7.9	1.8	1.2
Moth bean				
below the canopy	2.8	2.1	1.9	1.4
away from canopy	3.0	2.6	2.0	1.6
Cluster bean				
below the canopy	2.5	1.8	0.9	0.8
away from canopy	2.9	2.1	1.1	1.0
Sesame				
below the canopy	1.4	2.1	1.9	-
away from canopy	1.9	3.9	2.6	-
Green gram				
below the canopy	1.4	0.9	-	-
away from canopy	2.1	0.1	-	-

Table 6.8 Fuel wood and top feed production from traditional agroforestry systems in arid region of western Rajasthan

Agroforestry systems	Plant density (nos ha^{-1})	Fuel wood (t ha^{-1} year^{-1})	Leaf fodder (t ha^{-1} year^{-1})
P. cineraria-A. nilotica based	31.4	1.23	0.29
P. cineraria based	14.2	0.60	0.20
Ziziphus spp - *P. cineraria* based	91.7	0.62	0.90
Ziziphus spp - *P. cineraria* - *Salvadora* spp based	17.2	0.18	0.14

Lower Transect

Arable crop production: Croplands are nucleus of the traditional agroforestry system. The major rainfed (*Kharif*) crops grown in association of trees/shrubs were pearl millet, green gram, moth bean, cluster bean, and sesame (Table 6.9). On an average, across all the villages studied in entire region, it was recorded that 60 % area was under pearl millet. The pearl millet grain yield exhibited clear declining trend with decrease in total annual rainfall. The grain yield of pearl millet was found to be comparable both in open field and under the tree canopy cover in all the systems. However, under the tree canopy cover yield of green gram decreased 20.0–29.83 %, moth bean 15.69–32.43 %, cluster bean 20.0–36.61 % and sesame 21.05–23.53 %. The data clearly indicated that tall pearl millet crop production did not have any negative effect of tree canopy cover, however, production of grain in all short-statured leguminous crops were adversely affected under the tree canopy cover.

Fuel and leaf fodder production from woody component: Fuel wood and leaf fodder are main products of woody component in the traditional agroforestry systems, which farmers collect by way of lopping and or/pollarding the trees and shrubs. The maximum fuel wood (4.9 t

Table 6.9 Pattern of crop grain production (kg ha^{-1}) in different identified traditional agroforestry systems

System/ crop	BC/AC/ % decrease	*Azadirachta indica- A. nilotica* var. *cupressiformis*	*P. cineraria - A. indica*	*P. cineraria –T. undulata – Salvadora* spp	*P. cineraria- S. oleoides – T. undulata*
Pearl millet	BC	940	600	360	351
	AC	915	622	375	339
	% decrease	2.66	3.54	4.00	3.42
Green gram	BC	96	57	120	130
	AC	120	81	175	182
	% decrease	20.00	29.63	31.43	28.57
Moth bean	BC	75	86	100	105
	AC	111	102	162	135
	% decrease	32.43	15.69	38.27	22,22
Cluster bean	BC	107	104	90	62
	AC	75	130	125	92
	% decrease	29.91	20.00	28.00	32.61
Sesame	BC	95	104	120	–
	AC	121	136	152	–
	% decrease	21.49	23.53	21.05	–

AC Away from canopy; *BC* Below canopy
See text for scientific names of crops

Table 6.10 Fuel wood and leaf fodder production potential of different identified traditional agroforestry systems

Systems	Fuel wood		Leaf fodder	
	Total (t ha^{-1} $year^{-1}$)	Contribution of system farming species (%)	Total (t ha^{-1} $year^{-1}$)	Contribution of system farming species (%)
A. indica – A. nilotica sub sp. *cupressiformis*	3.10	39.7	0.71	54.8
P. cineraria – A. indica	1.20	74.8	0.43	68.9
P. cineraria – T. undulata – Salvadora spp	4.73	52.3	1.70	68.9
P. cineraria – S. oleoides - T. undulata	4.90	70.0	1.78	73.5

$ha^{-1}year^{-1}$) and leaf fodder (1.78 t $ha^{-1}year^{-1}$) production was recorded in *P. cineraria–S. oleoides-T. undulata*-based agroforestry system, which was prevalent (Table 6.10) in western part of Barmer district (rainfall zone >200–300 mm). The fuel wood (1.2 t $ha^{-1}year^{-1}$) and fodder (0.43t ha^{-1} $year^{-1}$) production was minimum in *P. cineraria – A. indica*-based system, which was dominant in 300–< 400 mm rainfall zone. It appeared that with decrease in rainfall to a certain limit, the woody component in the system are protected by the farmers as they provide substantial amount of livelihood in environmentally more stressed areas.

Improved Agroforestry Systems

In general, throughout the hot Indian arid regions native tree species are very few in number and that too are very slow growing, and more over, inhospitable environmental conditions also do not support much required natural regeneration of trees (Tewari et al. 1993). The problems of hot arid regions of the country attracted the attention of planners, policy makers and even thinking politicians, soon after independence and realizing the importance of plantation forestry for such areas, a Desert Afforestation Station was established at Jodhpur in 1952 and this station was elevated to full fledged multidisciplinary institute, named Central Arid Zone Research Institute (CAZRI) in the year 1959. Thus, this marked the beginning of R & D efforts in the frontiers of arid zone forestry in India. Introduction of fast growing species from iso-climatic regions of the world and also from other drier regions for different landforms of arid tract was given priority and the process of evaluation and screening of more than 200 tree species have been accomplished so far and most adaptable ones have been identified (Tewari and Harsh 1998).

Tree growing in hot arid region is basically concerned with the management of trees for conservation and for limited production objectives like wood for fuel, poles and fencing material; leaves for livestock fodder; and pod/seeds for many types use in human diet. The role of trees to conserve the fragile ecosystems of hot arid regions has been well recognized (Mann and Muthana 1984). The trees also provided so many services to mankind, which make them an intricate part of man-livestock-agriculture continuum, the lifeline of hot arid regions (Saxena 1997).

In the light of increasing pressure on land resources, CAZRI, Jodhpur initiated systematic studies on agroforestry systems in late 1970s. Since then a number of improved agroforestry practices in order to enhance overall productivity and economic returns of the farming communities have been developed and standardized. Following improved practices have been found promising and remunerative, and easy to fit in existing traditional agroforestry systems.

Fruit Tree-Based Cropping Systems

When Fruit Trees are Grown with Arable Crops

Leguminous crop (green gram) sown under *ber* (*Ziziphus mauritiana* cv. Seb) plantation produced 0.2 t ha^{-1} of grain and 0.8 t ha^{-1} quality *ber* fruits from same land unit even when seasonal rainfall was 200 mm (Table 6.11), thus rendering a drought proofing mechanism to the system. The density of *ber* plants were kept at 400 individuals per ha (Gupta 1997). The economics of this improved system indicated that in case of sole leguminous crop (green gram) farming, the net profit per hectare was Rs 4800, however, in case of *ber* intercropping, the profit was to a tune of Rs 8,000 per ha.

When Fruit Trees are Grown with Grasses

Ber (*Ziziphus* spp.)-based pastoral systems have proved highly remunerative in *Thar* desert on farmers' field. *Ber* trees were planted at a space of 6 m × 6 m and grass *Cenchrus ciliaris* was introduced between tree rows. On an average, the dry grass production was 1.55 t ha^{-1} year^{-1} (Tewari et al. 1999). The fruit, leaf fodder and fuel wood production from *ber* was 2.77, 1.87, 2.64 t ha^{-1} year^{-1}, respectively (Table 6.12).

Table 6.11 Improved fruit-based agroforestry practices: (*Ber* + green gram)

Treatment	Annual rainfall (mm)	Fruit yield (kg ha^{-1})	Grain yield (kg ha^{-1})	Net profit (Rs ha^{-1})
Sole crop	200	–	520	4,800
Intercropped *with Ber*	200	800	200	8,000

Table 6.12 Improved *(Ber + Cenchrus ciliaris)* system on farmers' field

Years	Dry grass (t ha^{-1})	Tree products		
		Fuel wood (t ha^{-1})	Leaf fodder (t ha^{-1})	Fruit yield (t ha^{-1})
1	1.50	3.10	2.13	3.10
2	1.67	2.74	1.80	2.74
3	1.42	2.46	1.71	2.49
4	1.66	2.49	1.63	2.85
5	1.48	2.43	2.01	2.80
6	1.57	2.63	1.94	2.67
			Average	1.55
2.64	1.87	2.77		

Forest Tree-Based Systems

When Trees are Grown with Grasses (Silvo-pastoral)

Arid region of Rajasthan is reported to have 89 species of grasses. The species like *Lasiurus sindicus* is efficient builder of biomass with energy use efficiency of 1.4–2.0 % (Harsh et al. 1992). In an improved silvi-pasture, *Hardwickia binata* was taken as tree component at 3 m × 3 m spacing with *Cenchrus ciliaris* grass. Results of 9 years study (Table 6.13) revealed that average carrying capacity of the system was 4.1 sheep ha^{-1} $year^{-1}$ against 3.7 for sole *C. ciliaris* pasture and 1.6 for sole *H. binata* plantation. In this type of improved silvi-pastoral system, in addition to grass + top feed production to the tune of 3.06 t ha^{-1} $year^{-1}$, a biomass of 0.26 t ha^{-1} $year^{-1}$ of fuel wood was also obtained.

When Forest Trees are Grown with Grasses and Arable Crops

Diversified production systems to appear very sustainable for hot arid regions. A recent study conducted on *Ailanthus excelsa*-based cropping system involving arable crops and grasses both revealed that the system has the potential to provide multitude of products in inhospitable environmental conditions of the region. Growth patterns of *A. excelsa* in association with pearl millet, moth bean, and *Cenchrus ciliaris* exhibited that wider spacing of 10 m × 10 m is highly suitable for overall growth of tree component in terms of height (Fig. 6.3) and basal area per tree (Fig. 6.4); and basal area of entire tree stand (Fig. 6.5). The production from herbaceous components was also superior in wider spaces.

On an average the system productivity of herbaceous component for food grain and fodder (in the form of crop straw and grasses) was found to be the best under wider spacing (10 m × 10 m) of *A. excelsa*. However, under closer (6 m × 6 m) and as well as moderate (8 m × 8 m) spacing of the species, the productivity of herbaceous component was satisfactory. This indicated that with proper management, diversified agroforestry systems have the potential to produce multitude of products in fairly good quantum. On an average

Table 6.13 Production potential of *Hardwickia binata* + *Cenchrus ciliaris* based improved system

Year after planting	Grass yield (t ha^{-1})	Leaf fodder yield (t ha^{-1})	Total fodder yield (t ha^{-1})	Carrying capacity (sheep ha^{-1} yr^{-1})
1	0.82	–	0.82	1.9
2	1.23	–	1.23	2.9
3	1.64	–	1.64	3.9
4	2.05	–	2.05	4.9
5	1.64	–	1.64	3.9
6	1.64	–	1.64	3.9
7	1.24	2.31	3.55	8.5
8	0.84	0.66	1.50	3.6
9	0.84	0.66	1.50	3.6
Average	1.33	0.40	1.73	4.1

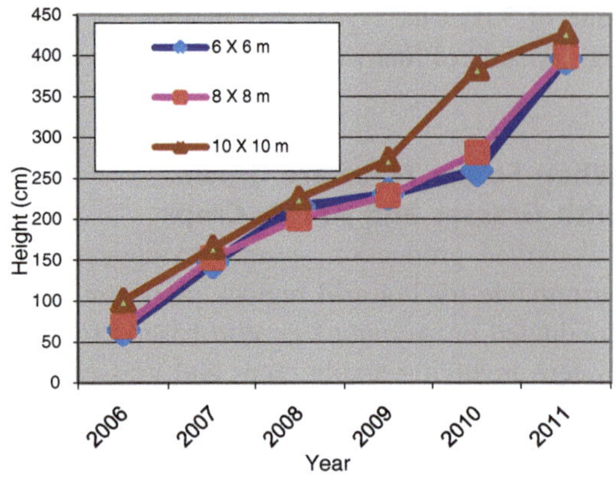

Fig. 6.3 Height (cm) of *Ailanthus excelsa* under different spacing in a cropping system having arable crops and grasses

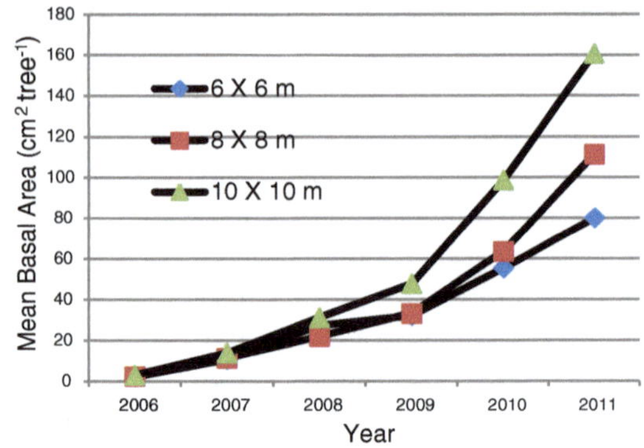

Fig. 6.4 Mean basal area (cm^2 $tree^{-1}$) of *A. excelsa* under different spacing in a cropping system having arable crops and grasses

the system productivity of herbaceous component for food grain (Table 6.14) and fodder (in the form of crop straw and grasses) was found to be the best under wider spacing (10 m × 10 m) of *A. excelsa*. However, under closer (6 m × 6 m) and as well as moderate (8 m × 8 m) spacing of the species, the productivity of herbaceous component was satisfactory (Table 6.15). This indicated that with proper management, diversified agroforestry systems have the potential to produce multitude of products in fairly good quantum.

As the system is still quite young, the amount of leaf fodder production (Table 6.16) during fifth year indicated that system is capable of providing leaf fodder in appreciable quantity in years to come.

Choice of Woody Species, Plantations, and Management in Hot Arid Landforms

As also mentioned earlier in physiography of hot arid regions of India, five major land forms, i.e., (1) sand dunes and hummocky terrain (2) sandy plains (3) shallow soil areas (4), rock-gravelly terrain, and (5) salt-affected soils areas are most common. Each of these landforms has unique features and also has unique problems. The native plant communities on these landforms are found in severe stages of degradation (Kaul and Ganguli 1964). Therefore, it is rare to find optional proportions of different components of natural plant communities on different landforms. Due to very slow growth of native plant species, especially tree species, it becomes necessary to introduce the species of relatively higher growth rate and at the same time tolerant to drought, frost, and salinity (up to certain level) (Tewari 1997). Efforts in this direction were initiated in early 1960s for different arid land forms (Bhimya et al. 1964; Kaushik et al. 1969; Mann and Muthana 1984) and the process is still continued (Tewari et al. 1989; Tewari and Harsh 1998). Till date, over 500 accessions of 200 tree species from isoclimatic regions of the world and other drier parts have been evaluated through long-term screening trials to assess their adaptability and growth performance on different landforms.

Species for Sand Dunes Stabilization

More than 34 % (11 million ha) of the total area of Indian hot arid regions is covered by drifting or semi-stabilized sand dunes, sometime up to 100 m in height, however, their intensities vary from place to place. Singh (1977) has identified seven types of sand dunes on the basis of their shape. There are parabolic, coalesced parabolic, longitudinal, transverse, burchan, obstacles, and shrub coppice type. Some of them are highly active and menace to inhabitants of arid regions.

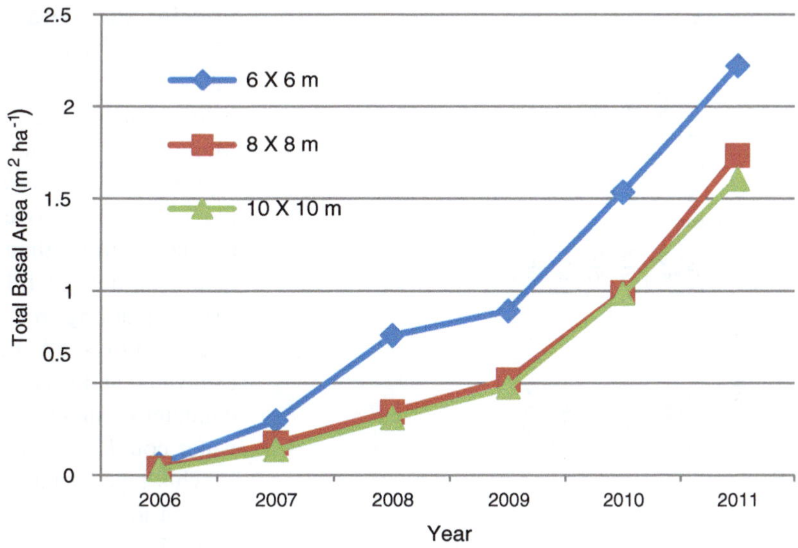

Fig. 6.5 Total basal area (m^2 ha^{-1}) in entire stand of *A. excelsa* under different spacing in a cropping system having arable crops and grasses

Table 6.14 Grain productivity (kg ha^{-1}) under different spacing of *Ailanthus excelsa* based cropping system

Years	Pearl millet			Moth bean		
	6 m × 6 m	8 m × 8 m	10 m x 10 m	6 m x 6 m	8 m x 8 m	10 m x 10 m
2007–2008	425	429	504	202	195	205
2008–2009	528	565	650	24	22	29
2009–2010[a]	–	–	–	–	–	–
2010–2011	1,312	1,408	1,552	233	250	256
Average	755	800.7	902	153	155.7	163.3

[a] Completely drought year

Some of the dunes are semi-stabilized but their crests and flanks remain active because of cultivation on them and mismanagement.

CAZRI, Jodhpur started efforts of sand dune stabilization since its inception in 1952 and within 10 years, i.e., by early 1960s successfully developed the technological package for their fixation. In brief, following steps are involved in sand dune stabilization process (Harsh and Tewari 1993):

- Protection against biotic interference.
- Treating the affected dunes by fixing micro-wind breaks across the wind direction in parallel strips or checker board design.
- Tree plantation on treated dunes.

A number of tree species were tried on sand dunes and the process of evaluation and screening of tree species for sand dune afforestation is still continuing. Table 6.17 incorporates the important features of tree species that performed well on sand dunes.

These species have been screened after long-term evaluation. In general, all these species are drought resistant both at seedling and mature stage. By adopting CAZRI developed technology, State Government of Rajasthan has stabilized more than 500 thousand ha of such area in arid western Rajasthan, the major hot arid zone of the country. Many times, native perennial grass species like *Cenchrus ciliaris*, *C. setigerus*, and *Lasiurus sindicus* are introduced on spaces between the tree species after 2 or 3 years of initial field out planting of tree seedlings. This provides vegetation cover on blank spaces, which further helps in better stabilization of active surface of sand dunes and also provide fodder for livestock.

Table 6.15 Crop straw/grass productivity (kg ha^{-1}) under different spacing of *Ailanthus excelsa* based cropping system

Years	Pearl millet crop			Moth bean crop			*Cenchrus ciliaris* grass		
	6 m × 6 m	8 m × 8 m	10 m × 10 m	6 m × 6 m	8 m × 8 m	10 m × 10 m	6 m × 6 m	8 m × 8 m	10 m × 10 m
2007–08	978	1,008	1,256	300	285	310	552	581	680
2008–09	1,368	1,292	1,665	260	231	267	562	708	890
2009-10[a]	456	508	677	212	227	296	411	507	666
2010–11	3,818	4,154	4,627	639	789	683	1,750	1,821	2,174
Average	1,655	1,740.5	2,056.3	352.8	383	389	818.8	904.3	1,102.5

[a] Completely drought year

Species for Sandy Plains

This type of landform is characterized by deep sandy soils with varying depth (70–150 cm), below which sometimes a Kankar pan is found. Majority of agricultural activities in hot arid regions are practised on this landform. Sandy plains are spread over in entire hot arid regions and cover more than 30 %of land area. In this landform, not much land preparation is required for out planting tree seedlings. For planting out tree seedlings 50 cm × 50 cm × 50 cm pit is excavated and a saucer shape basin of one meter diameter around the pit is made so that more rain water could be available to growing seedlings.

This type of landform occurs throughout the hot arid zone in rainfall gradient of 150–350 mm. The important feature of tree species which have been found suitable for this land form under varying rainfall situation are given in Table 6.18. The species for such landform must have good drought tolerance ability especially during first 2–3 years of growth. Some of these species are being used successfully in intensive agroforestry systems of hot arid regions (Hocking 1993).

Species for Shallow Soil Areas

Such sites are generally found in eastern and southeastern part of arid western Rajasthan and *Kutch* area. They generally consist shallow sandy loam soils having a depth of 30–45 cm. Below this depth a layer of calcareous nodules and/or 'Kankar pan' is found. In such site planting is done in pits of 60cm x 60cm x 60cm size.

The hard calcareous layer is first broken by crowbar to facilitate optimum root growth of tree species. If this operation is not done, the roots of growing trees are coiled once they reach to the depth of 'Kankar pan' and in turn their growth is retarded, ultimately resulting in drying of growing plants.

The shallow soil areas are also found in many pockets of Gujarat, Punjab, and Haryana. Through the introduction, evaluation, and screening of number of tree species for such land

Table 6.16 Canopy expansion and leaf fodder production by way of 70 % vertical lopping of *A. excelsa* trees in the cropping system

Spacing (m × m)	Mean crown cover (m^2 $tree^{-1}$)	Total canopy cover (% ha^{-1})	Green leaf fodder production (kg ha^{-1})
6 × 6	5.21	14.43	69.8
8 × 8	9.71	15.14	155.4
10 × 10	11.55	11.55	184.0

forms, some adaptable and promising tree species have been identified (Table 6.19). The choice of tree species for shallow soil areas is very limited. Only a very few tree species do well in such soils.

Species for Rocky-Gravelly Terrain

Such areas are scattered in pockets throughout the hot arid regions with exception in the east of Jaisalmer, where a large tract of lowly elevated plateau 150–325 m above mean sea level exists with number of rocky masses jutting out of the desert sand. In general, such areas consists barest rock without vegetation which forms small hills, however at the foot hills, shallow soils formed of finer particles and rock fragments are met with. For any plantation activity in such sites, land preparation is a prerequisite. Generally, some staggered counter trenches with a cross-section of 60 cm × 40 cm to minimize water erosion are constructed. Pits of 60 cm × 60 cm × 60 cm are dug out for out planting the seedlings. Good soil imported from other sites is used to refill the pits after out planting the seedlings of desired species. FYM @ 5 kg per pit is mixed with soil used for refilling the pits.

The selection of tree species for such rocky-gravelly terrains is of paramount importance for successful establishment of plantation. Over 40 years of R & D efforts of CAZRI revealed that choice of tree species is very limited for rocky-gravelly sites. *Acacia senegal*, *Prosopis juliflora*, *P. chilensis*, and *Wrightia tinctoria* perform best in such sites, however, species like *Ziziphus nummularia*, *Capparis decidua,* and *Grewia tenax* could also be planted successfully with better protection and management.

Table 6.17 Tree species for sand dune stabilization. In case of growth rate and uses, the higher the number (range 1–10), better it is in growth and higher is the quality for that use, respectively, relative to others

Species	Origin[a]	Growth rate		Utilization age (yrs)	Uses[b]
		Early	Late		
Acacia tortilis	I	7	10	8–35	F(7), Fd(4), T(5), G(4)
A. Senegal	N	3	6	5–20	F(3), Fd(5), Ft(6), G(8)
A. bivenosa	I	7	7	5–12	F(2), Fd(1)
Colophospermum mopane	I	8	7	5–20	F(8), Fd(9)
Dichrostachys nutans	I	5	6	3–25	F(2), Fd(7)
Prosopis alba	I	6	7	8–40	F(7), Fd(8), T(6)
P. chilensis	I	6	8	5–40	F(7), Fd(7), T(4)
P. cineraria	N	2	5	15–80	F(8), Fd(8), Ft(8), G(5)
P. pallida	I	5	8	5–45	F(7), Fd(7), T(7)
P. juliflora	I	8	9	8–50	F(8), Fd(5), T(6)
Hardwickia binata	O	3	7	15–45	F(4), Fd(9), T(8)
Ziziphus nummularia	N	7	6	4–20	F(3), Fd(9), Ft(5)
Z. rotundifolia	N	5	7	6–25	F(4), Fd(9), Ft(6)

[a] *N* Native; *I* Introduce from isoclimatic regions; *O* Introduced from other drier region

[b] *F* Fuel; *Fd* Fodder; *Ft* Fruits; *T* Timber; *G* Gum

Table 6.18 Tree species for sandy plains. In case of growth rate and drought resistant, the higher the number (range 1–10) better it is in growth and the more resistance to drought

Species	Origin	Annual rainfall requirement (mm)		Growth rate		Drough resistance		Main use
		Min	Max	Early	Late	Young tree	Mature tree	
Prosopis cineraria	N	75	700	2	5	8	10	Mp
Capparis decidua	N	100	1,000	3	6	9	10	Ft
Tamarix aphylla	I	100	400	3	6	8	10	F, Fd
Acacia tortilis	I	100	700	10	10	7	10	MP
A. salicina	I	125	550	5	8	6	8	T
Ziziphus nummularia	N	125	1,000	5	7	9	10	MP
Z. rotundifolia	N	125	1,000	5	8	8	10	MP
Z. mauritiana	N	150	2,000	4	6	7	9	MP
Prosopis juliflora	I	150	700	8	9	10	10	F, Fd
P. chilensis	I	150	700	6	8	10	10	F, Fd
P. pallida	I	150	750	5	8	9	9	MP
Acacia senegal	N	200	600	3	6	8	10	MP
Dichrostachys nutans	I	200	700	5	6	7	9	Fd
Acacia holosericea	I	200	1,500	5	8	8	8	F, Fd
Grewia tenax	N	200	800	4	7	5	7	Fd
Eucalyptus camaldulensis	I	250	2,000	4	8	7	8	F, MP
Hardwickia binata	O	250	1,500	3	7	6	8	MP
Acacia albida	I	300	1,500	2	6	6	8	Fd
Albizia lebbeck	N	300	2,250	7	9	8	8	MP
Azadirachta indica	N	300	1,100	5	8	7	8	MP
Cassia siamea	N	300	700	5	9	6	7	MP
Ailanthus excelsa	N	350	800	4	8	5	8	Fd, T
Emblica officinalis	N	350	2,500	3	6	5	8	Ft

N Native; *I* Introduced from isoclimatic region; *O* Introduced from other drier region; *MP* Multipurpose; *Ft* Fruits; *F* Fuel; *Fd* Fodder; *T* Timber

Table 6.19 MPTS which are found to grow satisfactorily in different rainfall zones

Category (<250 mm rainfall zone)	Category II (>250 mm rainfall zone)
Acacia salicina (I)	*Azadirachta indica* (N
Hardwickia binata (O)	*Dichrostachys nutans* (I)
Prosopis juliflora (I)	*Hardwickia binata* (O)
Prosopis alba (I)	*Prosopis juliflora* (I)
Grewia tenax (N)	*Acacia nilotica* (N)
Ziziphus nummularia (N)	*A. leucophloea* (N)
Cordia gharaf (I)	*Ziziphus mauritiana* (I)
Capparis decidua (N)	*Holoptelea integrifolia* (O)
Acacia jacquemontii (N)	*Eucalyptus camaldulensis* (I)
Acacia senegal (N)	*Pongamia pinnata* (N)

N Native; *I* Introduced from isoclimatic regions; *O* Introduced from other drier regions

Table 6.20 Important MPTs identified for saline and waterlogged areas in arid regions

Alkali soils	Saline soils	Bio-drainage species
Prosopis juliflora (I)	*Prosopis juliflora* (I)	*Eucalyptus camaldulensis* (I)
P. alba (I)	*Prosopis chilensis* (I)	*E. tereticornis* (I)
Terminalia arjuna (N)	*Azadirachta indica* (N)	*Acacia auriculaeformis* (I)
Dalbergia sissoo (N)	*Pithecellobium dulce* (I)	*Acacia ampliceps* (I)
Pongamia pinnata (N)	*Acacia nilotica* (N)	*Casuarina glauca* (I)
Acacia nilotica (I)	*Acacia tortilis* (I)	*Syzygium cuminii* (N)
Tamarix articulata (N)	*Salvadora persica* (N)	*Albizia lebbeck* (N)
Capparis decidua (N)	*Tamarix* species (N)	*Dalbergia sissoo* (N)

N Native; *I* Introduced from iso-climatic regions
For detailed see separate chapter in this publication

Species for Salt-Affected Soils

About 3.0 million ha area in hot arid region of India is suffering from soil salinity and brackish water problems. In such areas upper layers of soils often contain mainly silt and clay. The reclamation of such sites though is a costly proposition but not impossible. The soils containing soluble salts in root zone in quantities large enough to adversely affect plant growth are called salt-affected. Visually, they are recognized by the presence of a white or grayish white efflorescence of salts on the surface during dry months. Commonly they are devoid of good natural vegetation cover. Broadly, these soils are classified into two groups, alkali soils and saline soils (Yadav and Singh 2000). The detailed characteristics of these soils have been dealt in this volume in Chap. 9 by JC Dagar.

The promising MPTs which have been identified through long-term evaluation and screening for salty lands in arid regions are given in Table 6.20. More details are in Chap. 9.

Agroforestry for Livelihood Security

Agroforestry contributes to livelihood improvement in arid regions of India, where people have a long history of accumulated local knowledge. Arid regions are particularly notable for ethanoforestry practices and indigenous knowledge system on growing trees on farm lands. For example, the economics of previously discussed traditional agroforestry systems of upper transect arid western Rajasthan indicated that net Benefit: Cost (B:C) ratio of such systems is generally on positive sides (Table 6.21).

In fact, multipurpose trees in agroforestry systems provide a rational formula for drylands. In hot arid regions trees on farmlands are the organisms with an ecological investment strategy (Oldeman 1983). Matter and energy are immobilized in long-lived trees and majority of this accumulated production is harvested at longer interval than in case of arable crops.

Jujube (*Ber*) is a very useful agroforestry tree of arid regions. The leaves make excellent

Table 6.21 Economics of some traditional agroforestry systems of arid western Rajasthan

AF system	Expenditure (INR ha^{-1})	Returns (INR ha^{-1})			Gross returns (INR ha^{-1})	Net Returns (INR ha^{-1})	Net B:C ratio
		Crops	Fuel wood	Leaf fodder			
P. cineraria-A. nilotica based	1,850	4,103	1,230	870	6,203	4,353	2.3
P. cineraria based	1,550	3,670	600	420	4,690	3,140	2.0
Ziziphus. spp - *P. cineraria* based	1,550	1,506	620	600	2,726	1,176	0.7
Ziziphus spp - *P. cineraria* - *Salvadora* spp based	1,500	1,400	500	500	2,400	900	0.6

INR = Indian National Rupee (1 US$ = 54.34 INR in January, 2013)

Table 6.22 Economic returns (in terms of gross income) from a Jujube-based pastoral system

Particulars	Production (t ha^{-1})	Returns INR ha^{-1})
Dry grass	1.55	3,100
Fuel wood	2.64	2,640
Leaf fodder	1.87	7,480
Fruit	2.77	33,240
Total	–	46,640

fodder, crude protein 13–17 %, and fiber 15 %. Leaves are also food for silkworms. The branches and twigs are excellent fuel wood having calorific value 4,878 kcal kg^{-1}. Root bark and leaves have 7 % and 2 % tannin, respectively and are sometimes mixed with other sources for tanning leather. The fruits provided by all the improved cultivars are delicious and rich source of vitamin A, C, and B complex. Thus, cultivation of Jujube in arid regions offers solution to many problems. Tewari et al. (2001) demonstrated that agroforestry, particularly improved Jujube-based pasture system can be major source of livelihood in many parts of arid regions (Table 6.22). This system provides production in terms of human food, livestock feed, and cooking energy needs to communities. Such multifunctional agroforestry systems provide social wellbeing to the people in addition to goods and services.

The most spectacular evidence in context of livelihood improvement potential of agroforestry in arid region came through *Acacia senegal*-based localized traditional agroforestry system in parts of Barmer and Jodhpur districts (15 villages). *A. senegal* is source of gum Arabic which has very high commercial value. In addition to harvesting crop grain for food and crop straw for fodder, the farmers harvested substantial quantity of gum Arabic through small intervention using gum inducer provided by CAZRI. The data regarding production of gum and revenue earned by the farmers in past 3 years are given in Fig. 6.6. CAZRI gum inducer has been so popularized in the villages that farmers have made their own cooperatives for purchasing of gum inducer and sale of produce in the market.

During the year 2008–2009, 12,000 trees were treated through CAZRI gum inducer which resulted in exudation of 5.4 tons of gum Arabic and farmers earned Rs 2.7 million from the sale of gum. Subsequently in the year 2009–2010 and 2010–2011, 21,000 and 22,600 trees were treated, respectively resulting in production of 10.5 tons and 7.6 tons gum, respectively. In the process farmers earned INRs 5.2 million in the years 2009–2010 and Rs 3.8 million in the year 2010–2011. In this way, the farmers of 15 villages earned INRs 11.7 million from the sale of gum Arabic with the average modest sale price INRs 500 kg^{-1} in local market though at present, the rate of pure gum Arabic in the market is more than INRs 1,000 per kg.

Agroforestry in Adapting Climate Change in Arid Regions

Considering the role of agriculture in the social and economic progress of developing countries, the vulnerability of agricultural systems to the impacts of climate change has received considerable attention from the scientific community (Fischer et al. 2002; IISD 2003; Kurukulasuriya and Rosenthal 2003). Much of the available literature suggests that the overall impacts of climate change on agriculture especially in the tropics will be highly negative, although in a few areas there may be minor increases in crop yields in the short term (Maddison et al. 2007). Table 6.23 presents some of the projected changes in climate and their potential impacts on agriculture as summarized in the IPCC report (Parry et al. 2007).

Agroforestry, the integration of trees and shrubs with annual crops production, is an age old management system practised by farmers in arid regions of India. The system provides shade, a steady supply of food and/or income throughout the year, arrests degradation and maintains soil fertility, diversifies income sources, increases and stabilizes income, enhances use efficiency of soil nutrients, water and radiation, and provides regular employment.

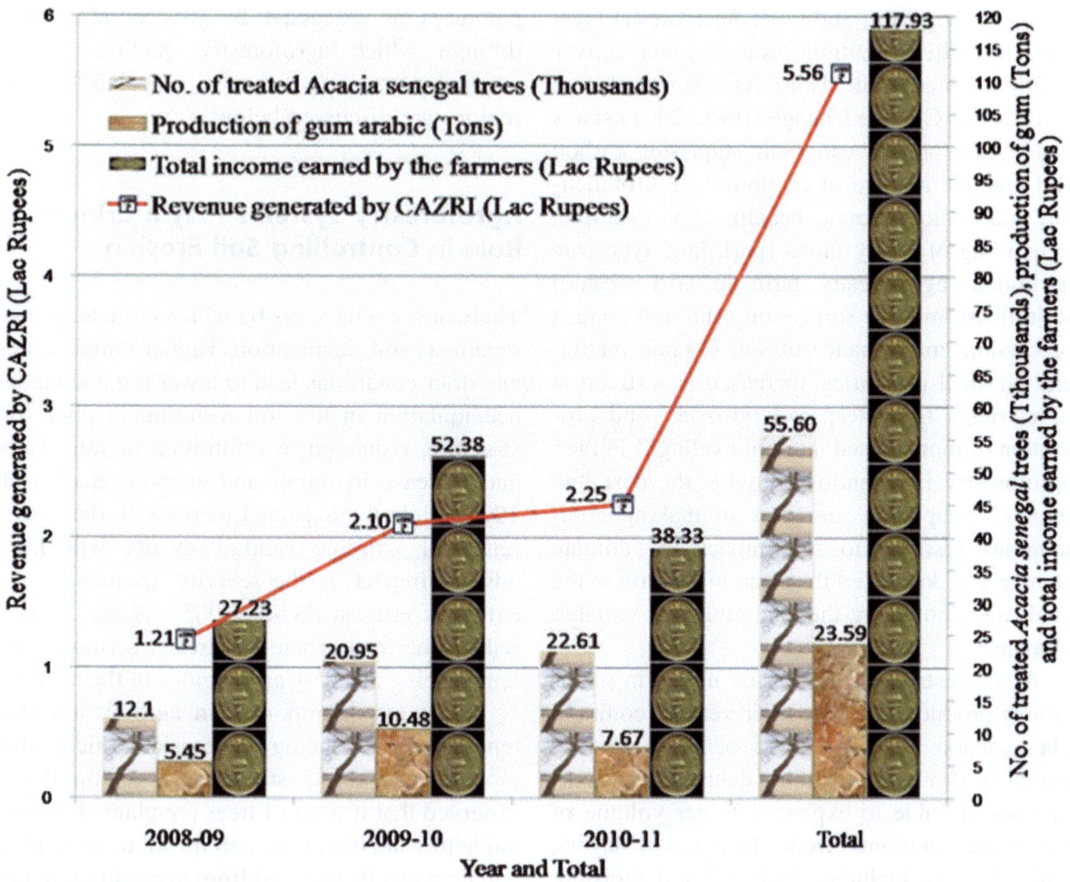

Fig. 6.6 Gum arabic production in certain villages of western Rajasthan (mainly Barmer and Jodhpur districts) INRs 1 lac = INRs 0.1 million

Table 6.23 Projected changes in climate and their impact on agriculture

Phenomena and direction of change	Likelihood of occurrence	Major projected impacts on agriculture
Warmer and fewer cold days and nights; warmer/more frequent hot days and nights over most land areas	Virtually certain (>99 % chance)	Increased yields in colder environments; decreased yields in warmer environments; increased insect outbreaks
Warm spells/heat waves: frequency increases over most land areas	Very likely (90-99 % chance)	Reduced yields in warmer regions due to heat stress; wild fire danger increase
Heavy precipitation events: frequency increases over most areas	Very likely	Damage to crops; soil erosion, inability to cultivate land due to water logging of soils
Area affected by drought: increases	Likely (66–90 % chance)	Land degradation, lower yields/crop damage and failure; increased livestock deaths; increased risk of wildfire
Intense tropical cyclone activity increases	Likely	Damage to crops; wind throw (uprooting) of trees; damage to coral reefs
Increased incidence of extreme high sea level (excludes Tsunamis)	Likely	Salinization of irrigation water, estuaries and freshwater systems

Recognizing the ability of agroforestry systems to address multiple problems and deliver multiple benefits, the Third Assessment Report of IPCC on Climate Change (IPCC 2001) states that "Agroforestry can both sequester carbon and produce a range of economic, environmental, and socioeconomic benefits. For example, trees in agroforestry farms (park land type true traditional agroforestry farm in arid western Rajasthan) improve soil fertility through control of erosion, maintenance of soil organic matter, and physical properties, increased N, extraction of nutrients from deep soil horizons, and promotion of more closed nutrient cycling." In fact, agroforestry interventions provide the best "no regrets" adaptation measures in making communities resilient to the impacts of climate change and do discuss the same in relation to the challenges posed by the changing and variable climate.

Agroforestry systems can be useful in maintaining production during drier years, a common phenomenon in arid regions of India. During complete drought situations, deep root systems of trees are able to explore a larger volume of water and nutrients from deeper soil layers, which help to maintain depleting soil moisture conditions to some extent. In drought prone environment of arid western Rajasthan, as a risk aversion and cooping strategy, the traditional agroforestry systems avoid long-term vulnerability as trees act as an insurance against drought, insect-pests outbreaks and other threats, instead of a yield-maximizing strategy aiming at short-term monetary benefits (Rathore 2004).

Adaptation to climate change is now inevitable. Research on agroforestry as an adaptation to climate change and as a buffer against climate variability is in the process of evolving. Many pathways as discussed by Roy et al. (2011) through which agroforestry qualifies as an adaptation to climate change, especially in arid regions are discussed below.

Agroforestry Systems Play a Critical Role in Controlling Soil Erosion

There are several ways by which climate change manifests soil degradation. Higher temperatures and drier conditions lead to lower organic matter accumulation in the soil resulting in poor soil structure, reduction in infiltration of rain water and increase in runoff and erosion (Rao et al. 1998) while the expected increase in the occurrence of extreme rainfall events will have adverse impact on the severity, frequency, and extent of erosion (WMO 2005). These changes will further exacerbate an already serious problem is being faced by arid regions of the country.

The woody component in agroforestry systems helps in reducing soil erosion which is the most harmful abiotic stress in arid region. It was observed that if rows of trees are planted at right angle to wind direction, a tremendous amount of soil drop can be checked from agricultural fields. The shelterbelts on agricultural fields form a type of agroforestry practice (Gupta et al. 1984) which is quite effective in controlling the soil erosion in arid regions. It is clear from the data (Table 6.24) that the soil loss was higher in agricultural field without trees as compared to field with agroforestry practices (shelterbelts).

Further, higher soil loss during Year II as compared to Year I was due to high wind speed and relatively more windy days during Year II. There was heavy flood (unprecedented) during Year I resulting in severe soil erosion and

Table 6.24 Effect of different type of shelterbelts on soil erosion

Type of shelterbelt	Total amount of soil loss (kg ha^{-1})		
	Year I	Year II	Mean
Prosopis juliflora	93.2	609.3	351.2
Cassia siamea	91.5	277.1	184.2
Acacia tortilis	106.0	494.1	300.0
Agricultural field without trees	262.7	831.0	546.8

loosening of soil in larger part, and perhaps this may also contributed to higher soil loss in Year II during the summer season when the study was conducted.

Agroforestry Systems Have Inbuilt Capacity to Enhance the Water Use Efficiency of Rainwater

Water is already a scarce resource and climate change is expected to make the situation worse. Climate change has both direct and indirect impacts on water availability. The direct impacts include changes in precipitation patterns, while the indirect ones are increases in losses through runoff and evapotranspiration.

There are several mechanisms whereby agroforestry may use available water more effectively than the annual crops. First, unlike in annual systems where the land lies bare for extended periods, agroforestry systems with a perennial tree component can make use of the water remaining in the soil after harvest and the rainfall received outside the crop season. Second, agroforests increase the productivity of rain water by capturing a larger proportion of the annual rainfall by reducing the runoff and by using the water stored in deep layers. Third, the changes in microclimate (lower air temperature, wind speed, and saturation deficit of crops) reduce the evaporative demand and make more water available for transpiration.

The tree canopies in agroforestry systems intercept the rain and reduce runoff (Khan et al. 1995). In a study at CAZRI, it was found that in *Acacia tortilis* silvi-pasture stand canopy interception was 21.4 %, whereas in *Colophospermum mopane* silvi-pasture it was 13.1 %. Rainfall interception was positively related with canopy cover and negatively related with throughfall. Average surface run off in *A. tortilis*-based silvi-pasture stand was 53 % higher than in *C. mopane*-based silvi-pasture. This indicated that hydraulic response to rain is dominated by plant species character, however, the percent annual runoff and soil erosion was very low in situations with trees on agricultural fields in comparison to bare soil condition. Thus, the enhanced use efficiency of rain water by woody species in agroforestry systems improves agricultural productivity.

Agroforestry Systems Provide Economically Viable and Environment Friendly Means to Improve Soil Fertility

Nutrient mining from continuous cropping without adequately fertilizing or fallowing the land is often cited as the main constraint to productivity increase in arid regions. It is observed that on average soils in Indian arid regions have poor nitrogen and phosphorus content. Organic carbon status in certain parts of arid regions has been depleted to an alarming state. While fertilizers offer an easy way to replenish the soil fertility, at the current prices it is very unlikely that there will be any change in the investments made by farmers of arid region in fertilizers. In this context, agroforestry systems have attracted considerable attention as an attractive and sustainable pathway to improve soil fertility. CAZRI, Jodhpur made substantial progress in the identification and promotion of agroforestry systems aimed at improving soil fertility.

Soil fertility under the tree species (*Prosopis cineraria* and *Acacia nilotica*) and in open field conditions was studied intensively (Singh and Lal 1969). In general, organic carbon content decreases with depth but the same was slightly higher under the canopy of both the tree species than in open field condition. The total nitrogen, available P and K were higher under the canopy of *P. cineraria* followed by *A. nilotica* and the values for the same were lowest in open field condition. The soil fertility build up under the 14–year-old stands of *P. cineraria* and *P. juliflora* were found to be of higher order as compared to open field condition. Similarly, status of micronutrients was also higher under tree canopies than in the open field (Table 6.25).

Table 6.25 Available macro and micronutrients under different tree-based agroforestry systems

Tree species	Macronutrients (kg ha^{-1})			Micronutrients (ppm)			
	N	P	K	Zn	Mn	Cu	Fe
Prosopis juliflora	231	7	333	0.89	9.3	0.58	3.3
Prosopis cineraria	221	11	479	1.44	10.8	0..89	2.8
Open field	199	6	3.2	0.19	7.0	0.38	3.5

Agroforestry Systems Play a Very Important Role in Moderating the Micro-climate in Arid Regions

The full genetic potential of many crops and varieties can only be realized when environmental conditions are close to optimum. Trees on farm bring about favorable changes in the microclimatic conditions by influencing radiation flux, air temperature, wind speed, saturation deficit of understorey crops all of which will have a significant impact on modifying the rate and duration of photosynthesis and subsequent plant growth, transpiration, and soil water use (Monteith et al. 1991). Following examples present microclimatic moderation effect in harsh climatic conditions of arid regions of western Rajasthan.

Air temperature exhibited appreciable variation under the canopy of *A. tortilis* during monsoon period (Tewari et al. 1989). A decline of 0.7 °C was evident beneath the canopy cover of *A. tortilis* than that of surrounding open areas at 07.00 h. Around 14.00 h, this decline was of the order of 2.0 °C. Similarly, it was found during monsoon season that the soil temperature just beneath the tree cover was lower by as much as 16 °C in top soil zone and 5 °C at 30 cm depth when compared to open field conditions, thereby indicating better soil-thermal regime. Seven-years-old *A. tortilis* trees provided more humidity on cluster bean grown with the trees. This finding proves the efficiency of agroforestry practices for maintaining favorable moisture status on arable lands (Ramakrishna and Shastri 1977).

Agroforestry Systems Improve Biodiversity

Continued deforestation is a major challenge for forests and livelihoods. Although agroforestry may not entirely reduce deforestations, particularly in arid regions, it may act as an effective buffer to deforestation. Trees in agro-ecosystems in Rajasthan have been found to support threatened cavity-nesting birds and offer forage and habitat to many other bird species. Biodiversity conservation may not be a primary goal of agroforestry systems. Nevertheless, in some cases traditional agroforestry systems in arid western Rajasthan were found to support very good species diversity and also act as a buffer to parks and protected areas (Pandey 2007).

In a study at Govardhanpura and Gokulpura cluster of villages in Bundi district having relatively little better rainfall regime, where a highly degraded silvi-pasture (45 ha) was revegetated and fenced to minimize abiotic and biotic disturbance, exhibited marked difference in biodiversity of plant species (Dixit et al. 2005). The area was divided into six different blocks. Though initially only five woody and some grass species were introduced with some water conservation structures in the area, after 5 years the fortified area with physical and social fencing exhibited the woody species richness to the tune of 20 species with a Beta (between habitat) diversity of 2.2. Maximum number of woody species (25 %) belonged to family Leguminosae/Mimosoideae. Rests of woody species were distributed among 11 other families/subfamilies. The herbaceous vegetation richness increased to

36 (number of species) of which 80 %were palatable. These 36 herbaceous species were distributed among 24 families. Cyperaceae family was represented by maximum percentage of the species. The Beta diversity of herbaceous species was 2.32 with index of general diversity of 3.437. The value of general diversity for woody and herbaceous species was very very low in adjoining unprotected area.

Dominance-diversity (d–d) curves were developed for herbaceous vegetation on the basis of relative biomass values for all the blocks and also for open degraded land. With exception of open degraded land site, the d–d curves on all the blocks exhibited more or less log normal distribution. This indicated that several herbaceous species were sharing relatively low importance values (biomass) or similar range of importance values. The log normal or curves approaching to log normal, in fact, indicate relatively stable population. The d–d curve of open degraded land site was closer to geometric series. Whittakar (1972) opined that communities having low diversity exhibit geometric series. In open degraded land ruthless exploitation of herbaceous vegetation has resulted in tremendous loss of biodiversity.

Success Stories

Transforming Subsistence Level, Rural Economy Through *ber* (*Ziziphus mauritiana*) Based Diversified Agroforestry System by Mr. Nand Kishor Jaisalmeria

In environmentally inhospitable hot arid tract, Shri N. K. Jaisalmeria of village Manaklao, located 25 km north of Jodhpur by adopting CAZRI technology package transformed the nature of undulating sandy unproductive landscape of 4.5 ha into a highly productive diversified agroforestry system is perhaps the best success story of CAZRI. Arable farming is by far the dominant pursuit of this densely populated principal arid zone of India. Though experience over the generations, the farmers have evolved a mixed livestock-crop-tree farming system. However, land productivity is low and permits no more than a subsistence living in most years and severe shortage of food and fodder in subnormal rainfall years such as occur in 2 out of 5 years on an average. Whereas for example in a good rainfall year, a hectare of land produces 500 kg of grains and 1.2 tons of fodder, a poor year gives hardly any grain and just 200–300 kg of fodder. Assets and cash reserves of the rural population are negligible. Serous indebtedness follows and despite governmental support, under nourishment of human population and livestock mortality is a common feature. Moreover, arable farming is somewhat over exploitative of land resources. With these considerations, Mr. Jaisalmeria started cultivation of *ber* fruits on his land with the technical assistance of CAZRI, Jodhpur.

He first established fairly large *ber* orchards with 750 trees under rainfed condition and second a *ber* nursery was established by him for supply of reliable budded *ber* plants. The periphery of entire 4.5 ha area was covered by MPTs like *A. tortilis, P. juliflora* and other woody species, which formed a very good shelter belt within 6 years. Superior biomass producing trees like *P. cineraria, Tecomella undulata* and *Azadirachta indica* growing within the orchards are protected. *Cenchrus ciliaris* grass has been planted between *ber* fruit trees. One hectare area is left to grow pearl millet which provided grain for human consumption and straw for livestock consumption. Eight goats are reared within the system without purchasing any kind of fodder. Farm waste is used to prepare compost within the system. The component and functioning of this diversified agroforestry system of Mr. Jaisalmeria is given in Fig. 6.7.

The system is producing *ber* fruits to the tune of 10.7t ha^{-1}, fuel wood (dry) 4t ha^{-1}, and fodder about 2.4 t ha^{-1} $year^{-1}$. This system is capable of providing employment to the tune of 331 man days per ha per year. The system is self-contained and the efforts of Mr. Jaisalmeria have been recognized in national and international forums. He was invited to Southeast Asia UNCCD head quarter at Bangkok, Thailand to

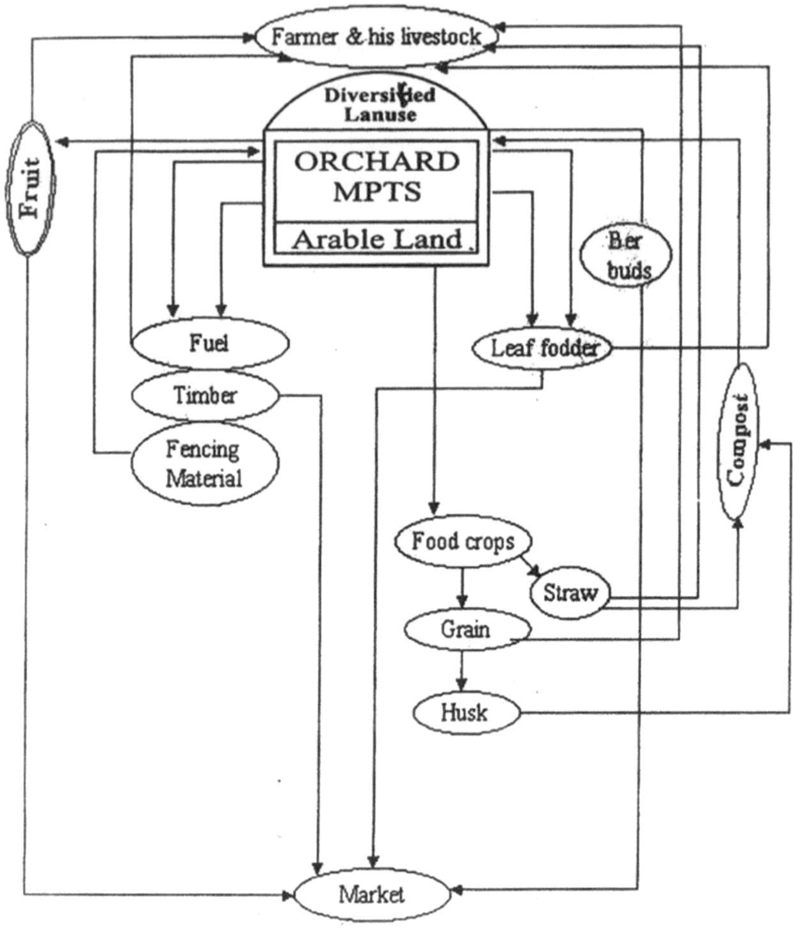

Fig. 6.7 Diversified agroforestry system developed by Mr. N. K. Jaisalmeria at village—Manaklao near Jodhpur: A schematic representation

present his success story of dry land development during 2001. In the year 2003, he was awarded Krishi Shiromani Award of Ministry of Agriculture, Govt. of India and N. G. Ranga Award of ICAR for diversified agriculture during the year 2005.

The success of Mr. Jaisalmeria's efforts can be better judged on the basis of following important issues:

Land use issue: The program was implemented on a sandy undulating land of hot arid zone having negligible productivity of rainfed crops. His efforts demonstrates how such unproductive lands in the desert areas can be transformed into an efficient production system, which in fact, is a sustainable land use in hot arid zone. This integrate land use having *ber* fruit tree plantation protected by shelterbelts of MPTs like *A. tortilis* and *P. juliflora* coupled with utilization of interspace by planting other drought hardy plant species of economic value not only provides sustainable production but also leads to soil conservation and improvement in the fertility level.

Social and economic issues: The land which was once not providing an income of Indian Rs 500/– (US$ 10) per ha per year is returning (net income) an average Indian INR 20,000 (US$ 400) per year. The cost: benefit ratio is 1:5.5, which is economically quite sound. Though his effort, the technological package to utilize such unproductive arid lands in an integrated manner of perennial crop-based system has been extended to 90,000 ha area from extreme north-west part of the country to southern states. At his farm the employment generation is in a tune of 1,200 man days per ha per year. Considering the same rate of man days requirement, it is estimated that these plantations are providing employment generation in order of 1.08 million man days per year in the country. The technological package is very simple and acceptable to the farmers with various levels of family economies, i.e., poor to rich ones. In country like India, especially in rural sector of arid areas, where huge labor force facing acute unemployment problems, the employment generation through *ber* based diversified agroforestry system could play vital role in improving the rural economy.

Social capital issue: A very high order of adoption of said technological package in hot arid areas of the country has been changed the economies of many poor farmers. Due to adoption of integrated *ber*-based diversified land use, the lands which have previously very poor or of no value, is now being considered worth million of Indian Rupees. This has direct impact on social status of the farmers in traditionally complex socioeconomic-cultural web of rural life of India, who have successfully adopted the technological package. Moreover, a high rate of employment generation in environmental inhospitable arid tracts, especially for women folk, as they are considered most efficient in harvesting and grading the *ber* fruits is playing significant role in the women empowerment

Policy-related issues: Once the success of his enterprise attracted the attention of farmers, subject matter specialists and policy makers, Horticulture departments of various state governments, especially those ones having area under hot arid zone, started giving serious thoughts to *ber* crop and in due course of time *ber* has been recognized as one of the important fruit crops. To promote the *ber* plantations, provision of subsidies and loans have been made by various state governments. Now many banks like National Bank for Agriculture and Rural Development, Gramin (village) banks, etc., and National Horticulture Mission have also made the provisions for short-term and long-term loans for growing *ber* based diversified production system.

Research Gaps and the Way Forward

Most of the land in arid regions has been under constant cultivation since past three decades regardless of its marginality. In many cases the farmer can hardly get back his seeds from the field he cultivates. In fact, in arid regions of India continuous cultivations and irrational utilization of resources under unbalanced man/resource ratio on the one hand and their inherently fragile nature on the other, have given rise to a critical deterioration of environment. Therefore, in arid tract combined protective-productive system and indirectly integrated system will be most suitable for enhancing tree cover in the landscape. Sand dune stabilization around cultivated sandy plains and tree-based pastoral system in village community lands, with proper management practices are the best examples of combined protective-productive system. In fact, combined protective-productive systems become much more important in critical areas, where they act as buffer strips to protect fringes of good lands for encroachment of desert conditions (Eren 1995). Increasing tree cover through farm boundary plantation of MPTs, plantation along roads, railway lines and canal, home gardens are some of the examples of indirectly integrated system. The above discussion clearly indicated that only integrated land use system could be a remedy to reverse the ruthless exploitation of woody biomass in one hand and will be helpful to restore the much required tree cover on the other.

The restorative forestry cannot be considered as the user of "residual land" and this is particularly true for arid land situation of western Rajasthan where even more than 70 % of sand dunes are farmers' owned. Faroda (1997) rightly pointed out that land tenure giving security to land owners would encourage long-term investment by the farmers. In any kind of plantation forestry and agroforestry program, component of tree tenure (i.e., right to own or inherit the trees, the right to use trees and tree products, and right to dispose tree) if included and followed in true spirit for larger interest of rural masses, the tree cover could be improved substantially in hospitable environmental condition of hot arid zone. Moreover, traditional agroforestry practised in entire arid western Rajasthan and other northwestern parts of country must be recognized as agroforest land use, as till date, such agroforests are classified as agriculture land use in revenue records. This will give a new dimension to agroforestry in principal hot arid region of the country.

References

Bhandari MM (1978) Flora of Indian desert. Scientific Publishers, Jodhpur, p 471

Bhimya CP, Kaul RN, Ganguli BN, Tyagi IS, Choudhari MD, Subbayyan (1964) Species suitable for afforestation of different habitats of Rajasthan. Annals of Arid Zone 2(2):162–168

Boffa JM (1999) Agroforestry parklands in sub-saharan Africa. Food and agriculture organization of the United Nations, Rome, p 230

CAZRI (2008) Annual progress report. central arid zone research institute (CAZRI). Jodhpur, India, p 166

Champion HG, Seth SK (1968) A revised survey of forest types of India. manager of publications. Govt. of India, New Delhi, p 404

Dhir RP (1982) The human factors in ecological history. In: desertification and development: dryland ecology in social perspective. Academic (in press), New York, pp 311–331

Dixit S, Tewari JC, Wani SP, Vineela C, Shaurasia AK, Panchal HB (2005) Participatory biodiversity assessment: enabling rural poor for better atural Resource Management. ICRISAT. Patancheru, A.P., p 16

Hocking D (1993) Trees for drylands. Oxford and IBH Publishing Co., Pvt., Ltd., New Delhi, pp 370

Eren T (1995) Agroforestry in arid and semi-arid zones. In: Sand Dune Stabilization, Shelterbelts and Afforestation in Dry Zones. FAO Conservation Guide 10, FAO, Rome, pp 169-174

Faroda AS (1997) Silvi-pastoral approach for arid and semi-arid areas: issues and strategies. In: Yadav MS, Singh M, Sharma SK, Tewari JC, Burman U (eds) Silvi-pastoral system in arid and semi-arid ecosystems (UNESCO: ICAR). CAZRI, Jodhpur, pp 1–10

Fischer G, Shah M, Van VH (2002) Climate change and agricultural vulnerability. International institute for applied systems analysis. Report prepared under UN institutional contract agreement 1113 for World summit on sustainable development. Laxenburg, Austria

Gupta JP (1997) Some alternative production systems and their management for sustainability. In: Gupta JP, Sharma BM (eds) Agroforestry for sustained productivity in arid regions. Scientific publishers, Jodhpur, pp 31–39

Gupta JP, Rao GGSN, Ramakrishna YS, Rao Ramana (1984) Role of shelterbelts in arid zone. Indian Farming 34(7):29–30

Harsh LN, Tewari JC (1993) Sand dune stabilization, shelterbelts and silvi-pastoral plantation in dry zones. In: Sen AK, Kar Amal (eds) Desertification in Thar, Sahara and Sahel Regions. Scientific Publishers, Jodhpur, pp 269-279

Harsh LN, Tewari JC, Burman U, Sharma SK (1992) Agroforestry in arid region. Indian Farming (Special issue on environmentally sound biotechnology) 45:32–37

IISD (2003) Livelihoods and climate change: combining disaster risk reduction, natural resource management and climate change. Adaptation in a new approach to the reduction of vulnerability and poverty, a conceptual framework paper prepared by the task force on climate change, vulnerable communities and adaptation

IPCC (2000) Special Report on Land Use. Land Use Change and Forestry. Summary for Policy Makers. Geneva, Switzerland, p 20

IPCC (2001) Climate Change 2001: The Scientific Basis. Contribution of the Working Group 1 to the Third Assessment Report of the IPCC. Cambridge, University (In Press) Cambridge

Kaul RN, Ganguli BN (1964) Afforestation studies in the arid zone of India. In Proceedings of symposium on problems of Indian arid zone. Ministry of Education, Goverment. of India, New Delhi, pp 183–187

Kaushik RC, Kaul RN, Subrahmanyam VP (1969) Homoclimes and introduction of trees and shrub species with special reference to the arid and semi-arid zones of India. Indian Forester 95(5):284–292

Khan MA, Tewari JC, Issac VC (1995) Hydrology of small forested catchments in the arid region of Rajasthan. Ann Arid Zone 34(4):259–262

Kurukulasuriya P, Rosenthal S (2003) Climate change and agriculture: a review of impacts and adaptations. climate change series paper 91. World Bank, Washington, p 106

Maddison D, Manley M, Kurukulasuriya P (2007) The impact of climate change on African agriculture: a ricardian approach. World Bank Policy Research Working Paper 4306

Malhotra SP, Trivedi HS, Mathur YN (1986) Agroforestry: a judicious use of desert ecosystem by man. CAZRI Publication No.27, Central Arid Zone Research Institute, Jodhpur

Mann HS, Muthana KD (1984) Arid Zone Forestry, CAZRI Monograph No. 23. CAZRI, Jodhpur, p 48

Monteith JL, Ong CK, Corlett JE (1991) Microclimatic interactions in agroforestry systems. For Ecol Manage 45:31–44

Oldeman RAA (1983) The design of ecologically sound agroforests. In: Huxley PA (ed) Plant research and agroforestry. International Council for Research in Agroforestry (ICRAF), Nairobi, Kenya, pp 173–207

Pandey DN (2007) Multifunctional agroforestry systems in India. Curr Sci 92(4):455–463

Parry ML, Canziani OF, Palutikof JP, van der Linden PJ, Hanson CE (2007) Climate change 2007: impacts, adaptation and vulnerability. contribution of working group II to the fourth assessment report of the intergovernmental panel on climate change. Cambridge University, Cambridge, p 1000

Pratap Narain, Tewari JC (2005) Trees on agricultural fields: a unique basis of life support in Thar Desert. In: Tewari VP, Srivastava RL (eds) Multipurpose trees in the tropics: management and improvement strategies. Arid Forest Res Inst, Jodhpur, pp 516–523

Ramakrishna YS, Shastri ASRAS (1977) Microclimate under *Acacia tortilis* plantation. Annual Progress Report, CAZRI, pp 69–70

Rao KPC, Steenhuis TS, Cogle AL, Srinivasan ST, Yule DF, Smith GD (1998) Rainfall infiltration and runoff from an alfisol in semi-arid tropical India. II. Tilled sys. Soil and Tillage Res 48:61–69

Rathore JS (2004) Drought and household coping strategies: a case of Rajasthan. Indian J Agri Econ 59(4):689–708

Roy MM, Tewari JC, Ram Moola (2011) Agroforestry for climate change adaptations and livelihood improvement in Indian hot arid regions. Int J Agri Crop Sci 3(2):43–54

Satyanarayan Y (1964) Habitats and plant communities of the Indian desert. In Proceedings of the Symposium on Problem of Indian Arid Zone. Govertment of India, Ministry of Education, New Delhi, pp 59–67

Saxena SK (1997) Traditional agroforestry systems of western Rajasthan. In: Sharma BM (ed) Gupta JP. Agroforestry for Sustained Production in Arid Regions Scientific Publishers, India, pp 21–30

Sharma AK, Tewari JC (2001) Arid zone forestry with special reference to Indian hot arid zone. In: Encyclopaedia of life support system (ELOSS), UNESCO, paris

Singh S (1977) Geomorphological investigations of Rajasthan desert. CAZRI monograph no. 7. CAZRI, Jodhpur, p 44

Singh KS, Lal P (1969) Effect of *Prosopis cineraria* and *Acacia nilotica* on soil fertility and profile characteristics. Ann Arid Zone 8:33–36

Suresh Kumar (1998) Vegetation resources in the Indian arid zone. In: Faroda AS, Manjit S (eds) Fifty Years of Arid Zone Research in India Central Arid Zone Research Institute, Jodhpur, pp 89–119

Tewari JC, (1997) Important fodder trees of arid zone and their management. In: Gupta JP, Sharma BM (eds) Agroforestry for Sustained Productivity, Arid Regions Scientific Publishers, Jodhpur, pp 147–153

Tewari JC, Harsh LN (1998) Forestry research in arid tract of India. In : Faroda AS, Manjit S (eds) Fifty Years of Arid Zone Research in India, Central Arid Zone Research Institute, Jodhpur, pp 307–322

Tewari JC, Harsh LN, Venkateshwarlu J (1989). Agroforestry research in arid regions - A review. In: Singh RP, Ahlawat IPS, Saran Ganga (eds) Agroforestry Systems in India: Research and Development, Indian Society of Agroforestry, IARI, New Delhi, pp 3-17

Tewari JC, Harsh LN, Venkateshwarlu J (1993) Some aspects of plantation forestry research in western Rajasthan. In: Dwivedi AP, Gupta GN (eds) Afforestation of arid lands. Scientific Publishers, Jodhpur, India, pp 62–72

Tewari JC, Bohra MD, Harsh LN (1999) Structure and production function of traditional extensive agroforestry systems and scope of agoroforestry in *Thar* desert. Indian J Agrofor 1(1):81–94

Tewari JC, Tripathi D, Narain Pratap (2001) Jujube : A multipurpose tree crop for arid land farming systems. The Botanica 51:121–126

Tod Col J (1832) Annals of Antiquites of Rajasthan (2 volumes). Routledge and Kegan Paul, London

UNEP (1992). World Atlas of Desertification. UNEP, Edward Arnold (A Division of Hodder & Stoughton), London–New York–Melbourne–Auckland, pp 69

Whittakar RH (1972) Evolution and measurement of species diversity. Taxon 28:213–251

WMO (2005) Climate and Land Degradation. WMO-No. 989 2005, World Meteorological Organization

Yadav JSP, Singh GB (2000) Natural resource management for agricultural production in india. International conference on managing natural resources for sustainable agricultural production in 21st century. Indian Society of Soil Science, New Delhi, p 1022

Agroforestry: A Way Forward for Sustaining Fragile Coastal and Island Agro-Ecosystems

7

J. C. Dagar, C. B. Pandey, and C. S. Chaturvedi

Abstract

India has 7,517 km total length of the coastline of mainland, Andaman and Nicobar Islands, and Lakshadweep Islands. Its peninsular region is bounded by the Arabian Sea on the west, the Bay of Bengal on the east, and the Indian Ocean to its south. It has two major island ecosystems, the Andaman and Nicobar Islands in the Bay of Bengal, and the Lakshadweep Islands in the Arabian Sea. The coastal and island ecosystems have a vide variability in climatic, topographical and edaphic conditions, and support diverse cultivated crops as well as natural vegetation ranging from tropical rainforests to coastal mangroves. The area is environmentally disadvantaged and at a great risk to the ill-effects of human activities and weather adversities. Frequent occurrence of strong cyclones causes colossal damage to the agricultural crops, human and animal lives, and other properties. Most of the cultivated area is predominantly monocropped mainly with rice cultivation. The coastal and island ecosystems offer vast scope of commercial use not only for vide varieties of fish, fruit and vegetable crops, but also plantation crops, spices, and medicinal plants. Multistorey plantation-based cropping systems and home gardens are dominant agricultural production systems. These are comprised of coconut, arecanut, oilpalm, cashew, cocoa, spices like cardamom, clove, black pepper, ginger, turmeric, and seed spices. Home gardens also consist of poultry, dairy, fishponds, and rice fields. Alley cropping, live fences, forest farming, or plantations under the shade of forest trees, and

J. C. Dagar (✉)
Krishi Anusandhan Bhawan–II, Indian Council of Agricultural Research, New Delhi, 110012, India
e-mail: dagarjc@gmail.com

C. B. Pandey
NRE Division, CAZRI, Jodhpur, Rajasthan 342003, India

C. S. Chaturvedi
Central Institute of Fisheries Education, Mumbai, 400061, India

J. C. Dagar et al. (eds.), *Agroforestry Systems in India: Livelihood Security & Ecosystem Services*, Advances in Agroforestry 10,
DOI: 10.1007/978-81-322-1662-9_7, © Springer India 2014

other agroforestry systems are found in the coastal and island regions. Marin aquaculture, brackish water aquaculture in vicinity of mangroves and sweet water aquaculture are prime fish production systems, which serve food and generate income to farmers. Most of these systems have been described here in this chapter.

Introduction

Coastal areas are commonly defined as the interface or transition areas between land and sea, including large inland lakes. These areas are diverse in function and form, dynamic and do not lend themselves well to definition by strict special boundaries. Unlike watersheds, there are no exact natural boundaries that unambiguously delineate coastal areas (FAO 1998). India with reported area of 328.73 Mha has land frontier of about 15,200 km and the total length of the coastline of the mainland, Lakshadweep Islands, and Andaman and Nicobar Islands is 7,517 km (India 2013) spread over 12 maritime states and Union Territories. Its economic jurisdiction in the seas for exploitation and management of the marine resources spreads over 202 Mha of an Exclusive Economic Zone (EEZ). The peninsular region of the country is bounded by the Arabian Sea on the southwest, the Laccadive Sea to the south, the Bay of Bengal on the southeast, and the Indian Ocean to its south. The Palk Strait and Gulf of Mannar separate India from Srilanka to its immediate southeast, and the Maldives are some 400 km to the southwest. It has two major island ecosystems, The Andaman and Nicobar Islands in the Bay of Bengal, and the Lakshadweep Islands in the Arabian Sea.

The Andaman and Nicobar Islands, some 1,200 km southeast of mainland, share maritime borders with Burma, Thailand, and Indonesia. Kanyakumari is the southernmost tip of the Indian mainland, while the southernmost point in India is Indira Point (earlier known as Pygmilian Point) on Great Nicobar Island. India's territorial waters extend into the sea to a distance of 12 nautical miles (22.2 km) from the coast baseline (http://en.wikipedia.org/wiki/Geography_of_India).

Coastal areas, mostly saline in nature due to periodical inundation with tidal water, and in case of lowlands having proximity to the sea, due to the high water table with high concentration of salts in it. Besides a number of soil and water-related factors limiting productivity of food grain crops, the entire shoreline is extremely fragile in nature. The areas are environmentally disadvantaged and are at a great risk to the ill-effects of human activities and weather adversities. Despite of problems and limitations, the coastal and island ecosystems provide excellent niche for rich biodiversity and various plantation crops like coconut (*Cocos nucifera*), arecanut (*Areca catechu*), cocoa (*Theobroma cacao*), cashew nut tree (*Anacardium occidentale*) and oil palm (*Elaeis guineensis*); spices like black pepper (*Piper nigrum*), cardamom (*Elettaria cardamomum*), clove (*Syzygium aromaticum*), ginger (*Zingiber officinale*), turmeric (*Curcuma domestica*); and seed spices like cumin (*Cuminum cyminum*), coriander (*Coriandrum sativum*), fennel (*Foeniculum vulgare*), chillies (*Capsicum acuminatum, C. annuum*); fruits such as banana (*Musa* x *paradisiaca*), pine apple (*Ananas comosus*), Sapota (*Achras zapota*), custard apple (*Annona squamosa*), and mango (*Mangifera indica*), and a variety of vegetables.

Agroforestry systems integrating woody perennials (including forest and fruit trees , bamboos, palms, and plantations) with agricultural crops, fishery, and livestock have a significant role to play in improving the agricultural and environmental scenario of the coastal and island regions through enhanced availability of food, fodder, timber, and fuel wood and also through effective conservation and amelioration of the natural resource bases. Plantation-based multistoreyed farming systems, home gardens, alley cropping, cultivation of fodders under coconut plantations (plantation-based silvo-pastoral systems), aqua-culture in association with mangroves, and multienterprise farming systems are some unique agroforestry systems of coastal and island regions. These play a vital role in sustaining the livelihood of coastal population and rendering ecosystem services. In scenario of climate change, all these systems can play a significant role in carbon sequestration and mitigation of climate change.

General Description of the Area

Situation, Agro-Ecology, Soil, and Land Use

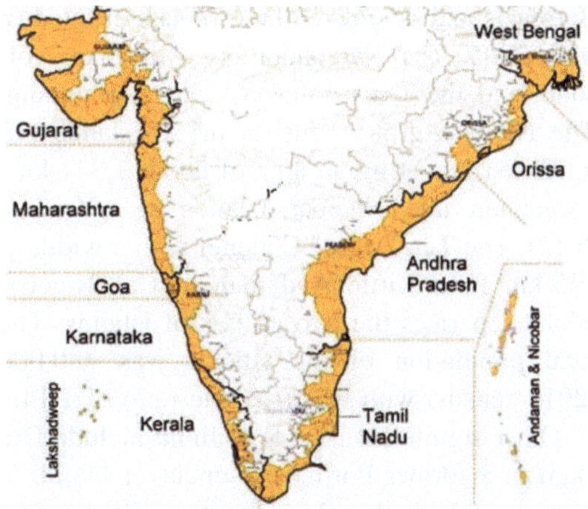

Fig. 7.1 Coastal and island regions of India

India has a total coast line exceeding 7,500 km stretched over about 32 Mha area. The Eastern Coastal Plain is a wide stretch of land lying between the Eastern Ghats and the Bay of Bengal. It stretches from Tamil Nadu in the south to West Bengal in the east. The Mahanadi, Godavari, Kaveri, and Krishna rivers drain these plains. The width of the plains varies between 100 and 130 km. The plains are divided into six regions—the Mahanadi delta, the southern Andhra Pradesh plain, the Krishna-Godavari deltas, the Kanyakumari coast, the Coromandel Coast, and sandy coastal areas. The Western Coastal Plain is a narrow strip of land sandwiched between the Western Ghats and the Arabian Sea, ranging from 50 to 100 km in width. It extends from Gujarat in the north and extends through Maharashtra, Goa, Karnataka, and Kerala (Fig.7.1). Numerous rivers and backwaters inundate the region. Mostly originating in the Western Ghats, the rivers are fast-flowing, usually perennial, and empty into estuaries. Major rivers flowing into the sea are the Tapti, Narmada, Mandovi, and Zuari. The Western Plain can be divided into two parts, the Konkan and the Malabar Coast.

The Lakshadweep and the Andaman-Nicobar Islands are India's two major island formations and are classified as Union Territories. The Lakshadweep is an archipelago located between 8° and 12° 13″ North latitude and 71–74° East longitude, about 220–440 km away from the coastal city of Kochi in Kerala, in the emerald Arabian Sea; consisting of 12 atolls, 3 reefs, and 5 submerged banks. It is single-district tiniest Union Territory of India and is the only coral island chain with an area of 32 km^2. It is comprised of 10 inhabited islands, 17 uninhabited islands attached islets, 4 newly formed islets, and 5 submerged reefs. The inhabited islands are Kavaratti, Agatti, Amini, Kadmat, Kiltan, Chetlat, Bitra, Andrott, Kalpeni, and Minicoy. Considering its lagoon area of about 4,200 km^2, territorial water of 20,000 km^2, and economic zone of 400,000 km^2; Lakshadweep is a large territory (http://Lakshadweep.nic.in). According to the 2011 Census, Lakshadweep has a population of 64,429 persons, with more than 93 % indigenous people. The main occupation of the people is fishing, coconut cultivation, and coir twisting·

The Andaman and Nicobar Islands comprise a chain of 572 islands, islets, reefs, and isolated rocks spread in the Bay of Bengal at a distance of about 1,200 km from the East Coast of maintained Indian. They lie between 6° and 14° N. Latitude and 92° and 94° E. longitude, extending between lower Burma and upper Sumatra (Indonesia), with a coastline of 1,962 km. The Andaman group consists of more that 325 islands (21 of them inhabited) with an area of 6,408 km^2. India's only active volcano,

Barren Island is situated here. It last erupted in May 2005. The Narcondum is a dormant volcano, and there is a mud volcano at Baratang. The Nicobar group consists of 28 islands (13 inhabited) covering an area of about 1,841 km^2 (Andaman and Nicobar Islands at a Glance 2012). The Ten Degree Channel with a width of 155 km from Little Andaman to Car Nicobar Island separates the two groups of islands. The total population of the islands was 380,000 (2011 census) with male: female ratio of 1.14.

Other significant islands in India include Diu Daman, a former Portuguese enclave; Majuli, a river island of the Brahmaputra; Elephant in Bombay Harbor; and Sriharikota, a barrier island in Andhra Pradesh. Salsette Island is India's most populous island on which the city of Mumbai (Bombay) is located. Forty-two islands in the Gulf of Kutchh constitute the Marine National Park.

The coastal ecosystems comprise of south-eastern coastal plains, extending from Kanyakumari to Gangetic delta in the eastern coast covering an area of about 8.5 million ha (Mha). The western coastal region comprises south Gujarat, western coastal plains of Maharashtra, Kerala, Karnataka states including Nilgiri hills of Tamil Nadu. The total area under this region is 22.8 M ha. The island regions comprising of Andaman and Nicobar Islands in Bay of Bengal and Lakshadweep Islands in Arabian Sea cover about 1 M ha area. There are about 3 Mha of coastal saline soils (Yadav et al. 1983) which includes about 466,000 ha mangrove areas (FSI 2011); occupying the narrow coastal plains, deltas, lacustrine fringes, lagoons, coastal marshes and creeks (Table 7.1).

Out of 15 agro-climatic regions of India, the coastal and island regions are distributed in East Coast Plains and Hill Region (regions of Andhra Pradesh-AP, Orissa, Tamil Nadu-TN, Pondicherry); West Coast Plains and Ghat Region (regions of Kerala, Karnataka, Maharashtra, Tamil Nadu); Gujarat Plains and Hill Region; Lower Gangetic Plain Region (coastal regions of West Bengal); and Island Region (Andaman and Nicobar Islands; and Lakshadweep, Minicoy and Amindivi Islands). To assess yield potentialities of different crops and to determine the crop suitability for optimization of land use in different ecological zones Sehegal et al. (1992, 1995) based upon the soil, bioclimatic type and physiographic situations grouped the country into 20 agro-ecological regions and 60 agro-eco-

Table 7.1 Extent of coastline, mangrove areas and saline soils of Maritime states and Union territories of India

State/Union territory	Coastal length (km)[a]	Mangrove area (000 ha)[b]	Saline area (000 ha)[c]
West coast region			
Gujarat	1214.7	105.8	714
Maharashtra	652.6	18.6	64
Karnataka	280.0	0.3	86
Kerala	569.7	0.6	26
Goa, Daman, and Diu	160.5	2.4	18
Pondicherry	30.6	0.1	1
East coast regions			
West Bengal	157.5	215.5	820
Orissa	476.4	22.2	400
Andhra Pradesh	973.7	35.2	276
Tamil Nadu	906.9	3.9	100
Island regions			
Andaman and Nicobar Islands	1962.0	61.7	15
Lakshadweep Islands	132.0	NA	NA
Total	7516.6	466.3	2,520 + 466 = 2,986

[a] India (2013), [b] (FSI (2011), [c] Yadav et al. (1983)

subregions. Further, Velayutham et al. (1999) classified each agro-eco subregion into agro-eco unit at district level for developing long-term land use strategies. The coastal regions can briefly be described (Sehgal et al. 1992, 1995; Mandal et al. 1999; Velayutham et al. 1999; Gajbhiye and Mandal 2006) under the following agro-ecological regions (Table 7.2).

Hot arid ecoregion with saline soils: The western plain, hot, and arid eco-region covers Katchh and part of Kathiawar peninsula in Gujarat State. The region is characterized by typical hot summer and moderate winter. The mean annual precipitation is usually less than 400 mm, which is just adequate to cover 15–20 % of annual potential evapotranspiration (PET) demand (1,500–1,900 mm). This results in large deficit of water throughout the year. The ecosystem represents *Aridic* soil moisture and *Hyperthermic* soil temperature regimes with an annual length of growing period (LGP) of less than 90 days (at times even < 60 days). The dominant soils representing the area are gently to very gently sloping Great Groups *Torripsamments, Camborthids*, and *Calciorthids* interspersed with level to very gently sloping *Salorthids* and *Natrargids* Great Groups as per Soil Survey Staff (1975) Taxonomy. Most of the soils are saline or alkaline in nature. The area is under rainfed traditional mono-cropping agriculture. The resistant and short duration rainy season crops such as pearl millet (*Pennisetum typhoides*), sorghum (*Sorghum* spp.) and pulses are grown on nonsaline soils. Castor (*Ricinus communis*) is very frequently grown.

In areas with available irrigation water cotton (*Gossypium arborium, G. barbadense, G. hirsutum*), sugarcane (*Saccharum officinarum*), mustard (*Brassica juncea, B. nigra*), gram (*Cicer arietinum*) and wheat (*Triticum aestivum*) are grown. The natural vegetation comprises sparse, sporadic tropical thorn forest.

Hot semi-arid ecoregion with medium and deep black soils: In coastal areas, this region covers Gujarat plains, Kathiawar peninsula and southern parts of Gujarat State. The climate is characterized by hot and wet summer and dry winter. The annual precipitation in the region ranges from 500 to 1,000 mm and covers 40–50 % of the annual PET demand (1,600–2,000 mm) resulting in gross annual water deficit of 800–1,200 mm. The LGP ranges from 90 to 150 days in a year. The dominant soil moisture regime in the region is Typic *Ustic* and soil temperature regime is *Isohyperthermic*. The dominant soils are gently to very gently sloping deep, loamy to clayey *Ustochrepts* and *Salorthids*. Dryland farming is common practice with pearl millet, sorghum, pigeon pea, groundnut, soybean, maize, and pulses in Kharif and safflower, sunflower and gram in Rabi season. Severe salinity and seasonal inundation by sea water in the Kathiawar coast result in crop failure.

Hot sub-humid to semiarid ecoregion with coastal alluvium-derived soils: The agro-ecoregion comprises the south-eastern coastal plain, extending from Kanyakumari to Gangetic Delta covering an area of about 8.5 Mha. The region experiences wide range of climatic conditions. The parts between Kanyakumari and south of Thanjavur (Tamil Nadu) and between north of Chennai and west Godavari (Andhra Pradesh) represent semi-arid conditions and receive 900–1,100 mm rainfall, of which about 80 % in October to December. The PET varies between 1,700 and 1,800 mm. Thus, the annual deficit of water is 800–1,000 mm. The LPG ranges from 90 to 150 days. The remaining part of eastern coast that lies between Nagapattinam and Chennai (Tamil Nadu) and extending to north-western part of coastal strip, including parts of western Godavari (Andhra Pradesh), Orissa and West Bengal, receives 1,200–1,600 mm rainfall of which about 80 % is received during June to September. The PET varies between 1,400 and 1,700 mm and the annual deficit of water is 600–800 mm. The LGP is much higher than the southern parts and varies from 150 to 210 days or more in a year. The region falls in sub-humid (moist) climatic type. The mean annual soil temperature of the entire region is higher than 22 °C and the difference between the mean summer and mean winter soil temperatures is less than 5 °C. Therefore, the soil temperature regime is *Isohyperthermic*. The two sub-regions have been grouped in one agro-ecosystem

Table 7.2 Eco-climatic characteristics, location, and major soil groups of coastal regions

Climate/Region description	Area Mha (LGP days)	Distribution States/Districts	Major soil groups	Rainfall (mm) {PET (mm)} [Mean temp °C]
• Katchh Peninsula, hot-arid ecosystem with saline and alkali soils • Katchh and Kathiawar Peninsula, hot-arid ecosystem with saline and alkali soils	2.0 (<60) 6.1 (60–90)	Gujarat: Lakhpat, Banni Great Runn of Katchh Gujarat: Bhuj, including Rapar, Adesar, Anjar, Kandla talukas, northern part of Jamnagar district	Salorthids, Natrargids, Camborthids, Torripsamments Camborthids, Calciorthids, Natargids, Halaquepts	<300 {1,800–1,900} [25–27] 400–500 {1,800–1,900} [26–27]
• Kathiawar Peninsula, hot semi-arid ecosystem, with deep black soils and medium black soils as inclusion • Kathiawar Peninsula (coast) with deltaic alluvium-derived soils	2.7 (90–120) 0.9 (120–150)	Gujarat: Northern part of Junagarh, Amrelli, Rajkot and western part of Bhavnagar Gujarat: Coastal parts of Junagarh, Amrelli and Bhavnagar; Diu (Daman and Diu)	Ustorthents, Ustochrepts, Chromusterts, Salorthids Halaquepts, Haplaquepts, Ustropepts, Chromusterts	600–700 {1,700–2,000} [24–25] 500–800 {1,800–1,900} [26–27]
• East (TN)Coastal Plain, hot semi-arid ecosystem, with coastal and deltaic alluvium-derived soils • East Coastal (TN)Plain, hot moist semi-arid ecosystem, with coastal and deltaic alluvium-derived soils • Eastern Coastal (AP) Plain, hot dry sub-humid, with coastal and deltaic alluvium-derived soils • Eastern Coastal (Utkal) Plain, hot dry sub-humid, with coastal and deltaic alluvium-derived soils • Eastern Coastal (Ganga) Plain, hot moist sub-humid, with coastal and deltaic alluvium-derived soils	0.5 (90–120) 1.6 (120–150) 2.0 (150–180) 3.2 (180–210) 1.2 (240–270)	Tamil Nadu (TN): coastal plains of Pudukkottai, Ramnathapuram, Toticorin, Tirunelveli, Kanyakumari TN: Madras, coastal plains of Chengal Pattu, Cuddalore, Thanjavur, Karaikal and Pondicherry (UT) AP: Coastal plain of W.Godavari, Krishna and Guntur, Prakasham and Nellore AP: Srikakulam, Coastal plains of E.Godavari (Kakinada) Vishakhapatnam, Vizianagaram Orissa: Coastal plain of Ganjam, Puri and Cuttack Orissa: Coastal plain of Baleshwar; WB: Coastal plain of Mediniopur and South 24-Parganas (including Sundarban Sagar Island)	Haplustalfs, Rhodustalfs, Ustorthents, Ustifluents, Ustipsamments, Chromusterts, Ustropepts Ustropepts, Chromusterts, Paleusterts, Ustifluents, Ustipsamments Ustifluents, Ustropepts, Chromusterts, Paleusterts, Rhodustalfs, Ustorthents, Haplustalfs Halaquepts, Fluvaquents, Haplaquepts, Ustifluvents, Ustochrepts, Haplustalfs, Chromusterts Halaquepts, Fluvaquents, Eutrochrepts, Ochraqualfs, Tropaquepts	900–1,000 {1,800–1,900} [27–28] 1,200–1,400 {1,600–1,800} [28–29] 900–1,100 {1,700–1,800} [28–29] 1,200–1,500 {1,600–1,700} [26–27] 1,600–1,800 {1,400–1,600} [26–27]

(continued)

Table 7.2 (continued)

Climate/Region description	Area Mha (LGP days)	Distribution States/Districts	Major soil groups	Rainfall (mm) {PET (mm)} [Mean temp °C]
• Western Ghat (Sahyadris), hot humid ecosystem with mixed red and black soils • Western Ghat (Sahyadris), hot, moist subhumid ecosystem with red and lateritic soils • West Coastal plain, hot perhumid ecosystem with coastal alluvium-derived soils	2.2 (180-210) 6.9 (210–270) 2.0 (240–270)	Guj: Southern parts of Surat, Dang, Valsad; Daman (Daman and Diu) and UT of Dadra Nagar Haveli; Mharashtra (Thane, Bombay, Alibagh-Kulaba) Maharashtra: Ratnagiri, Sindhudurg, Dang, Parts of Kohlapur; Goa: Panaji; Karnataka: Western parts of Uttar Kannad (Karwar), Shimoga, Dakshin Kannad (Mangalore), western parts of Chikmangalore and Kadagul (Madikari); Kerala: Cannanore (hilly part), Wayanad (Kottapadi), upland of Kozhikode, highlands of Malappuram, Palghat, Ernakulum, Pattanamtitta, Quilon,Trivandrum, Idukki; TN: Udagamandalam, Uplands of Trichur Maharashtra(narrow coastal strip of Ratnagiri, Sindhu-durg); Goa; Karnataka (narrow coastal strip of Karwar, Mangalore); Kerala (western Cannanore, narrow coastal strip of Melapuram, Calicut, Trichur, Ernaculum, Aleppy, Quilon, Trivandrum	Ustothents, Ustropepts, Chromusterts, Halaquepts, Fluvaquents, Ustipsamments Haplustalfs, Ustropepts, Rhodulstalfs, Tropofluvents, Udothents, Haplusdalfs, Dystropepts, Haplorthox Ustropepts, Tropaquepts, Troposamments, Haplustalfs, Haplusdalfs, Toporthents, Dystropepts	1,600–2,000 {1,400–1,600} [27–28] 2,000–3,000 {1,400–1,800} [27–28] >3,000 {1,400–1,700} [27–28]
• Eastern highlands (Andaman-Nicobar Islands), hot perhumid ecosystem, red loamy soils • Western highlands, hot humid ecosystem, sandy and littoral soils	0.8 (>300) 0.03 (240–270)	Andaman and Nicobar groups of Islands Lakshadweep group of Islands	Troporthent, Eutropepts, Tropaquents, Haplichy-draquents, Humitropepts, Troposamments, Ustifluvents Ustipsamments, Udifluvents	>3,000 {1,400–1,600} [27–28] 1,600–2,000 {1,400–1,600} [26–27]

Source Modified from Sehgal et al. (1992, 1995), Velayutham et al. (1999)

because of maritime climatic influences and limited area (Gajbhiye and Mandal 2006). The soils belong to *Heplaquents, Halaquepts*, and *Ustifluents* Groups occurring on level to very gently sloping topography and *Ustropepts* series. Motto (*Haplaquepts)* and Kalathur (*Pellusterts*) series are slight to moderately sodic. Both are clayey in nature but have marked differences in clay mineralogy. The later soils have high swell-shrink potential. Both rainfed and irrigated agriculture are prescribed in the region. Rice is main crop both in Rabi and Kharif seasons. Coconut is a dominant plantation crop of the region. In some parts, pulses, such as black gram and lentil; and oil seed crops such as sunflower and groundnut are cultivated after rice on residual moisture (Mandal et al. 1999). Besides agriculture, coastal and brackish water fisheries are important economic activities of the coastal people.

Hot humid perhumid ecoregion with red, lateritic and alluvium-derived soils: The agro-ecoregion comprises Sahyadris, western coastal plains of Maharashtra, Karnataka, and Kerala states. The climate is characterized by hot and humid summer and warm winter. The mean annual daily temperature varies between 25 and 28 °C. The mean summer and winter soil temperatures by less than 5 °C representing *Udic* moisture and *Isohyperthermic* temperature regime. The mean annual rainfall exceeds 2,000 mm in most of the area and exceeds PET demand (1,400–1,600 mm) in most of the months, except seasonal deficit of 300–400 mm during February to mid-April. The region is represented by a longer LGP ranging between 150 and 210 days, some places even exceeding 210 days in a year. The major soils of the region include red and laterite soils along the leeward flank of Sahyadris and the alluvium-derived soils in the coastal plains. These belong to the Great Groups of *Dystropepts, Eutropepts, Hapludults, Haplaquepts* with localized *Haplorthorx* as per soil taxonomy of Soil Survey Staff (1975). The soils of Trivandrum and Kunnamangalam series, typifying *Dystropepts* and *Haplorthox* are very deep, clayey, strongly to moderately acidic in nature and poor in base saturation. Because of the dominance of Kaolinite clay mineral, the soils are low in retentive capacity, suggesting poor inherent fertility. The area is intensively cultivated for rice, tapioca, coconut and spices. The natural vegetation comprises tropical moist deciduous forests. Excessive leaching leading to depletion of plant nutrients and bases; water-logging due to imperfect drainage; and inundation of land area resulting in localized saline marshes are some constraints.

Hot humid perhumid island ecoregion with red loamy and sandy soils: This agro-ecoregion comprises the group of islands of Andaman and Nicobar in the east and that of Lakshadweep in the west covering an area of about 0.8 Mha. The climate is characterized by tropical conditions with little difference between mean summer and mean winter temperatures. The annual rainfall of these two far-seated areas varies from 1,600 mm (Lakshadweep Islands) to 3,000 mm (Andaman and Nicobar Islands) representing humid and perhumid climate, respectively. The rainfall covers the entire annual PET demands, except for small seasonal water deficit of 300–400 mm during the post-monsoon period (January-March). The LPG is more than 210 days which is long enough to support double cropping system and plantation crops grown in the area. The area experiences the *Udic* soil moisture and *Isohyperthermic* soil temperature regimes. The soils of Andaman and Nicobar Islands are medium to very deep, red loamy soils including marine alluvium-derived soils along the coast qualifying for the Great Groups of *Hapludalfs, Dystropepts, Eutropepts*, and *Sulfaquents* (along the coast). The soils of Lakshadweep Islands, on the other hand, are highly calcareous and sandy in nature (*Udipsamments*).

The soils of the Bay Islands have developed under the dominant influence of vegetation and climate over diverse parent materials. Singh et al. (1988) broadly classified the soils of the Bay Islands into three orders, seven suborders and nine series. The Entisols are mostly in valleys and are very young and devoid of diagnostic horizons. The Inceptisols are characterized by altered horizons that have lost materials by leaching and contain some witherable minerals.

The Alfisols have an argillic horizon with achric epipedon having moderate to high base saturation. The soils show considerable variation in profile characteristics. These are shallow to deep alluvial, heavy to coarse textured and most of the time with coralline deposits. The pH of surface soil varied from 4.7 to 6.6, showing a generally acid nature. The presence of potential acid sulfate materials within 50 cm of the surface, high organic matter, persistent anaerobic conditions, and ingress of sea water rich in sulfates are common features in coastal areas. Arable soils are mostly of alluvial and colluvial origin. Available moisture storage is generally low owing to the high intensity of rainfall. Uncovered soils are subject to severe soil erosion and are poor in plant nutrients. A well-known characteristic of these humid tropical soils is their ability to hold phosphate ions strongly.

Vegetation

Based on habitat, the coastal ecosystem may be classified into beach, dunes, gravel shores, tide flat, marsh land, and upland. Characteristic groups of organisms including micro-organisms and vascular plants (vegetation) live on each types of shore/habitat. There is big heritage of species and genetic strains in these coastal areas. The biota of coastal ecosystem includes a great variety of plants, birds, fish, mammals, and invertebrate organisms. The natural vegetation comprises tropical evergreen rain forests, littoral, and swamp forests. Forests occur predominantly on the upland hill slopes though many plantations also constitute important part of coastal ecology. The climatic climax vegetation in most of the islands is tropical rain forests. The moist evergreen and semi-evergreen forests are luxuriant and have a variety of plants. Species such as *Dipterocarpus griffithii, D. turbinatus, Planchonia andamanica, Sideroxylon longipetiolatum*, and *Hopea odorata* reach 40–60 m height. Among deciduous forests, *Pterocarpus dalbergioides* supported by huge buttresses along with *Pterocymbium tinctorium, Bombax insignis* and species of *Terminalia* form important associations of timber trees. Tall trees are covered by many woody climbers, i.e., lianas. Mangrove forests are represented by 35 exclusive mangrove species and constitute a large number of associate species. Dagar (1982) and Dagar et al. (1991) have given a detailed account of mangroves in India. The Bay Islands are very rich in biological diversity and of about 2,400 species of vascular plants 292 species are endemic (Dagar 1989). The utilization of indigenous plants by the tribal people in medicine, food, fiber, and other purposes is of great interest. Dagar (1993, 1995b, 1998, 2003, 2005a), Dagar and Dagar (1991, 1999), Dagar and Singh (1999, 2007), Chakraborty et al. (2012) and many others have described the coastal vegetation of India and its utility among indigenous people at length. On western and eastern coast particularly along Orissa beaches *Acacia auriculaeformis, Anacardium occidentale, Azadirachta indica, Barringtonia acutangula, Borassus flabellifera, Caesalpinia bonduc, Calophyllum inophyllum, Cassia occidentalis, Casuarina equisetifolia, Clerodendrum inerme, Cocos nucifera, Hibiscus tiliaceous, Pandanus fascicularis, Phoenix paludosa, P. sylvestris, Pongamia pinnata, Salvadora persica, Tamarix troupii, Terminalia catappa, Thespesia populnea*, and *Vitex negundo* are important tree species. *Croton bonplandianum, Hydrophylax maritima, Ipomoea pes-caprae, Launaea sermentosa, Mimosa pudica, Opuntia stricta, Pedalium murex, Phyla nodiflora* and *Salicornia brachiata* are important herbaceous species found on sandy substrata. Mangroves of Andamans are among the most luxurious forests of the world. Dagar (1982, 2003, 2012) and Dagar et al. (1991, 1993) gave extensive account of mangrove ecology of Andaman and Nicobar Islands. About two-third area of the Andamans is under native forest and agriculture is confined only to specific areas around habitations, where the dominant crop grown is rice. In general, the land use is dominated by plantation crops, such as coconut, areca nut, red oil palm with or without intercultivation of pineapple, clove, cardamom, tapioca, and black pepper. In

Nicobar group of islands (particularly Car Nicobar Island), coconut is dominant plantation crop. In Lakshadweep, rice is mainly grown under lowland conditions and coconut is the main plantation crop with high yield. Besides agriculture, marine fishery is an important means of subsistence throughout the islands.

Degradation of the tropical rainforest ecosystem leads to severe soil erosion hazard. Inundation of coastal areas and denudation of mangroves leads to saline marshes and consequently formation of acid sulfate soils. Under this scenario, plantation-based agroforestry is most important part of coastal and island agriculture.

Promising Agroforestry Systems

A range of agroforestry systems and practices suitable for coastal and island regions have been studied and identified by many (Nair and Sreedharan 1986; Dagar 1991, 1994, 1995a, 1996, 2000; Kumar 2005, 2011; Kumar et al. 1998, 2001a, b, 2005; Pandey et al. 2007; Kumar and Kunhamn 2011). Some of these such as home gardens, plantation-based cropping systems, live fences and hedges, shelterbelts and shore protection, grazing under plantations (silvopastoral systems) and aqua-silviculture are already being practiced at large while others such as alley cropping, and fish/shrimp culture with mangroves, can be adopted by farmers after they receive suitable trainings. Some site specific practices such as rehabilitation of saline areas are used by different stake holders. Some important agroforestry systems/practices for coastal areas have been briefly described here. Many multipurpose tree species of coastal and island regions used in different systems along with their utilization have been shown in Table 7.3.

Home Gardens

Home gardens also known as homesteads, are an integral part of farming system, an adjunct to the house, where selected trees, shrubs and herbs are grown for edible products and cash income (Kumar and Nair 2004). It is an operational farm unit in which a number of crops, including food and cash crops, vegetables and tree crops, are grown with livestock, poultry and/or fish production, mainly for the purpose of satisfying the farmer's basic needs. The components of a homestead are so intimately mixed in horizontal and vertical strata as well as in time that a number of complex interactions exist among soil, plants, water, other components and environmental factors in the farmer's plot where he lives and manages the unit (Nair and Sreedharan 1986; Nair and Dagar 1991; Krishnakumar 2004). There is long history of home gardens in coastal areas. The travelogue of the Persian traveler Ibn Battuta (1,325–1,354) mentioned that in the densely populated and intensively cultivated landscapes of Malabar Coast, coconut and black pepper were prominent along the houses (Randhawa 1980). The writings of the colonial period (Mateer 1883; Logan 1906; Nagam 1906) that provide evidences of home gardening are other examples in this respect. Home gardens offer food and nutritional security to the subsistence farmers, besides acting a source of cash income. These also help getting a buffer to food insecurity during lean season, providing habitat protection, soil and water conservation, environmental services and high rate of carbon sequestration. The home garden farmers are perpetual experimenters and are constantly trying and testing new species and varieties and their management (Ninez 1984; Kumar and Nair 2004). Over centuries, they have selected specific species and manipulated their physical and ecological locations in a home gardens planting for maximizing space and production (Gillespie et al. 1993), based on their accrued wisdom and insights acquired through interactions with environment without access to exogenous inputs, capital or scientific skills (Kumar and Nair 2004). The crop combinations found in the home gardens of a region, however, are strongly influenced by the biophysical and sociocultural factors besides the specific needs and preference of the household and nutritional complementarities

Table 7.3 Important uses of prominent multipurpose trees and shrubs found in coastal and island regions

Species	Major uses	Distribution		
		WC	EC	IS
Acacia auriculaeformis	fu, ta, ti, nf	+	+	+
A. catechu	ti, re, fu, ch, nf, pw, ta	+	+	-
A. mangium	ti, fu, nf, ch	+	+	-
A. nilotica	ti, fu, ch, gum, fo, m, nf	+	+	+
Achras zapota	fr, b, fu	+	+	+
Adenanthera pavonina	fu, m, re, ti	+	+	+
Aegle marmelos	fr, m, fo, ti	+	+	+
Ailanthus excelsa	pw, fo, re, fu, m	+	+	–
Ailanthus triphysa	ch, fu, po, rep	+	–	–
Albizia amara,chinensis, lebbeck	fi, fu, ta, nf	+	+	+
A.odoratissima, procera	ti, fo, fu, ta, nf, o(es), m, pw	+	+	–
Alstonia scholaris	ti, po, rep, pw, m	+	+	–
Anacardium occidentale	n, b, gu, m, d, o	+	+	+
Annona squamosa, muricata	f, b	+	+	+
Anogeissus latifolia	ti, fu, ch, re, m	+	+	–
Anthocephalus chinensis	pw, fu, avenue	+	+	+
Artocarpus chaplasha, gomeziana, heterophyllus, incisa, lakoocha	v, ta, fo, source of floride	+	+	+
A. altilis syn A. communis syn. A. incisa	v, pw, fi, fo	+	+	+
Averrhoa carambola, bilimbi	fr, v, pi, ch, fo	+	–	+
Avicennia marina, officinalis	fu, fo, v, sp, ti, h	+	+	+
Azadirachta indica	m, ti, fu, ch, ta, sp, avenue	+	+	+
Bambusa spp.	pw, th, be	+	+	+
Barringtonia asiatica, racemosa	po, ch, boat making	+	+	+
Bauhinia acuminata,malabarica, purpurea, racemosa, variegata	v, fu, nf, fo, avenue	+	+	+
Bischofia javanica	ti, ta, fu	+	+	+
Bixa orellana	d, fu, m	+	+	+
Bombax malabaricum	fi, pw, ti	+	+	+
Borassus flabellifer	th, b, fi, radicles eaten roasted, avenue	+	+	–
Bridelia retusa, roxburghiana	ti, fo, fu, ta	+	–	–
Bruguiera cylindrica, gymnorrhiza, parviflora	fu, ch, ta, fo, sp, h	+	+	+
Buchnania lanzan, splendens	fo, fu, ch, ta, po, v	+	–	+
Butea monosperma	fo, fu, nf	+	–	–
Caeslpinia bouduc, crista, sappan	ta, gu, m, sp	+	+	+
Calophyllum inophyllum	boat making, fu, m, sp	+	+	+
Carallia brachiata	fo, ti, boat making	+	–	+
Careya arborea	ti, fu, b	+	–	–
Carica papaya	fr, b, m.	+	+	+
Carissa carandas	pi, fr, m.	+	+	+
Caryota urens	fu, ti, fo, fi	+	–	–
Casearia tomentosa	fish poison, fu, pw, m	+	–	+

(continued)

Table 7.3 (continued)

Species	Major uses	Distribution		
		WC	EC	IS
Cassia fistula	ti, nf, ti, fu, ta	+	+	+
C. siamea	ti, nf, fu, ta, sp	+	+	+
Casuarina equisetifolia	ti, nf, ti, fu, ta, sp	+	+	+
Ceiba pitandra	fi, ti, ornamental	+	+	+
Cerbera odollam	ch, po, fu,sp	+	–	+
Ceriops tagal	ti, ta, fu, fo, sp	+	+	+
Chrysophyllum cainito	fr, b, fu	+	–	–
Chukrasia tabularis	ti, ta, ch, po, pw, gum, sp	+	–	+
Cinnamomum malabatrum	spice, fu	+	–	–
Cinnamomum zeylanicum	Bark spice	+	–	+
Citrus spp.	fr, b	+	+	+
Clerodendrum inerme	m, sp, lf, ornamental	+	+	+
Cocos nucifera	n, oil, th, fi	+	+	+
Coffea arabica	B (coffee), fu, ch	+	+	+
Cordia dichotoma, rothii	fo, pi, gu, v, m, fu.	+	+	+
C. subcordata	boat making, ti, m, sp	+	–	+
Cullenia exarillata	ti, fu	+	–	–
Cynometra ramiflora	ti, fu, medicinal oil, sp	+	–	+
Dalbergia latifolia	ti, fo, fu, nf, ta	+	–	+
D. sissoo	ti, fo, fu, nf, ta	+	+	+
Delonix regia	ti, fu, gm, ornamental, h	+	+	+
Dendrocalamus asper, brandisii	th, poles, pw	+	+	+
Diospyros buxifolia, ebenum	ti, fu, ch, re, m	+	–	–
Dipterocarpus alatus, D. turbinatus, D. odorata	ti,re	+	–	+
D. bourdillonii, indicus	ti, re	+	–	–
Elaeocarpus serratus, tectorius	ti, fu, ch, fo, re, gm, m, waxes	+	–	–
Elaeis guineensis	Oil, b, th, feed for cattle	+	–	+
Embilica officinalis	pi, fr, gm, essential oil, ch, ta, m	+	+	+
Erythrina indica, variegata	ti, fu, fo, nf, support to black pepper	+	–	+
Eucalyptus tereticornis	ti, pw, sp	+	+	+
Euodia luna-ankaenda	Ti, fu, essential oil, pw	+	–	–
Excoecaria agalocha	fu, ta, ch, sp, latex poisonous, match stricks	+	–	+
Ficus gibbosa, hispida,racemosa, relegiosa, retusa, rumphii, virens	fu, ch,fo	+	+	+
F. ilastica, exasperata, tinctoria	fu, fo, m	+	–	–
Flacourtia inermis, montana	fu, ti, bi	+	–	–
Garcinia cowa, microstigma	fr, b, jam, d	+	–	+
G. gummi-gutta, mangostana, G. morella	fr, b, jam, essential oil	+	–	–
Garuga pinnata	ti, pw, canoe making, d, ta, fo, re	+	–	+
Gliricidia sepium	gm, fo, support for black pepper, lf, mu	+	+	+
Gmelina arborea	pw, ti, avenue	+	–	+

(continued)

Table 7.3 (continued)

Species	Major uses	Distribution		
		WC	EC	IS
Grewia tiliaefolia	ti, fu, fo, avenue	+	–	–
Heritiera fomes, H. littoralis	ti, fu, ch, sp	+	+	+
Hernandia peltata	ti, fu, ta, ch, sp, m, seeds edible	+	–	+
Hevea brasiliensis	Rubber, ti, fu, sp	+	–	+
Hibiscus tiliaceus	ti, fo, pw, fi, sp	+	+	+
Holarrhena pubescens	ti, fu, m	+	–	–
Holigarna arnottiana	pw, re, m, fu	+	–	–
Hydnocarpus alpina, pentandra	ti, fu, m, essential oil	+	–	–
Knema attenuata	fu, ti, re, m	+	–	–
Jatropha curcas, gossipifolia	Biofuel, fi(bark), m, mu, lf	+	+	+
Lagerstroemia hypoleuca	ti, avenue	+	–	+
L. microcarpa, parviflora	ti, fu, ch	+	–	–
L. reginae, speciosa	ti, fu, avenue	+	–	–
Lannea coromandelica	ti, fu, pw, fo, ta, m, ornamental	+	–	+
Lawsonia inermis	Dye, lf	+	+	+
Leucaena leucocephala	ti, fu, pw, fo, ta, lh, nf	+	+	+
Lumnitzera littorea, racemosa	sp, fu, ta, ch, ta	+	–	+
Macaranga peltata	pw, fo, ta, fu, m	+	–	+
Mallotus philippensis	,pw, ti, dye, ta, m, sp	+	–	+
Mangifera indica	fr, pickle, fu, v, ta, b	+	+	+
Manilkara littoralis	ti, pw, bows, d, fo, sp	+	–	+
Melia azedarach	ti, pw, medicinal oil, m	+	+	+
Memecylon edule	ti, ch, fu, d, fruit edible, m	+	–	+
M. molestum	Fu, ch, d, m	+	–	–
Michelia champaca	ti, d, essential oil, m, avenue	+	+	+
Mimusops elengi	ti, fruit edible, fo, dye, ta, saponin,m, oil	+	–	+
Morinda citrifolia/coreia	ti, d, fruit edible, fo, cleansing, m, sp	+	–	+
Moringa oleifera	v, fo, m	+	+	+
Morus alba	fr, silkworm rearing, fo, utencil making	+	+	+
Murraya koenigii, paniculata	flavor, lf, m	+	+	+
Musa paradisiaca	fr, v, b, fo, m	+	+	+
Mussaenda frondosa, macrophylla	v, m, fo	+	–	+
Myristica fragrans	Mace (fruit spice),m	+	–	+
Nypa fruiticans	th, mats, b, ta, alcohol, fi, m	+	+	+
Oroxylum indicum	ti, fo, d, ta, fruits for vegitable	+	–	+
Pajanella longifolia	ti, canoe, m	+	–	+
Pandanus fasicularis, leram, tectorius	fr, v, th, fi, sp	+	+	+
Paraserianthes falcataria	ti, pw, fo	+	–	–
Peltophorum pterocarpum	ti, ta, shade, fo, m	+	+	+
Persea macrantha	ti, pw, fu	+	–	–

(continued)

Table 7.3 (continued)

Species	Major uses	Distribution		
		WC	EC	IS
Phoenix paludosa, sylvestris	th, fi, fruits edible, sp	+	+	+
Pithecellobium dulce	fo, fr, v, fu, ti, d, m	+	+	+
Planchonella longipetiolatum, obovata	ti, pw, fo, m	+	−	+
Polyscias acuminata	fu, gm, m, po	+	−	−
Pongamia pinnata	ti, fo, biofuel, nf, avenue	+	+	+
Prosopis juliflora	ti, fu, pods (fo), nf	+	+	−
P. spicigera	ti, fu, m, fo, nf	+	+	−
Psidium guajava	fr, juice, b, fu, m	+	+	+
Pterocarpus dalbergioides	ti, fo, ta, nf	+		+
P. marsupium, santalinus	ti, fo, ta, nf	+	−	−
Pterocymbium tinctorium	ti, pw, dugout, bark fi, gum, poison	+	−	+
Punica granatum	fr, juice, b, m	+	+	+
Quassia indica	ti, fu, re, m	+	−	−
Rhizophora apiculata, mucronata	fu, ti, fo, ta, sp, h, wild life	+	+	+
Ricinus communis	seed oil, fo, fu, m	+	+	+
Salvadora oleoides/persica	seed oil, wax, fo, fu, sp	+	+	+
Samadera indica	fu, m	+	−	−
Samanea saman	ti, fu, nf, fo, gum, m	+	−	+
Sapindus trifoliatus	essential oil, soap, cleaning agent, m	+	−	−
Saraca asoca	sacred, fo, ti, avenue, m	+	+	+
Santalum album	ti, fo, essential oil, m	+	+	−
Scaevola sericea	fu, m, sp	+	+	+
Schleichera oleosa	ti, fu, ch, d, siri/lac culture, h, m	+	+	−
Semecarpus anacardium	ti, fu, ch	+	−	−
Sesbania grandiflora, javanica	fo, gm, nf,fu	+	+	+
Sonneratia alba, apetala, caseolaris	fu, ch, fo, ta,sp,fr,eaten by tribals	+	+	+
Spondias indica, pinnata	pw, fu, b, m	+	−	+
Stereospermum personatum	ti, fu	+	−	−
Sterculia guttata	ti, fu, bark fiber, gum	+	−	+
Strychnos nux-vomica	ti, poison, d, m	+	−	+
Swietenia macrophylla, mahagoni	ti, fu, ch, ta, gu, oil soap making	+	−	+
Syzygium aromaticum	spice, essential oil, m, fu	+	−	+
S. cuminii	fr, vinegar, b, ti, d, ta, m	+	+	+
S. samarangense	ti, canoe, fr, m, sp	+	−	+
Tabernaemontana crispa	fu, b, ornamental, m, lh, sp	+	−	+
Tamarindus indica	fr, chutneys, oil, ta, ti, fo, fu, ch, m, nf, h	+	+	+
Tamarix articulata	ti, fu, sp	+	+	+
T. troupii	fu, ta, sp	+	+	+
Tectona grandis	ti, fu, ta, d, fatty oil, m	+	+	+
Terminalia bellirica	ti, fu, ch, m, pw	+	+	−

(continued)

Table 7.3 (continued)

Species	Major uses	Distribution		
		WC	EC	IS
T. bialata	ti, ornamental wood, ta, kernel edible,m	+	–	+
T. catappa	oil, ti, fu, ch, seri-,lac-culture, h, d, m, sp	+	+	+
T. chebula	ti, fu, ch, fo	+	–	–
T. crenulata	ti, fu, ch	+	–	–
T. manii, T. procera	Ornamental timber, fu, ch, d	+	–	+
T. paniculata, tomentosa	ti, fu, ch	+	–	–
Tetrameles nudiflora	ti, canoe, fu, m	+	–	+
Theobroma cacao	b (cacao),chocolates, o, soap, d, m	+	–	+
Thespesia populnea	B(cacao)m,o	+	+	+
Tournefortia ovata	fu, sp, fo, bark fi, d, m	+	+	+
Trema tomentosa, orientalis	fo, ti, fu, ch	+	–	+
Trewia polycarpa	ti, fu, ch, essential oil, pw	+	–	–
Vateria indica, macrocarpa	ti, fu, re, ta, m	+	–	–
Vatica chinensis	ti, fu, resin	+	–	–
Vitex altissima	ti, fu, d	+	–	–
Vitex negundo, trifoliata	ta, m, boiled seed edible, lh, sp	+	+	+
Wrigtia tomentosa, tinctoria	v, ti, fu, oil, fi, lh, lf, m	+	+	+
Xylia xylocarpa	ti, fu, ch, green manure, ta, essential oil	+	–	–
Xylocarpus gangeticus, granatum	ti, ta, fu, ch, sp	+	+	+
Zanthoxylum limonella	Bark pi, v, oil, ti, fu, m	+	+	+
Ziziphus mauritiana	fr, ti, fo, fu	+	+	+

Source Kumar et al. (1994), Dagar (1995), Pattanaik et al. (2008), Kumar (2011)
Letters for uses depict as a- alley cropping, b-beverages, ch-charcoal, cs-crop support, csh-crop shade, d-dye, f-food, fi-fibre, fo-fodder, fr-fruit, fu-fuel wood, gm-green manure, gu-gum, h-honey bee forage, ir-insect repellent, lf-live fence, lh-live hedge, m-medicinal, mu-mulch, n-nuts, nf-nitrogen fixing, o-oil, p-pole, pi-pickle, po-poison, pw-pulp wood, re-resin, sp-shore protection, sw-silk worm food, ta-tannin, ti-timber, th-thatching, and v-vegetables

with other major food sources. As a consequence, home gardens vary greatly in species, species richness, structural complexity and size, but general principles are broadly similar (Gillespie et al. 1993). From the ground layer to the upper canopy, the gradient of light and humidity determine different niches that species exploit according to their own requirements (Fernandes and Nair 1986). The position, height and shade tolerance of plants are important traits that are acquired with time (evolutionary). Therefore, study of these parameters gives an idea of the temporal and spatial positioning of plants, species interaction, and mixed species silviculture that are pertinent for designing a multistrata agroforestry and management of its productivity (Gillespie et al. 1993).

Home garden in addition to plantation and fruit trees include poultry, livestock, and fish-pond. These components being an integral part of the home garden function in an interactive manner. In India, the home gardens of Kerala and Bay Islands are most popular because of the equatorial climate, which is optimal for growing most of the crops including plantations. These homesteads are unique being more or less coconut based with an array of inter or mixed crops resulting in multistory cropping system, thereby efficiently harnessing solar radiation and using soil moisture and nutrients. Dagar (1995a) and Pandey et al. (2007) have studied the home gardens of Andamans. It has been found that cow is the major animal of livestock averaging two cows per home garden; it ranges from 1 to 6 per homestead. Forty percent of the home gardens are not having livestock. Some farmers use the he-buffalo as a draft animal. Live stock feeding totally depends on grazing. They are trained in such a way that they go for grazing in morning in nearby forest after milking and comeback in evening at a fixed time. In Andamans, on an average, one poultry shade is found in the 80 % home gardens. Poultry birds are comprised of hen, cock and duck and the hen ranging from 2 to 55 (average 13 birds) per home garden. Sixty-seven percent home gardens are having poultry. An average 2,340 eggs are produced by hen and 633 eggs by ducks in a year for each home garden. Nearly 25 % of the total eggs are consumed by households whereas 75 % are sold. Though duck produces relatively greater number of eggs and cost of eggs are also high but numbers of duck are lower because all farmers do not own pond. All the poultry birds feed among the trees in home garden. Rice fields are located quite close to coconut-arecanut orchards in the Andamans. Smallholders grow mostly long duration (6 months) rice to be harvested in December. Generally between the two rice crops they grow vegetables in the rice fields. Though, the vegetable crops are common among the islands, their proportions as well as acreages vary. Proportion of tomato, brinjal, and chilli is always higher in Little Andaman than that in any other islands. However, proportion of cucumbers, gourds, cowpea, okra and amaranths is nearly equal among the islands. Maximum 0.5–2 ha per household are cultivated for vegetables in Middle Andaman followed by 0.25–1 ha per household in North Andaman. Acreages of vegetable cultivation are lowest (0.02–0.25 ha per household) in South Andaman. Cole crops are grown from November to January, whereas, the remaining crops round the year. Nicobaries in Nicobar do not know vegetable and rice crop husbandry. Paddy cultivation in Nancowry is similar to that in the Andamans, but the vegetable cultivation is negligible.

Taking into account the importance of home gardens, Agro-ecosystem Coastal Directorate initiated a project funded by National Action Taken Program (NATP) to collect basic data and restructuring the homesteads of Kerala and Andaman and Nicobar Islands on farmer participatory approach. In Kerala, the Central Plantation Crops Research Institute (CPCRI), Ksaragod, one among the five centers implemented this project. After collecting basic data from 815 homesteads of the four districts in 83 *Panchayats* (25 % of the total *Panchayats*) based on the resources (land, water and capital) availability as well as willingness of the farmers for restructuring their homesteads through farmers' participation 5 homesteads in each districts were shortlisted in different agro-climatic subzones for implementation of interventions. The important interventions included (Krishnakumar 2004): Planting hybrid (WCT × COD) high yielding coconut, planting tissue culture banana (G 9), planting rooted cuttings of pepper (cultivar Sreekara), cultivating high value vanilla as a mixed crop, mixed cropping tree spicies (clove and nutmeg), planting hybrid mango (H 87) grafts and sapota grafts, growing vegetables, planting vetiver grass on raised bunds, balanced fertilizer management for coconut, introduction of pineapple, establishment of vermicompost unit, strengthening of bunds for soil conservation, and introduction of chicks (cultivar Gramalakshmi). The performance of all the interventions implemented in the project is being monitored on regular basis and the results are encouraging. The basic data collected indicate that most of the farmers, in general, are cultivating a number of crops in their homesteads adopting all scientific knowledge regarding manuring, plant protection operations and other cultural practices. About 26 % farmers in Kozhikkode and Kannur districts and 32 % farmers in Malappuram and Kasaragod districts carryout their cultivation utilizing their family labor. While majority of the farmers (56 %) in Kazhikkode district sell their farm produce by direct marketing and all the surveyed homesteads keep livestock and maintain other enterprise to meet the domestic demand (Krishnakumar 2004).

Most of the home gardens in Kerala and Andaman-Nicobar Islands are rich in species composition and productivity. These have been worked out by many (Nair and Sreedharan 1986; Ismail 1986; Jose 1992; Kumar et al. 1994; Dagar 1995a and Pandey et al. 2007) and are described below.

Species Composition and Structure

Earlier studies of Kerala home gardens have traditionally emphasized species inventories (Nair and Sreedharan 1986; Babu et al. 1992) with little quantitative data. Jose (1992), however, calculated Simpson's diversity indices for different size-class categories and components. After analyzing 80 households, she concluded that the introduction of rubber (*Hevea brasiliensis*) into the home gardens resulted in a reduction of species diversity. Krishnankutty (1990) performed survey of 30 villages in Kerala and observed that just 10 species accounted for 74 % of the total homestead growing stock and 85 % of the total wood volume.

Kumar et al. (1994) conducted a survey in 17 selected *taluks* (revenue sub-divisions) among 252 farmers and analized density, structure, and standing stock of wood in the home gardens of Kerala. They found tremendous variability both in number of trees and shrubs present and species diversity of the selected homesteads in different provinces. The small (<2 ha), medium (2–5 ha), and large (>5 ha) sized holdings also exhibited profound variability in the number of woody taxa and the individuals present. In total, 127 woody species (girth at breast height >15 cm) were encountered. The mean number of woody taxa found in home gardens ranged from 11 to 39. The floristic diversity was higher in the smaller homesteads and it decreased with increasing size of holdings. Mean Simpson's diversity index ranged from 0.25 to 0.74 (mean value being 0.50) and that of small, medium, and large holdings was 0.61, 0.44, and 0.46, respectively (Table 7.4) suggesting that floristic diversity was moderate to low compared to a value over 0.90 for the species rich evergreen forests of the Western Ghats.

Farmers tend to prefer timber trees such as *Ailanthus triphysa* (highest frequency), teak (*Tectona grandis*), *Erythrina indica*, and *Bombax ceiba* besides fruit trees such as mango (*Mangifera indica*), jack (*Artocarpus heterophyllus, A. hirsuta*), cashew (*Anacardium occidentale*), *Tamarindus indica, Psidium guajava*, and goose berry (*Emblica officinalis*). Among other MPTs *Macaranga peltata, Thespesia populnea, Gliricidia sepium, Swietenia macrophylla, Casuarina equisetifolia, Delonix regia, Leucaena leucocephala, Azadirachta indica, Ceiba pentandra, Annona squamosa, Syzygium cuminii*, and *Dalbergia latifolia* are important. It is evident that both naturally occurring wild plants and deliberately introduced plants occur in the home gardens. A new species may be introduced exclusively based on its properties, i.e., food, vegetables, medicine, timber, ornamental, religious, etc. Standing stock of timber

Table 7.4 Density of woody perennials found in home gardens of different land holdings in Kerala

Holding size classes	Land holding (ha)	Number of species encountered	Number of individuals	Simpson's diversity index
Small	1	21	269	0.61
Medium	2	26	1,173	0.44
Large	5	20	561	0.46
F test (P)	<0.01	<0.05	<0.01	<0.01
LSD ($p = 0.05$)	1.73	4.7	308	0.08

Source Modified from Kumar et al. (1994)

and firewood ranged from 6.6 to 50.8 m^3 ha^{-1} (average 25 m^3 ha^{-1}) and 23–86 m^3 ha^{-1} (average 50 m^3 ha^{-1}), respectively (Kumar et al. 1994).

In Andaman and Nicobar Islands, Dagar (1995b) reported 66 species commonly grown in home gardens categorized as vegetables (29 species), fruits (16), pulses (6), nuts/oils (7), spices and condiments (6), and others (2). The average yield of economic product of these commodities has been reported in Table 7.5. Jacob (2004) also reported the yield of some vegetables and fruits in home gardens of Lakshadweep Islands. He observed varieties of chillies (*Capsicum* sp.) to produce 6.7–24.0 t ha^{-1}, brinjal (*Solanum melongena*) 20.3–33.7 t ha^{-1}, snake gourd (*Trichosanthes anguina*) 22.5 t ha^{-1}, cucumber (*Cucumis sativus*) 30.3 t ha^{-1}, tomato (*Lycopersicon esculentum*) 2.1 t ha^{-1}. Among fruits, varieties of banana are reported to yield 12.5–13.5 kg per plant, papaya 15.7 kg per plant, moringa (*Moringa oleifera*) 15 kg pods per plant and watermelon 2.4 t ha^{-1}. Among tuber crops 5 kg per plant was harvested from tapioca (*Manihot esculenta*) and 2 kg per plant from elephant foot yam (*Dioscorea* sp.) while sweet potato (*Solanum tuberosum*) could produce 1.6 t ha^{-1}.

In a comprehensive study, Pandey et al. (2007) analyzed the composition and structure of different components in both Andaman and Nicobar groups of islands. In total, 34 woody plant species are reported to be found in the home gardens of Andaman and 12 in Nicobar that are planted, cared, and harvested. They are classified as palms, fruit trees, spice trees, and agroforestry/forestry trees. Though the species are common, their density varies within and among the sites. Arecanut is dominant in the home gardens invariably in all the islands of Andaman. However, coconut dominates in the home gardens of Nicobar. Dominance of most of the fruit and spice trees is higher in Andaman compared to that in Nicobar's home gardens. Among the Andamans, almost all the fruit trees found there are encountered in the home gardens, but their proportions are different. Proportion of mango, banana, cashew nut, papaya, and sapota is highest in home gardens of South Andaman and lowest in that of the Little Andaman. However, proportion of custard apple and pineapple is highest in Little Andaman and citrus in North Andaman. Proportion of mango in North Andaman is nearly equal to that found in South Andaman. Like fruit trees, proportion of spice trees is also highest in South Andaman. Forest trees are found in almost all the home gardens of the studied islands, but their numbers are higher in the home gardens of Andaman compared to that found in Nicobar district. Plant species found in the studied home gardens are similar to that found in Kerala, India (Nair and Sreedharan 1986), Bengladesh (Millate-E-Mustafa et al. 1996) and Thailand (Boonkird et al. 1984). On the basis of cluster analysis, six home garden types namely arecanut-coconut-banana-pineapple, arecanut-coconut-banana-mango, areanut-coconut-mango-banana, arecanut- coconut- pineapple-banana, coconut-banana-arecanut, and coconut-arecanut-banana are found that correspond the home gardens of South Andaman, North Andaman, Middle Andaman, Little Andaman, Car Nicobar and Nancowry, respectively.

Diversity in different categories of trees differs among the home gardens across the islands. Palm trees registered lowest diversity but highest concentration of dominanace in the home gardens of all the islands. Among the Andamans, fruit trees record the highest diversity in Middle Andaman and lowest in Little Andaman. Similarly, spice trees register the highest diversity in Little Andaman and lowest in North Andaman. All species diversity is the highest in the home gardens of Nicobar and lowest in that found in the Andamans. In general, species richness increased whereas evenness declines with increasing diversity in the home gardens. In the home gardens, arecanut contributes maximum (54–76 %) followed by coconut to the density in the home gardens. Arecanut: coconut ratio is higher in the homesteads indicating that farmers prefer arecanut most as it provides economic security to household, whereas coconut serves the subsistence. Arecanut was a highly remunerative crop when free trade was not affected. The arecanut is distributed in closer spacing (60 cm) because of its natural growth from fallen seeds (nuts) always

Table 7.5 Yield of some fruit and vegetable species grown with plantations in Andaman Islands

	Vernicular name	Economic produce	Average yield
Vegetables			
Abelmoschus esculentus	Okra, Ladies' finger	Fruit	8–10 t ha^{-1}
Amaranthus gracilis, A. blitum var oleracea	Amaranthus, Chulai	Entire plant	10–15 t ha^{-1}
Artocarpus spp.	Jack fruit	Fruit	250–300 fruits (each 2–5 kg) tree^{-1}year^{-1}
Basella rubra	Poi	Leaves	15–20 t ha^{-1}
Capsicum annuum, C. frutescens	Green pepper	Fruit	20–25 t ha^{-1}
Carica papaya	Papaya	Fruit	13–15 t ha^{-1}
Colocasia esculenta	Taro, Arvi	Rhizome	8–12 t ha^{-1}
Cucurbita maxima, C. pepo, C. moschata	Pumpkins	Fruit	20–25 t ha^{-1}
Dioscorea spp.	Dioscorea	Tuber	20–25 t ha^{-1}
Eryngium foetidum	Burmese coriander	Leaves	7–10 t ha^{-1}
Ipomoea batatas	Sweet potato	Root	30–35 t ha^{-1}
Lagenaria vulgaris	Bottle gourd	Fruit	35–40 t ha^{-1}
Luffa acutangula, L. cylindrical	Ridge/smooth gourd,tori	Fruit	15–20 t ha^{-1}
Lycopersicon esculentum	Tomato	Fruit	20–30 t ha^{-1}
Manihot esculenta	Cassava	Tuber	15–20 t ha^{-1}
Moringa oleifera	Moringa, Drum's strick	Fruit	10–15 kg tree^{-1}
Momordica charantia	Bitter gourd	Fruit	25–30 t ha^{-1}
Musa paradisiaca	Banana	Fruit	1,500 bunches (each of 18–22 kg)
Raphanus sativus	Radish	Root	25–30 t ha^{-1}
Solanum melongena	Brinjal	Fruit	20–35 t ha^{-1}
Spinacea abracea	Spinach	Leaves	15–20 t ha^{-1}
Irichosanthes cucumbrina	Snake gourd	Fruit	25–30 t ha^{-1}
Pulses			
Cajanus cajan	Pigeon pea	Fruit	500–600 kg ha^{-1}
Cyamopsis tetragonoloba	Guar, Cluster bean	Fruit	600 kg ha^{-1}
Dolichos uniflorus	Horse gram	Fruit	500–700 kg ha^{-1}
Vigna radiata	Green gram	Seed	1–2 t ha^{-1}
V. mungo	Black gram	Seed	2–2.5 t ha^{-1}
V. unguiculata	Cowpea	Fruit	6–7 t ha^{-1}
Fruits			
Achras zapota	Sapota	Fruit	30–35 kg tree^{-1}year^{-1}
Aegle marmelos	Bael	Fruit	300–350 fruits tree^{-1}year^{-1}
Ananas camosus	Pine apple	Fruit	2–3 kg per plant
Annona squamosa, A. ramosa	Custard apple	Fruit	30–40 kg tree^{-1}year^{-1}
Artocarpus spp.	Jack fruit	Fruit	200–250 fruits tree^{-1}year^{-1}
Averrhoa carambola	Kamrakh	Fruit	75–125 kg tree^{-1}
Carica papaya	Papaya	Fruit	30–35 kg tree^{-1}

(continued)

Table 7.5 (continued)

	Vernicular name	Economic produce	Average yield
Citrulus lanatus	Watermelon	Fruit	25–35 t ha^{-1}
Citrus spp.	Citrus	Fruit	500–700 fruits $tree^{-1}year^{-1}$
Emblica officinalis	Goose berry	Fruit	15–20 kg $tree^{-1}$
Mangifera indica	Mango	Fruit	150-200 kg $tree^{-1}year^{-1}$.
Musa paradisiaca	Banana	Fruit, stem	25–30 kg $tree^{-1}$
Psidium guajava	Guava	Fruit	30–50 kg $tree^{-1}$
Syzygium cuminii	Jamun	Fruit	60–80 kg $tree^{-1}$
Tamarindus indica	Tamarind	Fruit	200–250 kg $tree^{-1}year^{-1}$
Nuts/oil and fats			
Anacardium occidentale	Cashew tree	Nuts	5–8 kg $plant^{-1}$
Arachis hypogea	Ground nut	Kernels	1.5–2.0 t ha^{-1}
Areca catechu	Areca nut	Nuts	10–12 kg $palm^{-1}$
Brassica spp.	Mustard	Seed oil	200–300 kg ha^{-1}
Cocos nucifera	Coconut nuts	Nuts	100–150 nuts $palm^{-1}$
Helianthus annuum	Sunflower	Flower/seed	400–500 kg ha^{-1}
Sesamum indicum	Sesame	Seed oil	800–900 kg ha^{-1}
Spices and condiments			
Cinnamomum zeylanicum	Cinnamon	Bark	1.5–2 kg quills $tree^{-1}$
Curcuma longa	Turmetic	Rhizome	3–5 t $ha^{-1}yr^{-1}$
Myristica fragrans	Nutmeg	Nuts	750–1,200 nuts $tree^{-1}year^{-1}$
Piper nigrum	Peper	Fruit	2–2.5 kg $vine^{-1}$
Syzygium aromaticum	Clove	Dry flower buds	2–3 kg buds $plant^{-1}year^{-1}$
Zingiber officinalis	Ginger	Rhizome	10–12 t ha^{-1}fresh ginger
Others			
Saccharum officinarum	Sugarcane	Culm(juice)	20–25 t ha^{-1}
Zea mays	Maize	Grains	130–150 kg ha^{-1}

Source Dagar (1995)

supplements its population resulting into high density. Generally arecanut is distributed under coconut, but in bigger size home gardens a separate block of arecanut is grown mainly for maximization of household income. Banana particularly var. champa, locally known as "*Cheena kela*," is most common, contributing about 85 % to the total banana population. Pineapple is another fruit crop found relatively more common in the home gardens with quite high density but the frequency was found to be low indicating its uneven distribution across the home gardens. Only few farmers (6 %) are found to grow the fruit species for commercial purpose but maximum (94 %) for household consumption. In maximum home gardens labor-intensive cultivation of pineapple is avoided because the major source of household income is from Governmental services in South Andaman and at least one person from each family is engaged in Government job. The perennial ligneous species like tamarind (*Tamarindus indica*) and *Ceiba pentendra* are found generally in each home garden. The former serves as food whereas later provides flosses, which are used for making beds and pillows. Tree spices like clove, nutmeg, and cinnamon are found in the home garden, but relative frequency of cinnamon is the lowest perhaps due to higher labor input, in harvesting, debarking, and drying and comparatively low return (INR 200 per kg) at 2 years interval. Rotation cycle of cinnamon is generally 2 years, but few farmers make delayed harvesting for higher yields.

Spatial and Temporal Arrangements of Plant Species in the Orchards

The palm, fruit, spice, and forest trees form three-storey structure in the home gardens of the Andamans (Fig. 7.2). Top storey (12–16 m) is always occupied by arecanut, coconut and forest trees. The coconuts are planted generally at wider spacing in the beginning and some arecanuts in their interspaces. Trees like mango, jackfruit, neem, tamarind, and *Ceiba pentendra* form the second storey (4.5–9.5 m) and the fruit trees are planted mostly closer to houses. The pineapple, seed spices, and many vegetables form the ground layer while many climbers such as black pepper and gourds as vegetables are quite frequent in almost all the home gardens.

Spice trees like nutmeg (*Myristica fragrans*) and cinnamon (*Cinnamomum zeylanicum*), and fruit trees like papaya and lemon occupy the first storey. They are grown mostly under the coconut and occasionally under the arecanut. Clove is found in both the first and the second storey as well. Cinnamon is grown commonly under the arecanut. Banana is grown always relatively in open where water from the house drains. Distribution of the houses has no specific pattern but house construction is preferred on uplands to avoid stagnation of rainwater. Ground cover, generally, is not cultivated for annual crops, but occasionally it is cultivated for pineapple. A few smallholders (2–3 %) grow *Curcuma longa*, *Zingiber officinale*, *Manihot esculenta* and *Amorphophallus campanulatus* in the home gardens in South Andaman and only 8–10 % in North and Little Andaman mainly for household consumption. Flowers like *Crossandra infundibuliformis*), tuberose (*Polianthes tuberosa*), marigold (*Tagetes erecta*), and jasmin (*Jasminum grandiflorum*) are grown a little in front of the houses and *Ixora parviflora* on the boundary of home gardens. Unlike Andaman, coconut forms the second storey whereas banana the first storey in Nicobar. A wild tuber (*Tacca leontopetaloides*) and grasses constitute the ground flora in these home gardens.

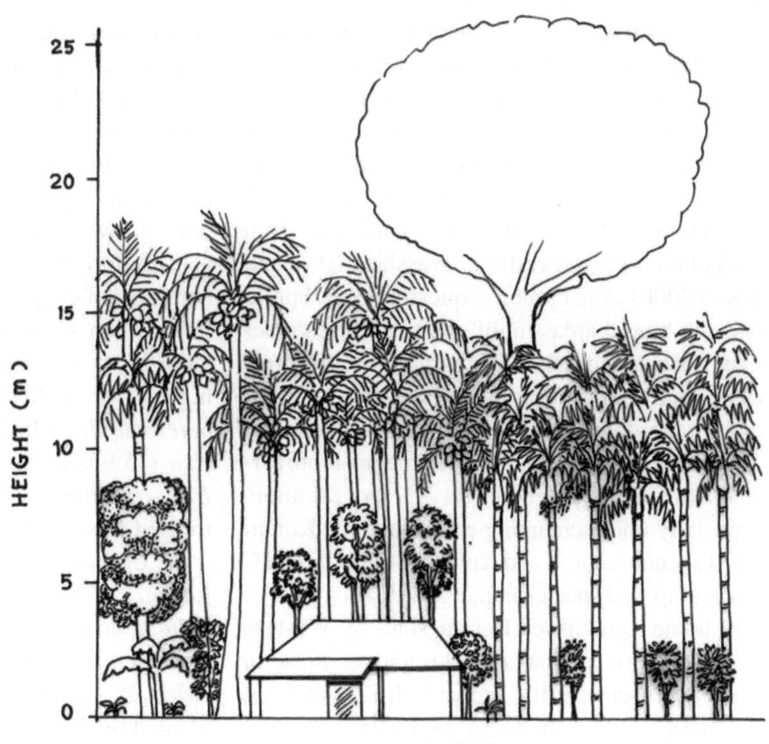

Fig. 7.2 Pictorial view of a typical home garden in the Andaman and Nicobar Islands

Temporal organization of crops is an important feature of homestead agroforestry. Temporal organization ensures round the year production in the system because of the difference in the biological cycle in the crops (Table 7.6). Plantation crops like coconut and arecanut flower and bear nuts round the year. Harvesting in arecanut generally occurs from September to December but maximum yields are found from October to November. Harvesting in coconut occurs from January to February for copra, but tender coconuts are harvested round the year.

Carbon Sequestration in Home Gardens

Change in land use pattern and management practices greatly influence C sequestration in soil. Large amount of soil organic matter and carbon are lost from soil following deforestation (20–50 % of stored carbon) and subsequent conversion to grazing and agricultural land (Eswaran et al. 1993; Reicosky and Lindstorm 1994; Reicosky et al. 1997). It has also established that the soil C can be increased to new higher equilibria with sustainable management practices and adoption of appropriate farming systems such as conservation agriculture (Reicosky 2002) and agroforestry (Nair et al. 2010). The rate of average global C sequestration was estimated to be 33.8 and 33.2 g C m^2 $year^{-1}$, respectively, due to changing land use from agriculture to agroforestry or grassland (Post and Kwon 2000). The carbon sequestration in humid and sub-humid areas in different land uses/practices such as conservation agriculture, agroforestry, and afforestation ranged 0.3–0.8, 0.2–3.1, and 4.0–4.8 t C ha^{-1} $year^{-1}$, respectively. Tropical home gardens with high agro-biodiversity have high potential for C sequestration, especially under changing environments. Kumar and Takeuchi (2009) and Kumar (2011) studied C stock of mixed species home gardens in a total of 839 home gardens of Kerala state of Western Coast and found that aboveground C stock of trees (>20 cm girth) ranged from 16.3 to 35.2 t ha^{-1} with a mean of 24.3 t ha^{-1}. Size of gardens was a major factor affecting C stocks per unit area and it decreased with increase in size. In home gardens of smaller size, the richness of species per unit area is richer than the larger gardens. Saha et al. (2010) also reported similar results for soil organic carbon stocks in the home gardens of Thrissur, Kerala.

Farming in Forests (Cultivation of Commercial Crops Under the Shade of Natural Forests)

Growing of commercial crops under the shade of trees of natural forests of the Western Ghats of peninsular India is very common. Cardamom hill reserves (CHR) in the high altitude regions of Kerala constitute a traditional agroforestry system. It involves growing of small cardamom (*Elettaria cardamomum*), a sciophytic commercial crop, under the shade of trees in natural forest. Dominant trees in the evergreen and semi-evergreen forests selectively retained by the growers provide shade to the cardamom crop in this age-old cultural system. Shade trees also vary from place to place depending on local preferences and the silvicultural characteristics of trees. Kumar et al. (1995) compared floristic diversity indices of trees in three CHR and a natural forest in the Western Ghats of Kerala and recorded the highest floristic diversity index for CHRs at Pampadumpara (Table 7.7) despite having lowest tree density.

The floristic spectrum of Pampadumpara site consisted of 40 tree species (522 trees per ha). The trees which showed more than 10 % importance value index (IVI) included *Vernonia arborea* with highest IVI (94.6 %), followed by *Artocarpus heterophyllus* (46.3 %), *Actinodaphne malabarica* (15.9 %), *Persea macrantha* (13.6 %) and *Erythrina lithosperma* (10.6 %). Other common trees at this site included *Cinnamomum malabatrum, Cedrela toona, Prunus ceylanica, Bischofia javanica, Chionanthus malabarica, Macaranga peltata*, and *Mallotus albus*. As many as 20 species were, however, represented only once or twice. A significant portion of shade trees was planted by the crop growers, presumably to offset the ill-effects of larger canopy gaps. The common

Table 7.6 Calendar of events of economically important plant species in the home gardens of Bay Islands

Sl No.	Species	Month											
		January	February	March	April	May	June	July	August	September	October	November	December
1.	*Plantation crops*												
	Coconut	FH	FH	FH	FH	FH	FH	FH	FH	FH	FH	FH	FH
	Arecanut	FH	F	FH	F	F	F	F	F	F	FH	FH	F
2.	*Fruits crops*												
	Cashew nut	F	–	–	H	H	H	–	–	–	–	–	F
	Mango	F	F	–	H	H	H	–	–	–	–	–	F
	Lemon		–	–	F	F	F	–	H	H	H	–	–
	Banana	FH	FH	FH	FH	FH	FH	FH	FH	FH	FH	FH	FH
	Papaya	FH	FH	FH	FH	FH	FH	FH	FH	FH	FH	FH	FH
	Pineapple	F	F	–	–	H	H	H		–	–	–	–
	Jack fruit	–	–	H	H	H	H	–	–	–	–	F	F
3.	*Spices*												
	Clove	H	–	–	–	–	–	F	F	–	–	–	–
	Cinnamon	–	–	–	–	–	–	–	–	–	H	H	–
	Nutmeg	FH	FH	FH	FH	FH	H	FH	FH	FH	FH	FH	FH
	Black pepper	H	–	–	–	–	F	F	–	–	–	–	H
	Bay leaf	H	H	H	H	H	H	H	H	H	H	H	H
4.	*Others*												
	Rice	H	–	–	–	–	–	T	T	–	–	–	H
	Vanilla	–	F	F	–	–	–	–	–	–	–	H	H

F Flowering, *H* Harvesting, *T* transplanting
Source Pandey et al. (2007)

Table 7.7 Floristic diversity indices of trees (>10 cm girth at breast height) in the cardamom hill reserves (CHR) and evergreen forests of high ranges in the Western Ghats of Kerala (plot size 5,000 m^2)

Site	Number of species	Number of individuals	Number of individuals per ha	Simpson's index
Pampadumpara (CHR)	40	261	522	0.860
Kumili (CHR)	22	352	704	0.697
Devikolam (CHR)	25	350	700	0.701
Ayyappancoil (Forest)	42	988	1,976	0.933

Source Kumar et al. (1995)

species found on other two CHR sites namely Kumili (with 704 trees per ha) and Devikolam (with 700 trees per ha) included *Cullenia exarillata, Palaquium ellipticum, Trema orientalis, Erythrina indica, Mesua nagassarium, Canarium strictum, Macaranga peltata, Artocarpus heterophyllus, A. hirsutus, Ficus hispida, Bischofia javanica, Cedrela toona, Mangifera indica, Myristica dactyloides*, and *Garuga pinnata*. Other species were present occasionally on both or either of the two sites. The evergreen natural forest site at Ayyappancoil recorded a much higher tree density (1976 trees per ha) containing 42 tree species. *Aporusa lindleyana, Hydnocarpus pentandra, Alstonia scholaris, Litsea stocksii, Clerodendron viscosum, Antidesma bunis, Vernonia arborea, Cullenia exarillata, Mesua nagassarium, Holigarna arnottiana, Coreya arborea, Buchanania axillaris, Artocarpus hirsutus, Palaquium ellipticum, Wrightia tinctoria*, and *Vitex altissima* are important tree species found in evergreen natural forests. Presence of heliphilic components (*Aporusa lindleyana, Hydnocarpus pentandra, Alstonia scholaris*, and many others) suggest that these forests are not altogether free from anthropogenic disturbances, and that, in turn, may help the regeneration and survival of such components. The other commercial crops such as cacao (*Theobroma cacao*), coffee (*Coffea* spp), tea (*Camellia sinensis*) and spices are also cultivated under shade trees.

"Farming in forests" has also been prevalent since long in the Andaman and Nicobar Islands. Young regenerating forest trees are retained or suitable multi-purpose woody perennials planted at regular intervals. These also support climbers such as black pepper or betel vine. Nutmeg (*Myristica fragrans*), cinnamon (*Cinnamomum verum*), clove (*Syzygium aromaticum*), coffee (*Coffea arabica, C. indica, C. robusta*), and cacao (*Theobroma cacao*) are planted between the retained or planted trees as their root system is not competitive with deep rooted trees. Betel vine and black pepper are planted near trees. *Erythrina variegata, Cieba pentendra, Areca catechu*, and *A. triandia* are frequently used as support trees for climbers in the islands. The interspaces on sloping lands are also used for fruits like pineapple (*Ananas comosus*) but fodder grasses such as *Pennisetum purpureum, Trepsicum laxum, Panicum maximum, Pennisetum polystachion*, and *Setaria anceps* and fodder legumes such as *Stylosanthes guienensis, S. hamata*, and *Calopogonium mucunoides* are considered more suitable as ground cover to check soil erosion from sloping lands and meet fodder requirements in the islands.

The Jirikatang Farm of the Department of Agriculture, Andaman and Nicobar Islands is a unique example of this type of farming system. The tall forest trees, including commercial timber trees (e.g., *Dipterocarpus* spp. *Pterocarpus dalbergioides, Terminalia* spp., *Bombax insigne, Legerstroemia hypoleuca, Artocarpus chaplasha, Albizia lebbeck, Canarium euphyllum*, and others) have been retained as canopy with the above-mentioned spice species as middle storey crops, and pineapple or fodder crops as a ground storey crops. Farmers may get economic benefits at regular intervals from this system.

Plantation-Based Multiple Cropping Systems (Plantations Other Than Home Garden Trees)

Contrary to the popular belief, a sizeable percentage of the total production of most plantation crops (coconut, red oil palm, rubber, cashew, etc.) in developing countries comes from small holdings, and for these, multiple cropping systems are of great importance and relevance. In India, there are about 5 million coconut holdings with 985 of them occupying only <2 ha (Thampan 1996). The hilly tracts in the Western Ghats are generally dominated by monospecific plantations such as rubber, red oil palm and cashew nut trees. Such plantations offer opportunities for integrating various herbaceous crops. For example, intercropping pineapple (*Ananas camosus*), banana (*Musa* spp), and cassava (*Manihot esculenta*) in combination with rubber in Kerala (Rajasekharan and Veeraputhran 2002); and turmeric, cassava and banana intercropped with cashew nut, coconut and rubber in Andamans (Dagar 1995a) are quite successful. Ismail (1986) and Dagar (1995a) described the livestock component as most important and advocated to integrate it with most of the plantations particularly coconut and rubber plantations.

Rubber and red oil palm plantations of the Forest and Plantation Development Corporation of Andamans cover 1,003 and 1,593 ha, respectively. But Singh et al. (1988), Dagar (1994) and Dagar et al. (1995) showed that the soil under these plantations has deteriorated in terms of pH, organic carbon and available nutrients (Table 7.8) particularly under red oil palm, which was raised after removal of rain forest and deteriorated soil after 13 years of plantation. So monocultures are to be discouraged in favor of sustainable agroforestry systems, well blended with leguminous species.

Agroforestry interventions in coconut plantations are another interesting area. Studies carried out in Kerala and elsewhere reviewd by Thampan (1996) who concluded that intercropping was of great economic advantage to the small and marginal farmers. Among different combinations of perennials like cacao, pepper, cinnamon, clove, nutmeg, coffee, and mulberry under coconut,

Table 7.8 Physico-chemical characteristics of soil in upper 15 cm layer under different types of plant cover in Andamans

Plant cover → Soil parameters ↓	Evergreen forest	Teak (24 years)	Oil palm (13 years)	Rubber (23 years)	Coconut (23 years)	Cashew nut (21 years)	Paddy
pH (1:2)	6.1	7.2	5.5	5.7	5.4	5.4	5.5
Bulk density (g cm^{-3})	1.11	1.47	1.55	1.35	1.49	1.40	1.48
Water storage (cm m^{-1})	32.3	18.0	23.0	27.1	ND	ND	ND
Cumulative intake rate of water (cm in 3 h)	26.6	13.0	13.8	16.0	ND	ND	ND
Organic C (%)	3.2	1.7	1.0	2.7	2.1	1.7	1.1
Extrct. P (mg/100 g)	0.7	0.4	0.1	0.6	0.4	0.4	0.5
Exchangeable K (mg/100 g)	14.0	9.3	6.3	8.3	2.0	5.8	9.6
Exchangeable Ca (mg/100 g)	13.0	6.0	7.8	5.2	6.4	4.3	7.5
Exchangeable Mg (mg/100 g)	5.6	4.7	8.9	3.2	5.3	3.7	3.6

Source Dagar et al. (1995)

cacao was found to be the most ideal mixed crop in a 50-year old coconut plantation. Besides producing good yields, cacao improved the productivity of coconut as well. When cacao was grown in a single row (300 plants per ha), the post-treatment average annual yield of coconut for 6 years was 64 nuts per palm as against 57.6 nuts for the corresponding pre-treatment period. When cacao was grown in double rows (600 plants per ha), the post-treatment average annual yield of coconut was 67 nuts per palm as against 60 nuts for the pre-treatment period. The average yield of coconut in the absence of cacao for both the periods was 63 and 60 nuts, respectively. The average yield of cacao beans in single and double row planting was 165 and 378 kg per ha per year, respectively. Among the coconut-based farming systems, mixed farming was found to be more efficient in employment generation than either of the cropping patterns such as intercropping, mixed cropping, and multistoreyed cropping. With properly organized mixed farming involving fodder cultivation and maintainance of milch animals (4–5 animals per ha) created a labor potential of 1,000 mandays annually assuring full time employment to three adult members of a farmer family. Further, when irrigation was available, the net income from integrated coconut farming could be increased to as much as 400 % compared to that of monocrop of coconut.

There are many success stories of coconut farmers representing Andhra Pradesh, Karnataka, Kerala, Tamil Nadu, and Andaman-Nicobar Islands, which include:

Introduction of livestock and poultry in the system; diverse crop combinations including flower plants of commercial value (orchids, crotons, species of *Anthurium, Alpinia, Heliconea, Dieffenbachia, Aglonema*, etc.); black pepper trained on coconut palm; cultivation of high value crops like nutmeg (*Myristica fragrans*), guava (*Psidium guajava*), mangosteen (*Garcinia indica, G. cowa*), rambutan (*Nephelium lappaceum*), durian (*Durio zibethinus*), etc.; developing nursery blocks of improved high yielding germplasm; maintaining nursaries of shade loving decorative palms such as *Areca catechu, Areca lutescens, Livistonia rotundifolia*, etc.; integration of miscellaneous trees and shrubs such as arecanut palm, citrus species, banana, goose berry, ginger, turmeric, tuber crops, etc. and livestock; besides fruit trees and tuber crops manufacturing of vermicompost; implementation of organic farming involving livestock and high value crops; coffee, cardamom and pepper as mixed crops in coconut; coconut based mixed farming involving a combination of diverse crops with milch animals, apiary, poultry, etc.; cover crops for suppressing weeds and paddy in lowlying areas; introduction of green manure crops, turmeric and banana in hybrid coconut; and integrated farming systems. The synthesis of data from these farmers indicate that many progressive farmers are adopting organic farming using the raw material from the farms and getting annual income ranging from INR 50–200 thousand per ha (at times even more).

In Andaman-Nicobar Islands, coconut and arecanut-based multistoreyed agroforestry systems are quite common and banana, yams, pineapple, and grasses are cultivated in interspaces while black pepper is grown along

Fig. 7.3 Coconut-based multistoreyed agroforestry system in Andamans (Photo Courtesy: Dr. RK. Gautam)

Fig. 7.4 Arecanut plantations having grasses as ground cover and black pepper as climber. Photo: Dr. RK. Gautam

arecanut palm or *Gliricidia* as support (Figs. 7.3, 7.4). Although banana, colocasia (*Colocasia esculenta*), greater yam (*Dioscorea alata*) and elephant foot yam (*Amorphophallus campanulatus*) are popular understorey crops in coconut gardens, growing multipurpose tree species (MPTs) in the interspaces needs adequate attention. A wide spectrum of MPTs is used in tropical plantation programs. For example, *Ailanthus triphysa* is prominent in traditional land use systems of Kerala and widely used in match industry and yields a highly viscous aromatic resin used in medicines. The tree is commonly used as shade tree for commercial crops like ginger (*Zingiber officinale*).

Kumar et al. (2001a, b) observed the yield parameters of ginger under different densities of *Ailanthus* tree and when given different doses of fertilizers. They recorded highest rhizome yield (5.0 t ha^{-1}) in the 2,500 trees per ha stocking level under 3-year old plantation, which was optimum for below 5-year old stands on good sites. It represented 25 % mean daily photosynthetic active radiation (PAR) flux or 73 % midday PAR flux. The rhizome yield was 3.7, 3.6, 4.0 and 3.5 t ha^{-1} in the 3,333, 1,660, 1,111 trees per ha stocking level and control (without trees), respectively. Though application of fertilizer dose (N: P_2O_5: K_2O:: 50:25:25 kg ha^{-1}) gave highest yield (4.4 t ha^{-1}) as compared to other doses (100:50:50, 150:75:75 kg ha^{-1} and control without any fertilizer) the yields were not significantly different. The interaction of tree density and fertilizer was also not significant. The system improved the site nutrient capital when ginger was adequately fertilized, despite treeless ginger having relatively higher initial soil nutrient availability. In another study, Kunhamu et al. (2008) observed that the removal of two-third trees of 8-year-old *Acacia mangium* promoted understorey ginger productivity producing 787 kg ha^{-1} ginger rhizome as compared to 439 kg ha^{-1} under unthinned tree stand (1,600 trees ha^{-1}) while 2,473 kg ha^{-1} ginger was produced without any tree. However, the net present value in INR was obtained highest in *A. mangium* + ginger (INR 4,81,772; BC ratio 4.26) unthinned stand followed by one-third removed (1,066 trees per ha), 50 % removed (800 trees per ha) and two-third removed (533 trees per ha), respectively while from sole ginger only INR 1,58,488 could be obtained (Table 7.9) showing that agroforestry system is more sustainable and remunerative.

In Andamans, fruits such as banana (*Musa* x *paradisiaca*), papaya (*Carica papaya*), mango (*Mangifera indica*), guava (*Psidium guajava*), pineapple (*Ananas comosus*), sapota (*Achras zapota*), lemon (*Citrus limon/Citrus medica*), jack fruit (*Artocarpus* spp.), and custard apple (*Annona squamosa*) are frequently grown. The MPTs such as *Ceiba pentendra*, *Erythrina variegata, Areca catechu*, and *Gliricidia sepium* are grown as support for black pepper (*Piper nigrum*) in coconut plantations. Clove (*Syzygium aromaticum*), cinamon (*Cinnamomum zeylanica*), and coffee (*Coffea arabica*) are grown as middle storey cash crops. Dagar (1995a) successfully established coconut-based multitiered cropping system on highly degraded sloping land. Besides coconut, small tree-spices such as cinnamon and clove as second storey crops; and grasses such as lemon grass (*Cymbopogon fluxuosus*) for commercial aromatic oil, *Pennisetum purpureum* (hybrid napier) and leguminous stylo (*Stylosanthes guianensis*) for fodder; and Kudzu vine (*Pueraria lobata*) and *P. phaseoloides* as ground cover crops in the interspaces of coconut plantations were established.

Black pepper vine was cultivated at the base of coconut trees. After the age of 8–10 years, coconut produced 60–150 (average 135) nuts per plant; cinnamon started yielding bark (1.2–3.0 kg quills per shrub after third year; clove produced 2–3 kg buds per shrub; and

Table 7.9 Yield and economics of ginger intercropping with *Acacia mangium* as influenced by thinning regime

Thinning intensity (trees remaining)	Ginger rhizome dry weight (kg ha^{-1})	Ginger*		*A. mangium* + ginger**		
		BCR	NPV (INR)	BCR	NPV (INR)	IRR (%)
Unthinned (1,600 trees ha^{-1})	439^{c}	0.98	1,090	4.26	4,81,772	30.7
One-third removed (1,066 trees ha^{-1})	538^{bc}	1.18	10,055	3.96	4,35,607	30.4
50 % removed (800 trees ha^{-1})	613^{bc}	1.22	13,730	3.66	3,93,642	29.6
Two-third removed (533 trees ha^{-1})	787^{b}	1.54	33,305	3.36	3,45,339	28.8
Sole ginger (treeless control)	2,474^{a}	2.15	1,58,488	–	–	–

BCR Benefit cost ratio, *NPV* net present value in INR at 5 % discount rate, *IRR* internal rate of return
Source Kunhamu et al. (2008). Incolumn 2 values followed bysamesuperscript donot differsignificantly

hybrid napier, stylo and lemon grass produced up to 98, 57, and 127 tons fresh biomass per ha, respectively. Tuber crops such as tapioca, turmeric, and foot yam and vegetable crops such as okra, brinjal, cucurbits, and beans are cultivated as is the case in home gardens.

Many MPTs are being improved and domesticated for their multipurpose use. For example, during last one decade, Noni (*Morinda citrifolia*) has been widely collected and cultivated as sole tree and in combination with several plantations such as coconut, cashew nut, and fruit trees in Andamans. Farmers have been trained to utilize its fruit as raw, in pickels and making juice and jellies and now has become a tree of every household (Singh et al. 2009). The Bay Islands have been blessed with several trees in natural stands with several varieties, which may be domesticated for high value crops; collected and grown for improving their germplasm. For example, varieties/ecotypes of nutmeg (*Myristica fragrans*), Rudraksha (*Elaeocarpus sphaericus*), betel vine (*Piper betle*), banana, mango, and many others are found distributed wild in natural forests.

In Lakshadweep Islands, coconut is the main crop of economic importance. Due to rapid fragmentation of land holdings, farmers grow coconut trees in close spacing, which has resulted low yield. It has been estimated that on an average, 400–500 palms of all ages are found growing in one hectare of land as against 170–200 normally recommended for optimum yield. Average per palm annual yield was found to be 135 in normal spacing and 40 in double the density of planting (Jacob 2004). As the organic content of soil in the islands is very low and water and nutrient holding capacity is very poor, to increase productivity vermin-composting of coconut waste has been found successful. Experiments carried out at Minicoy have revealed that 98 % nitrogen and 28 % each of potassium and phosphorus could be substituted by growing sun hemp (*Crotolaria juncea*) in the interspaces of coconut (Jacob 2004). Cultivation of *Gliricidia sepium* as green manure crop is also practical method of increasing fertility of the soil. Further, selection of high yielding palms is also very important to increase productivity. Hybrid combinations produced by crossing the locally available cultivars (Laccadive Ordinary, Orange, Green, and Yellow dwarfs) performed better in nut-yield and early bearing in contrast to their parents/hybrids introduced from outside (CPCRI 2002). Production of palm jaggery and vinegar from the neera/coconut water, charcoal from coconut shell, compost from coir pith, vermin-compost (using *Eudrillus* earthworm) from coconut waste, and coir from coconut husk have been proved useful for increasing income of farmers. Banana (vars. Nendran, Kunnan, Robusta), papaya, watermelon, tuber crops (tapioca, sweet potato, and foot yam), mushroom and vegetables (tomato, chilli, snake gourd, bitter gourd, cucumber, pumpkin, brinjal, annual moringa, and amaranthus) cultivation has been found profitable and feasible to increase the productivity of coconut gardens in the islands (Jacob 2004; Jacob et al. 2002).

For the past two decades, several attempts were made in the cropping system research to include aromatic and medicinal plants with the conventional food, commercial, and plantation (including MPTs) crops. This not only increased the land use efficiency, productivity, but also the net returns from the particular piece of land (Maiti and Raju 2004). Mixed cropping, alley cropping, and multitier cropping with these plants proved to be quite successful with increased nutrient recycling having reduced pest and disease load on the crops. Among important medicinal and aromatic species suitable for coastal ecosystem include *Abelmoschus moschatus, Acorus calamus, Adhatoda vasica, Aloe barbadensis, Alpinia galanga, Andrographis paniculata, Asparagus racemosus, Bacopa monnieri, Cassia angustifolia, Catharanthus roseus, Centella asiatica, Clitoria ternatea, Coleus forskohlii, Curculigo orchioides, Curcuma longa, C. aromatica, Cymbopogon flexuosus, C. martini, Gloriosa superba, Kaempferia galanga, K. rotunda, Mucuna pruriens, Piper longum, Plumbago zeylanica, Pogostemon cablin, Sida cordifolia, Solanum surattense, Tinospora cordifolia, Tribulus terrestris*, and *Zingiber officinale*.

Silvo-Pastoral Systems (Cut and Carry Systems/Fodder Farming)

Though silvo-pastoral system refers to land use system in which pasture (grazing land) and livestock production are integrated with woody perennials on the same land management unit and grazing is major component; but here the concept covers broadly "cut and carry" fodder production practices also. These land use systems are generally characterized by higher productivity on account of the vertical stratification of the shoot and root systems of different components. The trees in managed species have a great potential for efficient cycling of plant nutrients. Growing of nitrogen fixing trees has additional advantage as these help in fixing the atmospheric nitrogen into the soil which in turn is utilized by the associated field crops.

The results of a field trial conducted by Mathew et al. (1992) in Kerala revealed that growth and yield of fodder species significantly influenced by tree components only after tree canopy formation.The fodder species such as *Pennisetum purpureum, Panicum maximum, Brachiaria ruziziensis* and *Euchlaena mexicana* grown in association with *Casuarina equisetifolia* and *Ailanthus malabarica* recorded comparatively higher forage yield even after canopy formation. However, forage yield in association with *Acacia auriculaeformis* and *Leucaena leucocephala* was relatively lower. The forage grasses performed in order *P. purpureum* > *P. maximum* > *B. ruziziensis* > *E. mexicana* producing mean biomass 74.5, 59.0, 42.5 and 23.9 t ha^{-1}, respectively. Kumar et al. (2001a, b) also studied performance of understorey herbage production of four forage grasses under *Acacia auriculaeformis, Ailanthus triphysa, Casuarina equisetifolia* and *Leucaena leucocephala* and observed that herbage production increased until 3 years in all tree + grass combinations, but declined subsequently, as the tree crown expanded after 3 years. Overall *Casuarina* among above-mentioned trees and hybrid napier (*Pennisetum purpureum*) and guinea (*Panicum maximum*) among forage crops (other forage crops were congo signal-*B. ruziziensis*, and teosinte-*Zea mexicana*) performed better than others. In another experiment, the same trees after 7 years could produce average total above ground biomass of 73.4, 7.8, 13.5, and 25.4 kg per tree and 183.5, 19.4, 33.7, and 63.5 t ha^{-1}, respectively with an annual mean increment of 26.2, 2.8, 4.8, and 9.1 t ha^{-1} year^{-1}, respectively from *A. auriculaeformis, A. triphysa, C. equisetifolia* and *L. leucocephala*, (Kumar et al. 1998).

Nutrient removal at harvest from the site depends on both nutrient concentration of different tissue fractions and biomass yield. Nutrient concentrations were found to vary markedly among the species, age classes and tissue types. In general, mineral element (N, P, K) concentration of tissue types decreased in the order foliage > branches > roots > bole. Although no species had clearly higher or lower nutrient concentrations for all tissue types, generally nitrogen fixing leguminous trees had markedly higher foliar nitrogen levels.

There are many opportunities of growing salt-tolerant fodder trees in situations like wetlands of West Bengal. Besides fodder trees, forage grasses such as *Coix lachryma-jobi, Brachiaria mutica*, and *Echinochloa* spp. can successfully be cultivated giving 3–5 cuts in a season. Based on the average of 2 years data these grasses could produce 41.3, 31.1, and 24.4 t ha^{-1}, respectively forage biomass from five cuts during *Kharif* season. When applied 60, 80, 100, and 120 kg nitrogen *Coix lachryma-jobi* could produce 39.2, 43.5, 48.4, and 50.0 t ha^{-1}, respectively forage biomass (Biswas 1994) showing that 100 kg nitrogen per ha is sufficient to get optimum yield from this grass, which grows well in stagnant water.

In Andamans, there are more that 8,300 ha of fallow lands and permanent pastures which can be upgraded or used for fodder cultivation and about 12,000 ha of barren and culturable wastes which can be managed as high quality pasture lands. The soil cover under sizeable area of coconut and rubber plantations could also be developed for perennial fodder cultivation. The productivity of local grasslands is very low. Though the practice of animal husbandary in the islands is just a century old and a spontaneous increase in ruminants has resulted in an acute shortage of green fodder especially during the lean period of the year, from December to April. There is hardly any piece of fallow land where cows, buffaloes, and goats are not seen grazing and browsing. This has caused tremendous pressure on these lands, resulting in the deterioration of vegetal cover, adversely affecting the physical and chemical properties of soil, and also exacerbating soil erosion. Sharma et al. (1990) advocated that by protecting these grazing lands from animals their productivity can be increased significantly. In one experiment the performance of exotic species of grasses and legumes was assessed in rainfed conditions without any irrigation on a sloping land producing substancial biomass (Sharma et al. 1991, 1992) during the year through multiple cuts (Table 7.10). These forages also perform well in wider spaces of fodder trees like *Trema tomentosa, Morinda citrifolia* and *Leucaena leucocephala. Morinda citrifolia* has been identified, adopted and domesticated as very useful MPTs by local people and they prepare several products from its fruit.

Moreover, the grasses such as thin-napier and guinea and legume stylo have high potential to propagate naturally through seeds once introduced. These may also be raised after protection of grazing lands and under old coconut or arecanut plantations. In wider spaces of these plantations nitrogen fixing trees like *Gliricidia sepium* can be grown, which may also be used as support for black pepper and improved fodder grasses such as napier or guinea or *Panicum* grass can be cultivated successfully (Fig. 7.5). Leguminous species such as *Calopogonium mucunoides, Clitoria ternatea, Phaseolus atropurpureus,* and *Stylosanthes* spp. are good cover crops and may also be cultivated as interspace crops in red oil palm and rubber plantations. These not only provide nutricious fodder but also increase soil fertility and protect it from erosion.

Multipurpose trees such as *Aegle marmelos, Artocarpus* spp., *Bauhinia variegata, Erythrina variegata, Grewia glabra, Hibiscus tiliaceous, Moringa oleifera, Pitchecelobium dulce, Pongamia pinnata, Samanaea saman, Sesbania grandiflora* and *Trema tomentosa*, which are growing successfully in these islands, may also be raised as fodder banks.

Table 7.10 Forage yield of some exotic perennial fodders without irrigation

Fodder species	Vemacular name	Oven dry biomass (tons $ha^{-1}year^{-1}$)	Number of cuts
Grasses			
Andropogon gayanus	Sadabahar	28.7	9
Brachiaria mutica	Para grass	28.1	7
B. ruzizensis	Congo signal	33.2	6
Panicum antidotale	Blue panic	18.0	6
P. maximum	Guinea	34.0	6
Paspalum plicatulum	Bryan	30.3	6
Pennisetum purpureum	Hybrid napier	30.5	8
Setaria anceps	Kazungula	20.5	8
Tripsacum laxum	Guatemala	20.6	7
Legumes			
Calopogonium mucunoides	Calopo	10.9	2
Clitoria ternatea	Condofan pea	8.7	2
Phaseolus atropurpureus	Siratro	4.5	2
Sytlosanthes guianensis	Stylo	23.4	2
S. scabra	Stylo	10.5	2

Source Sharma et al. (1992), Dagar (1995)

Fig. 7.5 Arecanut-*Gliricidia* (support to black pepper)-fodder grasses based agroforestry system in Andamans. Photo: Dr. RK. Gautam

Hedge Row (Alley) Cropping

These practices are usually adopted for sloping lands where forage shrubs are planted across the slope and forage grasses and legumes or crops in the interspaces. *Gliricidia sepium, Leucaena leucocephala, Cassia siamea, Morus alba*, and *Pithecelobium dulce* are trained as hedge row crops. Two parallel rows usually one meter apart of these woody perennials are raised across the slope in close spacing. Further, these rows are repeated leaving 4–6 m wide space for growing intercrops. The hedges are frequently cut at about 1 m height from the ground and the sticks are usually used as fuel wood and foliage as fodder or mulch. In one experiment conducted in high rainfall area (Andamans), *Gliricidia sepium* was established from cuttings of mature plants, which were planted in alleys on a gravely sloping land (across the slope). Four herbaceous fodder species, hybrid napier (*Pennisetum purpurium),* Kazungula (*Setaria anceps*), guinea (*Panicum maximum*) and stylo (*Stylosanthes guianensis*) were raised in interspaces (Dagar and Kumar 1992). *Gliricidia sepium* yielded 24.5 t ha^{-1} of fresh lopped biomass in two cuts after 1 year of growth. During the second year, the yield of lopped *Gliricidia* and intercrops increased abruptly owing to gap filling and greater branching (Dagar 1995a). Forage crops produced 17.6–46.8 t ha^{-1} fresh forage biomass but due to root competition the biomass was reduced from 13.5 % in guinea grass to 49.8 % in hybrid napier grass (Table 7.11).

Taungya

Taungya, a traditional system of establishing commercial forest plantations such as *Tectona grandis*, *Eucalyptus* spp., and *Ailanthus triphysa* in which agricultural crops are cultivated on a temporary basis between regularly arranged rows of trees, has been widely practiced in Western Ghats till recently. The greatest disadvantage of taungya and other systems established on sloping lands, however, has been soil erosion caused by bed preparation for cultivation of arable crops (Alexander et al. 1980; Moench 1991). In a taungya system involving cultivation of cassava (*Manihot esculenta*) with forest tree *Eucalyptus*, Gopinathan and Sreedharan (1989) reported that 10 % substitution of cassava with grass strips reduced soil erosion by 41 %.

Table 7.11 Performance of forage crops and *Gliricidia sepium* in alley cropping system (t ha^{-1})

Fodder crop	Fresh forage yield			Fressh weight of *Gliricidia* loppings		
	Without *Gliricidia*	With *Gliricidia*	Reduction in yield (%)	Stick	Foliage	Total
Pennisetum purpureum (Hybrid napier)	73.3	36.8	49.8	12.8	25.6	38.4
Setaria anceps (Kazungula)	48.9	35.0	28.5	34.6	75.0	109.6
Panicum maximum (Guinea)	54.1	46.8	13.5	22.0	47.6	69.6
Styloanthes guianesis (Stylo)	26.8	17.6	34.4	22.4	49.6	72.0
CD at 5 %	18.0	16.5	–	–	–	–

Source Dagar (1995)

Many of the nitrogen fixing species can convert substantial quantities of atmospheric nitrogen into a combined form (Danso et al. 1992) and can successfully be utilized to improve taugya plantation stands. The growth of teak was reported to increase in Java when planted in association of *Leucaena* (Van Noordwijk et al. 1996) and growth of *Eucalyptus saligna* was significantly more when it was interplanted with *Albizia falcataria* (DeBell et al. 1989; Binkley et al. 1992). *Leucaena* has been reported to fix 100–150 kg of nitrogen $ha^{-1}year^{-1}$ (Dommergues 1987). A significant portion of this nitrogen is released into the rhizosphere through leaf litter, fine roots and nodule turnover. Associated crops may utilize the nitrogen so released (Mathew et al. 1992).

Effects of intercropping of teak with *Leucaena* were studied by Kumar et al. (1998) in their experiments conducted in Kerala and observed the significant increase in tree growth and the system also modified soil characteristics. Fourty-four months after planting, teak in the 33 % teak −67 % *Leucaena* (2 rows of *Leucaena* for every row of teak) mixture were 45 % taller and 71 % larger in diameter at breast height (DBH) than those in pure stands. The increase in height and DBH was 20 and 26 % when teak and *Leucaena* were planted in alternate rows. Total N content of the soil increased with increasing relative proportion of *Leucaena*; available P levels were heighest in the 1:1 teak-*Leucaena* mixture, while available K was heighest in the 1:2 mixture. Despite the favorable effects of intercropping *Leucaena* on teak growth, increasing the relative proportion of *Leucaena* substantially (>50 %) may be counter-productive, as it would substantially reduce teak density. A 50 % mixture (alternate rows of teak and *Leucaena*), therefore, is considered optimal. Therefore, if *Leucaena* is planted in alternate rows with teak, intercropping with agronomic crops might be profitable and a substantial quantity of firewood could be produced in these improved taugya agroforestry systems.

Woodlots

These days high biomass production is an important consideration in all tropical tree planting programs. Biomass productivity of the MPTs, however, differs enormously with species, site characteristics and stand management practices. Nonetheless, it is always useful to know the stocks of carbon as biomass per unit area, not only to facilitate choice of species but also to assess the impact of deforestation and re-growth rates on the global carbon cycle (Deans et al. 1996). These days agriculture has become labor-oriented and private industrial forestry has become a recent phenomenon. The farmers with medium to large holdings are found to integrate trees with field crops and/or animal production. At many places, they raise MPTs as woodlots to have enhanced income and avoid labor investment. Plantations of *Eucalyptus, Casuarina, Bambusa,* and *Acacia auriculaeformis* are quite frequent all along coastal regions. Commercial plantations (woodlots) such as of cashew nut (*Anacardium occidentale*) are quite common

Table 7.12 Mean biomass accumulation (t ha^{-1}) in MPTs of 8 years and 10 months old trees in Kerala

Species	Bole	Branch	Foliage	Roots	Total aboveground biomass	Mean annual increment (t ha^{-1} $year^{-1}$)	Number of trees sampled
Acacia auriculaeformis	274.93 (109.97)	42.55 (17.02)	8.95 (3.58)	17.73 (7.09)	326.43 (130.57)	37.09	31
Ailanthus triphysa	28.78 (11.51)	7.68 (3.07)	4.08 (1.63)	7.40 (2.96)	40.54 (16.21)	4.61	30
Artocarpus heterophyllus	54.38 (21.75)	19.85 (7.94)	7.78 (3.11)	10.13 (4.05)	82.01 (32.77)	9.32	32
Artocarpus hirsutus	32.15 (12.86)	14.83 (5.93)	11.95 (4.78)	11.15 (4.46)	58.93 (23.57)	6.70	28
Casuarina equisetifolia	73.25 (29.30)	16.63 (6.65)	5.70 (2.28)	5.60 (2.24)	95.58 (38.23)	10.86	26
Emblica officinalis	46.20 (18.48)	18.23 (7.29)	4.43 (1.77)	12.63 (5.05)	68.86 (27.54)	7.83	17
Leucaena leucocephala	15.28 (6.11)	6.25 (2.50)	1.28 (0.51)	3.23 (1.29)	22.81 (9.12)	2.59	18
Paraserianthes falcataria	141.18 (56.47)	37.25 (14.90)	5.05 (2.02)	13.78 (5.51)	183.48 (73.39)	20.85	19
Pterocarpus marsupium	52.60 (21.04)	10.08 (4.03)	3.43 (1.37)	7.30 (2.92)	66.11 (26.44)	7.51	30

Values in paranthesis are kg per tree
Source Kumar et al. (1998)

near Goa on West Coast and Puri, East Coast. Despite of raising woodlots of MPTs, very limited efforts have been made for systematic studies of these woodlots regarding their biomass production and nutrient use efficiency.

Kumar et al. (1998) estimated tree biomass of nine MPTs at 8 years and 10 months of their growth. Biomass accumulation showed wide variations and the aboveground biomass was highest (326.4 t ha^{-1}) in *Acacia auriculaeformis* followed by *Paraserianthes falcataria* (183.5 t ha^{-1}) and minimum (22.8 t ha^{-1}) was from *Leucaena leucocephala* (Table 7.12).

Live Fences and Hedges

Many trees are found grown on field boundaries, which are used as multipurpose trees by the farmers. *Acacia nilotica, Ailanthus excelsa, Bambusa* spp., *Borassus flabellifer, Casuarina equisetifolia, Cocos nucifera, Carissa carandas, Cordia rothii, Dalbergia sissoo, Ficus* spp., *Leucaena leucocephala, Moringa oleifera, Prosopis juliflora, Syzygium cuminii, Tamarindus indica* and *Ziziphus mauritiana* are very frequently found on bunds or farm boundaries. Many of these in association with shrubs are trained as live fences or hedges.

In coastal areas of Gujarat region, *Prosopis juliflora* is commonly trained as protected hedge on farm boundaries along with species of *Agave, Capparis* and cactus such as *Cactus indicus, Cereus peruvianus,C. triangularis, C. hexagonus, Opuntia dillenii, O. monocantha, O. tuna,* and many others. All along Orissa coast *Casuarina, Pandanus,* and *Acacia auriculaeformis* are very common. In many areas several plants are grown as live fence and hedges around farms and home gardens. *Bambusa* spp., *Vitex trifoliata, V. negundo, Jatropha gossipifolia, Ficus rumphii, Agave sisalana, Ehretia microphylla, Clerodendrum inerme, Duranta repens, Erythrina variegata, Lawsonia inermis, Pithecellobium dulce,* and *Gliricidia sepium* are the most common species grown but there has been no systematic research in this field. In one trial with *Leucaena leucocephala* about

23.3 t ha^{-1} dry biomass was obtained in six cuts round the year when planted densely (50 cm × 50 cm) as hedge crop. *Carissa carandas, Gliricidia sepium* and *Pithecellobium dulce* also perform well in the islands. Many plants coppice well and are mentioned under multipurpose woody perennials but more research is needed to find out more suitable species for live fences and hedges.

Bamboos occur extensively in the managed ecosystems both as plantations and in agroforestry as scattered clumps as well as hedgerows on farm boundaries. Now the bamboo is considered to be an important livelihood strategy of rural people and is being elevated from a raw material known as the "poor man's timber" to the status of "timber of the 21st century." Among the ~ 130 wild and cultivated bamboo species reportedly occurring in India (Sharma 1987), thorny bamboo (*Bambusa bambos*) is considered the most important in the life of rural people along West Coast of India. Chandrashekara (1996), Shanmughavel and Francis (1996) and Shanmughavel et al. (2001) studied the ecology and aboveground biomass accumulation in bamboo stands of Kerala.

Kumar et al. (2005) have studied aboveground biomass production and nutrient uptake of hedgerow-raised 20 years-old thorny bamboo (*B. bambos*). They measured the aboveground biomass of bamboo clumps averaged 2,417 kg per clump with an average accumulation of 242 t ha^{-1}. Highest biomass accumulation (82 %) was observed in live culms followed by thorns + foliage (13 %) and dead culms (5 %). Nutrient (NPK) export at harvest was also in the same order. Average N, P, and K removal was 9.22, 1.22, and 14.4 kg per clump, respectively. Litter accumulation on the forest floor averaged 9.1 t ha^{-1} accounting for 482, 367, and 430 kg ha^{-1} of N, P and K, respectively.

Shelterbelts and Shore Protection

In coastal areas, high winds also carry salt with them and damage crops. Many trees and shrubs such as *Casuarina equisetifolia, Acacia auriculaeformis* and *Gliricidia sepium* may play very important role in reducing the speed of these winds and may protect the crops from injury. These not only protect the crops but also help in soil amelioration. Most of the coastal areas are prone to damage caused by cyclones and even Tsunamies. Mangroves have very dense root systems and protect the shore from the damage caused by these natural desasters. We must protect and conserve all the present stands and all afforestation programs must be at place to restore the mangrove degraded areas by planting suitable species. Besides mangroves littoral species such as *Pandanus* spp., *Thespesia populnea, Scaevalia taccada, Tournefortia ovata, Hibiscus tiliaceus*, and *Salvadora persica* may also play important role in protecting the shores and beaches. MPTs such as *Calophyllum inophyllum, Pongamia pinnata, Heriteria littoralis, Terminalia catappa*, and *Manilkara littoralis*, which are found growing luxuriously along beaches of Andamans, may be raised on degraded low lying areas. These belts protect the shores/beaches, provide valuable forest products and also give shelter to wild life.

Multienterprise Farming Systems/ Aqua-Silviculture Systems

In coastal areas, aqua (shrimp and fish) culture in association with paddy cultivation or in denuded mangrove areas is age-old practice. On the bunds of fish ponds, plants of coconut palm and banana are quite frequent. Many farmers also grow vegetables on dykes of fish ponds. Despite of the fact that most of the small families are dependent on this system of cultivation very little research efforts have been made to improve this cultivation. During last two decades, efforts have been made to develop integrated farming systems particularly in waterlogged areas involving fish culture in fishpond, livestock, food and forage crops, vegetables, fruit trees on dykes, paultry/duckry, piggery (if feasible) and plantation crops.

The concept of farming system has been interpreted and adopted through Agricultural Universities in coastal states. Components like horticulture, sericulture, forestry, fish culture, and live stock production have been considered most viable to integrate with annual cropping. For wetland situation, model farming systems have been developed integrating components like fishery and poultry with cropping. To get rid of uncertain yield or very low yield from the traditional paddy cultivation in coastal saline lands, brackish water fish (*Peneous monodon*) and fresh water fish-like *Tillapia mossambica, Mugil passia* and *Mugil tade* have been found most remunerative in situations like West Bengal and Orissa. Integrated farming system approach combining field, horticulture, and plantation crops; livestock (dairy), biogas and goatery; and silviculture proved to be a viable preposition for marginal and small farmers under dryland/rainfed situation (Mahapatra and Panda 1994).

An integrated farming system was evaluated in farmers'field of Cuttack district of Orissa (Mohanty et al. 2004). Out of 2.5 ha water-logged area, 1.64 ha was converted into fish-pond, while vegetables, flower, and fruits were grown on raised embakment. During stocking, a density of 7,500 fingerlings per ha was maintained in fish pond with species composition of 30: 40: 15: 15 (*Catla catla*: *Labeo rohita*: *Cirrhinus mrigala*: *Cyprinus carpio*). In addition to this, prawn post-larvae of *Macrobrachium rosenbergii* were also stocked in the main pond for polyculture with Indian major carps (@ 15,000 per ha). Poultry sheds were also constructed for rearing about 4,000 birds in such a way that droppings could fall into pond as organic manure and feed for fish. The average productivity of fish and prawn culture alone was 8.1 t ha^{-1} per annum and gross and net return from fish and prawn in 2002 was INR 3,76,317 ha^{-1} and 2,01,868 ha^{-1}, respectively. This accounted for INR per m^3 of water productivity in the pond system alone. The gross and net returns from the whole system of 2.47 ha during the year were INR 6,51,110 (INR 2,63,607 ha^{-1}) and INR 3,62,515 (INR 1,46,767 ha^{-1}), respectively. Thus, the system was found to be most profitable and sustainable.

Most of the coastal aquatic animals including fish utilize the mangrove water as nursery and breeding grounds. A variety of these animals associates with mangroves and make complex but interesting food web. Mangroves contribute substantially the nutrients to fisheries in the adjacent coastal waters. There are many forms of aquaculture, such as oyster, crab, fish, and prawn culture enclosed either in pans or cages, which may be undertaken in mangrove swamps without destruction of the habitat. In many coastal areas, shrimp culture is practised by means of pond construction behind mangrove areas. The ponds are connected to coastal waters through channels and during daily high tides the water flows into the pond. The outflow of water and shrimps is controlled by sluice gates. Species of *Avicennia, Sonneratia, Rhizophora, Bruguiera, Ceriops,* and *Cynometra* are considered good fodder trees and may be raised in paired rows (1–2 m apart) in mangrove swamps. On maturity, the alternate rows may be harvested for fuel and fodder. Some low lying areas which are partially reclaimed are being used in Andamans for coconut cultivation. Andaman Tall and Katchal Tall varieties perform well yielding 30–65 nuts per palm per year. These may be planted on dykes of fishponds along with other MPTs. On raised bunds, fodder grasses such as guinea (*Panicum maximum*), thin napier (*Pennisetum* sp.), Guatemala (*Tripsicum laxum*), and stylo (*Stylosanthes guianensis*) could perform well.

Afforestation of Coastal Saline Areas (Site Specific Systems)

On the coastlines wherever sulfur-containing sediments accumulate in tidal marshes or swamps or mangrove ecosystems, acid sulfate soil formation takes place. These formations are quite frequent in southern India and Andaman-Nicobar Islands). Rice is a major crop in the tropical and sub-tropical areas where acid sulfate soils are formed. Rice is otherwise a quite tolerant crop, but it requires submerged conditions, which help in controlling the pH. However, under such conditions it also suffers from an excess of water-soluble iron and particularly of aluminum and toxic effect of these is a consequence of the strongly acidic pH. The application of lime mitigates the adverse and harmful effects on plants, but it is a costly preposition. Therefore, alternative land use systems need to be evolved.

From the management purposes, the salinity-related problematic soils of the coastal areas may be classified as: (i) land impregnated with high salinity and flooded with sea water, (ii) acid sulfate soils, (iii) land impregnated with high salinity and waterlogging but not flooded with sea water, (iv) land with low salinity and shallow water table (at 0.5–4.0 m depth) with good quality water but saline water beneath, and (v) waterlogged and saline soils caused by seepage in canal command areas. Suitable agroforestry systems offer scope for increasing the income and employment generation for small farmers, meeting the local needs of fodder and fuel, conserving biodiversity, improving the coastal environment, protecting the soil from erosion, and creating environment for the wild life. Nair and Sreedharan (1986), Dagar (1991, 1994, 1995a, b, 1996, 2000, 2009, 2012), Dagar and Tomar (1998), Pandey et al. (2007), and Kumar and Kunhamn (2011) have explained several agroforestry practices for coastal and island regions in detail.

Afforestation of land impregnated with high salinity and flooded with sea water

The areas lying closer to the sea are flooded regularly with seawater, therefore, have high salinity. The tidal areas protected against high wind velocity and waves of high intensity (as in case of many creeks, lagoons, and estuaries) form a suitable situation for mangroves. Dagar (1982, 2003, 2005b, 2008) and Dagar et al. (1991, 1993) gave an illustrative account of distribution, zonation pattern, importance and management of mangrove forests. For cultivation of mangroves, we need seedlings of appropriate size. Many mangrove genera such as *Rhizophora, Ceriops, Aegialitis, Bruguiera, Kandelia, Aegiceras*, and *Cynomitra* are viviparous and seed germinate when the fruit is intact. The radical falls when mature and develop roots when meets the muddy substratum. These mature radicals can be collected from the mangrove stands and planted directly in mangrove habitats or along protected shores as nursery. When seedlings are of proper size, these may be planted directly in tidal zone. Earlier attempts at restoration, regeneration, and afforestation have been undertaken with direct planting of mangrove propagules or seeds (Karim et al. 1984; Hamilton and Snedaker 1984; Kogo 1985; Kogo et al. 1986). In an attempt to rehabilitate denuded mangrove areas, Untawala (1993) attempted to raise mangrove nursery in tidal zone on slightly raised platforms supported with split bamboo on the sides to prevent possible drifting of polybags. The polybags filled with mangrove soil are planted with mangrove propagules/seeds. The seedling are available throughout the year hence may be planted any time of the year. The area with freshwater influence gives better results. As found in natural zonation pattern, *Rhizophora* may be planted facing sea followed by belts of *Bruguiera, Kandelia, Ceriops, Avicennia, Sonneratia*, and *Excoecaria* in middle zone toward land; and species of *Ceriops, Aegiceras, Aegialitis,* and *Camptostemon,* grown toward border with associate mangroves such as *Thespesia populnea, Pongamia pinnata, Terminalia catappa,* and *Calophyllum innophyllum.* The natural zonation pattern gives perfect understanding where to plant a particular species (Dagar 1982, 2003; Dagar et al. 1991, 1993).

While planting the mangrove seedlings, the distance between the rows and seedlings may be maintained at 1.0–1.5 m. After 10–15 rows, a gap of 8–10 m may be left for future forestry operations. Species such as *Rhizophora mucronata, R. apiculata, Avicennia marina, A. officinalis, Bruguiera gymnorrhiza*, and *B. parviflora* prefer sandy clay substratum and can be grown in highly saline substratum; while species such as *R. stylosa, Ceriops tagal, Aegiceras corniculatum,* and *Sonneratia alba* are found more predominantly on silty clay substratum and in middle zone while *Sonneratia casealaris, Xylocarpus granatum* and *Excoecaria agallocha* prefer low salinity and silty substratum. *Avicennia marina* is most tolerant to biotic stress and may be planted widely in all kinds of mangrove habitats. *Nypa fruticans*, a mangrove palm is more predominant in muddy substratum along creeks. This can be propagated from suckers. *Terminalia catappa* (coastal almond), *Pandanus* spp., *Calophyllum inophyllum* and *Pongamia pinnata* are useful oil-yielding trees and can be grown bordering mangroves as commercial plantations but their nursery cannot be raised in tidal zone. *Salvadora persica* and *Salicornia* are useful oil yielding bushes and may be raised in highly saline swamps behind mangroves. In Rann of Katchh area *Salvadora persica* has been found a life line among many small farmers (Rao et al. 2003).

Several projects have been initiated for rehabilitation of mangroves in Indian sub-continent and quite sizeable area has been planted with mangroves particularly in Korangi-Phitti creek and Indus Delta in Pakistan; and Goa and Pichavaram in Tamil Nadu. Some denuded areas have been rehabilitated with suitable mangrove species along Goa and Tamil Nadu coasts of India. The sandy beaches along Orissa coast have been planted successfully with *Casuarina glauca, C. equisetifolia, Pandanus*, cashew nut (*Anacardium occidentale*), coconut and *Acacia auriculaeformis.*

Brackish water aquaculture

In December 2004 due to Tsunami about 4,000 ha of agricultural land was inundated in Andaman and Nicobar Islands causing significant alteration to the agricultural production. This effect was predominantly visible in South Andaman and Sipighat, Chouldari, Teylarabad, Badmaspahar, Port Mout, Methakhadi, Dunduspoint, and Namunaghar sites were the worst affected areas. Hitherto, the land, which was used for paddy cultivation, turned into degraded due to increase in salinity. The soil became acid sulfate (pH varying between 3.5 and 6.5, the bulk density of surface soil from 1 gm to 1.4 gm cm^{-3} and organic carbon from 1.5 to 1.8 %). As a result, the farmers were forced to search for a viable alternative livelihood.

Reclamation of these brackish water acid sulfate soils is not easy as during drying process exposed mud becomes severely acidified. Any dyke construction from this mud results in acid leaching during heavy rain to pond water reducing its pH causing distress or even death to the cultured stock. The iron carried into pond water is oxidized and later precipitated out as insoluble ferric hydroxide that clogs the gill filaments of aquatic organisms (fish) leading to asphyxiation (Dam Roy and Rai 2005). The iron in pond interferes with smooth osmotic exchange of gases between organisms and water resulting in poor excretion. The carbonates and bi-carbonates in pond water deplete as they react with acid resulting in soft shelling and poor moulting response of shell fish. These chemical reactions also negatively affect bacteria-plankton-meiobenthos- microphytes composition affecting the pond food chains. The reclamation of saline acid sulfate soils for brackish water aquaculture can be achieved only through a model, which nullifies or reduces the above-mentioned problems. A procedure involves drying and filling of the soil to oxidize pyrite, filling the pond with water and holding till water pH drops to below 4, and then draining the pond with water, repeating the procedure until the pH stabilize over and above 5. Several experiments on farmers' field were conducted (Dam Roy and Krishnon 2005, Dam Roy et al. 2005) at several sites with an idea to tackle these problems and to achieve a better productivity. There was notable improvement in soil and water pH during the culture period.

Shrimp or fish farming is one of the viable commercial alternatives to agriculture in these areas. The development of brackish water aquaculture especially shrimp farming has been one and it has been found to have substantial economic gains. Estimated tiger shrimp brood stock availability in Andaman and Nicobar

Islands reveals that 9,320 brood stock of tiger shrimp available in the islands out of which as per conservative estimate 2,976 brood stock spanners can be collected annually for seed production. Prior to Tsunami, 608 ha of marshy wetland water areas were identified as suitable for shrimp culture in the islands. Besides this, about 1,206 ha of coastal agricultural lands in South Andaman are inundated with seawater, which cannot be used for productive agriculture. Pre-Tsunami reports (Kohli1989) revealed the ability of air-breathing fishes such as *singhi* and *magur* to adopt better in their natural habitat, as they breed in paddy fields naturally. Several studies conducted by CARI (1995–1996 to 1999–2000) revealed the performance and suitability of these fishes. But the post-Tsunami scenario has shown that these fishes are badly affected by the change in environment due to the advent of sea water in paddy fields or irrigation ponds (Dam Roy et al. 2005).

In a series of experiments conducted in Andamans, it was found that there is the feasibility of culturing mudcrab (*Scylla serrata*) in brackishwater ponds filled during tide (at stocking density from 1,000 to 5,000 per ha). The highest production of 878 kg ha^{-1} during eight months was obtained at a stocking density of 5,000 ha^{-1}. The fattening of milkfish (*Chanos chanos*) was done for a period of 1 year and 4 months and the net production at harvest was 1,030 kg ha^{-1}. The mullet (*Liza tade*) seeds were stocked at stocking densities ranging from 6,000 to 30,000 per ha and the production rate varied from 111 to 342 (average 232) kg ha^{-1}. The production of seabass (*Lates calcarifer*) at 66 % survival was found to be 3 t per ha. Two species of prawns namely the tiger (*Penaeus monodon*) and banana prawn (*Penaeus merguensis*) were cultured at a stocking density varying from 30–100,000 per ha and with 60 % survival it was estimated to produce 1.3 t ha^{-1} in 120 days. Tilapia (*Oreochromis urolepis*) with 87 % survival could produce 1,036 kg per ha in 6 months. Thus, the vast inundated wetlands of South Andaman can be looked upon as a site for promising innovative shrimp farming, which will be an alternate livelihood source for the farmers affected by Tsunami. On dykes of the ponds plantations such as multipurpose Noni (*Morinda citrifolia*) or coconut can successfully be grown.

Carp culture in fresh water has been an important activity in Bay Island and about 90 % fish farmers in North and South Andaman are frequently culturing mixture of Rohu, Catla and Mrigla. The indigenous cultivated species of carps include *Catla catla, Labeo rohita,* and *Cirrhinus mrigala* and the market value of these carps is very high (INR 200 per kg). Cat fish (Magur) also called "walking cat fish", is commonly found in fresh water mainly as swampy water ditches, ponds and paddy fields in rainy season. Another carp *Clarias batrachus* commonly called as "Indian Magur" is commercially important fresh water species fetching high price (INR 300–400 per kg). *Channa striatus* (snaked head fish because of its flavor and bone less nature and *Channa marulius* (Giant murrel) are also quite popular in the islands. The air-breathing fishes such as Singhi and Magur can successfully be cultivated in paddy fields of Bay Islands during rainy season (Dam Roy et al. 2005). Integrating fish culture with one or two live stocks at a time is quite often practiced. The concept of integration of fin and shell-fish culture with agriculture (paddy), horticulture, piggery, duckery, poultry, and livestock is quite old traditional way of farming in the islands that has been refined over the years.

Aquaculture keeping mangroves intact is most feasible and sustainable option for promotion of aquaculture in inundated areas (Fig. 7.6). After creating imbankment coconut and multipurpose tree Nauni (*Morinda citrifolia*) can be grown on raised bunds and in channels as well as mangrove creeks fish or shrimp culture is quite feasible.

In a preliminary study, mullet, prawn tilapia, and fish culture could be made feasible and poof table connecting these culture-ponds with brackish water behind mangroves particularly in association with *Avicennia* communities. The yield of shrimp fry (*Penacus* and *Metapenaeus*) was up to 690 kg ha^{-1} year^{-1} and during rainy season 3 t ha^{-1} of rice could also be produced.

Fig. 7.6 Aquaculture keeping mangroves intact (coconut in background)

Nypa fruticans, a mangrove palm (frequent in Sunderbans and Andaman-Nicobar Islands) is cultivated in Philippines and Bangladesh as a commercial crop. The alcohol production was reported about 15,000 l ha^{-1} $year^{-1}$ (Vannucci 1989). Fish/prawn culture, keeping mangrove intact is very viable and useful system particularly in areas where fresh water streams merge with seawater. Bee humps are natural in Sunderbans hence bee-keeping and duckry/poultry can be blended with fish culture associated with mangroves particularly along creeks. The agri-silvi-aquaculture system of *tumpang sar*-approach of Indonesia is the ideal solution for producing shrimps with the least disturbance to the mangrove ecosystem. In the post Tsunami scenario, in South Andaman alone, due to the subduction of the land by about 1.25 m, the level of submergence due to tidal influence has also increased. A survey conducted revealed that approximately 4,000 ha area of agricultural farmlands was submerged, out of which 630 ha was found suitable for coastal aquaculture.

Rehabilitation of acid sulfate soils

Most of these areas are reclaimed for rice cultivation. Coconut has been found suitable for cultivation on raised (1–2 m) bunds. In some rice fields, palmyrah palm (*Borassus flabellifer*), *Sesbania sesban, Gliricidia sepium*, and *Casuarina* are found grown on bunds or as live fences. Lime application leads to high yield of rice but more research attempts are needed for developing more salt-tolerant crops and improving sub-surface and surface drainage. Besides coconut, other littoral MPTs along creeks and bunds of aquaculture ponds may be useful for the local rural population. In Andaman-Nirobar Islands and Kerala, coconut has been grown as a successful crop on reclaimed mangrove areas. Some varieties like Andaman Tall and Katchal Tall are doing well on raised bunds in brackish water yielding 19 nuts per palm (at soil ECe 22.5 dS m^{-1}) to 63 nuts per palm (at EC 11.4 dS m^{-1}) in a year. Arecanut is also frequently grown. Fodder species such as *Pennisetum purpureum, Tripsicum laxum*, and *Panicum maximum* could be raised in inter-spaces of raised platforms yielding 20–34 t ha^{-1} fresh forage and the channels could be utilized for fish culture (Dagar 1995a). In some areas on bunds of rice fields, alongwith coconut, *Acacia auriculaeformis, Sesbania sesban, Casuarina equisetifolia, Vitex negundo*, and *Borassus flabellifer* palm are also grown. *Glircidia sepium* is useful live fence in high rainfall areas. In the low lying acid sulfate areas of Southeast Asia forest tree species such as *Casuarina junghuniana* and *Melaleuca leucadendron* are grown successfully. About 10,000 ha of *M. leucadendron* stands have been established in Mekong Delta and other species such as *M. acacoides, M. viridiflora, Eucalyptus camaldulensis*, and *E. microtheca* have been found the most tolerant for several acid soils. In the central plain near Bangkok (Thailand), farmers plant areca nut (*Areca catechu*) in low lying areas, in combination with fish raised in wide ditches under *Casuarina junghuniana* and vegetables grown on raised beds, which may also be adopted for Indian conditions. These situations are similar to Indian coastal areas, hence may be tried.

Afforestation of Land Impregnated with High Salinity but not Flooded with Seawater

Most of the area in Rann of Kutchh along Gujarat coast comes under this category. Because of low rainfall and high evapotranspiration the problem becomes more severe. In many areas, natural salt is prepared in evaporation saltpans. MPTs such as *Prosopis juliflora, Salvadora persica, Tamarix articulata, T. troupii*, and many halophytes are found growing naturally in these areas with stunted growth. A slivo-pastoral system may be developed incorporating suitable salt-tolerant forages such as species of *Atriplex, Kochia indica, Aeluropus lagopoides, Dichanthium annulatum, Leptochloa fusca*, and *Sporobolus helvolus*. Oil yielding species such as *Salvadora persica, Salicornia bigonie, Pongamia pinnata*, and *Terminalia catoppa* and firewood trees like *P. juliflora, Acacia nilotica,* and *Casuarina glauca* can be raised in furrows and above-mentioned grasses in interspaces. The grasses such as *Leptochloa fusca, D. annulatum,* and *Eragrostis* sp., when planted on 45 cm high ridges, could produce 3.17, 1.85, and 1.09 t ha^{-1} forage, respectively. When planted in furrows, these could yield 3.75, 1.76, and 0.54 t ha^{-1}, respectively, showing their potential for these highly degraded lands. In coastal sandy areas particularly along beaches of Orissa *Casuarina equisetifolia* is successfully grown. At many places, plantations of *Eucalyptus*, cashew nut (*Anacardium occidentale*), soapnut (*Sapindus trifoliatus*), *Acacia leucophloea, A. auriculaeformis*, and *Tamarindus indica* are raised successfully.

Agroforestry on Waterlogged Saline Soils Caused by Intrusion of Sea Water (biodrainage)

Cramer et al. (1999) showed that *Eucalyptus camaldulensis* intercepted deep ground water while *Casuarina glauca* relied on shallower unsaturated zone. Keeping this fact into consideration Roy Chowdhury et al. (2011) conducted experiments in coastal deltaic Orissa where problem of waterlogging was both due to sea water intrusion and due to topographical depression. They planted *Casuarina glauca* and *Eucalyptus camaldulensis* at two sites each. *Casuarina* was also found to be more efficient in discharging saline ground water hence was used for bio-drainage plantation at the sites having more salinity. The deltaic Orissa on an average experiences annual rainfall more than 1,400 mm, during monsoon. As a result, for about 10–12 weeks period, the plantations experience above ground waterlogged condition and remain so till end of monsoon. Therefore, in present scenario, the scope for assessment of efficacy of biodrainage plantation has been limited to only in post monsoon season. It is worthwhile to assess how fast the plantation is able to bring the field to cultivable condition so that *rabi* crops can be raised as early as possible. The effect of planted tree species on underlain water table was monitored by them through observation wells monitoring systems. The mean of first 2 years (April 2004 to March 2006) lowest water level at Patna (Orissa) was 102.0 cm below ground which declined to 117.7 cm in 3 years (i.e., 2006–2007 to 2008–2009). Similarly decline at Baghadi (Orissa) site was from 127.0 to 152.3 cm; at Alishibindha from 168.5 to 185.3 cm, and at

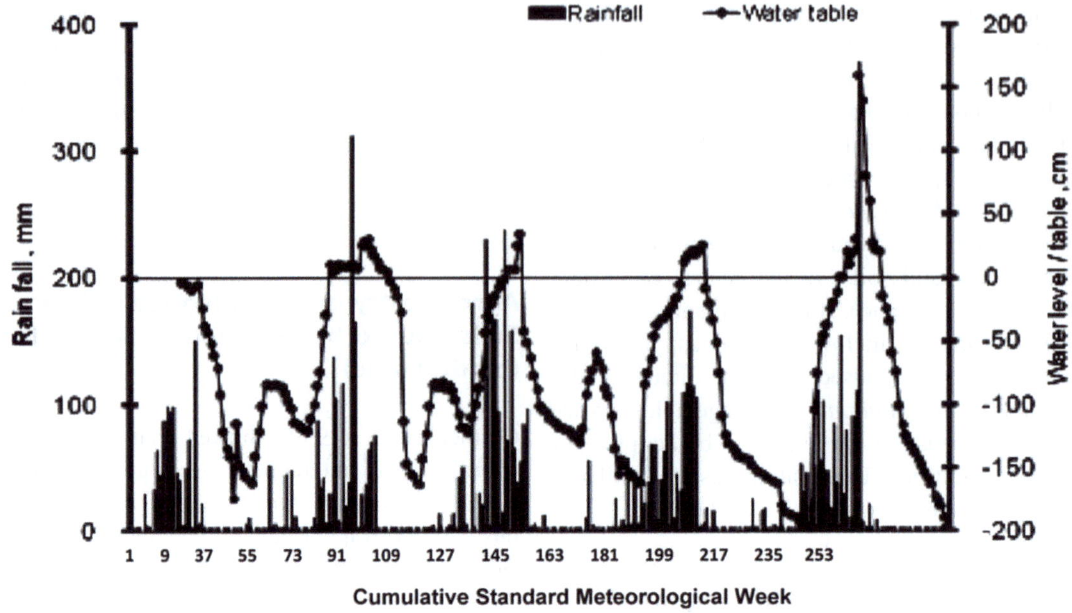

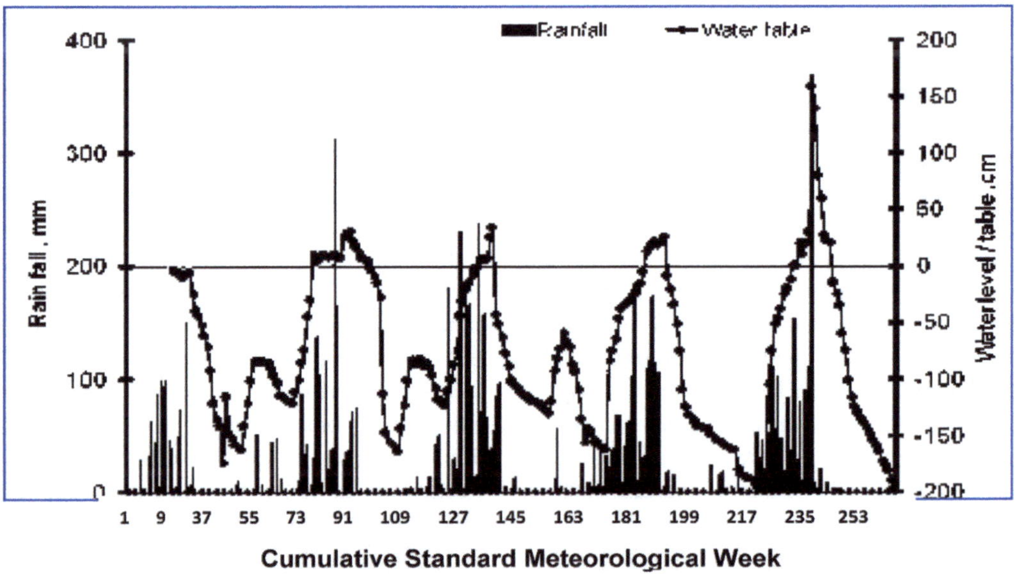

Fig. 7.7 The changes in level of water inside observation bore during experimental period and weekly rainfall underneath (*top*) *Eucalyptus* at Alshibindha and (*bottom*) *Casuarina* at Baghadi in deltaic Orissa from 2004–2005 to 2009–2010 (*Source* Roy Chowdhury et al. 2011)

Ambapada the decline was up to 150 cm (Figs. 7.7 and 7.8). Thus, from the data, it is evident that at phreatic surface there has been a clear draw down in level of water table underneath bio-drainage vegetation.

This accelerated drainage has helped the farmer to advance *rabi* cultivation by a period of 15–20 days. Through this process, the cultivation of watermelon as intercrop inside *Casuarina* vegetation could get additional benefit of

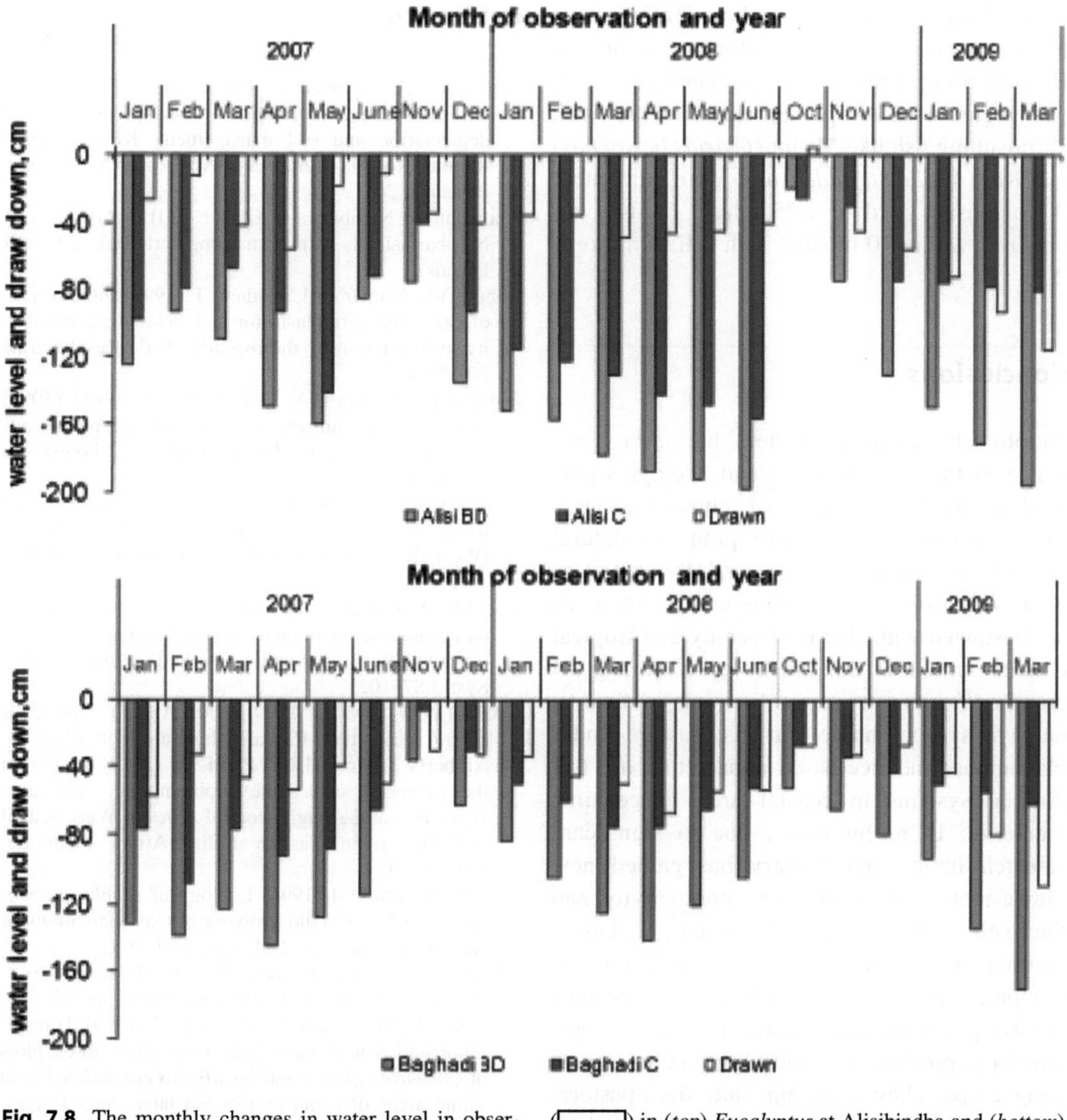

Fig. 7.8 The monthly changes in water level in observation bore wells inside plantations (grey), outside plantations (black) and the difference in water level (white) in (*top*) *Eucalyptus* at Alisibindha and (*bottom*) *Casuarina* at Baghadi, deltaic Orissa (*Source* Roy Chowdhury et al. 2011)

about INR 15,000 per ha for the farmer due to better market price of the crop as well as avoiding the market glut. In *kharif* season, rice was taken as intercrop inside *Casuarina* vegetation at one site. The final average yield of the paddy obtained during 4 years was 1.75 t ha^{-1}. The yield under *Eucalyptus* ranged from 2.3 to 3.5 t ha^{-1} (average 2.6 t ha^{-1}) during the same period. At another site, the net return of watermelon under *Casuarina* plantation in *rabi* season was INR 30,000 with B:C ratio of 2.14. Similarly under *Eucalyptus* from groundnut, net return was INR 21,000 with a B:C ratio 2.10 and from watermelon, net return was INR 62,500 and with B:C ratio of 3.67. Many other crops were also tried and showed promise as intercrops with these plantations. Aquaculture intervention in the bio-drainage field was also initiated during first week of June 2007 using a dug out pond of 400 m^2 of water surface area at Baghadi along with *Casuarina* plantation. After carrying out standard pond preparation protocol, air breathing fish like Magur (*Clarias batrachus*) and Koi (*Anabas testudeneous*) were cultivated. A composite yield of 1.25 t ha^{-1} of fish was obtained within 10 months with a B: C ratio of 2.5.

Conclusions

Agroforestry land use system has great relevance to the coastal and island ecologies particularly in the scenario of climate change. These ecosystems are more prone to natural calamities (such as cyclones and Tsunamis) as well as anthropogenic interferences. These areas are bestowed with rich biodiversity and tropical climate which is congineal for a variety of cropping/farming systems. Farming in forests, taungya system, home gardens, pasture under plantations, and rice-based aquaculture are traditional systems in coastal areas since time memorial. In recent times, due to significant research inputs, agroforestry has gained new dimensions in improving the productivity and the livelihood dimensions of coastal population. Agroforestry systems such as multistoreyed cropping systems; integrated farming systems involving several components like food crops, livestock, poultry, fish, and high income generating crops; alley cropping; and silvo-pastoral systems particularly under coconut plantations have been improved a lot for their sustainability, profitability, and adaptability. These are contributing toward livelihood security of marginal and small farmers of coastal and island ecologies. Mangrove ecosystem has been denuded and needs special attention for its rehabilitation. More research efforts are needed in domestication of valuable multipurpose tree species, value addition, carbon sequestration, root interaction among different components, popularization of integrated farming systems, and it requires separate agroforestry policy both at state and country level.

References

Alexander TG, Sobhana K, Balagopalan M, Mary MV (1980) Taungya in relation to soil properties, soil depreciation and soil management. KFRI Reserch Report No. 9. Kerala Forest Research Institute, Peechi, p 24

Andaman & Nicobar at a Glance (2012) Andaman & Nicobar Islands Administration, Port Blair, (A & N Islands)

Babu KVS, Kumar BM, Mathew T (1992) Field testing of *Leucaena* germplasm for their relative susceptibility to infestation by the psyllids. Agric Res J Kerala 30:135–137

Binkley D, Dunkin KA, DeBell D, Ryan MG (1992) Production and nutrient cycling in mixed plantations of *Eucalyptus* and *Albizia* in Hawaii. Forest Sci 38:393–408

Biswas CR (1994) Studies on feasibility and yield of forage crops in coastal saline areas of Sundarbans (West Bengal). J Indian Soc Coastal Agric Res 12:91–94

Boonkird S, Fernandes E, Nair P (1984) Forest villages: an agroforestry approach to rehabilitating forest land degraded by shifting cultivation in Thailand. Agrofor Syst 2:87–102

CARI (1995–1996 to 1999–2000) Annual reports of central agricultural Research Institute, Port Blair

Chakraborty T, Mondal AK, Parui SM (2012) Studies on the prospects and some problems of sand dune vegetation at the fragile coastal zones of West Bengal and Orissa, in Eastern India. Afr J Plant Sci 6(2):48–56

Chandrashekara UM (1996) Ecology of *Bambusa arundinacea* (Retz.) Willd growing in teak plantations of Kerala. India For Ecol Manage 87:149–162

CPCRI (2002) Annual report. Central Plantation Crops Research Institute (CPCRI), Kasaragod, pp 120

Cramer VA, Thornburn PJ, Fraser DW (1999) Transpiration and ground water uptake from farm forest plots of *Casuarina glauca* and *Eucalyptus camaldulensis* in saline areas of south east queensland. Aust J Agric Water Manage 39:187–204

Dagar HS, Dagar JC (1991) Plant folk medicines among the Nicobarese of Katchal Island, India. Econ Bot 45:114–119

Dagar JC (1982) Some ecological aspects of mangrove vegetation of Andaman and Nicobar Island. Sylvatrop Philippines For Res J 7:177–216

Dagar JC (1989) Endemic plant species of Bay Islands. J Andaman Sci Assoc 5:161–168

Dagar JC (1991) Agroforestry practices for coastal Bay Islands. J Indian Soc Coastal Agril Res 9:433–439

Dagar JC (1993) Structure of vegetation and litterfall in tropical rain forests of Andaman and Nicobar Islands, India. Asia Life Sci 2(1):43–70

Dagar JC (1994) Strategies for sustainable use of Island resources of India. J Indian Soc Coastal Agril Res 12:187–191

Dagar JC (1995a) Agroforestry systems for Andaman and Nicobar Islands in India. Int tree Crops J 8(2&3):107–128

Dagar JC (1995b) Ecology of halophytic vegetation in India: a review. Int J Ecol Env Sci 21:273–296

Dagar JC (1996) Rehabilation of coastal saline lands by planting with suitable species. J Indian Soc Coastal Agril Res 14:185–191

Dagar JC (1998) Vegetation of salt affected soils and its scope for agroforestry interventions. Int J Eco Environ Sci 24:49–57

Dagar JC (2000) Agroforestry systems for coastal and island regions. Indian J Agrofor 2:59–74

Dagar JC (2003) Biodiversity of Indian saline habitats and management and utilization of high salinity tolerant plants with industrial application for rehabilitation of saline areas. In: Alsharhan AA, Wood WW, Gouie AS, Fowler A, Abdellatif EM (eds) Desertification in the third millennium. Swets and Zeitlinger Publishers, Lisse, p 151–172

Dagar JC (2005a) Salinity resources in India: an overview. Bull Nat Inst Ecol 15:69–80

Dagar JC (2005b) Ecology, management and utilization of halophytes. Bull Nat Inst Ecol 15:81–97

Dagar JC (2008) Indian mangroves: status, management and their plausible benefits for livelihood security. J India Soc Coast Agril Res 26(2):121–128

Dagar JC (2009) Opportunities for alternate land uses in salty and water scarcity areas. Int J Ecol Env Sci 35(1):53–66

Dagar JC (2012) Utilization of degarded lands/habitats and poor quality water for livelihood security and mitigating climate change. Indian J Agrofor 14(1):1–16

Dagar JC, Dagar HS (1999) Ethnobotany of Aborigins of Andaman-Nicobar Islands. Surya International Publications, Dehra Dun, p 203

Dagar JC, Kumar V (1992) Agroforestry for Bay Islands. Indian For 118(6):411–415

Dagar JC, Mongia AD, Bandyopadhyay AK (1991) Mangroves of Andaman and Nicobar Islands. Oxford and IBH Publishing Co. Ltd, New Delhi

Dagar JC, Mongia AD, Singh NT (1995) Degradation of tropical rain forest soils upon replacement with plantations and arable crops in Andaman and Nicobar Islands in India. Trop Ecol 36(1):89–101

Dagar JC, Singh G (2007) Biodiversity of saline and waterlogged environments: documentation utilization and management. National Biodiversity Authority, Chennai pp 762

Dagar JC, Singh NT (1999) Plant resources of Andaman and Nicobar Islands. Bishen Singh Mahendra Pal Singh Publishers, Dehra Dun

Dagar JC, Singh NT, Mongia AD (1993) Characteristics of mangrove soils and vegetation of Bay Islands in India. In: Lieth H, Al Masoom A (eds) Towards the rational USE of high salinity tolerant plants vol. 1. Kluwer Academic Publishers, The Netherlands, p 59–80

Dagar JC, Tomar OS (1998) Suitable agroforestry systems for coastal saline regions: case study. J Range Manage Agrofor 19:165–172

Dam Roy S, Krishnan P (2005) A road map for marine fisheries development in Andaman & Nicobar Islands. Discussion organized by Fishery Survey of India

Dam Roy S, Rai RB (2005) Impact of Tsunami on fisheries and on the ecosystem of Andaman and Nicobar Islands. Paper in seminar strategies for improved farming and ecological securities of coastal region held in 2005

Dam Roy S, Soundararajan R, Krishnamoorthy V, Krishnan P (2005) Tsunami impact on fisheries sector, rehabilitation required and strategies to be followed in Andaman and Nicobar Islands. Paper in national symposium "The Seventh Indian Fisheries, Forum" 8–12, November, 2005, Bangalore, India

Danso SKA, Bowen GD, Sanginga N (1992) Biological nitrogen fixation in trees in agroecosystems. Plant Soil 142:177–196

Deans JD, Moran J, Grace J (1996) Biomass relationship for tree species in regenerating semi-deciduous tropical moist forest in Cameroon. For Ecol Manage 88:215–225

DeBell DS, Whitesell CD, Schuber TH (1989) Using N_2 fixing *Albizia* to increase growth of *Eucalyptus* plantations in Hawaii. For Sci 35(1):64–75

Dommergues YR (1987) The role of biological nitrogen fixation in agroforestry. In: Steppler HA, Nair PKR (eds) Agroforestry-a decade of development. international council for research on agroforestry, Nairobi, Kenya, pp 245–271

Eswaran HE, van Den Berg, Reich P (1993) Organic carbon in the soils of the world. Soil Sci Soc Am J 57:192–194

FAO (1998) Integrated coastal area management and agriculture, Forestry and Fisheries, FAO Guidelines. Food and Agricultural Organization, Rome

Fernandes ECM, Nair PKR (1986) An evaluation of the structure and function of tropical home gardens. Agrofor Syst 21:279–310

FSI (2011) India state of forest report 2011. Forest Survey of India (FSI), Ministry of environment & forests, Goverment of India, Dehra Dun p 286

Gajbhiye KS, Mandal C (2006) Agro-ecological zones, their soil resource and cropping systems. Status paper. In: Status of farm mechanization in India. Indian agricultural Statistics research institute, New Delhi, pp. 1–32 (http://agricoop.nic.in/farm%20-mech.20%pdf/05024_pdf.)

Gillespie AR, Knudson DM, Geilfus F (1993) The structure of four homegardens in the Peten, Guatemal. Agrofor Syst 24:157–170

Gopinathan R, Sreedharan C (1989) Soil erosion as influenced by rainfall erosivity under different agroforestry systems. In: Reifsnyder WS, Darnhofer TO (eds) Meteorology and agroforestry: proceedings international workshop on the application of meteorology to agroforestry systems planning and management. International Center for Agroforestry, Nairobi, Kenya, pp 407–418

Hamilton LS, Snedaker SC (1984) Handbook of mangrove area management. UNEP and East-West Centre, Environment and Policy Inst., Hanolulu, Hawai

India (2013) India 2013-A reference mannual. Publication division, Ministry of Information & Broadcasting.Government of India, New Delhi, p 1206

Ismail T (1986) Integration of animals in rubber plantations. Agrofor Syst 4:55–66

Jacob PM (2004) Strategies for enhancing production and productivity in the coastal ecosystem of Lakshadweep Islands. J Indian Soc Coastal Agric Res 22:152–155

Jacob PM, Bhat KS, Vasanthasenan R (2002) Increasing production and productivity in the existing coconut gardens of Lakshadweep Islands. Indian Coconut J 33(5):2–4

Jose D (1992) Structure and productivity of homegardens of Kerala. In: Nair CGR (ed) Proceedings 4th Kerala science congress, Thrissur, pp 17–19. Science Technology & Environment Department, Government of Kerala,Thiruvananthapuram, India

Karim A, Hossain Z, White KJ (1984) Study of the growth of mangrove plants in relation to edaphic factors in coastal afforestation plantation of Chittagong. In: Proceedings Symposium Mangrove Environment Research & Management, Kuala Lumpur, Malaysia

Kogo M (1985) Mangrove research & recommendations of afforestation in Pakistan. A report for UNDP/UNESCO Mangrove Project by Al Gurm Research Centre, Tokyo

Kogo M, Kaminura D, Miyagi T (1986) Research for rehabilitation of mangroves in Truck Islands. In: Mangroves of Asia & the Pacific-Status and monument. Technical report. UNDP/4NESCO Regional Mangrove Project, New Delhi

Kohli MPS (1989) Natural breeding of air- breathing fishes in Andaman and Nicobar islands. J Andaman Sci Assoc 5:96–97

Krishnakumar V (2004) Management of homegardens and their economic prospects under coastal agroecosystem. J Indian Soc Coastal Agric Res 22:137–141

Krishnankutty CN (1990) Demand and supply of wood in Kerala and their future trends. KFRI research report 67. Kerala Forest Research Institute, Peechi, p 66

Kumar BM (2005) Land use in Kerala: changing scenerios and shifting paradigms. J Tropical Agric 43(1–2):1–12

Kumar BM (2011) Species richness and and aboveground carbon stocks in the homegardens of central Kerala, India. Agric, Eco Environment 140:430–440

Kumar BM, George SJ, Chinnamani S (1994) Diversity, structure and standing stock of wood in the homegardens of Kerala in peninsular India. Agrofor Syst 25:243–262

Kumar BM, George SJ, Suresh TK (2001a) Fodder grass productivity and soil fertility changes under four grass + tree associations in Kerala, India. Agrofor Syst 52:91–106

Kumar BM, Kumar SS, Fischer RF (1998) Intercropping teak with Leucaena increases tree growth and modifies soil characteristics. Agrofor Syst 42:81–89

Kumar BM, Kunhamu TK (2011) Quarter century of agroforestry research in Kerala: a compendium of research publications. Kerala Agriculture University, Kerala 404

Kumar BM, Nair PKR (2004) The enigma of tropical homegarden. Agrofor Syst 61: 135–152 areas in the Western Ghats of peninsular India. Agrofor Syst 31:275–289

Kumar BM, Sajikumar V, Mathew T (1995) Floristic attributes of small cardamom (*Elettaria cardamomum* (L.) Maton) growing areas in the Western Ghats of peninsular India. Agrofor Syst 31:275–289

Kumar BM, Sudheesh KG, Rajesh G (2005) Standing stock of thorny bamboo [*Bambusa bambos* (L.) Voss] in the homegardens of Thrissur. Kerala J Trop Agri. 43:61–66

Kumar BM, Takeuchi K (2009) Agroforestry in the Western Ghats of Peninsular India and the Satoyama landscapes of Japan: a comparision of two sustainable land use systems. Sustain Sci 4:215–232

Kumar BM, Thomas J, Fisher RF (2001b) *Ailanthus triphysa* at different density and fertilizer levels in Kerala, India: tree growth, light transmittance and understorey ginger yield. Agrofor Syst 52:133–144

Kunhamu TK, Kumar BM, Viswanath S, Sureshkumar P (2008) Thinning promotes understorey ginger productivity in *Acacia mangium* Willd. Stands of Kerala. In: Yahya AZ, Ngah ML, Md.Ariff FF, Rahman ZA, Abdullah MZ, Philip E, Chik SW (eds) Proccedings of the International Agroforestry Conference. Trans. of the Malaysian Society of Plant Physiology, vol. 15. Forest Research Institute, Kuala Lumpur, Malaysia, pp 47–60

Logan W (1906) Reprint (1989) malabar manual (2 vols). Asian education services, 2nd edn. Madras, India, p 772

Mahapatra IC, Panda M (1994) Farming systems research and development in the eastern coast of India. J Indian Soc Coastal Agric Res 12:67–77

Maiti S, Raju S (2004) Problems and prospects of medicinal and aromatic plants cultivation in the coastal ecosystems of India. J Indian Soc Coastal Agric Res 22:185–190

Mandal C, Mandal DK, Srinivas Cv, Sehgal J, Velayutham M (1999) Soil climatic database for crop planning in India. Technical Bulletin No. 53. National Bureue of Soil Survey & Land Use Planning, Nagpur, India p 1014

Mateer S (1883) Native life in Travancore. WH Allen & Co, London, p 450

Mathew T, Kumar BM, Suresh Babu KV, Umamaheswaran K (1992) Comparative performance of some multi-purpose trees and forage species in silvopastoral systems in the humid regions of southern India. Agrofor Syst 17:205–218

Millate-E-Mustafa MD, Hall JB, Teklehaimanot Z (1996) Structure and floristics of Bangladesh homegardens. Agrfor Syst 33:263–280
Moench M (1991) Soil erosion under a succesional agroforestry sequence: a case study from Idukki District, kerala, India. Agrofor Syst 15:31–50
Mohanty RK, Sahoo N, Jena SK, Roy Chowdhury S, Verma HN (2004) Water productivity and economic evaluation of aquaculture-based integrated farming system approach in waterlogged ecosystem. J Indian Soc Coastal Agric Res 22:298–303
Nagam A (1906) Travancore State Manual, Vol III. Travancore Government Press, Trivandrum, p 700
Nair MA, Sreedharan C (1986) Agroforestry farming systems in the homesteads of Kerala, Southern India. Agrofor Syst 4:339–363
Nair PKR, Dagar JC (1991) An approach to developing methodologies for evaluating agroforestry systems in India. Agrofor Syst 16:55–81
Nair PKR, Nair VD, Kumar BM, Showalter JM (2010) Carbon sequestration in agroforestry systems. Adv Agron 108:237–307
Ninez VK (1984) Household gardens: theoretical considerations on an old survival strategy. International potato center. Potatoes in Food Systems Research Series Report No. 1. Limá. Peru
Pandey CB, Rai BR, Singh L, Singh A (2007) Homegardens of Andaman and Nicobar, India. Agric Syst 92:1–22
Pattanaik C, Reddy CS, Dhal NK (2008) Phytomedicinal study of coastal sand dune species of Orissa. Indian J Tradit Knowl 7(2):263–268
Post WM, Kwon KC (2000) Soil carbon sequestration and land use change-process and potential. Global Change Bio 6:317–327
Rajasekharan P, Veeraputhran S (2002) Adoption of intercropping in rubber smallholdings in Kerala, India: a Tobit analysis. Agrofor Syst 56:1–11
Randhawa MS (1980) A History of Indian agriculture. Vol 2, 8th to 18th Century. Council of Indian Agricultural Research, New Delhi, p 358
Rao Gururaja G, Nayak AK, Chinchmalatpure AR (2003) *Salvadora persica*: A life support species for salt affected black soils. Technical Bulletin 1/2003. CSSRI, Karnal, India
Reicosky DC (2002) Conservation agriculture: Global environmental benefits of soil carbon management. In: McKinnon D, Hamilton-Manns M (eds) Proceedings Seminar Monsanto Conservation Agriculture, held at Taupo & Ashburton, New Zealand, Caxton Press, Christchurch
Reicosky DC, Dugas WA, Torbert HA (1997) Tillage induced soil carbon dioxide loss from different cropping systems. Soil Tillage Res 41:105–118
Reicosky DC, Lindstrom MJ (1994) Fall tillage methods: effect and short-term carbon dioxide flux from soil. Agron J 85:1238–1243
Roy Chowdhury S, Kumar A, Brahmanand PS, Ghosh S, Mohanty RK, Jena SK, Sahoo N, Panda GC (2011) Application of Bio-Drainage for Reclamation of Waterlogged Situations in Deltaic Orissa. Research Bulletin 53. Directorate of Water Management (ICAR), Bhubaneswar, India, p 32
Saha S, Nair PKR, Nair VD, Kumar BM (2010) Carbon storage in relation to soil size-fractions under some tropical tree-based land-use systems. Plant Soil 328:433–446
Sehgal J, Mandal DK, Mandal C, Vadivelu S (1992) Agro-ecological regions of India,2 Edn. Technical bulletin No. 24, National Bureue of soil survey & land use planning, Nagpur, India p 130
Sehgal J, Mandal DK, Mandal C (1995) Agro-ecological subregions of India (map). National Bureue of Soil Survey & Land Use Planning, India
Shanmughavel P, Francis K (1996) Aboveground biomass production and nutrient distribution in growing bamboo (*Bambusa bambos* (L.) Voss). Biomass Bioenergy 10:381–391
Shanmughavel P, Peddappaiah RS, Muthukumar T (2001) Biomass production in an age series of *Bambua bambos* plantations. Biomass Bioenergy 20:113–117
Sharma AK, Dagar JC, Bandyopadhyay AK (1990) Fodder resorces of Bay Islands. CARI research bulletinno.3. Central Agricultural Research Institute, Port Blair, p 71
Sharma AK, Dagar JC, Pal RN (1991) Comparative yield performance and water use efficiency of eleven exotic fodder grasses in the humid tropics. Trop Ecol 33:245–254
Sharma AK, Dagar JC, Pal RN (1992) Performance of perennial fodder grasses under continuous growth in tropical islands. Trop Sci 32:383–388
Sharma YML (1987) Inventory and resources of bamboo. In: Rao AN, Danarajan G, Sastry CB (eds) Recent Research on Bamboos. Chinese Academy of Forestry and International development Research Centre, pp 14–22
Singh HP, Singh DR, Srivastava RC (2009) *Morinda citrifolia* L—the divine noni. Central Agricultural Research Institute, Port Blair, p 241
Singh NT, Mongia AD, Ganeshmurthy AN (1988) Soils of Andaman and Nicobar Islands. CARI research bulletinno.1, Central Agricultural Research Institute, Port Blair
Soil Survey Staff (1975) Soil taxonomy, a basic system of soil classification for making and interpreting soil surveys. USDA agriculture handbook No. 436. Soil Conservation Service, Washington DC
Thampan PK (1996) Coconut for Prosperity. Peekay Tree Crops Development Foundation, Gandhi Nagar pp 271
Untawale AG (1993) Development of an intertidal mangrove nursery and afforestation techniques along the Indian Coast. In: Lieth H, Al Masoom A (eds) Towards the rational use of high salinity tolerant plants vol. 1. Kluwer Academic Publishers, The Netherlands pp 371–378
Van Noordwijk M, Lawson G, groot JJR, Hairiah K (1996) Root distribution in relation to nutrients and competition. In: Ong CK, Huxley PA (eds) Tree-crop

interactions-a physiological approach. CAB International, Wallingford, pp 319–364
Vannucci M (1989) The Mangroves and US a synthesis of insight. Indian Association of the Advancement of Science, New Delhi
Velayutham M, Mandal DK, Mandal C, Sehgal J (1999) Agro-ecological subregions of India for planning and development. NBSS & LUP Publ. No. 35. National Bureue of Soil Survey & Land Use Planning, Nagpur, p 372
Yadav JSP, Bandyopadhyay AK, Bk Bandyopadhyay (1983) Extent of coastal saline soils of India. J Indian Soc Coastal Agric Res 1(1):1–6

Agroforestry for Wasteland Rehabilitation: Mined, Ravine, and Degraded Watershed Areas

8

O. P. Chaturvedi, R. Kaushal, J. M. S. Tomar, A. K. Prandiyal, and P. Panwar

Abstract

Wasteland is defined in various ways by different agencies. However, in general it represents degraded, unused, and uncultivated lands. These lands have utilized in recent past to bridge the gap between demand and supply of food, fodder, timber, and also for resource conservation. Area under mines in the country is about 0.19 m ha and ravine lands 4 m ha. Though mining is important for industrial growth, it also has negative impact on the environment and renders the land unproductive. Rehabilitation of such degraded areas requires systematic and scientific approach which includes proper survey, choice of species, and techniques for establishment of plant species. Rehabilitation of ravine lands involves treatment of table and marginal lands contributing runoff to the gullies and proper gullies/ravines on watershed basis. It requires an integrated approach of using gullies according to land capability classes, soil, and water conservation measures and putting land under permanent vegetation cover involving, afforestation, agroforestry, horticulture, pasture, and energy plantations. Watershed development has become the major intervention for managing natural resources. Majority of the watersheds in the country are degraded and suffer from poor productivity, biotic pressure, acute fodder shortage, poor livestock productivity, poverty, water scarcity, and poor infrastructure. A multitier ridge to valley sequenced approach is required to treat the watersheds for enhancing productivity and resource conservation. This

O. P. Chaturvedi (✉) · R. Kaushal · J. M. S. Tomar
Central Soil and Water Conservation Research and Training Institute (CSWCRTI), Dehradun, Uttarakhand 248195, India
e-mail: chaturvediopc@gmail.com

A. K. Prandiyal
Regional Centre CSWCRTI, Kota, Rajasthan 324002, India

P. Panwar
Regional Centre, CSWCRTI, Chandigarh, 160019, India

J. C. Dagar et al. (eds.), *Agroforestry Systems in India: Livelihood Security & Ecosystem Services*, Advances in Agroforestry 10,
DOI: 10.1007/978-81-322-1662-9_8,

chapter deals with the aapproaches for development of mined, ravine, and degraded watersheds through bioengineering, afforestation, and agroforestry along with success stories.

Introduction

The wasteland represents degraded, unused, uncultivated, and common land as (a) lands not available for cultivation, barren, and unculturable wastes, (b) other uncultivated land excluding fallow, culturable waste, permanent pastures, and land under miscellaneous trees, (c) fallows under wastelands. Different departments define the wasteland according to their land use pattern. Agriculture land lying fallow for more than 2 years can be termed as agricultural wasteland. Lands under the control of Revenue Department not fit for agriculture lying barren can be termed as Revenue wasteland. Similarly, grasslands and lands under the control of Forest Department which do not have tree cover can be termed as forest wasteland (Luna 2006). As per the Report of the Technical Task Group constituted by the Planning Commission, the wastelands are the degraded lands which can be brought under vegetative cover with reasonable efforts and which are currently under utilized and the land which is deteriorating for lack of appropriate water and soil management or on account of natural causes (http://dolr.nic.in/wasteland_division.htm). National Wastelands Development Board (NWDB) defined wasteland as "that land which is degraded and is presently lying unutilized (except as current fallow) due to different constraints." NWDB also suggested that any land which is not producing green biomass consistent with the status of soil and water must be treated as wasteland. Wastelands primarily consist of culturable and unculturable wastelands. Culturable wastelands are those lands which have potential for the development of vegetative cover or may be reclaimed at later stage. It includes gullies and or ravinous lands, undulating upland surface, water logged and marshy areas, salt-affected land, shifting cultivation area, degraded forest land, sandy area, mining and industrial wasteland, strip land, pasture and grazing land. Unculturable wastelands are those lands which cannot be developed for vegetative cover. These lands are barren and cannot be put to significant uses. However, some of the areas of such lands can be converted/ changed into pasture land. It includes barren rocky area, steep slopes, and snow covered glacial areas. Although no consensus have yet been arrived at definition of wasteland but it is largely accepted that wastelands are the areas which are underutilized and which produce less than 20 % of its biological productivity (Mishra et al. 2013).

Extent of Wastelands

About 2 billion ha area in the world is affected by various forms of human-induced land degradation (Oldeman 1991). The productive lands, in the country are in the constant process of various degrees of degradation and are fast turning into wastelands. Depending upon varying definitions of land degradation, data sources, classification systems, methodologies and scales, various estimates of wastelands have been given by different central and state agencies (Table 8.1).

Thirteen categories of wastelands viz., gullied and/or ravenous land, underutilized/degraded notified forest land, mining/industrial wasteland, barren rocky/stony waste/sheet rock area, land with or without scrub, steep sloping area, snow covered and/or glacial area, degraded pastures/ grazing land, degraded land under plantation crops, sands inland/coastal, water logged and marshy land, land affected by salinity/alkalinity-coastal-inland, and shifting cultivation area have been identified in the country by National Remote Sensing Agency (NRSA), which constitute about 20.17 % of total geographical area (NRSA 2000). According to the Wasteland Atlas of India, the spatial extent of different classes of

Table 8.1 Land degradation assessment by different organizations

Agency	Estimated extent (million ha)	Criteria for degradation (year).
National Commission on Agriculture	148	Based on the secondary data (1976).
Ministry of Agriculture (Soil and Water Conservation Division)	175	Based on the NCA's estimates. No systematic survey was undertaken (1978).
Society for Promotion of Watershed Development (SPWD 1984)	129	Based on the secondary estimates (1984).
NRSA	53	Mapping on 1:1 million scale based on the random sampling techniques (1985).
Ministry of Agriculture	173	Land degradation statistics for states (1985).
Ministry of Agriculture	107	Estimation of duplication of area. Area reclaimed counted (1994).
NBSS and LUP	187	Mapping of 1:4 million scales based on the Global Assessment of Soil Degradation (GLASOD) guidelines (1994).
NBSS and LUP	146	1:1 million scale soil map (2004).
Department of Environment	95	(1980).
National Wasteland Development Board	123	(1985).

Source Gautam and Narayan (1988), Maji et al. (2010), Mishra and Rath (2013)

wastelands was 4,67,021 sq.km. in 2008-09 (NRSC 2011). Nearly 83 % of the wastelands are in Andhra Pradesh, Gujarat, Karnataka, Madhya Pradesh, Maharashtra, Orissa, Rajasthan, Tamil Nadu, and Uttar Pradesh (Sreedevi et al. 2007).

The area reported by different agencies show wide variation in the extent of wastelands. Keeping above in view, a harmonized classification comprising wasteland classes and soil degradation was prepared by National Rainfed Area Authority (NRAA) in 2008. The methodology adopted for harmonization of data sets consisted of collection of information available with NRSA, National Bureau of Soil Survey & Land Use Planning (NBSSLUP), and Central Arid Zone Research Institute (CAZRI) while examining the definitions adopted by them, scope for harmonizing the classes in the legend of the maps, and availability of maps (Ramakrishna et al. 2007).

The comparison of legends between wastelands and degraded soil indicates that the common categories between wasteland maps and soil degradation maps are gullied and/ravinous lands, semi-stabilized to stabilized sand dunes, waterlogged and marshy lands, and land affected by salinity/alkalinity. However, there are some exclusive categories such as land with/without scrub, shifting cultivation, degraded forest-scrub dominated, degraded pasture/grazing land, agriculture land inside notified forest, degraded land under plantations, steeply sloping area loss of top soil, terrain deformation, over blowing, and loss of nutrients in soil degradation map. After thorough deliberations on the data sets of wastelands and soil degradation, a legend comprising wasteland classes and soil degradation was prepared, and the statistics were generated on degraded lands of India. The wasteland classes were compared with soil degradation classes to arrive at common classes and mutually exclusive classes (Sreedevi et al. 2007).

According to harmonized area statistics, total degraded area comes to 120.8 m ha out of which 73.3 million ha is estimated to suffer from water erosion, 12.4 m ha from wind erosion, 17.4 m ha from chemical degradation, and

Table 8.2 Harmonized area statistics of degraded and wastelands of India

Degradation type	Arable land (m ha)	Open forest (<40 % canopy) (m ha)	Data source
Water erosion (>10 tons/ha/yr)	73.27	9.30	Soil loss map of India—CSWCR&TI
Wind erosion	12.40	–	Wind erosion map of India—CAZRI
Subtotal	85.67	9.30	
Chemical degradation			
Exclusively salt-affected soils	6.70	–	Salt-affected soils map of India, CSSRI, NBSS&LUP, NRSA, and others
Salt-affected and water eroded soils	1.20	0.10	–
Exclusively acidic soils (pH < 5.5)#	5.09	–	Acid Soil Map of India NBSS&LUP
Acidic (pH < 5.5) and water eroded soils#	5.72	7.13	–
Subtotal	17.45	7.23	–
Physical degradation			
Mining and industrial waste	0.19	–	Wasteland map of India—NRSA
Waterlogging (permanent surface inundation)$	0.88	–	–
Subtotal	1.07	–	–
Total	104.19	16.53	
Grand Total (arable land and open forest)	120.72		

Source Maji et al. (2010)

1.1 m ha from physical degradation (Table 8.2). This figure however exclude degraded lands under dense forest, unculturable wastelands, barren rocky/stormy waste land and snow covered ice-caps which are source of water, and building material and are not treated as wastelands. The above land use classification, however, does not say anything about the extent of land degradation or loss in productivity over time due to various natural and man-made causes. NRSC (2011) reported spatial changes in wastelands between 2005-06 and 2008-09 using 3-season Resourcesat-1 LISS-III data on 1:50000 scale. Data revealed that the spatial extent of different classes of wastelands was 4,67,021 sq.km. in 2008-09, compared to 4,72,262 sq.km. in 2005-06. It was observed that during this period, a total area of 32,340 sq.km. of different wasteland categories has changed into non-wasteland category. Bulk of area (20,000 sq.km.) from different wasteland categories has changed into 'cropland' (including 'fallow') class. A considerable area change in case of 'degraded forest-scrub dominant' has been noticed getting converted into 'forest-dense/open' & 'forest plantation' classes (9,600 sq.km.). About 800 sq.km. of wasteland has been converted into other plantations. During the same period, an area of 27,098 sq.km. of non-wasteland has been converted into different wasteland classes.

Increasing population and biotic pressure on the fragile ecosystem, unplanned urbanization and rural poverty, break down of traditional institutions for managing Common Property Resources, and lack of appropriate management practices are some of the key reasons for the land degradation. Other activities such as mining, deforestation, and faulty land use practices also contributes to the problem of land degradation. Wind and water are the major agents responsible for the land degradation. They cause erosion due to which about 5334 m tons of soil is lost (16 t ha^{-1}) annually highest in black soils (24–112 t ha^{-1}) followed by Shiwaliks (80 t ha^{-1}) and northeastern region by shifting cultivation with 27–40 t ha^{-1} soil loss (Mandal et al. 2008). The land degradation has both on-site and off-site impacts which need to be managed properly to arrest further degradation. This chapter discusses the integrated approaches for rehabilitation of mined, ravine areas, and degraded watersheds.

Approaches to Rehabilitation of Degraded Lands

There are many approaches to land and vegetation rehabilitation. These depend on severity of damage to the land resource, the goals of rehabilitation, and availability of resources for repairing the damage. Improvement of degraded lands implies restoration or reclamation or rehabilitation. The IUCN guidelines for the restoration of degraded ecosystems define restoration as process bringing a degraded ecosystem or landscape back to same prescribed, productive condition in short process of revitalization (Lamb 1988). Ecological restoration concentrates on processes such as persistence of species through natural recruitment and survival, functioning food webs, systems wide nutrient conservation via relationships among plants, animals, and the detrivore community. Reclamation on the other hand is referred as means that the site will be similar in ecological functioning after disturbance and will be habitable by similar but not necessarily the same organisms. Rehabilitation means that the land will be made useful but with different land use and usually with different species. The most intensive rehabilitation is sometimes termed reclamation, because the damage is so severe that soils have to be replaced and landscapes may have to be reshaped (Ang 1994; Bradshaw 1997; Singh et al. 2002). An important goal of ecological rehabilitation is to accelerate natural succession so as to increase biological productivity, reduced rates of soil erosion, increase soil fertility, and increase biotic control over biogeochemical fluxes within the recovering ecosystem (Mishra 2013).

With rising global population and ever increasing pressure on agricultural land for industrial and other infrastructural projects, waste land reclamation has become a priority to ensure sustainable food production and environmental conservation. Wasteland reclamation and development in our country falls under the purview of Wasteland Development Board. Degraded lands can be suitably reclaimed for agriculture or some alternate uses through afforestation and agroforestry. Both these measures are not only helpful in meeting the diverse needs of fuel, fodder, timber, etc., but are also able to improve environment condition, biodiversity, and socioeconomic conditions of the stakeholders. Besides, these measures are simple, cost-effective, and can be helpful in relieving the pressure on traditional cultivated lands and forests. The development of degraded mined, ravine areas, and degraded watersheds through afforestation, agroforestry, and bioengineering approaches are discussed below:

Mined areas

Extent

The country is having a well-developed mining sector, which has vast geological potential with over 20,000 known mineral deposits. The distribution of minerals in the country varies from region to region (Table 8.3).

India produces as many as 84 minerals comprising 4 fuel, 11 metallic, 49 nonmetallic industrial, and 20 minor minerals. The mining leases numbering 9,244 are spread over 21 States on about 13,000 mineral deposits occupying about 0.7 million hectares which is 0.21 % of the total land mass of the country (Tata Energy Research Institute 2001). Himalayan region also contributes significantly to the mineral wealth of the country (Table 8.4).

Table 8.3 Distribution of mineral belts in the Country

Mineral Belt	Location	Minerals found
Northeastern Peninsular belt	Chota Nagpur plateau and the Orissa plateau covering the states of Jharkhand, West Bengal, and Orissa	Coal, iron ore, manganese, mica, bauxite, copper, kyanite, chromite, beryl, apatite, etc. This region possesses India's 100 % kyanite, 93 % iron ore, 84 % coal, 70 % chromite, 70 % mica, 50 % fire clay, 45 % asbestos, 45 % China clay, 20 % limestone, and 10 % manganese
Central belt	Chattisgarh, Andhra Pradesh, Madhya Pradesh and Maharashtra	Manganese, bauxite, uranium, limestone, marble, coal, gems, mica, graphite, etc., exist in large quantities and the net extent of the minerals of the region is yet to be assessed. This is the second largest belt of minerals in the country
Southern belt	Karnataka plateau and Tamil Nadu	Ferrous minerals and bauxite. Low diversity
Southwestern belt	Karnataka and Goa	Iron ore, garnet, and clay
Northwestern belt	Rajasthan and Gujarat along the Aravali Range	Non-ferrous minerals, uranium, mica, beryllium, aquamarine, petroleum, gypsum, and emerald

Table 8.4 Distribution of mined lands in Himalayan States of India

State	Mining area (00 ha)	Mineral
Uttarakhand	68.3	Limestone, phosphorite, limestone, dolomite, copper, silica, gypsum, magnesite, chromite
Meghalaya, Assam and West Bengal	114.7	Dolomite, silica, coal, limestone, China clay, quartz, mica
Jammu and Kashmir	8.9	Limestone, China clay, gypsum, magnesite, bauxite, sapphire
Himachal Pradesh	4.4	Limestone
Total	250.6	

Source Juyal et al. (2007)

Varieties of minerals such as limestone, dolomite, phosphorite, magnesite, gypsum, etc., are found in this region. Limestone is the most prominent mineral available in almost the entire Himalayan belt.

Production and Growth of Mining Sector in the Country

Mining industry in the country contributes significantly toward the socioeconomic condition. The GDP contribution of the mining industry varies from 2.2 to 2.5 % only but going by the GDP of the total industrial sector, it contributes around 10–11 %. The value of mineral production during 1999–2000 was estimated at INR 452.3 billion of which the contribution from public sector was INR 378.4 billion (Khullar 2006). The mineral production in the country is maximum from coal followed by limestone. Among metallic mineral, iron contributes toward maximum production (Table 8.5).

The growth of mining industries indicates that area under coal and metal mines have decreased. The values of these minerals however, have shown an increasing trend (Table 8.6).

Mining and Land Degradation

Land degradation is considered as an unavoidable by-product of mining. Due to overexploitation and mismanagement of natural resources by mining, land degradation has reached

Table 8.5 Production of some selected minerals in India (by mineral groups) 1999–2000

Fuel minerals		Metallic minerals		Non-metallic minerals	
Minerals	Production	Minerals	Production	Minerals	Production
Coal	300.0	Bauxite	6.8	Limestone	127.9
Lignite	21.9	Chromite	1.7	Dolomite	2.9
Crude oil	32.0	Copper ore	3.1	Gypsum	3.3
Natural gas[b]	26.4	Iron ore	73.5	Diamond[a]	41.0
		Manganese ore	1.6		
		Lead and zinc ore	2.7		
		Gold ore	0.7		

Units Million tons; [a] Unit in 10^3 carats; [b] Unit in BCM
Source Tata Energy Research Institute (2001)

Table 8.6 Growth of mining activities in India

Years	No. of reporting mines			Value (INR[a] in 10^3 million)		
	Coal	Metal	Oil	Coal	Metal	Oil
1951	893	1810	–	0.50	0.23	N.A.
1961	848	2323	–	1.14	0.48	N.A.
1971	781	1995	13	2.54	1.08	0.75
1981	496	1768	8	18.11	3.62	2.74
1991	561	1787	24	79.79	19.07	18.53
2001	568	1907	43	261.08	54.03	106.74
2007	567	1770	49	419.27	235.35	256.94

[a] *INR* Indian National Rupee (1 US$ = 54.34 INR in Jan 2013)
Source Ministry of Labor and Employment, Govt. of India

alarming proportions. During initial stages of development, mining leads to loss of biodiversity due to forest land diversion, economic, and livelihood loss due to displacement and encroachment of agricultural land, loss of local water resource due to reduction in catchment area, and destruction of streams. Some of the land-related impacts such as water pollution, loss of agricultural productivity, depletion of ground water, and siltation of river are more pronounced at the operational stage. Postoperational impacts on land are most prominent when reclamation is not carried out properly. For instance, a capacity of 10 million tones opencast coal mine in 20 years has potential to destroy around 800 ha of land. According to one estimate, around 4 ha of land is damaged for every million tons of coal mined by the surface mining (http://www.terienvis.nic.in). In underground mining, risks of land subsidence are very common due to which engineering structures such as highways, buildings, bridges, and drainage and ground water regime are seriously damaged.

Mine Ecology

Mining destroys vegetation and causes extensive damage to the soil and biodiversity (Fig. 8.1). Mine areas are usually characterized by high temperature and low humidity of soil moisture which affect the physical, chemical, and biological properties of the soil. Due to excessive use of explosives in mining operations, shear resistance of soil mass is reduced and shear stresses are increased which result in development of gullies and exposure of rock/hard pan on the surface. In hilly terrains, mining increases slope height thereby resulting in slope instability (Juyal et al. 2007). Stony nature of mine wastes further aggravates the situation for establishment of vegetation by developing low infiltration rates

Fig. 8.1 A view of area affected by limestone mining

and water retention due to compaction (Grunwald et al. 1988; Norland 1993). Surface drainages are disturbed and blocked due to surface mining which reduces recharge of aquifers.

Mining results in changes in soil texture and loss of soil structure. Due to absence of topsoil, mine soils have low pH, low organic matter, and low levels of plant nutrients. The soils also contain high concentration of Al, Mn, Fe, Zn, and Cu which are toxic to the plants. Moreover, shortage of soil micro-flora restricts the decay of plant material due to which the natural biodiversity is invaded by inferior species (Corbett et al. 1996; Singh et al. 2002). Some mineral overburdens release salt, heavy metal, and radioactive pollutants, effluents like ammonia, arsenic, asbestos, cadmium, copper, cyanide, lead, mercury, etc., which on leaching contaminates the land and water resources. The magnitude of impact intensifies when mine lease is close to the river and drainage flow is toward the river (Chaterji 1988; Verma 1988; Dadhwal et al. 1992).

Overburdened dumps change the natural land topography thereby affecting the drainage system and preventing natural succession of plant growth which ultimately results in acute problems of soil erosion and environmental pollution (Singh et al. 1996). Emission of flume gases suspended particulate matter like silica, fluorides, etc., and gaseous pollutants like sulfur dioxide, nitrogen dioxide, carbon monoxide, etc., also causes considerable air pollution and health hazard to animals and people inhabiting the adjoining areas (Chaterji 1988).

Rehabilitation of Mined Areas

Rehabilitation measures aim at returning the damaged ecosystem to productive use which is environmentally and socially acceptable. The various steps for rehabilitation of mined areas are discussed below:

Preliminary Survey and Proper Configuration of Disposal Dump

Before taking any mining operation, preliminary survey of the mining site should be taken up and topographic, vegetative, and soil maps should be prepared. The survey should detail out the limiting site factors, origin of the waste, size of the waste particles, etc. The overburden if any, to be generated should be analyzed for pH, acid base balance, and heavy metals like Fe, Mn, Al, etc., and their seepage should be checked. Mining muck, low grade ore, tailing, etc., should be disposed in areas which are not productive. The dumped over burden should be graded so that it can be used for carrying out afforestation work. If possible overburden should be filled back in same order as was found in nature. Weathering material and soils should be kept on top whereas; coarse/toxic material can be dumped sufficiently deep. The top soil (5–10 cm) is the most important for plant growth as most of the soil seed reserves are found in the surface and thus should be separately preserved and replaced on the top of overburden material (Roberts 1981; Iverson and Wali 1982; Putwain and Gilllham 1990). However, the collection, storage, and use of topsoil for restoration of mine areas is difficult and therefore, recent reclamation strategies are focused on creating soil that can support short-term establishment of native plant species to sustain long-term successional development (Noyd et al. 1997; Pflegar et al. 1994; Singh et al. 2002).

Site Preparation

In mined affected areas, establishment of vegetation is difficult due to absence of fertile soil, high runoff/debris movement during monsoon, and lack of soil moisture during summer. Site preparation in mined areas consists of reshaping the waste areas, covering the whole area with topsoil, filling in cavities, and leveling. In monsoon, uncontrolled runoff results in slope failures. Mechanical measures are therefore very important before starting any plantation work for stabilizing slopes and channels, control the runoff water and create conditions favorable to plant growth by arresting fine soil, and improving soil moisture status. These measures are followed by vegetative measures so that both of them act in union as a bioengineering measures, supporting and supplementing each other (Juyal et al. 2007). Slope stabilization measures are mainly used to provide mechanical stability to the eroded slopes and conserve soil moisture for establishment of permanent vegetation while channel stabilization measures are used to guide the flow and store and recharge the ground water. The important slope and channel stabilization measures used for rehabilitation of mine lands along with their functions are given in Table 8.7.

Role of Vegetation in Rehabilitation

Vegetation is the most appropriate and cost-effective long-term remedy to encounter the majority of underlying problems of derelict-mined land (Bradshaw 1997). Vegetation helps in ameliorating microclimatic conditions and has a marked catalytic effect on succession on severely degraded sites (Parrotta 1992). It plays a critical role in restoring productivity, ecosystem stability, and biological diversity through numerous processes, including maintenance or increase of soil organic matter, biological nitrogen fixation, uptake of nutrients from deep soil layer, increase water infiltration and storage, reduce loss of nutrients by erosion and leaching, improve soil physical properties, reduce soil acidity, and improve soil biological activity (Schaller 1993; Filcheva et al. 2000). The development of a permanent vegetation covers on mine waste aims at establishment of plant community that can maintain itself indefinitely without attention or artificial aid and support native fauna. However, establishment of vegetation is very difficult due to lack of top soil, altered pH, lack of organic matter, coarse rock fragments, and many other adverse biological and chemical factors. It takes about 50–100 years to attain a satisfactory vegetation cover. Redevelopment of advanced communities may even take a millennium or more (Dadhwal and Juyal 2012; Singh et al. 2002; Dobson et al. 1997).

Choice of Species

Different species behave differently in their capacity to stabilize soils. The rate of rehabilitation largely depends on the selection of appropriate species for revegetation (Parrotta 1992). Selection of plant species for revegetation of overburden dumps depends on various parameters such as climate, physical, and chemical properties of dump materials, topography, viability, and surrounding vegetation (Singh and Jha 1992). Some ecological variables must be considered while selecting the species for plantation. Plant species chosen should be preferably native of that place and should be fast growing so that it can provide quick and immediate vegetative cover and accumulate biomass rapidly on affected sites to avoid further damage. Local species are more suitable, better adapted as well as economically and ecologically viable for reclamation (Soni and Vasistha 1986). Species choice is also guided by seed and seedling availability, local uses for the species and economic aspects. Species that represent lower successional stages and maintains itself indefinitely without attention should be preferred. The species selected should have high root soil binding characteristics and high esthetic, economic, and ecological values. Nitrogen fixing trees help in improving the soil fertility and should be preferred. The species selected should improve the soil organic matter, available soil nutrients, and soil microbial biomass. Grasses help in soil and water conservation and can tolerate drought, low soil nutrients, and other climatic stresses and should be encouraged in rehabilitation program. A list of

Table 8.7 Important slope and channel stabilization measures used for reclamation of mined lands

Measures	Function
Slope stabilization measures	
Diversion drains	• To divert runoff water away in order to protect the downstream area and discharge it safely into protected waterway
Contour trenching	• Break the velocity of the runoff
	• Store whole or part of runoff
	• Moisture conservation in plantations and grassland development
Crib structures	• Slope stabilization
Retaining walls	• Stabilizing precipitous hill slopes/*nala* banks
Gunny bag/*katta-crate*/structures	• Construction of barrier for slope stabilization
Bench terracing with earthen/stone-cum-earthen shoulder bunds	• To break slope length and reduce degree of slope
	• Conserve soil moisture for better plant growth
Geotextiles	• Protection of land surface from splash erosion and surface runoff
	• Holding the seed/vegetation in place from washing away by runoff
	• Conserving fine soil and moisture particle for plant germination and establishment
	• Mechanical strength to land surface
Live check dams	• Stabilization of rills/small gullies
	• Check runoff and soil loss
	• Retain fine soil and conserve moisture
Wattling	• Breaking length of slope by providing contour wattles at 3–5 m interval
	• Reduce runoff velocity
	• Check soil loss in form of debris deposition
Channel stabilization measure	
Temporary check dams	• Stabilization of first-order gullies
	• Improve moisture conditions
	• Store and recharge ground water
Gabion check dams	• Flow guidance
	• Grade stabilization of channel bed
	• Store and recharge ground water
	• Checking runoff velocity in steep and broad gullies
	• Retention of soil/debris
Spurs	• Guide the stream flow along a desired alignment and to prevent soil erosion by aiding sedimentation along it

important trees and shrubs suitable for rehabilitation of different mined areas is given in Table 8.8.

Planting and Aftercare

The process of redevelopment in mine spoils begins naturally through colonization by species from the surrounding plant communities (Jha and Singh 1991). For afforestation, direct seeding is cost-effective and easier. Artificial seeding of quick growing grasses with short life cycle, legumes and forage crops accelerate development of vegetation; helps in improving site fertility; moisture retention capacity; stabilizing sloping areas and encourage natural invasion of tree and shrub seedlings (Vogel 1973; Dadhwal and Juyal 2012). Plantation of mixed species of economic importance should be done after 2–3 years of growing grasses (Singh 2004). Direct seeding of tree species for 3 years with grasses and leguminous forbs has been found to

Table 8.8 Plant species suitable for revegetation of mine spoils

Mine spoil category	Suitable plant species
Bauxite mined area of Madhya Pradesh (MP)	*Eucalyptus camaldulensis, Grevillea pteridifolia, Pinus* spp, *Shorea robusta*
Coal mine spoils of MP	*Acacia auriculaeformis, Acacia nilotica, Dalbergia sissoo, Eucalyptus camaldulensis, Eucalyptus* hybrid, *Pongamia pinnata*
Lime stone mine spoils of outer Himalayas, Uttarakhand	*Acacia catechu, Agave americana, Arundo donax, Bauhinia retusa, Buddleja asiatica Chrysopogon fulvus, Dalbergia sissoo, Erythrina suberosa, Eulaliopsis binata, Ipomoea carnea, Leucaena leucocephala, Mimosa himalayana, Pennisetum purpureum, Rumex hastatus, Salix tetrasperma, Vitex negundo*
Rock—phosphate mine spoils of Mussoorie, Uttarakhand	*Acacia catechu, Buddleja asiatica, Dalbergia sissoo, Leucaena leucocephala, Mimosa himalayana, Pennisetum purpureum, Rumex hastatus, Saccharum spontaneum, Salix tetrasperma,* and *Vitex negundo*
Lignite mine spoils of Tamil Nadu	*Acacia* spp, *Agave* spp, *Eucalyptus species, Leucaena leucocephala*
Mica, copper, tungeston, marble, dolomite, limestone, etc., minespoils of Rajasthan	*Acacia tortilis, Acacia senegal, Cenchrus setigerus, Cymbopogon* spp, *Cynodon dactylon, Dichanthium annulatum, Grewia tenax, Prosopis juliflora, Salvadora oleoides, Sporobolus marginatus, Tamarix articulata, Ziziphus nummularia,*etc
Iron ore wastes of Orissa	*Leucaena lencocephala* and local plant species, etc
Hematite/magnetite, manganese spoils from Karnataka	*Albizia lebbeck*, local plant species, etc

be useful (Juyal et al. 2007). Overtime, self-organizing biotic communities develop to optimize the flow of nutrients and energy within the ecosystem. The different techniques for the establishing the plantation in mined areas are given in Table 8.9.

Pits of 45–60 cm^3 have been found optimum for planting tree seedlings. Mine spoils are very poor in nutritional status and thus fertility management should be given due importance for the quick establishment of vegetation. Studies conducted for the limestone mine spoil revealed that 50 kg Nitrogen ha^{-1} and 20t FYM or leaf litter ha^{-1} to *Eulaliopsis binata* grass gave better results. Similarly, 10 g di-ammonium phosphate per sapling gave encouraging results in the first year of establishment of MPTS. Soil toxicity in terms of pH, metals, and salts should also be reduced by buffering toxic elements by the use of soil, peat, humus, clay, etc., with a high exchange capacity. Plantation of resistant species and cultivars can also reduce soil toxicity. Strong acid or alkaline soils should be neutralized by the use of lime, sulfur, or waste of the opposite reaction. Saline soils should be leached by means of rainwater or by fertilizing with organic or green manures or with chemical fertilizers. Placing normal soil up to a depth of 30 cm in the initial year of establishment has been found good for *Eulaliopsis binata* and *Chrysopogon fulvus* grass whereas, in second year, mixing treatment up to the same depth was found better over placing.

Mixing 50:50 (soil: debris) has been found suitable for the establishment of various tree, shrub, and grass species. Appropriate soil moisture conservation measures should be adopted for better survival and growth of the trees and grasses. Mulching with organic matter and leaf litter helps in soil moisture conservation and can be incorporated in mine spoil/debris. Dadhwal et al. (1991) reported that 20 t ha^{-1} of mulch was found to be the ideal dose which gave better results for *Eulaliopsis binata*. In Amarkantak, attempts were made to afforest the Bauxite mined area, which after open cast mining was devoid of top soil, and organic matter was full of laterite boulders and bauxite rejects. The area was leveled and filled with detritus in the month of May. The pits of the size 45 cm^3 at

Table 8.9 Afforestation of mining wastelands in India-few case studies

Mine and its location	Species tried	Technique	Results and References
Singrauli Coal field, Sidhi and Shandol (M.P.), Mirzapur District of U.P.	*Acacia nilotica, Prosopis juliflora* in Bina and Kakri mines	Planting of seedlings in pits	During 1982–85, density of cover achieved 0.5 (Jha 1987)
Dhanpuri coal mine, Shandol District, M.P.	*Eucalyptus* hybrid, *E. camaldulensis, Emblica officinalis, Pongamia pinnata, Acacia nilotica, A. auriculaeformis, A. catechu, Dalbergia sissoo,* bamboo etc.	60 cm^3 in 2 × 2 m grid pits filled with surface soil of natural sal forest along with 5 kg FYM	Survival rate 70–94 %, *Eucalyptus, A. auriculaeformis* and bamboo very successful (Prasad and Shukla 1985)
Dolomite mined area, Bilaspur District, Chattisgarh	*Gmelina arborea, Acacia auriculaeformis, A. nilotica, Eucalyptus* spp, *Albizia procera, Pongamia pinnata,* bamboo	45 cm^3 pits in 2 × 2 m grid filling as above	Survival 54–90 %, *Gmelina, A. auriculaeformis, Eucalyptus, P. pinnata* most successful (Prasad and Chadhal 1987)
Iron ore mine, Dalli Rajhara in Durg District, Chattisgarh	*Dalbergia sissoo, Pongamia pinnata, Acacia procera, A. aurriculiformis, Albizia lebbeck, Eucalyptus* spp *Emblica officinalis, Azadirachta indica, Terminalia arjuna,* bamboo, etc.	45 cm^3 pits in 3 × 2 m grid filling as above with 2.5 kg FYM per pit	Survival 73–100 %, *Dalbergia sissoo, Eucalyptus,* Bamboo, *Pongamia, Albizia, Emblica officinalis* most successful (Prasad 1989)
Limestone quarries, Sahastradhara, Mussoorie Hills, Uttaranchal	*Acacia catechu, Dalbergia sissoo, Leucaena leococephala, Bauhinia* spp, *Salix tetrasperma, Grewia* spp (trees); *Vitex negundo, Ipomoea carnea* (shrubs); and *Chrysopogon fulvus, Eulaliopsis binata, Saccharum* spp (grasses), etc.	Seedlings planted in contour trenches/pits filled with good soil mixed with minespoil. Geojute technique to stabilize unstable and degraded mine spoil	*Acacia catechu, Bauhinia, Dalbergia sissoo, Salix* most successful trees (Dhyani et al. 1988; Juyal et al. 1998)
Limestone quarries, Nandini near Bhilai, Durg District, Chattisgarh	*Dalbergia sissoo, Pongamia pinnata, Acacia aurriculiformis, Eucalyptus* spp, *Emblica officinalis, Leucaena,* Bamboo.	45 cm^3 pits n 3 × 2 m grid filling with surface soil and 2.5 kg F M per pit	Most of the species well established except bamboo (Prasad 1989)
Rock phosphate and Limestone mined area, Mussoorie hills, Uttaranchal	*Acacia catechu, Dalbergia sissoo, Leucaena, Salix, Pinus roxburghii, Robinia pseudoacacia, Populus* and many shrubs, grasses, and sedges.	Director sowing, seedling and stump planting, cuttings and rootstocks in contour trenches	Indigenous species have higher (>60 %) survival than exotics (Soni et al. 1990)

Source (Dhyani et al. 2007)

a spacing of 2 x 2 m were dug in June, and imported soil from the nearby sal forest was used to fill up these pits along with cow dung manure @ 0.5 kg pit^{-1}. The results of 5 years old plantations revealed that survival and growth of *Grevillea pteridifolia* and *Eucalyptus camaldulensis* was superior with mean height of 5.01 and 7.76 m, respectively. *Pinus caribaea* and *Acacia auriculaeformis* were also found to be suitable. Studies also revealed that the biomass production efficiency of *Grevillea pteridifolia* was comparable to many other species (*E. camaldulensis* and *P. caribaea*) and its plant nutrient requirements were less, which makes it adapted to orphaned sites (Parsad and Chadhar 1987; Parsad and Shukla 1988).

Legal Framework for Rehabilitation

Management of mineral resources is the responsibility of the Central and State Governments in terms of Entry 54 of the Union List (List I) and Entry 23 of the State List (List II) of the Seventh Schedule of the Constitution of India. Management of mineral resources has therefore to be guided by long-term national goals and perspectives. According to Forest conservation Act, open cast and underground mining are non-forestry activity and therefore, prior approval of the Central Government is essential before a mining lease is granted in any forest area. No mining activity is permitted in National Parks and Wildlife Sanctuaries and adjoining National Monuments, areas of cultural heritage, ecologically fragile areas, areas rich in biological diversity, genepool, etc. All proposals involving forest land more than 20 ha in plains and more than 5 ha in hills must be accompanied by a cost-benefit analysis to determine whether diversion of the forest land for non-forestry use is in the overall public interest. No mining lease can be granted to a private or a public party without a proper mining plan including environmental management plan approved by Indian Bureau of Mines. The mining plan is reviewed after 5 years and a mining scheme for next 5 years is submitted. Environmental Impact Assessment (EIA) for development projects have been made compulsory. As per National Mineral Policy, preference is given to the schedule tribes for the grant of mineral concessions of small deposits. If the project involves displacement of people, a detailed rehabilitation plan is to be submitted along with the proposal. The SC and ST population should be separately considered and a plan for their rehabilitation should be in cognisance with their socio-economic, cultural lifestyle.

Five Acts viz., The Water (Prevention and Control of Pollution) Act 1974, The Water (Prevention and Control of Pollution) Cess Act 1977, The Air (Prevention and Control of Pollution) Act 1981, The Environment (Protection) Act 1986, and The Public Liability Insurance Act 1991 are directly applicable to the management of the environment in mining areas. The common objectives of these legislations are to arrest further damage to the environment and ecosystem resulting from mining, to take positive steps toward conservation of the environment, to take measures to restore the environment in areas damaged including measures such as reclamation of degraded land, to create authorities to administer the policy and contents of the legislation, and to provide penalties and prosecution for violation of laws. For mineral companies, general standards for discharge of effluents from mining and mineral processing activities and ambient air quality standards in respect of noise for industrial, commercial, residential areas, and silence zones have been formulated.

Compensatory afforestation is one of the most important conditions stipulated by the Central Government while approving proposals of dereservation or diversion of forest land for non-forest uses. It must be done over equivalent area of non-forest land close to reserved forest or protected forest to enable the forest department to effectively manage the newly planted area. The identified non-forest land has to be transferred to the ownership of the State Forest department and declared as Protected Forests so that the plantation raised can be maintained permanently (TERI 2001).

Table 8.10 Status of afforestation and trees survived in mined areas in India (up to 1999–2000)

Minerals	Mines covered	Area covered (ha)	Trees planted (millions)	Trees survived (millions)	Survival rate (%)
Bauxite	53	1451	5.0	3.9	77
Chromite	14	338	1.5	0.9	60
Copper	7	331	1.3	0.8	62
Dolomite	50	278	0.5	0.3	68
Gold	5	412	0.9	0.6	70
Iron Ore	130	7714	21.5	15.2	70
Iron and Manganese	31	172	0.5	0.4	74
Lead and Zinc	7	1332	0.7	0.7	88
Limestone	260	7714	12.6	9.2	74
Manganese	57	1940	5.0	3.2	64
Magnesite	16	485.25	0.43	0.30	68.37
Pyrite	1	7	0.02	0.013	69.03
Others	270	1382.99	2.43	1.58	65
Total	901	23556.72	52.32	37.03	70.76

Source Ministry of Statistics and Programme Implementation (2001)

Status of Mine Rehabilitation in the Country

At present, mine rehabilitation is completely ignored area in the country. According to report of the Union ministry of mines (Ministry of Statistics and Programme. Implementation, Govt. of India), 53 abandoned mines covering an area of about 660 ha were reclaimed/rehabilitated in 2007–08 and about 1135 mines covering 11200 ha area were still under rehabilitation program. No information about the kind of reclamation/rehabilitation undertaken has been provided in these reports, nor is there any information on the whereabouts of these rehabilitated mines. A total of 901 mines covering an area of 23556 ha were planted with 52 million trees out of which 70 % survived (Table 8.10).

Success Stories

Rehabilitation of Limestone Mined Areas in Shahastradhara

Indian limestone in general is high in silica and magnesia, therefore almost equal quantity of waste and overburden is generated due to mining of this mineral. Limestone mines are confined to 40–50 m depth and can be categorized as shallow operations. Dehradun-Mussoorie belt in the Himalayan region contains limestone of high grade for steel making. There was a time when the production of limestone from this area was 1.5 million tons per annum which used to be contributed by 50–60 small mines and medium mines. Public litigation was filled in the Supreme Court, which gave a ruling to close all mines in the area after which number of successful rehabilitation programs in the region were carried out by the Forest Research Institute and CSWCRTI, Dehradun and Ecotask Force (TERI 2001). One such project was undertaken in Shahastradhara, Dehradun by CSWCRTI (Juyal et al. 2007). Biological measures with proper scientific technologies were undertaken to restore the productivity of land and to maintain esthetic beauty and visual impact on ecology. Results indicated that planting of slips of *Eulaliopsis binata* was ideal on the degraded, steep mine spoil areas. Planting of leguminous species, such as *Leucaena leucocephala*, and *Peuraria hirsuta* in the mine spoil provided foliage rich in N which served as fodder, organic manure, mulch, etc. Species, such as *Thysonoleana maxima*, *Saccharum munja*, *Pennisetum*

Table 8.11 Changes in minespoil characteristics overtime (average of 14 locations)

Properties	Years						
	1985	1991	1992	1994	1995	1997	2002
pH (1:2.5)	8.1	7.9	7.7	7.6	7.6	7.5	7.40
Organic C (%)	0.13	0.18	0.26	0.39	0.4	0.42	0.45
Total N (%)	0.01	0.02	0	0	0	0.05	0.05
$CaCO_3$ (%)	54.6	34.6	34	32	31.8	31	29.50
Bulk density (Mg m^{-3})	1.63	1.53	1.5	1.48	1.48	1.47	1.45

Source: Juyal et al. (2007)

purpureum, *Eulaliopsis binata*, *Ipomoea carnea*, and *Vitex negundo* performed well under geotextiles. With continued biotic protection together with rehabilitation measures the retrogression and erosion came to stand still. The vegetal cover increased from 10 to >90 % due to reforestation over a period of 20 years. On an average, after 18 years, pH of the spoil came down from 8.1 to 7.4, organic carbon increased from 0.13 to 0.45 %, whereas $CaCO_3$ decreased from 54.6 % to 29.5 % and bulk density from 1.63 to 1.45 Mg m^{-3} (Table 8.11).

The equivalent slope in the treated hilly terrain reduced from 38 % to about 19 % over a period of 14 years. Runoff and soil loss also decreased considerably over the time with mechanical and biological measures. Because of the various treatment measures, the main drainage channels which used to dry up by the end of October, have now become perennial. The dry weather flow recorded in the main drainage channel was to the tune of 265 and 100 cum per day in the months of November and February, respectively. The augmented flow is now being used for irrigation purposes by the farmers. Runoff water from treated mine site was found within the permissible water quality limit (Table 8.12).

Neyveli Lignite Corporation (Tamil Nadu), India

The achievement of Neyveli Lignite Corporation (NLC) in land reclamation of mines spoils and afforestation have been noteworthy. The mined out area or the de-coaled area is refilled with overburden, since this does not contain plant nutrients or have the proper texture. The dumped soil is improved in stages through modern techniques to bring back its original fertility and the agricultural operations are carried out by adding nutrients, like organic, inorganic and biofertilizers. Today crops and vegetables of various varieties are being constantly raised on 250 ha of rehabilitated land.

Massive afforestation of *Cassuarina equisetifolia, Leuceana leucocephala, Dalbergia sissoo, Acacia auriculaeformis, A. nilotica, Albizia lebbeck*, and *Eucalyptus citridora* have been carried out around the mine spoil area, industrial units, and township. The technology consists of planting seedlings in pits after filling 2/3 depth with red soil, FYM, and sand after leveling. From period of 1970–86, the survival was 80 % (Narayana 1987). The total plantation in Nlc is of around 17 million trees of various species on an area of 2750 ha. Lakes and ponds were formed in the afforested areas and a picnic spot was also created with boating facilities, along with a mini zoo. Neyveli has now become a bird sanctuary and number of bird species visit during the different seasons.

Issues

Mineral development, while increasing employment and incomes create serious environmental damage and can undermine other socio-economic development opportunities of local communities. It is, therefore, necessary to address both the environmental damage and the socio-economic constraints created by mining

Table 8.12 Impact of rehabilitation measures

Particulars	Pre-treatment	Post-treatment
Minespoil outflow (t ha^{-1} yr^{-1})	550	6
Surface runoff (%)	57	37
Lean period flow (days)	60	>240
Water quality	Not portable	Portable
Vegetal cover (%)	10	>80
Cost of debris clearance (INR yr^{-1})	0.1 million	Nil

Source: Juyal et al. (2007)

operations. A multidimensional approach that integrates a number of activities that range from the technical to the social, to training and capacity building is, therefore, urgently required. The following issues need to be taken up for rehabilitation of mined areas:

- Identification of stress-tolerant plant species having a positive influence on soil fertility and design of management systems for maximizing ecosystem productivity under a wide range of degraded site conditions.
- Research on ecological aspects of reclaimed areas, such as role of organic matter in productivity and nutrient relationships, biodiversity restoration potential need to be taken up.
- Improving the social and environmental knowledge base to allow local communities to control more actively their own environments, in particular women community members.
- Improve or maintain community relation programs and sustainable environmental management systems.
- Capacity building of the stakeholders.
- Policies for prevention of further degradation of and involvement of local community.

Ravine Lands

The word 'ravine' means a deep gorge and represents the last stage of water erosion. Ravines are the systems of gullies running almost parallel to each other and draining into a river after a short distance with the development of deep gorges (Fig. 8.2).

Fig. 8.2 A view of ravine land

A ravine is a very small valley—almost like a canyon but narrower—which is often the product of stream cutting erosion. Ravines are channel of ephemeral flow, denuded and guided essentially by the process of rejuvenated streams, and having steep sides and head scarps with a width and depth always greater than a gully. Ravines may or may not have active streams flowing along the down slope channel which originally formed them. Ravine cuts in the alluvial deposits are quite common and are the most stunning topographic features along most of the river systems of the Indian subcontinent. These may extend up to 2–3 km into the tablelands and vast areas go out of cultivation due to this land degradation. Ravines are classified as larger in scale than gullies, although smaller than valleys. The depth of small, medium, and large gullies are <1 m depth, 1–5 m depth, and 5–10 m depth, respectively. Ravines are usually >10 m depth.

Cause of Ravine Formation

Ravines are formed when the upper layer of vegetal cover is not strong enough and the roots are unable to hold and bind the soil together. Constant rainfall erodes the soil and washes away the crust of the earth due to which, the water flow turns into drains, creating cracks. In due course of time, these cracks are further eroded, and become large ravines. The devastating effect of rain drops striking bare soil by detachment of soil particles is the principle factor for soil erosion. In the initial stages of rainfall, splashing effect of the raindrops detach the soil particles. But as the rainfall continue on the bare soil, a sheet of water begins transporting loose soil down slope. As the water accumulates in low areas, it takes shape into a stream, the deeper it becomes, the faster it flows, and higher its erosive power becomes, cutting gullies while flowing down. Gullies erode away entire soil profiles and often cut into unconsolidated material beneath the soil (Luna 2006).

Extent of Ravine Lands

Ravines and gullies occur all over India, the largest incidence is found in Madhya Pradesh, Uttar Pradesh, and parts of Rajasthan. The National Remote Sensing Agency, based on the Land sat data, has estimated nearly 4 m ha of ravine land. The extent of ravine area in the different states of the country is given in Table 8.13.

The ravines occur along the river Yamuna and its tributaries; river Gomati and Kholas of Ganga in Uttar Pradesh; along river Chambal and its tributaries; Assan, Kalishindh, and Kshipra in Madhya Pradesh; river Chambal and its tributaries like Kalishindh, Mej, Parvati, Banas, Parwan etc.; river Banganga and Sabi in Rajasthan; and river Mahisagar, Narbada, Sabarmati, Watrak, Saraswati, Dhadhar, etc., in Gujarat (Tejwani et al. 1975). There are other areas of substantial ravine erosion in Chota Nagpur, Bihar, the Mahanadi, and Upper Son Valley of Mirzapur, all on the southern fringes of the Gangetic Plain, and the Shiwalik and *bhabar* tract of the western Himalayan foothills (Sharma 1980).

Ravine Ecology

The climate, topography, and soils of the ravine region are typically harsh and thus adversely affect the composition and growth of the vegetation. The climate of the ravine region in Rajasthan and Madhya Pradesh is semi-arid to subhumid. The average rainfall (250–750 mm with almost 9 dry months annually) is both low and erratic. Ravine areas also experience frost and drought which cause a severe damage to the vegetation. Deficiency or rainfall and excessive evaporation due to more exposed surface area results in concentration of soluble salts on the surfaces.

The soils of the ravinous regions are poor in fertility and physical conditions (Dagar and Mall 1980) as almost all the soil profiles are eroded leaving behind rocky, sandy, or stiff clay with poor water retention or water absorbing power. The process of soil formation is also very slow due to the fact that little interaction takes place due to non-availability of water and vegetation roots, sometimes the lime develops into *kankar* pan (deposits of calcium carbonate) which further prevents the roots of the plants to penetrate and get water from deeper layers. The ravine soils vary in pH from 7.5 to 8.0. In the zones of lighter soils in Uttar Pradesh, Madhya Pradesh, and Gujarat, the water holding capacity is low. Soils of these regions are deep alluvial deposits having sandy to sandy loam texture. The soils of Chambal ravine regions are less deep alluviums as compared to those in Yamuna and Mahi ravine region in Uttar Pradesh, Kshipra river (Madhya Pradesh), and Gujarat. They have silty clay loam to clay texture and calcium carbonate content increases with increasing depth. These soils are prone to water logging when wet and to

Table 8.13 Extent of ravines in India

S No.	State	Area (m ha)
1.	Uttar Pradesh	1.23
2.	Madhya Pradesh	0.68
3.	Rajasthan	0.45
4.	Gujarat	0.40
5.	Maharashtra	0.02
6.	Punjab and Haryana	0.12
7.	Bihar	0.60
8.	Tamil Nadu	0.06
9.	West Bengal	0.10
10.	Odisha	0.11
11.	Others	0.19
12.	Total	3.98

compaction when dry. The infiltration rates in these soils are low to moderate and soils are highly erodable in the upper layers.

The land feature of ravines is rugged and characterized by three distinct parts namely, ravine top or hump (irregular in size and shape), slope and bottom/bed. In general, the depth of ravines ranges from 1 to 20 m but may reach up to 55 m. Generally, ravines of the Chambal river region are 'V' shaped where the sub-soil is comparatively resistant to erosion than the top-soil. Ravines of 'U' shaped are found in Yamuna and Mahi, where thickness of alluvium increases and the soils are light textured. All types of gullies and ravines have been reported to occur along Kshipra river in Madhya Pradesh (Dagar and Mall 1980; Dagar and Singh 1980; Dagar 1995).

The ravines are characterized by dominance of thorny vegetation. Champion and Seth (1968) have classified the natural vegetation of ravines as northern tropical ravine thorn forests. Sajwan (1975, 1976) and Prajapati et al. (1995) have described the floristic wealth of Yamuna ravines. Parandiyal et al. (2000) also studied the floristic diversity of Chambal ravines. The typical vegetation consists of low growing type of coarse grasses like *Aristida adscensionis, Eremopogon faveolatus, Heteropogon contortus, Themeda triandra, Eragrostis* spp., *Eleusine* spp. and rarely occurring *Dicanthium annulatum* among grasses, bushy shrubs of *Ziziphus nummularia., Capparis zeyalanica, Capparis decidua, Crotolaria burhia, Cassia tora, Tephrosia purpurea, Xanthium strumarium, Calatropis procera, Clerodendron phlomides, Grewia pilosa, Grewia tenax, Lantana camara,* etc., and xerophytic trees like *Dichrostachys cineria, Acacia senegal, Acacia leucophloea, Acacia nilotica, Prosopis juliflora, Balanites aegyptiaca*, etc. Dagar and Mall (1980), Dagar and Singh (1980), and Dagar (2001) gave extensive account of vegetation ecology of Kshipra ravines and its tributaries including the successional trends of the vegetation under biotic pressure. Over-hardy grasses such as *Sehima nervosum, Heteropogon contortus, Dichanthium annulatum, Chrysopogon fulvus, Cynodon dactylon* and legumes *Alysicarpus rugosus,* and *Rhynchosia minima* withstand biotic pressure while species such as *Themeda triandra, Iseilema laxum, I. prostratum, Apluda mutica,* and legumes like *Indigofera glandulosa, Rhynchosia capitata,* and *Alysicarpus longifolius* appear when the area is protected from grazing. Ecological studies have revealed that in most ravenous tracts, the secondary succession takes 10–15 years to reach a stable tree stage while in ravine bottom, it takes 6–7 years only.

Rehabilitation and Management of Ravine Lands

Barring the marginal and shallow gullied lands, which can be rehabilitation economically and can be utilized for agriculture, most of the ravinous lands are not fit for cultivation of agricultural crops and are classified as non-arable lands. Crop cultivation in these lands is not desirable as it leads to accelerated erosion thus adding to their fast degradation. The best scientific land use for these lands is to place them under permanent vegetation involving forest and fruit trees along with forage grasses and energy plantations. For the rehabilitation of ravine areas, the whole gullied watershed is to be considered which consists of two important components a) table and marginal lands contributing runoff to the gullies, and b) gullies proper. Rehabilitation of gullies and ravines involves treatment of these two components (Verma and Chinamani 1990). Ravine rehabilitation requires an integrated approach of using gullies according to land capability classes and using soil and water conservation measures and productive utilization of land through agriculture and afforestation measures which are discussed below:

Using Land According to Gully Reclamability Classification

In India, classification of gullies has been made in Uttar Pradesh, Madhya Pradesh, and Gujarat. The classification differed from agriculture department to forest department from state-to-state. The classification is linked with treatment of gullies for different land uses like agricultural and non-agricultural ones. It takes into account the influence of soil characteristics and gully dimensions on suitability of reclaiming gullies for cropping, horticulture, forestry or growing grasses, etc. Tejwani and Dhruvanarayana (1961) classified ravines into four classes based on depth, bed width, and side slopes of the gullies. In Rajasthan, the forest department uses the classification based on depth of gullies to classify them as G_1 (gully depth<1 m), G_2 (gully depth 1–5 m), and G_3 (gully depth > 5 m). Bali and Karale (1977) have suggested criteria for reclamation of gullies. Verma (1981) proposed a detailed classification of ravines for optimum utilization in which gullies have been classified into six reclaimability classes. Verma and Bhushan (1986) also proposed gully reclamability ratings by classifying them into following six classes:

Class 1: Shallow gullies which require minor leveling work on gently sloping sides on bed for reclamation for agriculture. They have favorable soil texture for most of the crops. The bed may be wide enough to be converted into agricultural fields with minor shaping. Good management practices are required

Class 2: Gullies which require moderate leveling work for reclamation for agriculture. The gullies are 1.5–3.0 m deep and sides are more sloping than in Class 1. Maintenance of reclaimed plots requires more care of terrace faces

Class 3: It has got more limitations than Class 2 of reclaimability class gullies. Soil texture may be lighter or heavier than loam and the gully may be deeper, presenting more limitation for reclamation and management of reclaimed gully for cropping

Class 4: It has severe limitations for reclamation for agriculture in the soil texture or gully dimensions. Steeper side slopes and more gully depth constitute a borderline case for reclamation for cropping. Such gullies may be put under agri-horticulture after constructing narrow terraces

Class 5: This category includes gullies described in Classes 1–4, but they cannot be gainfully reclaimed for agriculture as they are prone to seasonal backflows from a nearby river or have developed water logging, salinity problems, etc., due to irrigation of the adjoining table land or any other adverse factors. Under such very adverse conditions, gullies may be put under perennial vegetation like suitable fuel and fodder trees and grasses (silvipasture). Grass cutting for stall-feeding may be allowed. Fuel trees may be exploited on rotation basis

Class 6: Gullies presenting limitations more severe than Class 4 in soil characteristics like texture and gully features like depth and side slopes are included in this category. They may be put under perennial vegetation.

In the gullies belonging to reclaimability Classes 1, 2, 3; all local crops such as sorghum (*Sorghum durra*), maize (*Zea mays*), pearl millet (*Pennisetum typhoides*), wheat (*Triticum aestivum*), green gram (*Vigna radiata*), black gram (*V. mungo*), pigeon pea (*Cajanus cajan*), gram (*Cicer arietinum*), mustard (*Brassica juncea*), and tobacco (*Nicotiana tobacum*) can be grown in gully beds and on bench terraces made on gully sides. Class 4 may be put under arable crops along with fruit trees such as lemon (*Citrus limon*), mango (*Mangifera indica*), ber (*Ziziphus mauritiana*), and aonla (*Emblica officinalis*) at required spacing in agriculture crops in gully beds. Classes 5 and 6 may be planted under perennial vegetation like fuel and fodder producing trees and grasses.

Soil and Water Conservation Measures

Measures for Checking Extension of Gullies

(i) ***Peripheral bunds***

The uninterrupted high velocity flow of surplus runoff water from the arable lands into the adjoining nalah/gullied lands leads to the extension of gullies into the arable lands. It is essential that the runoff may be allowed to reach the gullies at controlled velocity and at designated points only where safety measures for controlling the damages are installed. Peripheral bunds are erected around the arable lands at a distance from the gully heads that is equal to double the depth of gullies. For the clay soils, peripheral bunds of 2.0–2.5 m cross section with 0.1–0.2 % grade are recommended. These bunds are reinforced by planting of grasses. The construction of these bunds also leads to moisture conservation to the tune of 15–20 % in the adjoining lands.

(ii) ***Permanent Drop Structures for Runoff Disposal in Narrow and Deep Gullies***

At the places where the excess runoff is allowed to enter the gullies from the peripheral bunds, spillways like drop spillways and chute spillways are recommended depending upon the fall of gully, contributing area of the catchment. The spillways are constructed for safe disposal of excess runoff from the adjoining arable lands through the peripheral bunds to the gully beds. Drop spillways are recommended where the runoff contributing area is large and gully fall is not more than 3 m. In case of higher fall (3–6 m) chute spillways are used for large flows. These spillways are designed based on the peak rate of runoff, vertical fall of gully, and width of the gully. These spillways also help in grade stabilization in the gully beds thus helping in their restoration.

Gully Control Structures for Controlling Gully Erosion

(i) ***Brush Wood Check Dams***

These are erected for controlling runoff in small- and medium-deep gullies, where permanent protection is sought by developing vegetative cover. These temporary check dams hold certain amount of soil to facilitate the establishment of vegetation and are thus known as soil savers. In the small gullies (1–1.5 m depth), brush wood check dams covering the width of gullies (2.5–6 m) are erected. These brush wood check dams may be single row or double row. For this, 8–10 cm diameter poles of 1.5 m height are fixed across the gully into the soil at a depth of 50 cm at a spacing of 40 cm from each other. These poles are then interwoven with the locally available brushwood or small branches (Fig. 8.3).

The brushwood check dams reduce the velocity of runoff and help in reduction of soil loss while also checking the velocity of runoff. A substantial quantity of soil gets deposited behind these structures and together with the improved moisture availability, it facilitates the raising of economically useful vegetation and leads to better survival and growth of vegetation. The silt deposited behind these structures leads to stabilization of gully beds. Construction cost of single row brush wood check may vary from INR 80–100 per running meter length depending on location and availability of material.

(ii) ***Wattling***

Wattling is a structure that holds the earth flow reducing by the raindrop impact. Wattles are laid out at a vertical interval of 3 m up to 66 % slope and 5–7 m up to 33 % slope. These are laid with poles of 8–12 cm in diameter, 1.5 m in length of easily sprouting species, buried half in the ground, 1–1.25 m apart in trenches packed with earth and local soil, etc. Above ground portion of the poles acts as shut, across which brush wood is woven in a way that it holds the soil. In due course, the poles sprout, form roots, firmly hold the soil and intercept rainfall. Intervening space between each such wattle is covered with about 10 cm thick bundles of grass held firmly by 20 gage G.I. wire.

(iii) ***Permanent (Gabion/Masonry) Gully Control Structures***

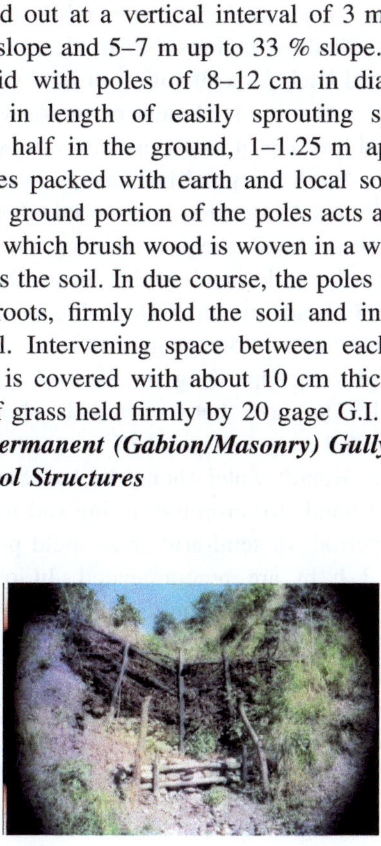

Fig. 8.3 Temporary brushwood check dam

For the control of deep gullies permanent gully control structures may be needed. These are designed depending on the catchment's area, peak flow, and grade of gullies. The fortification of all kinds of mechanical structures with planting of vegetation strengthens them and acts as second line of defense initially and with passage of time stabilizes the gullies while ensuring optimum utilization of these lands. The cost of these structures varies from INR 25 thousand to 80 thousand depending on the runoff contributing area, gully width, and depth. Permanent gully control structures are erected only when their benefits outweigh the costs involved. While these structures prevent the extension of gully, they also help in retaining large quantity of runoff and silt behind them. These structures thus improve the ground water, allow more water to infiltrate into the soil resulting in higher water table in the downstream wells.

(iv) ***Live Vegetative Check Dams***

In the small gullies where runoff from the catchment is not high, live vegetative checks of locally available shrubs and perennial vegetation are effective for retention of soil and reducing the grade of gully bed. These live vegetative checks are low cost and permanent in nature. Once established, they add to the productive value of ravines by providing fuel, fodder, and/or other usable biomass. Two or more rows of vegetation is planted or raised by seeding across the gully in staggered manner to develop an effective live check dam. Depending on the moisture availability, shrubs like *Agave americana, Capparis decidua*, and tall grasses like *Arundo donax* and *Erianthus munja* having protective as well as economic value or *Ipomoea carnea, Lantana camara, Grewia pilosa, Clerodendron phlomoides,* etc., having protective value may be utilized for live vegetative check dams usually in a combination of more than one species. Seeding of grasses behind these check dams may improve fodder availability. Brush wood like *Lantana camara* should be avoided because these have negative effect on fodder grasses. Some of the recommended gully treatments for different sized gullies are given in Table 8.14.

Table 8.14 Recommend gully treatments

Name	Depth (m)	Recommended techniques	
		Waterways	Gully areas
Rills	<0.30	Fascine (community lands), minimum tillage (agricultural lands)	Planting grasses
Small	0.3–1	Brushwood check dam	Planting/brush layering/wattling
Medium	1–5	Brushwood and loose stone check dams	Planting/brush layering/wattling/ retaining walls
Large	5–10	Loose stone and gabion check dams	Planting/brush layering/wattling/ retaining walls
Ravines	>10	Loose stone and gabion check dams	Planting/brush layering/wattling/ retaining walls

(v) ***Trenching***

As moisture may be the most important limiting factor affecting the survival and growth of planted trees as well as their production from the ravine lands, measures for optimum utilization of rainwater are essential for ravine reclamation. Continuous or staggered contour trenches are dug up for *in situ* rainwater harvesting on the non-arable lands. Depending on the land slope and the tree species to be planted. These staggered trenches are dug up at horizontal interval of 1–3 m. Bhushan et al. (1986) reported that staggered contour trenches proved very effective in preventing gully head extension, soil loss, and interception of runoff in Yamuna ravines at Agra. Sethy et al. (2011) reported that staggered trenching for trapping 75 % runoff from the watershed resulted almost doubling the productivity of 6-year-old fruit tree-based pastoral system consisting of *Emblica officinalis* + *Cenchrus ciliaris* grass in Chambal ravines over the no moisture management treatment. In order to maintain proper moisture in and around the roots of planted trees, the land between two trees is shaped into V shape with 5 % slope. This facilitates the movement of water to the root zone of trees, which results in improved survival, growth, and production in respect of the fruit trees. Covering the upper part of V-shaped catchment with polythene further improves the runoff yield.

Productive Utilization of Ravine Lands

Arable Lands near Head of Gullies/ Ravines

The catchments of gullies or the tablelands located near the ravines need special attention, as these lands are highly vulnerable to degradation. Shallow ravines (<3 m deep) can be reclaimed back for cultivation of field crops by putting earthen embankment or masonry check dam and leveling of side slopes into steps bench terraces. Earth marginal bund helps in preventing gully head extension into the table/marginal land. These bunds help in leading the runoff water to the gully heads which are made stable with gully head structures. Desirable mechanical measures like contour or graded bunds need to be employed, depending on the soil type and rainfall. The area between bunds should be leveled and proper disposal of excess runoff be ensured. Runoff water should be harvested into dug out ponds for crop use during soil moisture stress period. In semi-arid areas, field ponds of about 2 h m are recommended. In order to ensure optimum economic returns from these areas simple agronomical practices like contour cultivation, green manuring, mulching for reducing moisture loss, ensuring optimum plant population, and proper crop geometry should be followed. For reducing erosion from cultivated areas, sod-strip checks, cover crops and intercropping of cereals with legumes are widely recommended (Dev Narayan and Bhushan 2002). On gentle sloping beds sod-strip checks with a minimum width of 30 cm are used at a space of 2–3 m for arresting silt. Intercropping provide soil cover, check erosion and is capable of providing better returns as compared to sole cropping. Several profitable intercropping combinations have been identified for the ravinous regions mainly depending on soil type and moisture availability.

Non-arable Lands

Fencing

One of the prerequisites for reclamation of ravine lands is to effectively protect these areas by fencing from biotic agencies which will result in substantial increase in yield of grasses. Closure should be accompanied by soil conservation measures such as construction of check dams and disposal of runoff. The gullies can be brought into good grass cover in 2–5 years, good tropical thorn forest in 25–50 years, and dry deciduous forest in 100–200 years with protection. Chinnamani (1986) reported that 30 years of protection at Soil Conservation Centres, gully systems of Yamuna, Chambal, and Mahi rivers have developed dense natural forests with trees, shrubs, grasses, and leaf litter with almost no runoff and soil loss, high infiltration and enriched wildlife. Fencing should always be accompanied by soil conservation measures.

Choice of Species

The selection of tree species is of important factor in ravine areas. The highly eroded soils coupled with extremes of temperature and limitations of moisture hamper the survival and growth of trees. Selection of the tree species is guided by the compatibility of tree crop soil interactions, favorable tree architectures, ecological suitability of tree species, economic value of tree species, and farmer's perception of the tree species. Fast growing herbaceous plants, grasses, shrubs, trees which are easy to establish, and have fibrous roots are suitable for ravine lands. The species selected should have the ability to endure unfavorable site conditions and provide the quickest cover for protection against the erosion. Species like *Acacia catechu* and *Dalbergia sissoo* with a dormant period during the dry summer months should be preferred for gully slopes and humps to overcome moisture deficiency in the soil. In the ravine beds where moisture availability is better than ravine top or side slopes, valuable species like *Dalbergia sissoo*, *Tectona grandis*, *Eucalyptus,* and *Dendrocalamus strictus* provides good economic returns. Species like *Mours alba*, *Broussonetia papyrifera, Dendrocalamus strictus, Gmelina arborea* may be introduced in the gully beds. For saline and alkali patches, species like *Acacia nilotica, Azadirchta indica, Albizia* spp, and *Prosopis juliflora* should be preferred. For salt-affected ravines, *Acacia tortilis, Albizia amara,* and *Dichrostachys cinerea* are promising species (Luna 2006). For energy plantation *Acacia nilotica, Acacia catechu, Acacia tortilis, Acacia senegal, Albizia lebbeck, Azadirachta indica, Leucaena leucocephala*, and *Prosopis juliflora* are some of the suitable species for ravine top and bottom which can provide 2–3 t ha^{-1} $year^{-1}$ fuel wood in a rotation of 15 years. *Acacia nilotica, Ailanthus excelsa, Albizia amara, Bauhinia purpurea, Leucaena leucocephala, Kigelia pinnata, Morus alba, Prosopis cineraria,* and *Soymida febrifuga* planted on the ravine top and ravine bottom are good source of leaf fodder and can produce 2–4 t ha^{-1} $year^{-1}$ leaf fodder from 5th year of planting. Fruit trees can also be selected for plantation. However, care should be taken that selected fruit species are capable of withstanding moisture stress. Species like ber (*Ziziphus* spp.), aonla (*Emblica officinalis*), sharifa (*Annona squamosa*), pomegranate (*Punica granatum*), phalsa (*Grewia subinaequalis*), lasora (*Cordia dichotoma*), imli (*Tamarindus indica*), karonda (*Carissa carandas*), bael (*Aegle marmelos*), etc., can be good choice for ravine habitats. Among grasses,

Dicanthium annulatum, *Cenchrus ciliaris*, *Cenchrus setigerus*, *Panicum antidotale*, *Panicum maximum*, *Pannisetum purpureum*, and *Bracharia mutica* are some of the important species suitable for improving the fodder availability in ravine regions.

Afforestation

Planting of fuel, fodder, or multipurpose trees on ravine lands can mitigate the scarcity of fuel and fodder for rural households while ensuring adequate protection to these lands against land degradation. Most of the tree species are sown in one or two lines on the ridge but species which are best raised by planting are raised either by entire planting or stump planting. Trees can also be raised by sowing of seeds. Seeding is carried out in contour furrows of 10 × 10 cm cross section at 1 m horizontal interval just before onset of monsoons. The dug out soil of contour furrows is placed on the down streamside of slope.

For afforestation of waste lands near ravines, interrupted (in case of slope, on contour) trenches 1.5 or 3 m long, 60 cm wide, and 45 cm deep are made. The trenches are staggered and are 3.5 m to 4.5 m apart in the same contours or lines which are 4.5 m apart. In every fifth line, i.e., at an interval of 18 m, there is a continuous trench. In order to prevent water from the barren areas to go toward the head of the ravine, two continuous trenches are made at an interval of 4.5 m all round the three sides of the ravine with the first continuous trench at about 2.5 m from the bank of the ravine. In order to prevent these continuous trenches from becoming deep channels (*nalas*), 30 cm soil is left as such without any earth-work after every 3 m (Khanna 1996).

Tall and well-grown seedlings have better chances of success under ravine conditions. Seedlings raised in containers are preferred to naked root plants, as the rich nursery soil serves to improve the degraded site as well as hold water for impending drought. Long narrow tubes may be preferred in order to induce longer root system in the seedlings. Before transporting seedlings to the site, they should be hardened off in the nursery. For planting of nursery raised seedlings in ravines, pits of 45 × 45 × 45 cm size are dug at ravine top and ravine bottom. In the areas having mild slopes half slanting pits are dug. On steep side slopes, seeding in the contour furrows is the only practicable method. Use of insecticides is very essential in low rainfall areas to prevent termite attack. Soil mulching by weeds also helps in moisture conservation.

Agroforestry Approach/Agrisilviculture system

The most important measure for reducing the risk of degradation of catchments of gullies and marginal lands along the ravines is introduction of trees in cropping lands. The planted trees provide cover and security against the uncertainties of crop failure in the harsh conditions of ravinous environment which also assist in reducing the soil and nutrient loss from these lands. Apart from the benefit of product diversification and multi-strata farming, agroforestry has the potential of amelioration of less productive lands. Depending on the problems and needs of the area, trees may be introduced as alley, boundary plantation, or scattered tree plantation in the field. Though introduction of trees on cropped lands leads to loss of agricultural production due to belowground and aboveground competition among the woody and non-woody species, the overall economic returns have been found better than the sole cropping of agricultural crops in these habitats (Prasad et al. 1985, 1989; Prasad 1994a, b; Parandiyal et al. 2006).

Parandiyal et al. (2006) reported that the yield reduction in castor and pigeon pea in the vicinity of all tree species were compensated well from income of the fuel wood produced by the trees (Tables 8.15, 8.16).

Cereal crops are usually more compatible in tree crop associations as compared to legumes in

Table 8.15 Reduction of castor yield under boundary plantation system in Chambal Ravines

Tree Species	% Reduction of castor yield ha^{-1}				Av. reduction ha^{-1} (%)	Yield reduction (INR ha^{-1})[a]
	North	South	East	West		
Acacia nilotica	4.04	4.32	4.20	4.08	4.16	481
Azadirachta indica	4.55	4.87	4.45	3.80	4.42	511
Albizia lebbeck	5.34	5.85	5.07	5.24	5.37	679
Overall	4.64	5.02	4.57	4.37	4.65	538

[a] the value of castor is calculated by the price of INR 1300 qtl^{-1}

Table 8.16 Value of different tree species after 12 years of age

Tree species	Expected yield per tree in kg			Value in INR ha^{-1a}
	Final	Intermediate	Total	
Acacia nilotica	390	150	540	10980
Azadirachta indica	360	222	582	11480
Albizia lebbeck	690	371	1061	21116

[a] the price of final product and intermediate produce is taken as INR110 and 80 per 100 kg

rainfed farming. Prasad et al. (1987) and Prasad (1994a) reported that under alley cropping sorghum had better compatibility with *Leucaena* as compared to pigeonpea (*Cajanus cajan*) and black gram (*Vigna mungo*). Sorghum was also found to be the most compatible with *Eucalyptus tereticornis* raised under boundary plantation system (Prasad et al. 1985). Prasad (1994b) reported similar results with *Lucaena* shelterbelt. Improvement of soil fertility through addition of 371 kg ha^{-1} N through *Lucaena* leaf litter was also reported in this study.

In Yamuna ravines at Agra, Prajapati et al. (1993) observed that *Dendrocalamus strictus* produced 30–33 harvestable culms every 3 years after proper establishment at a spacing of 3 × 3 m to 8 × 8 m bamboo and gave average bamboo yield of 4000 poles ha^{-1}. Similarly *Eucalyptus tereticornis* planted at 2 × 2 m spacing at the ravine bottom was observed to yield 26–37 t ha^{-1} wood for house hold construction and fuel wood in a rotation of 10–12 years.

Agrihorticulture system

Land Class IV may be put under agri-horticulture. Fruit trees such as lemon (*Citrus limon*), mango (*Mangifera indica*), ber (*Ziziphus mauritiana*), and aonla (*Emblica officinalis*) can be grown with agricultural crops in humps and gully beds (Verma et al. 1986). Depending on the tree architecture and species, spacing of planting may vary from 2 × 2 m to 8 × 8 m. Pits of 1 × 1 × 1 m size are dug up during summer and the dug up soil is exposed to bright sun thus helping in eradication of pests and soil borne pathogens. Planting is completed soon after onset of monsoons. With a view to utilize the interspaces in widely planted trees planting of palatable grasses is a better option. If possibilities of developing irrigation facilities exist, the installation of drip irrigation facilities may be beneficial. The development of drip irrigation for trees spaced at 6 m x 6 m distance costs about INR 30 thousands to 35 thousands per ha. The farmers can also avail the subsidies provided for the adoption of drip irrigation by state governments. In Rajasthan, the state government provides a subsidy of up to 90 % of cost for installation of drip irrigation in the fruit plantations. Life saving irrigation may be essential for the fruit cultivation in ravines at least during summers.

Silvo-pastoral system

Cultivation of grasses with the multipurpose trees could increase production of fuel, fodder, or other goods, and services also added to revenue generation proving a viable option of economic utilization of ravines. Medium and shallow gullies can be utilized under forest and fruit tree-based pastoral system. Grasses provide desired levels of protection to various mechanical structures erected for reclamation of ravines. *Dicanthium annulatum, Cenchrus ciliaris, Cenchrus setigerus, Panicum antidotale, Panicum maximum, Pannisetum purpureum,* and *Brachiaria mutica* are some of the important grass species suitable for improving the fodder availability in ravine regions. For raising of grasses, palliated grass seeds are sown in 10 cm deep furrows dug along the contours at the onset of monsoon. Rooted slips of grasses are also planted at a spacing of 50 × 50 cm to achieve quick establishment of grasses. Erecting water spreaders not only reduces rill erosion but also ensures improved distribution of water to planted grasses on mildly sloping lands (Prajapati et al. 1982). The grasses planted on ravine top usually have better compatibility with trees due to better light availability than at ravine bottom.

By planting and protection of grasses, good green fodder yield is achieved in 2–5 years. Dhaman (*Cenchrus ciliaris*) and Kala Dhaman (*Cenchrus setigerus*) grasses could provide 5–7t ha^{-1} yr^{-1} air dry fodder which is equivalent to 18–20 t ha^{-1} yr^{-1} of green fodder. This practice also reduces runoff and soil loss considerably to the tune of 6–10 t ha^{-1} yr^{-1}. Planting of grasses further leads to improvement of soil structure and organic matter addition in these highly eroded soils. Prakash et al. (1987) recorded significantly lower runoff and soil loss from steeply sloping lands under silvipastoral system than relatively less sloppy lands under agricultural use. Sharda et al. (1982) reported superiority of mixed vegetation of trees and grasses in arresting the runoff and soil loss in ravinous watersheds. Studies in Chambal and Yamuna ravines reveal that even in the natural condition also ravines are able to support rearing of four goats ha^{-1} without causing erosion hazards under controlled grazing system (Chinamani et al. 1985, Prajapati et al. 1989).While comparing conservation values for several species, Dagar (1987b, 1995) reported that grass species such *Dichanthium annulatum, D. caricosum,* and *Heteropogon contortus* have great conservation and forage value. Dagar (1987a, c) and Dagar and Pathak (2005) reported the ecology and management of grazing lands in India. Ravinous catchments of Chambal at Kota when planted with *Acacia* + *D. annulatum* and *D. annulatum* alone generated 5.8 and 2.6 % of runoff and 1.26 and 0.62 t ha^{-1} of soil loss, respectively compared to 14.7 % of runoff and 3 t ha^{-1} of soil loss from agricultural catchments. Production of 4.5 t ha^{-1} of air dry grass + firewood from such degraded lands proved the effectiveness of grasses and trees as an alternative land use for protection and productive utilization of degraded ravine lands (Sharda and Venkateswarlu 2007). Prajapati et al. (1993) reported 1.5–2 t ha^{-1} yr^{-1} mean yield of grasses and top feed yield of 4.2 t ha^{-1} yr^{-1} from a 14-year-old *Acacia nilotica* and *Acacia tortalis* + *Cenchrus ciliaris* silvipastoral system in Yamuna ravines.

Success Stories

Bamboo and Anjan Grass-Based Agroforestry System for Enhancing the Productivity of Ravines (http://www.icar.org.in/node/5101)

An indigenous Bamboo and Anjan grass (*Cenchrus ciliaris*)-based silvo-pastoral system system for enhancing productivity of ravines has been developed by Central Soil and Water Conservation Research and Training Institute's Research Centre Vasad in Anand district of Gujarat. It was proved that the ravine lands, under unproductive use can be successfully reclaimed by planting bamboo, a fast growing plant species, on the gully beds and Anjan grass on the side slopes, and the interspaces of gully bed for economic utilization of gullied land. The cost of planting bamboo and sowing grass slips in ravine is about INR 22 thousands per hectare. During the initial years, green fodder of the grass is harvested. Planting of grass also protect the ravine slopes and reduces soil loss. Initial results revealed that about 7.1 t $ha^{-1}year^{-1}$ of green fodder could be obtained from the stabilized slopes. The grass yield from interspaces of bamboo planted on bed portion was about 10 t $ha^{-1}year^{-1}$ during initial 5 years. The grasses fetched an income of INR 3000–6000 per hectare over the period of 5 years. The system also yield about 300 clumps of bamboo with 3000–4000 old and 1000–1500 new culms per hectare bamboo plantation in 7 years old plantation. About 30 % of old culms can be harvested easily per year. At 2008 prices, the bamboo planted on the ravine bed fetched INR 6 thousands to INR 27 thousands ha $^{-1}year^{-1}$.

The watershed under this system absorbs more than 80 % of rainfall that is either utilized by the plant or percolated deep, to recharge the ground water. Due to less runoff, soil loss is reduced to less than one ton $hectare^{-1}$ $year^{-1}$ only from about 20 ton per hectare per year from degraded ravines prior to plantation. Organizations including Anand-based Foundation for Ecological Security (FES), Gujarat State Watershed Management Agency (GSWMA), Gujarat State Land Development Corporation (GSLDC), Forest, and Agricultural departments and other users' agencies have taken up this technology for stabilizing the degraded ravine lands and improving livelihood of primary stakeholders through reclamation and productive utilization of ravine lands. All the agencies put together, have utilized this technology for reclamation of nearly 1000 hectare of community and wastelands in Mahi river stretch in Gujarat.

Productive Utilization of Ravines Through Fruit-Based Agroforestry

A 6 years study was conducted by Sethy et al. (2011) at CSWCRTI, Research Centre, Kota from 2006 to 2011 for evaluation of production potential of ravine lands for sustainable production. Soils of the study site were very deep clay loam in texture; low in organic carbon, available nitrogen and phosphorus; and medium to high in potash with slightly alkaline pH. Four ravinous micro-watersheds. (RW1 = 0.4, RW2 = 1.4, RW3 = 1.1 and RW4 = 1.00 ha) were treated with 4 different trenching densities (i.e., 0, 25, 50 and 75 % runoff trapping potentials), i.e., no trenches, 139 trenches ha^{-1}, 278 trenches ha^{-1}'and 417 trenches $ha^{-1}$with uniform size of 3.0 × 0.60 × 0.45 m. A uniform land use management system was applied in all the micro-watersheds. Aonla (*Emblica officinalis*) was planted on ravine humps, bamboo on gully beds, and interspaces were planted with *Cenchrus ciliaris* grass. The numbers of plants were kept 278 ha^{-1} with spacing of grass slips at 0.5 × 0.5 m. After 6 years of planting, the total productivity of ravinous lands was 3.90, 4.52, 6.31, and 8.53 t ha^{-1} yr^{-1}aonla fruit equivalent under RW1, RW2, RW3, and RW4, respectively. The production of aonla fruit ranged from 1.31 t ha^{-1} in control (no trenches) to 6.61 t ha^{-1} in treatment involving trenching for trapping 75 % runoff, the production of grass ranged from 7.71 to 9.91 t ha^{-1}. The average number of harvestable clums/clump in bamboo ranged from 8 to 16 under different treatments. The cost and return economics indicated that RW4, RW3, RW2, and RW1 started yielding positive return at the end of 3rd, 4th, and 5th years, respectively after imposition of treatments. The trenching at 75 % runoff trapping potential density proved superior to others in terms of moisture conservation (87.3 %), reduction in runoff (78.23 %) and soil loss (90 %) and improving production potential (218.7 %) of ravinous lands under medium and deep ravinous lands.

Issues

During the past three decades research efforts in agroforestry in ravine areas can be rated as commendable which has helped in better understanding of ravine lands. Research efforts have lead to development and adoption of new technologies which has clearly demonstrated significant impact on production and protection. Future areas of R&D in agroforestry need to include the following issues:

- Treatment of gullies in accordance with gully reclamability classification.
- Survey, collection, assembling, and domestication of wild trees and shrubs suitable for ravine areas.
- Refinement of already developed technologies to make it more useful to the stakeholders.
- Develop agroforestry models for different site conditions for optimizing multiple output production, carbon sequestration, and biodiversity conservation.
- Development and distribution of quality planting material.
- Managing below ground interaction among tree crop associations.
- Capacity building of the stakeholders through trainings on nursery production, tree growing, soil and water conservation techniques, biofuel production and other income generating avenues Value addition through processing and marketing of tree products.
- Enhancing transfer of technology to the farmers field.

Degraded Watersheds

India is hydrologically demarcated into six major Water Resource Regions, 35 River Basins, 112 Catchments, 500 Sub-catchments, and 3237 watersheds. Watershed is a naturally occurring geohydrological unit draining to a common point and hence it is an ideal unit or development.

Watershed development aims at integrated development of non-arable land, arable land, rainwater, vegetation, livestock, local materials, common property resource, and human resource in a participatory process (Samra and Dhyani 2001). The Government of India has accorded highest priority to the holistic and sustainable development of rainfed areas and wastelands through watershed development programmes (Wani et al. 2008; Joshi et al. 2005).

The objectives of watershed management are:

1. Integrated land development and prevention of accelerated erosion.
2. Conservation of soil and water.
3. Control and/or moderate droughts.
4. Sustained production of crops, timber, firewood, fodder, fruits, and minor forest produce.
5. Sustained water supply for hydroelectric, irrigation, domestic, and industrial consumption.
6. Conservation of vegetation and biodiversity (with regard to trees and crops).
7. Improvement in the economic status and standard of living of rural population.

Various Watershed Development Programmes are being implemented in the country by different agencies for development of degraded lands. Since inception up to Tenth-Five-Year Plan (2002–07), an area of 51 million hectare has been developed on watershed basis (Sharda et al. 2006). The experience of watershed development projects in the country, however, has not been very encouraging. The few successful projects are outnumbered by the many unsuccessful ones. Many watershed projects, designed to conserve rainwater to improve irrigation, have tended to ignore communities primary need of access to drinking water. On similar lines, some projects have neglected to develop forests, pasturelands, and propagate soil-moisture conservation practices mainly through engineering structures. Majority of the water sheds in the country are degraded and suffer from poor productivity, biotic pressure, acute fodder shortage, poor livestock productivity, poverty, water scarcity, and poor infrastructure. The challenges in these watersheds are to enhance productivity, income, livelihood security, and conserve the resources on sustainable basis through integrated and participatory approach.

Role of Vegetation in Watersheds

Vegetation in the form of trees, shrubs, and herbs plays an important role in the watershed. The presence or absence of vegetation in watersheds has a profound impact on the hydrological process. Dense vegetation through interception of rainfall by the canopy, the subsequent flow of water from stems (called stem flow) and leaves (called leaf drip), slows down water movement, reduces runoff, soil loss, and facilitates water infiltration into the ground. Large number of twigs, branches, and fallen leaves form numerous microdams on the ground surface. Forest litter reduces rainfall impact and helps water absorption into the ground. It also prevents runoff and allows the water to percolate into the soil thereby helping ground water recharge.

Vegetation roots create macropores in the soil which increase water infiltration. Litter and other organic residue change soil properties which affect the capacity of soil to store water. The surface cover, debris, and roots trap sediments stop their down slope movement. In the absence of adequate vegetative growth, most of the rainfall gets converted into runoff and surface flows and cause soil loss. Through their canopy, trees provide shade which prevents the soil to become too dry and help in maintaining microclimate.

Rehabilitation of Degraded Watersheds

For rehabilitation of watershed, a multitier ridge to valley sequenced approach is followed. Prior to start of rehabilitation work, bench mark survey and resource inventorization should be done which is helpful in proper planning and execution of watershed management plan. Watershed problem should be identified using participatory rural appraisal (PRA) exercise as it helps in building rapport with the villagers. Using the basic resource information of the watershed (with regard to tree and grass species), decisions should be taken with regard to the area under cropping and new areas which can be brought under cultivation by employing suitable soil conservation techniques, type of tree species to be planted, the location and area where it is to be planted, the acceptability of these species by the farmers. Conceptually watershed has three main parts: catchment area, command area, and submergence area. These parts have specific role to play and are hence managed accordingly. Following are some of the important measures which can be taken for rehabilitation of degraded areas in different parts of watershed:

Catchment Area

Upper Catchment

Majority of forests and degraded lands are located in the catchment area. Treatment of these areas is often ignored, as the areas are typically under the control of the forest department, which does not permit other departments to operate on their lands. The onus of implementation in this part of watershed therefore lies with the Forest Departments and the Joint Forest Management Committees (JFMCs). As per the guidelines, degraded forest lands can be developed in two phases, viz. natural regeneration phase and new plantation phase using soil and water conservation measures.

Soil and Water Conservation Measure

The different soil and water conservation measure in this part of the watershed aims at checking runoff, soil loss soil, and increase the infiltration of water into soil profile so that it can be harvested in the reservoirs downstream. For slope stabilization, low cost gunny bags (*kattacrate* structure) should be used. First-order gullies/channels receiving small quantity of runoff should be stabilized by brush wood check dams or loose stone masonry check dams. For larger gullies, masonry check dams should be provided with aprons on lower side and their wing wall should be properly embedded in nala sites. On silting masonry dams, the area should be planted with *Agave, Ipomea, Vitex, Arundo,* etc. Geotextile should be used in the area where mass erosion is a major problem. Trees such as *Leucaena leucocephala, Melia azedarach, Sesbania grandiflora, S. sesban, Gliricidia sepium, Calliandra* sp, *Grevillea robusta*, etc., are added to earth structures employed for soil and water

Fig. 8.4 A series of check dams for *nala* stabilization

Fig. 8.5 Staggered contour trenching

conservation. Drainage line treatment should be done in this part of the watershed to reduce that the runoff velocity within permissible limit for which series of check dams are usually made (Fig. 8.4). Diversion drain should also be provided at appropriate places in order to protect the downstream area and discharge the flow safely.

Natural Regeneration

Securing adequate regeneration from already existing mother trees in the watershed can be helpful in improving the degraded condition at low cost within reasonable time. Cultural operation like weeding, cleaning, cutting, improvement felling, piling of debris, control burning helps in assisting and completing the regeneration and should be given due importance. Weeding should be done immediately after the germination of seed as they compete for light, nutrient, and moisture. Care should be taken to avoid weeding beyond October as it helps in protecting the dormant seedlings from frost. At least three number of weeding should be done during the first year, two during the second, and one in the third year. Hoeing is done along with the weeding and mulching is done along with the last weeding of the year. Once the regeneration has reached sapling stage, cleaning of climbers and inferior growth should be done periodically at an interval of 3–5 years.

Afforestation/Plantation

For sustainable results plantation of new species should be taken up only where first phase of natural regeneration through social fencing has been successfully completed. (Sreedevi et al. 2007). Before start of any plantation work, a reconnaissance survey should be conducted to know the topographic variations, aspect, slope, and elevation. Information on presence of weeds, rocks, crops, and drainage pattern should also be collected. The site to be afforested should be made free from the bushes for proper layout. Prior to start of plantation work, the area should be fenced or should be managed under cut and carry system to check the grazing. Proper soil working is important for the establishment of plantation. In areas having low rainfall, staggered contour trenches should be made to enhance the survival and growth of the seedlings. Trenches should be designed to store 60–70 % of runoff and accordingly the number and size of trenches should be decided. The numbers of trenches are usually equal to number of trees to be planted. Size of trenches are usually $3 \times 0.30 \times 0.30$ m (Fig. 8.5). However, if trees are to be closely spaced, size of trench should be reduced. The dug up soil is placed on the downhill slope along the trench. The seeds are either sown or seedlings are planted on bream of the trenches or in diagonally half filled trenches. The successive lines of contour trenches are kept at 2–4.5 m depending on slope. If soils are shallow, spacing should be ignored and direct sowing of seeds or planting should be done in patches or pits. Planting should be done after a good rain when moisture penetrates to a depth of 30 cm in the profile pit. Care should be taken that planting operations are carried out in a short time. The seedlings should be kept free from the weeds (Sharda et al. 2006).

Fire is one of the major factors responsible for large-scale mortality in both natural and artificial regenerated areas in the watershed. It makes the soil dry, accelerates soil erosion, and depletes water table and affects the water supply in the lower part of the watershed. For proper rehabilitation, fire needs to be controlled effectively which can be done by making fire lines and control burning prior to fire season. Care should be taken to cut the grasses before it dries. The dead seedlings must be replaced immediately.

Second Tier/Intermediate Slopes

In catchment area, the second tier is the intermediate tier or the slopes, which are just above the agricultural lands. The slopes can be used for agriculture after following mechanical measures like bench terracing and bunding (Fig. 8.6).

Stonewalls should be constructed across the hill slopes at predetermined spacing for developing land for cultivation. Diversion drains should be made to divert the runoff water away in order to protect the downstream area and discharge it safely into a protected waterway. These slopes can also be sustainably used for promoting agri-horticulture, agroforestry, silvipasture, and other alternate land use system.

The terrace risers of agriculture field should be planted with suitable grasses as it helps in binding the soil, checking the erosion, and

Fig. 8.6 Terracing to check soil erosion and productivity enhancement

increasing crop productivity. Grass species such as, Hybrid napier (*Panicum x purpurium), Panicum maximum, Chrysopogon fulvus, Vetivera zizanoides, Eulaliopsis binata,* etc., have been found suitable in the Shiwaliks and lower hills. Maize (*Zea mays*) and wheat (*Triticum aestivum*) yield was found increased by 23–40 % and 10–20 %, respectively when planted with vegetative grass barriers. In addition, grass yield of 0.6–1.7t ha^{-1} yr^{-1} was also observed. Vegetative barriers were also found to reduce the runoff and soil loss by 18–21 % and 23–68 %, respectively on slopes varying from 2 to 8 % (Ghosh 2010). On steep slopes the trees are also maintained in the form of hedges for checking soil loss. A study conducted at Dehradun indicated that sediment deposition along hedge rows during a period of 3 years and the same from the tree-rows in 9 years varied from 184 to 256 t ha^{-1} which is equivalent to 15–20 mm of soil depth. *Eucalyptus tereticornis* trees and *Eulaliopsis binata* (Bhabar grass) planted in Shiwalik foothills (@ 2500 trees ha^{-1}) in paired rows under storey grass planted at 50 × 50 cm spacing allowed no soil loss with an annual return of about INR 4000 ha^{-1} yr^{-1} from commercial grass alone besides additional returns from *Eucalyptus* and proved to be more remunerative than traditional rainfed crop (Sharda and Venkateswarlu 2007).

Fruits like citrus (*Citrus* spp.), banana (*Musa paradisiaca*), mango (*Mangifera indica*), apple (*Malus pumila*), walnut (*Juglans regia*), plum (*Prunus domestica*), peach (*P. persica*), cherry (*P. avium*), etc., should be grown on non-arable land with soil moisture conservation practices like half moon trenching, mulching, etc. Fruit trees should be grown on terrace risers in combination with vegetables such as potato (*Solanum tuberosum*), pea (*Pisum sativum*), capsicum (*Capsicum annuum*), cabbage (*Brassica oleracea* var. *capitata*), cauliflower (*B. oleracea* var. *botrytis*), etc.; spices such as ginger (*Zingiber officinale*), chillies (*Capsicum* spp), cardamom (*Elettaria cardamomum*), saffron (*Crocus sativus*), etc.; and many flowers including orchids, gladiolus, marigold, chrysanthemum, etc. Runoff water may be utilized for irrigating the fruit trees. Each tree depending upon species, age, slope, and intensity of rainfall should be provided with microcatchment for water harvesting. If possible, drip irrigation system should be installed for improving water use efficiency. Sharda and Venkateswarlu (2007) reported that intercropping of cluster bean (*Cyamopsis tetragonoloba*), cowpea (*Vigna unguiculata*) or pearl millet (*Pennisetum typhoides*) with peach (*Prunus persica*); turmeric (*Curcuma domestica*) with papaya (*Carica papaya*); *Chrysopogon fulvus* or Napier grass (*Pennisetum purpureum*) with aonla (*Emblica officinalis*) or ber (*Ziziphus mauritiana*) is an economically viable and eco-friendly system for rehabilitation of degraded lands. Aonla gave the highest yield of 86 kg fruits tree^{-1} in association with pigeon (*Cajanus cajan*) but the yield was reduced by 17 and 23 % respectively in association with *Chrysopogon* and Napier grass.

Community lands in intermediate slopes of the watershed should be used for promoting grasslands, or silvipasture system using appropriate soil and water conservation measures. Light harrowing, contour trenching, and furrowing are very effective for soil and moisture conservation and establishment of grasses. Small brushwood check dams should be constructed for gully plugging, stabilization, and improvement of gullied rangelands. Poor rangelands should be stocked with high forage yielding perennial grasses by reseeding or slip planting. Trees having high fodder values such as *Acacia nilotica, Acaia catechu, Ailanthus excelsa, Anogeissus pendula, Albizia lebbeck, Bauhinia variegata, Butea monosperma, Cordia dichotoma, Celtis australis, Ficus glomerata, Grewia optiva, Kydia calycina, Lannea coromandelica, Morus alba* should be given preference. A mixture of species is, however, always better. Pollarding and lopping of the trees should be done during the lean period. The left branches can be utilized for fuel wood. As community lands in the watershed are property of all and responsibility of none, therefore participatory approach is better in selecting, planting, tending, and protecting the trees and grasses. Group of villages, whose livestock utilize the common grazing lands should be jointly form societies to carry out the work efficiently (Dhyani et al. 2007).

Command Area

These are the plains and the flat areas, where typically, the farmers operate and large concentration of labor-intensive works are required. The main aim of this area is production of agriculture and livelihood security. The agronomical measures such as contour farming, intercropping, strip cropping, terracing, mixed cropping, land cover or canopy management, mulching, conservation tillage practices, and diversified cropping systems are generally recommended in this part of watershed for maximizing *in situ* rainwater conservation. Water productivity on farm lands can be increased manifold (2–4 times) by adoption of suitable cropping systems, good variety of seeds, integrated nutrient management (INM), integrated pest management (IPM), weed management, *in situ*, and inter-plot-water harvesting, etc.. Runoff water from stored tanks/ponds in submergence area should be utilized for irrigating the annual crops during moisture stress period.

Submergence Area

The lower part of watershed can be utilized for development of small water harvesting structures such as low cost farm ponds, check dams, percolation tanks, and ground water recharge measures which can be used for collecting and storing rainfall, runoff, and subsurface flow to meet increased water demand and overcome moisture stress conditions, and providing ecosystem services like ground water recharge. The harvested water can be used for drinking, irrigation, livestock, fisheries, ground water recharge, etc., to optimize the water productivity. The runoff water could be harvested and stored in suitable storage structures for supplemental irrigations during moisture stress periods for the crops. Samra and Narain (1998) reported an yield increases up to 200 % even with one supplemental irrigation of 5 cm.

Success Stories

Few community groups have improved their socioeconomic conditions by linking improved *in situ* moisture conservation with economic activities that build up social capital which shows that watershed development is a viable model for the economic development of poverty-stricken rural areas. Better-performing projects are based on promoting communities' traditional water harvesting and conservation practices. Some successful models are:

Ralegaon Siddhi

Ralegaon Siddhi Project, covering four watersheds in geographical area of about 892 hectares, is one of the success stories in Maharashtra. In a total project outlay of INR11.3 million, the State Government contributed 5.27 million, 4.7 million was borrowed from banks, 1.1 million was put together by villagers through Shramdan (volunteer labor), and the remaining 0.2 million was raised from other sources. Series of check dams, cemented *bandharas*, and *nullah* bunds were built at strategic locations through community participation. All these structures increased the infiltration of harvested water and recharged ground water. Today Ralegoan Siddhi has two percolation tanks, thirty nullah bunds, eighty-five wells, and eight borewells all of which are viable right through the year. The villagers have stopped grazing their animals on common lands. To take care of equitable distribution of water, village associations *pani puravatha* mandals was formed. Farmers are now growing two or three crops every year including fruits and vegetables.

Sukhomajri watershed

This project was undertaken by the CSWCRTI, Research Centre Chandigarh at Sukhomajri in Panchkula district of Haryana. Until 1975, Sukhomajri had no source of regular irrigation. The entire agricultural land (52 hectares) was under rainfed single cropping. An area of about 85 hectares of the worst affected topmost catchment area was treated with mechanical measures including construction of staggered contour trenches, stone and earthen check dams, brushwood check dams and grade stabilizers, supported by debris basin and vegetative measures. Grazing and illicit cutting of vegetation was stopped. The vegetative measures consisted of planting of tree species like khair *(Acacia catechu)* and shisham *(Dalbergia sissoo)* in pits, and *bhabbar* grass *(Eulaliopsis binata)* at mounds of trenches, and also *Agave americana* and *Ipomoea carnea* in critical areas to protect the soil against erosion. To promote agriculture and water availability in the area, four earthen dams were built. A network of underground pipeline system was laid in the entire command area of about 20 hectares to convey stored rainwater from the dam to the fields for providing supplemental irrigation. Risers were provided at suitable places and each outlet provided irrigation to 3–4 fields. To take care of watershed and equitable distribution of water Hill Resource Management Society (HRMS) was formed. Different interventions resulted in increased grass production from 3.82 to 7.72 t ha^{-1}. Canopy cover increased from 3 % to about 18 % in the catchment area of the dam. Barseem (*Trifolium alaxandrinum*) is now being grown over an area of 4 ha with an annual biomass production of 140 tons which have boosted milk production. With the availability of irrigation and introduction of improved agro-techniques, maize (*Zea mays*) productivity increased from 0.6 to 1.95 t ha^{-1} while sorghum from 8.0 to 14.0 t ha^{-1}. Productivity of wheat increased from 0.86 to 2.7 t ha^{-1} while that of sugarcane (*Saccharum officinarum*) from 15.0 to 25.0 t ha^{-1} (Tiwari et al. 2011).

Fakot Watershed

An integrated watershed management project (IWMP) located at Fakot watershed representing the lower and middle Himalayas in Tehri

Garhwal district was executed through demonstrational approach by Central Soil and Water Conservation Research and Training Institute, Dehradun during 1975–86 in 370 ha area. The main focus of the project was on adoption of participatory approach for the management of rain water, livestock, Horticulture development, crop production, livelihood, and environmental security. The watershed area was treated with various soil and water conservation measures during 1975–76 to 1987–88 after completing bench mark surveys of hydrology, soils, vegetation, and socio-economic conditions of the inhabitants. Irrigated area increased from 11.8 ha in 1974–1975 i.e., preproject period (PPP) to 23 ha by the end of active operation phase (AOP) in 1985–1986, which further increased to 25 ha during financial withdrawal phase (FWP) in 1994–1995. Net cropped area decreased from 80 ha to 74 and to 71.6 ha during the same period. Cropping intensity increased by more than 70 %. Farmers continued to adopt improved crop production technologies. Agricultural production from watershed increased 4.5 times during AOP and further increased to 8 times during FWP. Area under horticultural plantation increased from 0.5 to 18 ha by the end of AOP and to the level of 26 ha by FWP. Fruit production increased from almost negligible to 221tons. Many cash crops like gladiolus, ginger, tomato, cabbage, etc., are being cultivated by farmers on commercial lines and farmers are earning over INR 0.7 million out of these. Fuel fodder production of 24 ha area and protection of community land boosted fodder production from 13 to 36.8 t ha^{-1} and therefore, pressure on resource forest areas eased out. A large number of casual and regular employment opportunities were generated. Consequently, migration of work force declined from 26.7 % at PPP stage to 9.3 % in AOP and 0.7 % in FWP. The project proved to be economically viable with benefit cost ratio (B:C) of 1.93 and payback period of 10 years (Sharda et al. 2006).

Garhkundar–Dabar Watershed

The Garhkundar–Dabar watershed situated in the most economically backward Tikamgarh District, Madhya Pradesh, in the Bundelkhand region has an area of 850 ha, with 895 human and 2648 animal populations from three village panchayats. Soil and water conservation measures in the watershed included construction of eight check dams in series in approximately 10 km total length of third and fourth-order streams. In addition, 150 gabion structures with volume varying between 3 and 6.75 m^3 were laid across the first- and second-order streams to check gully formation and silt inflow in main water course. Three '*khadins*' (water spreaders) were constructed in depressions to check the concentrated flow of runoff. Bunding was carried out along the margins of the fields in 40 ha area for *in situ* water harvesting and erosion control. These bunds were provided with proper spillways (15 numbers) for safe disposal of excess water.

Adoption of integrated soil and water conservation measures in participatory mode not only reduced runoff, but also increased water availability in time and scale. The peak discharge from the treated watershed in 2009 was lower (0.018 $m^3s^{-1}ha^{-1}$) compared to the untreated watershed (0.028 $m^3\ s^{-1}\ ha^{-1}$). Similarly, total runoff recorded from the treated watershed was lower (41.5 mm or 5.4 % of annual rainfall) than the untreated watershed (77.3 mm or 10.1 % of annual rainfall). Runoff per unit area from the treated watershed (415 thousand 1 ha^{-1}) was 46 % lower than the untreated watershed. The average water column in the wells within the watershed increased steadily from 0.88 m in 2006, through 1.62 m in 2007 to 5.06 m in 2008, and 4.36 m in 2009. The water column was more than 6 m in 40 % of the wells and more than 4 m in 75 % of the wells. These wells sustained continuous pumping for 10–12 h with a 5 HP pump.

Furthermore, the surface water availability in water courses and groundwater availability in open dug wells in 2009–2010 increased round the year in the middle and lower reaches of the watershed, and for more than 5 months in the upper reaches compared to 4–5 months only in the lower reaches earlier. Soil loss from the untreated watershed was 76 and 73 % higher compared to the treated watershed in 2008 and 2009, respectively. N and P losses in 2008 from the treated watershed were 7.35 and 4.97 kg ha^{-1}, respectively, whereas the corresponding losses from the untreated watershed were 12.6 and 8.54 kg ha^{-1}, respectively. Similarly, the losses in 2009 were 3.29 and 1.17 kg ha^{-1} in the treated and 5.70 and 2.03 kg ha^{-1} in the untreated watersheds respectively (Palsaniya et al. 2011).

Issues

Watershed approach has been accepted as a means to increase agricultural production while arresting ecological degradation in rainfed and resource poor areas. However, despite of huge investment in this program, the results have not been very encouraging in the country which needs immediate attention. The immediate issues which need to be addressed are:

- Watershed programs need to be 'demand driven' and not 'target driven' as is usually the case.
- Effective vertical and horizontal coordination among various line departments and non-governmental organization.
- Involvement of landless and assetless people in the watershed.
- Postproject sustainability needs to be addressed adequately. Poor participation of people in many of the watersheds program.
- Strengthening of forestry component in watershed programs particularly in upper catchment areas by involvement of Forest department.
- Livelihood through integrated farming system involving livestock and fishery component.
- Involvement of institutions of multidisciplinary professionals.
- Scientific planning and capacity building of all functionaries and stakeholders involved in the watershed program for the effective implementation.
- Making available common lands and other common property resources (CPRs) physically for use by the village community including the landless.
- Gender empowerment to ensure equal partnership.
- In-built flexibility for midway corrections in the program for confidence building and increasing resource use efficiency.
- Sustainable development of common pool resources (CPRs) of land, water, forest, fisheries, wildlife, and agriculture.

Conclusions

Productive lands are in the constant process of various degrees of degradation and are fast turning into wastelands. Increasing population pressure, biotic pressure, unplanned urbanization, breakdown of traditional institutions for managing Common Property Resources, and lack of appropriate management practices are some of the key reasons for the land degradation. The land degradation has both on-site and off-site impacts which need to be managed properly. Degraded lands can be suitably reclaimed for agriculture or some alternate uses following afforestation, agroforestry, and bioengineering measures which are simple and cost-effective.

Mining activities and ravines areas contribute significantly to loss in productivity and land degradation. Rehabilitation measures aims at returning damaged ecosystem to productive use which is environmentally and socially acceptable. Mined areas can be suitably rehabilitated by proper configuration of disposal dump, using bioengineering measures, afforestation measures, and proper aftercare. Many studies in the country have shown successful rehabilitation of mined areas. However, sill the issues of, involvement of local people, environmental protection, and strict follow of laws and guidelines need to be taken up.

Ravine rehabilitation requires an integrated approach of treatment of table and marginal lands contributing runoff to the gullies, and gullies proper on watershed basis. Using gullies according to land capability classes, soil, and water conservation measures and putting land under permanent vegetation cover involving afforestation, forest, and fruit trees based pasture systems and energy plantations are some of the simple and cost-effective way of rehabilitation of ravine areas. Despite demonstrable technology of ravine land rehabilitation, their adoptions by the farmers have not been very encouraging. Thus, there is immediate need for developing of new technologies, refinement of existing technologies, capacity building, and technology transfer. In addition, intangible benefits due to ravine rehabilitation also need to be demonstrated.

Majority of the watersheds in the country are degraded and suffer from poor productivity, biotic pressure, acute fodder shortage, poor livestock productivity, poverty, water scarcity, and poor infrastructure. A multitier ridge to valley sequenced approach is, therefore, required to treat the watersheds for enhancing productivity and resource conservation. Integrated approach of soil water conservation, afforestation, improvement of existing forest, agroforestry, horticulture agronomical measures, and allied activities is required on community basis for rehabilitation of these watersheds. For achieving the best results, people should be sensitized, empowered, and involved in the program. The issues of lack of interdepartmental coordination, postproject sustainability, lack of scientific planning and capacity building also need to be addressed for the successful implementation of watershed programs.

Acknowledgments Authors like to acknowledge Director CSWCRTI for providing the necessary facilities and Dr J. C. Dagar, Assistant Director General (Agronomy/Agroforestry), NRM Division (ICAR) for critical suggestions. We are also grateful to all the Scientists of CSWCRTI for their kind help in compiling the information. Special thanks are due to Mr. Laxmi Kant Sharma for providing the photographs.

References

Ang LH (1994) Problems and prospects of afforestation on sandy tin tailings in peninsular. Malaysia. J Trop For Sci 7:87–105

Bali YP, Karale RL (1977) Reclamability classification of ravines for agriculture. Soil Conserv Digest 5(2):40–47

Bhushan LS, Tiwari AK, Mishra PR (1986) Contour trenching—an effective measure to control erosion on sloping lands. In: Proceeding national symposium on problems of arid and semi- arid regions and their remedial measures, Jodhpur, 08–09 Feb 1986, pp 83–85

Bradshaw AD (1997) The importance of soil ecology in restoration science. In: Urbanska KM, Webb NR, Edwards PJ (eds) Restoration ecology and sustainable development. Cambridge University Press, Cambridge, pp 33–64

Champion HG, Seth SK (1968) A Revised survey of forest types of India. Controller of Publication, New Delhi

Chaterjee PC (1988) Mine spoils and their rehabilitation. In: Mining and environment in India. H.R. Publishers, Nanital, pp 394–409

Chinamani S, Prakash C, Rathore BL (1985) Goat grazing in ravines and its effect on runoff and soil loss—a preliminary note. In: National seminar on soil conservation and watershed management, New Delhi, p 20

Chinnamani S (1986) Forestry, social forestry and agroforestry in ravine reclamation. Lecture delivered at the Summer Inst.,'Wasteland in Arid and Semi-arid zones and technologies for their Improved Utilization' CAZRI, Jodhpur

Corbett EA, Anderson RC, Rodgers CS (1996) Prairie revegetation of a strip mine in Illinois: fifteen years after establishment. Restor Ecol 4:346–354

Dadhwal KS, Juyal GP (2012) Agroforestry for rehabilitation of mines poils in Dehradun valley. In: Panwar P, Tiwari AK, Dadhwal KS (eds) Agroforestry system for resource conservation and livelihood security in lower Himalayas. New India Publishing Agency, New Delhi, pp 117–138

Dadhwal KS, Singh B, Narain P (1991) Effect of placing and mixing of a normal soil on the performance of two grasses in limestone minespoil. J Indian Soc Soil Sci 39(4):751–756

Dadhwal KS, Singh B, Narain P (1992) Characteristics of limestone minespoil/debris from outer Himalayas of Uttar Pradesh. Indian For 118(9):650–658

Dagar JC (1987a) Species composition and plant biomass of an ungrazed and a grazed grassland at Ujjain, India. Trop Ecol 28:208–215

Dagar JC (1987b) Responses of clipping treatments on nine palatable perennial grasses. Trop Ecol 28:216–221

Dagar JC (1987c) Studies on reclamation of Kshipra ravines. Indian J For 10(2):83–89

Dagar JC (1995) Soil conservation values of some grasses and forbs. Proc Indian Nat Sci Acad B 61:163–170

Dagar JC (2001) Biological spectrum and successional trends in Kshipra watershed areas. Indian J For 24(3):351–356

Dagar JC, Mall LP (1980) Studies on the vegetation of ravines and banks of the river Kshipra and its tributaries. J Indian Bot Soc 59:234–245

Dagar JC, Pathak PS (2005) Grassland dynamics and their management. Range Manag Agrofor 26(1):7–31

Dagar JC, Singh VP (1980) The ecological studies of the vegetation of the river Kshipra and its tributaries. Biologia 26(1):43–62

Narayan Dev, Bhushan LS (2002) Interrelationship between crop canopy and erosion parameters in alluvial soils. Indian J Soil Conserv 30(1):41–45

Dhyani SK, Dadhwal KS, Katiyar VS (1988) Biological measures for rehabilitation of abandoned minespoil areas. Indian Farm 38(7):15–18

Dhyani SK, Sharda VN, Juyal GP (2007) Training manual- soil conservation and watershed management : conservation forestry and watershed management. In: CSWCR&TI, Dehradun, p 315

Dobson AP, Bradshaw AD, Baker AJM (1997) Hopes for the future: restoration ecology and conservation biology. Science 277:515–522

Filcheva E, Noustorova M, Gentcheva-Kostadinova SV, Haigh MJ (2000) Organic accumulation and microbial action in surface coal-mine spoils. Pernik Bulgaria Ecol Eng 15:1–15

Gautam NC, Narayan LRA (1988) Wastelands in India. Pink Publishing House, Mathura, p 96

Ghosh BN (2010) Vegetative barriers for erosion control in western Himalayan region. Technology Brochure. CSWCRTI, Dehradun p. 8

Grunwald C, Iverson LR, Szafoni DB (1988) Abandoned mines in Illinois and North Dakota: towards an understanding of revegetation problems in rehabilitating damaged ecosystems. CRC Press, Boca Raton, vol 1, pp 40–56

Iverson LR, Wali MK (1982) Reclamation of coal mined lands: the role of Kochiascoparia and other pioneers in early succession. Relam Reveg Res 1:123–160

Jha AK, Singh JS (1991) Spoil characteristics and vegetation development of an age series of minespoils in a dry tropical environment. Vegetation 91:63–96

Jha MN (1987) Environmental management of mining operations in SingrauliCoalfieds—NCC. In: Dhar BB(ed) Proceeding national workshop on environmental management of mining operations in India. BHU, Varanasi, pp 47–53

Joshi PK, Jha AK, Wani SP, Joshi L, Shiyani RL (2005) Meta-analysis to assess impact of watershed program and people's participation. Research report 8, comprehensive assessment of watershed management in agriculture. International crops research institute for the semi-arid tropics and Asian development Bank, 21p

Juyal GP, Katiyar VS, Dadhwal KS, Joshie P (1998) Reclamation of mine spoils in outer Himalayas. In: Bhushan LS, Abrol IP, Rama MS Mohan Rao (eds) Soil and water conservation: challenges and opportunities. 8th ISCO conference. Indian Association Soil and Water Conservation, Dehradun, pp 1384–1392

Juyal GP, Katiyar VS, Dhadwal KS, Joshie P, Arya RK (2007) Mined area rehabilitation in Himalayas-Shastradhara experience. Central Soil and water Conservation Research and Training Institute, Dehradun, p 104

Khanna LS (1996) Principles and practice of silviculture. Khanna Bandhu, Dehradun, pp 377–380

Khullar DR (2006) "Mineral Resources", India: a comprehensive geography. Kalyani Publishers, New Delhi, pp 630–659

Lamb D (1988) IUCN guidelines for restoration of degraded ecosystems. draft- world conservation union (IUCN), Gland, Switzerland

Luna RK (2006) Plantation forestry in India. Offset Printers and Publishers, Dehradun, pp 361–386

Maji AK, Reddy, GPO, Sarkar D (2010) Degraded and Wastelands of India: Status and Spatial Distribution. Directorate of Information and Publications of Agriculture. Indian Council of Agricultural Research, New Delhi, p 158

Mandal D, Sharda VN, TiwariA (2008) Soil loss tolerance limits in planning conservation measures in different districts of India. In: CSWCRTI, Dehradun, India, p 112

Mishra CSK, Rath Monalisa (2013) Wasteland reclamation: novel approaches. Ecoscan 3:99–105

Ministry of Statistics and Programme Implementation (2001) Compendium of environment statistics. Central Statistical Organisation, Ministry of Statistics and Programme Implementation, Govt. of India

Narayana MP (1987) Neyveli open cast mine. A review of environmental management of mining operation in India. Banaras Hindu University, Varanasi, pp 54–62

Norland MR (1993) Bureau of mines information. US Department of the Interior, Washington

Noyd RK, Pfleger FL, Norland MR, Hall DL (1997) Native plant productivity and litter decomposition in reclamation of taconite iron ore tailing. J Environ Qual 26:682–687

NRSC (2011) Wastelands Atlas of India. Change Analysis Based on Temporal Satellite Data of 2005-06 and 2008-09. National Remote Sensing Centre, Hyderabad, p 268

Oldeman LR (1991) Global extent of soil degradation. Bi-ennial report, ISRIC, Wageningen, The Netherlands

Palsaniya DR, Ramesh S, Yadav RS, Tewari RK, Dhyani SK (2011) Now it is water all the way in Garhkundar-Dabar watershed of drought prone semi-arid Bundelkhand, India. Curr Sci 100(9):1287–1288

Parandiyal AK, Kumar A, Prasad A, Singh KD (2006) Study of tree crop association under boundry plantation system in south eastern Rajasthan. Abs.: National symposium for livelihood security, environmental protection and biofuel production, NRC Agroforestry, Jhansi, 16–18 Dec 2006

Parandiyal, AK, Samara JS, Singh KD, Singh Ratan, Rathore BL (2000) Floristic diversity of Chambal Ravines under varying levels of protection. Indian J. Soil Cons. 28 (2): 160-166

Parrotta JA (1992) The role of plantation forests in rehabilitating degraded tropical ecosystems. Agric Ecosyst Environ 41:115–133

Parsad R, Chadhar SK (1987) *Grevillea pteridifolia* successful introduction to mined out areas of Madhya Pradesh. Indian For 113(10 & 11):110–118

Pfleger FL, Stewart EL, Noyd RK (1994) Role of VAM fungi in mine land reclamation. In: Pfleger FL, Linderman R (eds) Mycorrhizae and plant health. American Phytopathological Society Press, St. Paul, pp 47–81

Prajapati MC (1995) Floristic wealth of Yamuna ravines. Indian J Soil Cons 23(1):4–61

Prajapati MC, Joshi P, Rathore BL, Dubey LN (1982) Surface water management for grass and tree land development in ravinous watersheds—a case study. Indian J Soil Conserv 10(2 and 3):30–38

Prajapati MC, Nambiar KTN, Puri DN, Singh JP, Malhotra BM (1993) Fuel and fodder production in Yamuna ravines at Agra. Indian J Soil Conserv 21(3):8–13

Prajapati MC, Tiwari AK, Nambiar KTN, Singh JP, Malhotra BM, Sharda VN (1989) Effect of goat grazing on vegetation, runoff and soil loss in a stabilized ravine watershed vis-à-vis animal growth. Indian J Soil Conserv 17(1):9–16

Prakash C, Rao DH, Verma B, Chinamani S, Prajapati MC (1987) Hydrological behaviour of small watersheds: under different land uses in Chambal ravines. In: National symposium on land and water management in ravines, CSWCRTI, Research Centre, Agra, 19–22 March, 10p

Prasad A (1994a) Alley cropping in *Leucaena lecocephala* with field crops for sustained productivity. In: Proceeding of the international conference on sustainable development of degraded lands through agroforestry in Asia and Pacific, vol 1, New Delhi, Nov 25–30 1994, pp 471–478

Prasad A (1994b) Shelterbelt of *Leucaena* and its effect on crop yield in rainfed agroforestry system. In: Proceeding of the international conference on sustainable development of degraded lands through agroforestry in Asia and Pacific, vol 1, New Delhi, Nov 25–30 1994, pp 499–506

Prasad R (1989) Removal of forest covers through mining and technology for retrieval. J Trop For 5:109–116

Prasad R, Chadhal SK (1987) Afforestation of dolomite over burden in Madhya Pradesh. J Trop For 3:124–131

Prasad R, Shukla PK (1985) Reclamation and revegetation of coal mine overburden in Madhya Pradesh. J Trop For 1:74–84

Prasad R, Shukla PK (1988) Restoration of ecological balance to the bauxite mined areas of Madhya Pradesh. In: Wasteland development for fuelwood and fodder production (a compilation). Forest Research Institute, Dehradun pp 25–33

Prasad SN, Srivastava AK, Narain P, Bhola SN (1985) Studies on the Compatibility of field crops with *Eucalyptus tereticornis* under agroforestry. Indian J Soil Conserv 13:41–46

Prasad SN, Verma B, Prasad A, Srivastava AK (1989) Alley cropping in *Leucaena* for higher productivity and profitability in South–East Rajasthan. Indian J Soil Conserv 17(1):30–34

Prasad SN, Verma B, Prasad A, Singh HB (1987) Competition studies on treecrop association with subabool on reclaimed lands of Kota region under rainfed conditions. Abstract no 20. In: National symposium on land and water mangement in ravines. CSWCRTI, Research Centre, Agra, 19–22 March 1987, p 17

Putwain PD, Gillham DA (1990) The significance of the dormant viable seed bank in the restoration of heath lands. Biol Conserv 52:1–16

Ramakrishna YS, Rao GGSN, Kausalya Ramachandran, Osman M (2007). Climatic variability, its impact on rain-fed agro-ecosystem and coping strategies. Paper presented in VIII science congress 2007 held at TNAU, Coimbatore, India, Feb 15–17 2007

Roberts HA (1981) Seed banks in soils. Adv Appl Biol 6:1–55

Sajwan SS (1975) Ground flora and their distribution in forest plantation in Jamuna ravines. Indian For 101(5):269–275

Sajwan SS (1976) Dominant grasses and their associates in ravinous tracts of Jamuna. Soil Conserv Digest 4(1):26–35

Samra JS, Dhyani SK (2001) Role of trees in watershed development. In: Puri Sunil et al (eds) Agroforestry. ICAR, New Delhi

Samra JS, Narain P (1998) Soil and water conservation. In: Singh GB, Sharama BR (eds) Proceeding 50 years of natural resource management research. NRM division, ICAR, New Delhi, pp 145–176

Schaller N (1993) The concept of agricultural sustainability. Agric Ecosyst Environ 46:89–97

Sethy BK, Parandiyal AK, Shakir Ali, Ashok Kumar, Singh RK (2011) Cost effective conservation measures for management of medium and deep ravenous lands. Annual report, CSWCRTI, Dehradun, pp 73–74

Sharda VN, Bhushan LS, Raghuvir (1982) Hydrological behaviour of ravinous watersheds under different land uses. In: Proceeding international symposium on hydrological aspects of mountainous watersheds, vol 6, Roorkee, Nov 04–06 1982, pp 14–18

Sharda VN, Sikka AK, Juyal GP (2006) Participatory integrated watershed management—a field manual. CSWCR&TI, Dehradun 366

Sharda VN, Venkateswarlu B (2007) Crop diversification and alternate land use systems in watershed management. In: Best-bet options for integrated watershed management. Proceedings of the comprehensive assessment of watershed programs in India, 25–27 July 2007, ICRISAT, Patancheru, pp 111–128

Sharma HS (1980) Ravine erosion in India. Concept Publishing Co., New Delhi, pp 1–96

Singh AN, Raghubanshi AS, Singh JS (2002) Plantations as a tool for mine spoil restoration. Curr Sci 82(12):1436–1441

Singh JS, Jha AK (1992) Restoration of degraded land: an overview. In: Singh JS (ed) Restoration of degraded land: concepts and strategies. Rastogi Publications, Meerut, pp 1–7

Singh JS, Singh KP, Jha AK (1996) Final technical report. Submitted to the Ministry of Coal, Govt. of India, p 116

Singh S (2004) Biological reclamation of degraded mined land a sustainability indicator. Eniviron News 10(1):1–3

Soni P, Vasistha HB (1986) Reclamation of mine spoils for environmental amelioration. Indian For 112(7):621–631

Soni P, Vasistha HB, Kumar Om (1990) Restoration technology for mined lands. In: Proceeding of the national seminar on technologies for afforestation of wastelands, FRI, Dehradun, 6–8 Nov

Sreedevi TK,Wani SP, Osman M, Tiwari S (2007) Rehabilitation of degraded lands Watersheds. In: Best-bet options for integrated watershed management. Proceedings of the comprehensive assessment of watershed programs in India, July 25–27 2007, ICRISAT, Patancheru, pp 205–220

Tata Energy Research Institute (2001) Overview of mining and mineral industry in India. Report no. 185. Tata Energy Research Institute, New Delhi

Tejwani KG, Gupta SK, Mathur HN (1975) Classification, control and reclamation of ravine lands. In: Soil and Water Conservation Research (1956–1971). Indian Council of Agricultural Research Publication, New Delhi, pp 254–310

Tejwani KG, Dhruvanarayana VV (1961) Soil conservation survey and land use capability planning in the ravine lands of Gujarat. J Indian Soc Soil Sci 9:233–244

Tiwari AK, Panwar P, Agnihotri Y (2011) Environment and livelihood security through participatory watershed management in Shivaliks. In: Sobti RC, Singh KP, Tiwari AK (eds) Shiwaliks at a glance-present status and future strategies. Punjab University Press, Chandigarh, pp 136–154

Verma B (1981) Twenty classification of ravines for optimum utilization. Central Soil and Water Conservation Research and Training Institute, Research Centre, Kota

Verma B, Bhushan LS (1986) Management of gullied watersheds. Indian J Soil Consev 14(3):105–110

Verma B, Chinamani S (1990) Rehabilitation of ravine lands. In: Abrol IP, Dhruva Naryana VV (eds) Technologies for wasteland development. ICAR, New Delhi, pp 263–272

Verma B, Chinamani S, Bhola S N, Rao D H, Prasad S N, Prakash C (1986). Twenty five years of research on soil and water conservation in ravine lands of Rajasthan. Central Soil and Water Conservation Research and Training Institute Publication, Research Centre, Kota, pp 1–215

Verma SK (1988) Environmental aspects of mining activities. In: Mining and environment in India. H.R. Publishers, Nanital, pp 388–393

Vogel W G (1973) The effect of herbaceous vegetation on survival and growth of trees planted on coal-mine spoils. In Proceedings of the Research and Applied Technology Symposium in Mined-Land Reclamation, National Coal Association, Pittsburgh, Pennsylvania: 197-207

Wani SP, Sreedevi TK, Reddy TSV, Venkateswarlu B, Prasad CS (2008) Community watersheds for improved likelihoods through consortium approach in drought prone rain-fed areas. J Hydrol Res Dev 23:55–77

Greening Salty and Waterlogged Lands Through Agroforestry Systems for Livelihood Security and Better Environment

9

J. C. Dagar

Abstract

Nearly, one billion hectares of arid and semi-arid areas of the world are salt-affected and remain barren due to salinity or water scarcity. In India, about 6.75 Mha lands are either sodic or saline in nature and 6.41 Mha land is degraded due to waterlogging. These lands constrain plant growth owing to the osmotic effects of salt, poor physical conditions leading to poor aeration, nutrition imbalances, and toxicities. To meet the requirements of food and other agricultural commodities for the burgeoning population is a big challenge for agricultural community. With the increasing demand for good quality land and water for urbanization and development projects, in future, agriculture will be pushed more and more to the marginal lands and use of poor quality water for irrigation is inevitable. With use of appropriate planting techniques and salt-tolerant species, the salt-affected lands can be brought under viable vegetation cover. Further, in most of the arid and semi-arid regions the groundwater aquifers are saline. Usually, cultivation of conventional arable crops with saline irrigation has not been sustainable. Concerted research efforts have shown that by applying appropriate planting and other management techniques (e.g., sub-surface planting and furrow irrigation), the degraded salty lands (including calcareous) can be put to alternative uses (agroforestry) and salt-tolerant forest and fruit trees, forage grasses, medicinal and aromatic, and other high value crops can be equally remunerative. Such uses have additional environmental benefits including carbon sequestration and biological reclamation. Agroforestry is not only a necessity for increasing tree cover and hence decreasing pressure on natural forests, but also a most desired land use especially for reclaiming

J. C. Dagar (✉)
Division of Natural Resource Management,
Indian Council of Agricultural Research,
Krishi Anusandhan Bhawan II, Pusa, New Delhi,
110012, India
e-mail: dagarjc@gmail.com

J. C. Dagar et al. (eds.), *Agroforestry Systems in India: Livelihood Security & Ecosystem Services*, Advances in Agroforestry 10,
DOI: 10.1007/978-81-322-1662-9_9,

and rehabilitating the degraded lands. In developing countries like India, there seems to be little scope for bringing the fertile lands under forestry cover. It may be emphasized here that we can bring unproductive wastelands and waterlogged areas under forest cover and take agroforestry tree plantation on nonforest community and farmlands. The long-term studies conducted show that salt-affected and waterlogged areas and saline waters can be utilized satisfactorily in raising forest and fruit tree species with improved techniques, forage grasses, conventional and nonconventional crops, oil-yielding crops, aromatic and medicinal plants of high economic value, petro-crops, and flower-yielding plants. Opportunities for raising salt-tolerant crops and alternate land uses through agroforestry on salty and waterlogged areas, especially in arid and semi-arid regions have been discussed in this chapter.

Introduction

Among the various degradative processes, the salinity afflictions of landscapes constitute a major threat. In fact, the salt-affected soils are not alien to land; these have been associated with mankind since the inception of agriculture. Civilizations in Mesopotamia, Nile Valley, Mohenjo-Daro and several other places and their subsequent fall are the testimony to the occurrence of salinity in agriculture. Salinity afflicted landscapes, which now occupy nearly a billion hectares (about 7 % of land area), have their origin either due to natural or man-induced causes. The salinity caused due to anthropogenic factors (secondary salinization) is related to clearing of natural deep-rooted vegetation and large-scale development of irrigated agriculture without adequate drainage. Due to secondary salinization, the world is losing at least 3 ha of arable land every minute (Bridges and Oldeman 1999) representing a serious threat to the expanding needs of food, fuel, fodder, and fiber production. Worldwide experiences show that though human-induced salinity problems can develop rapidly while the hydrological, agronomic, and biological solutions put forward for reclamation are very expensive and time consuming. Moreover, implementation of these solutions is constrained due to the socioeconomic and political considerations. Thus, despite the availability of technical know-how, the rehabilitation of the salty and waterlogged lands is progressing at a very slow pace. The use of agroforestry systems is now being put forward as a viable alternative. Though the salinity and waterlogging stresses can be as hostile for the woody tree species, these are known to tolerate these stresses better than the annual species.

The history of planting trees on salt-affected soils in India goes back to the follow-up of the recommendations of *Reh* Committee, which was set in 1879 to deliberate on the origin of salts in soils. While tracing the history of salt-affected soils in India, Singh (1998) mentioned about the scheme of experiments recommended by Reh Committee, which was directed toward preparing the soil for profitable cultivation by (i) removal of salts, (ii) drainage, (iii) silting, (iv) deep cultivation, (v) manuring, and (vi) ploughing of green crops. A conference was held in 1879 at Aligarh, during which it was decided to set up a series of experiments for the reclamation of salt lands and the decisions of the conference have been documented by Leather (1897). In one of the experiments near Aligarh, reclamation of *usar* (alkali land) was commenced with land and planting of trees. Plantations were later made at other locations in Aligarh and Kanpur districts. Various degrees of success was obtained when farm manure was used along with initial leaching through embankments of enclosed plantations and grassland. Greig (1883) visited one of the plantations at Pardilnagar near Sikandra Rao in the Aligarh district and seven plantations near A-wargarh and made some important observations on the heterogeneity in salt lands. He recorded that a simple and reliable way of finding good patches in an *usar* plot was to stop grazing for a year and then examine the ground. *Dub* (*Cynodon dactylon*) and other grasses will appear on good patches but on soil highly impregnated with *reh* only one kind of grass will grow that had not found on any other soil (we call these as indicator species). According to him, good *usar* was the soil that could grow *Acacia* tree without any special soil treatment. In real *usar,* the trees only did well if the pits were filled with good soil and when the pits were deep enough to let roots grown in good soil to extend beyond 90–120 cm of *reh* into a less-deteriorated soil below. These observations are forerunner of the auger hole method of planting trees in alkali soils advocated by the Central Soil Salinity Research Institute, Karnal (Singh 1998).

Though the records of plantations on alkali soils in India and elsewhere are available from 1874 (Oliver 1881; Leather 1897; Moreland 1901) no systematic experimentations were conducted to raise vegetation until recent past. Khan and Yadav (1962) observed that successful afforestation of alkali soils was not possible unless these are reclaimed to the desired level through appropriate techniques. Later on, it was suggested (Pande 1967; Yadav 1972; Yadav et al. 1975) that such soils could be planted using soil replacement technique in which the soil from $90 \times 90 \times 90$ cm^3 pit was replaced by a normal soil brought from out side. These measures, however, suffered from some intrinsic drawbacks like higher amendment needs, laborious and time consuming operation, leaving the $CaCO_3$ layer (which usually is located at about 0.5–0.7 m depth) untreated, and difficulties in preparing pits of this dimension in alkali soils. The difficulties were overcome through the development and perfection of the tree planting technique termed as the auger hole planting technique (Sandhu and Abrol 1981; Gill 1985; Singh and Gill 1992; Dagar et al. 2001a; Singh and Dagar 1998, 2005).

Moreover, the research efforts in the recent past have greatly enhanced the understanding of biology and management of forestry plantations on salt lands and with the use of saline waters (Lieth and al Masoom 1993; Lieth and Lieth 1993; Singh et al. 1993; Dagar et al. 2001b; Dagar 2003; Tomar et al. 2003a, b, Singh and Dagar 2005). Evidences are that subject to some of the obligatory changes in reclamation technologies, the salt lands can be successfully put to alternate land uses through agroforesty programmes. In addition to meeting ameliorative and long-term ecological goals on these landscapes, the alternate land use systems can be as economical as some cropping alternatives. Therefore, an attempt is made here to collate the existing information on afforestation technologies for the varied agroclimatic situations demanding site-specific solutions and agroforestry systems/ practices evolved for saline and waterlogged environments and utilizing saline waters. Some of the plausible benefits including environmental and socioeconomic impacts of rehabilitation of salt lands are also presented in this chapter.

Salt-Affected Soils and Saline/Sodic Waters

Salt-Affected Soils and Their Distribution

All soils contain a certain quantity of water-soluble salts, which are indeed essential for healthy growth of plants. If the quantity of soluble salts in a soil exceeds a certain threshold value (which in turn depends on the geochemical and environmental conditions, physicochemical properties of soil, and chemical composition of salts causing salinity), the growth of the most plants is adversely affected. Such soils are designated as *salt-affected.* These soils occur under different environmental conditions and have different morphological, physicochemical, and biological properties, but one common feature is the dominating influence of electrolytes on the soil-forming processes (Szabolcs 1979). Historically, the salt-affected soils have been referred to soils where growth of the most of the crops gets adversely affected either by the presence of excess soluble salts, sodium on the exchange complex or both. Thus, attempts have been made to classify the soils on the basis of total soluble salts measured in terms of electrical conductivity of the soil's saturation paste extract (ECe) or various dilutions (soil: water 1:2 or 1:5), exchangeable sodium percentage (ESP) or sodium adsorption ratio (SAR), and pH of the saturation paste (pHs) or other dilutions. The US Salinity Laboratory Staff in 1954 (USSL 1954) originally proposed the three categories of salt-affected soils on the basis of these parameters, i.e., saline, saline-alkali, and alkali soils.

The definitions in respect of these three categories were slightly modified later by Soil Science Society of America (SSSA 1987). It was described that owing to excess salts (ECe > 4 dS m^{-1}, or > 2 dS m^{-1} later on) and absence of significant amount of sodium (ESP < 15), saline soils are generally flocculated and as a consequence their conductivity is equal to or even greater than their nonsaline counterparts. A saline-alkali soil (ECe > 4 dS m^{-1}; ESP > 15) was described similar to that of saline soils as long as sufficient salts are present; whereas upon leaching, these soils become alkaline (pH > 8.5) leading to dispersion and their permeability reduces to levels those affect crop growth. The term "alkali" was discarded later on to be replaced with "sodic" and these soils contain sufficient exchangeable sodium (ESP > 15) to affect the physical behavior of soils and interfere with growth of the most of the crops.

The Indian classification of salt-affected soils is also based on the above criteria but in place of three categories of salt-affected soils (saline, sodic, and saline-sodic), these soils were classified into two groups based on the nature of plant responses to the presence of salts and the management practices desired for their reclamation. Abrol and Bhumbla (1978) concluded that "The so called 'saline sodic' soils" are in fact, of rare occurrence. It has been argued that usually recognized two categories of salt-affected soils "saline" and "saline-sodic" are no different from each other and that both should be categorized as "saline" because in these soils plant growth is not adversely affected due to the effect of excess exchangeable sodium or soil physical properties or lack of calcium. Soils containing sodium carbonate are necessarily sodic in nature. Bhumbla and Abrol (1979) argued that the so called "saline-sodic or saline-alkali" soils are rendered normal rather than sodic with lowering of SAR once the salts are leached. However, the term "alkali' has been invariably used for "sodic" in the Indian literature to represent soils which have both excess of exchangeable sodium and have salts capable of alkali hydrolysis, e.g., carbonate and bicarbonates of sodium or both. Therefore, Gupta and Abrol (1990) defined that the alkali soils in narrower context are the soils having (i) both high pHs (>8.2) and sodicity (ESP > 15) and containing soluble carbonates and bicarbonates of sodium such that $Na/Cl + SO_4 > 1$.

Looking at some of the recently modified classifications in the world literature, Rengasamy et al. (1984) have proposed a scheme based on the TEC concept where the soils have been classified into dispersive, potentially dispersive, and flocculated soils. The scheme has been further modified by Sumner et al. (1998) who have classified the salt-affected soils into nine classes based on EC and SAR of 1:5 (soil : water) extracts. Sodic soils have been differentiated into three categories based upon the dispersibility of soils, i.e., spontaneous dispersive, mechanically dispersive, and flocculated. Levels of ESP and SAR (1:5 extracts) used to describe corresponding soil sodicity are: no-sodic (SAR1:5 ≤ 3, ESP ≤ 6), sodic (SAR1:5 3–10, ESP 6–15), and very sodic (SAR1:5 > 10, ESP > 15). Each sodicity level has been combined with three levels of salinity (nonsaline, saline and very saline) with increasing limits of EC (1:5). A definite advantage of this scheme seems to be the use of EC and SAR of 1:5 extracts, while conventionally the saturation extract values have been used. It may be pointed out that the former gives little idea of salt release component, which is so important in maintaining the infiltrability in salt-affected soils containing sparingly soluble salts like gypsum and/or lime.

The salt-affected soils have been classified into three categories viz. nondispersive saline, dispersive saline-sodic, and sodic/alkali soils and based on the severity of the sodicity problem, further categorization has been proposed for different textured soils (Minhas and Sharma 2003).

Salt-affected soils are found distributed in all the continents covering about 954.832 million ha (Mha), which is ~ 10 %, of the total surface of dry land (Szabolcs 1989). As per recent FAO/UNESCO Soil Map of the World (FAO/AGL 2000), the total salt-affected area is 831 Mha, out of which 397 Mha are saline and 434 Mha sodic soils. Oldeman et al. (1991) reported that of the current 230 Mha of irrigated land, 45 Mha is salt-affected and of almost 1,500 Mha of dry land agriculture, 32 Mha are salt-affected to varying degrees by human-induced processes. Thus, globally almost 77 Mha of land is salty due to human-induced salinization (Oldeman et al. 1991; Bridges and Oldeman 1999). Recent figures indicate that in India salt-affected soils are distributed almost throughout the country (Fig. 9.1) over about 6.75 Mha (Table 9.1), 3.8 Mha being sodic and rest saline.

Saline and Sodic (alkali) Waters

Beneath many of the world's deserts are reserves of saline water. These include the Thar Desert of Indian sub-continent, the Arab Desert of the Middle East countries, the Sahara Desert in North Africa, the Kalahari Desert in Southern Africa, the Atacama Desert in South America, the California Desert in North America, and the West Australian Desert. Groundwater surveys in India indicate that poor quality waters being utilized in different states are 25–84 % of the total groundwater development (more in arid and semiarid regions); 84 % in Rajasthan, 62 % in Haryana, 47 % in Uttar Pradesh, 38 % in Karnataka, 30 % in Gujarat, 32 % in Andhra Pradesh, and 25 % in Madhya Pradesh (Minhas 1998). Many more areas with good quality aquifers are endangered with contamination as a consequence of excessive withdrawals of groundwater.

For assessing the quality of irrigation water, main parameters determined are: salt content (EC, dS m^{-1}), sodium adsorption ratio ($SAR = Na^+/\sqrt{[(Ca^{2+} + Mg^{2+})]}$, mol l^{-1}), residual alkalinity [$RSC = (CO_3^{2-} + HCO_3^-) - (Ca^{2+} + Mg^{2+})$ meq l^{-1}], divalent cation ratio ($DCR = \sum M^{2+}/\sum M^{n+}$), and presence of specific ions such as NO_3, F, B, and Se. Based on the characteristic features of majority of groundwaters in use with the farmers in different agroecological regions and the above indices, irrigation waters have been broadly grouped (Minhas and Gupta 1992) into good water ($ECiw < 2$ and $SAR < 10$), saline water ($ECiw > 2$ and $SAR < 10$), high SAR saline water ($ECiw > 4$ and $SAR > 10$), and alkali waters (ECiw variable, SAR variable, and $RSC > 2.5$). Globally, mass (1985, 1990) and Rhoades et al. (1992) produced the exhaustive data for the limits of salt tolerance of plants while in India very useful information has been generated on salinity limits of irrigation waters for different arable crops (Table 9.2), particularly for arid and semi-arid regions.

Natural Vegetation of Salt Lands

Saline habitats are usually characterized by sparse vegetation and highly saline soils are often barren "scalds". Based on their genetic potential to counter the defect of root zone salinity, the plants differ in their capacity to adapt to saline habitats. The capacity to lower the osmotic potential of cell sap, salt exclusion, salt secretion, and succulence are common but differentially expressed attributes of salt land vegetation. Thus, the plants, which are able to grow on saline habitats, possess special adaptive procedures and collectively are called halophytes. Based on the adaptability, Sen et al. (1982) classified the halophytes into the following three categories:

True (obligate) halophytes: Plants mainly attaining optimal growth on saline soil (above 0.5 % NaCl level) e.g. *Suaeda fruticosa, Cressa cretica, Aeluropus lagopoides, Salsola baryosma, Haloxylon recurvum* and *Zygophyllum simplex*.

Facultative halophytes: Those plants which can grow and achieve optimal growth on saline soil (at 0.5 % NaCl level) like true halophytes, as well as on nonsaline soils e.g., *Trianthema triquetra, Tamarix dioica, Salvadora persica,*

Fig. 9.1 Distribution of salt-affected soils in India (Based on the map prepared by NRSA Hyderabad, CSSRI Karnal, and NBSS & LUP, Nagpur 1997)

Launaea nudicaulis, Eragrostis pilosa, and many others.

Glycophytes or transitional halophytes: Plants of nonsaline habitats, which always grow and achieve optimal growth at nonsaline niches of the salt basins. For example, *Haloxylon salicornicum, Sporobolus helvolus, S. marginatus*, etc.

Dagar (1982, 1995, 2003, 2005b, 2006, 2008) described the halophytic vegetation inhabitating the mangrove swamps, littoral woodlands (behind mangroves), coastal salt marshes, saline

Table 9.1 Extent of salt-affected soils in India (000 h[a])

States	Sodic	Saline	Total
Andhra Pradesh	196.6	77.6	274.2
Andaman & Nicobar Islands	0	77.0	77.0
Bihar	105.9	47.3	153.2
Gujarat	541.4	1,680.6	2,222.0
Haryana	183.4	49.2	232.6
Jammu & Kashmir	17.5	0	17.5
Karnataka	148.1	1.9	150.0
Kerala	0	20.0	20.0
Madhya Pradesh	139.7	0	139.7
Maharashtra	422.7	184.1	606.8
Orissa	0	147.1	147.1
Punjab	151.7	0	151.7
Rajasthan	179.4	195.6	375.0
Tamil Nadu	354.8	13.2	368.0
Uttar Pradesh	1,347.0	22.0	1,369.0
West Bengal	0	441.3	441.3
Total	3,788.2	2,956.9	6745.1
		(Say 6.75 Mha)	

Source Mandal et al. (2010). (Based on NRSA data of 1996 and recon ciled during 2006 jointly by NRSA, CSSRI and NBSS & LUP, Nagpur)

[a] Exact figures may slightly differ because of rounding off the data

Table 9.2 Guidelines for saline irrigation waters (RSC < 2.5 me l^{-1}) in India

Soil texture (% clay)	Crop tolerance	Upper limits of ECiw (dS m^{-1}) in rainfall region (mm)		
		<350 mm	350–550 mm	550–750 mm
Fine soil (>30 %)	Sensitive	1.0	1.0	1.5
	Semi-tolerant	1.5	2.0	3.0
	Tolerant	2.0	3.0	4.5
Moderately fine soil (20–30 %)	Sensitive	1.5	2.0	2.5
	Semi-tolerant	2.0	3.0	4.5
	Tolerant	4.0	6.0	8.0
Moderately coarse soil (10–20 %)	Sensitive	2.0	2.5	3.0
	Semi-tolerant	4.0	6.0	8.0
	Tolerant	6.0	8.0	10.0
Coarse soil (<10 %)	Sensitive	–	3.0	3.0
	Semi-tolerant	6.0	7.5	9.0
	Tolerant	8.0	10.0	12.5

Source Minhas and Gupta (1992)

sand formations (beaches), inland salt marshes, and alkali/sodic lands.

Recently, it has been felt that these lands can effectively be utilized for salt-tolerant biological systems (Pasternak 1982; Le Houe'rou 1986; NAS 1990; Lieth and Al Masoom 1993; Dagar 1991, 1994, 1996, 2003, 2005a, b, 2006; Dagar and Singh 1998, 2007). For evaluation and identification of suitable salt-tolerant species for a specific situation, it is important to have a catalog of salt-tolerant species describing their distribution with respect to ecoclimate, habitat, and soil salinity.

Aronson (1989) set the criterion of ECe/ECiw (soil/irrigation water) of 7–8 dS m^{-1} salinity for a species to be designated as salt-tolerant, when plants were found growing well on this salinity without any significant yield reduction. Following the same criteria, Dagar (2003) and Dagar and Singh (2007) described the vegetation of saline habitats in India. Halophytes are not a single taxonomic group, but are represented by several thousand species of forbs, grasses, shrubs, and trees. These represent a wide range of habitats and there are about 1,560 species in which salt tolerance has already evolved (Aronson 1989).

Of the 500 halophytic genera of flowering plants listed in Aronson's database of 1989, almost 50 % belong to only 20 plant families. Chenopodiaceae (368 species) and Poaceae (136 species) have the largest number of species followed by Fabaceae (82), Asteraceae (62), Plumbaginaceae (58), Aizoaceae (52), Cyperaceae and Papilionaceae (46 each), Tamaricaceae (32), Zygophyllaceae (28), and Arecaceae and Mimosaceae (21 species each). In India, Dagar (2003) and Dagar and Singh (2007) have documented 1,140 species represented by 541 genera and 131 families of flowering plants distributed in saline and waterlogged including mangrove habitats.

The sodic soils support very restricted natural vegetation, comprising only a few species. *Prosopis juliflora* (an introduced species) has gregariously established widely especially on abandoned lands and along road sides. Among other species *Acacia nilotica, Capparis decidua, C. sepiaria, Salvadora oleoides, S. persica* and *Clerodendrum phlomidis* are prominent woody species on very high pH soils while *Acacia leucophloea, A. eburnea, Mimosa hamata, Prosopis cineraria, Butea monosperma, Diospyros tomentosa, Balanites roxburghii* and *Maytenus emarginata* are frequent on slightly low (up to 9) pH. Among herbaceous species *Sporobolus marginatus, S. coromandelianus, Desmostachya bipinnata, Chloris virgata, C. barbata, Leptochloa fusca, Cynodon dactylon, Kochia indica, Suaeda fruticosa, S. maritima, Pluchea lanceolata,* and *Portuleca oleracea* are important among herbaceous species.

In saline areas, *Suaeda fruticosa, Aeluropus lagopoides, Salsola baryosma, Haloxylon salicornicum, H. recurvum, Heliotropium curassavicum, Cressa cretica,* and *Sporobolus marginatus* are predominant in high salinity areas. Many salt-tolerant species such as *Sporobolus diander, S. tremulus, Eleusine compressa, Solanum xanthocarpum, Zygophyllum simplex, Portulaca oleracea, Fagonia cretica,* and *Dactyloctenium aegyptium* are frequently found on saline patches. Among woody species, *Prosopis juliflora* and *Salvadora persica* are dominant while *Ziziphus nummularia* is also quite frequent.

Along the Indian coast in tidal zone mangroves (species of *Avicennia, Aegiceras, Ceriops, Rhizophora, Bruguiera, Sonneratia, Heriteria, Excoecaria, Scyphiphora, Nypa,* and *Xylocarpus*); and mangrove associates such as *Terminalia catappa, Pongamia pinnata, Cynometra ramiflora, Pandanus spp., Salvadora persica, Arthrocnemum indicum, Acrostichum aureum, Barringtonia acutangula, Brownlowia tersa, Caesalpinia crista, Calophyllum inophyllum, Clerodendron inerme, Dalbergia spinosa, Hibiscus tiliaceus, Ochrosia oppositifolia, Salicornia brachiata,* and *Wrightia tomentosa* are predominant. Cultivaed species such as *Casuarina equisetifolia, Anacardium occidentale, Acacia auriculaeformis, Azadirachta indica, Cocos nucifera, Eucalyptus* spp., and *Pandanus* sp. are frequently found grown especially along the beaches of eastern coast.

Agroforestry for Salty Environments

For the successful establishment of agroforestry systems, site-specific system is required for management of salinity. This is because of the reason that the tree component of the system is (i) deep rooted, thus demanding characteristics of even deeper soil strata, and (ii) long-lived and thus major changes in salts may occur from establishment to maturity. Moreover, the systems on the most sites have to thrive on natural rainfall and other climatic conditions as against the preferential input of amendments, nutrients, water, etc. for crop production activities. The level of productivity is of course the most important criteria for assessing the suitability of agroforestry systems on the normal soils, but their environmental benefits and also the amelioration of soils are given due consideration for greening of salt lands. In addition to other site characteristics like climate and soil types, the nature of salts also defines the performance of different plant species. Therefore, the performance of species is discussed separately under sodic/alkali and saline conditions.

Developing Suitable Agroforestry Practices/Systems for Alkali/Sodic Soils

Alkali soils have a compact hard sub-surface layer or a caliche (calcite) bed (of nodulated or amorphous $CaCO_3$) in lower depths, which imposes physical impediment to root penetration/development and correspondingly poor aeration when wet (due to dispersion of clay colloids by sodium); nutrition imbalances including deficiencies of zinc, calcium, and magnesium (due to high pH); and toxicity of specifications (e.g., sodium and boron). Unlike soil reclamation for arable crops, where only plough layer is sought to be improved in the first instance, deep-rooted trees require reclamation of the soil to lower depths. The planting technique should further ensure efficient utilization of rainwater, and leaching of reaction products after interaction of amendments and to root development in the soil profile, soil structural improvement for increased water retention to encourage rapid root penetration in the vertical rather than horizontal direction, and minimize direct sodium toxicity hazards. Keeping this in view, several attempts were made to develop suitable techniques for planting trees on such lands. Several forest and fruit trees, grasses, medicinal and aromatic plants, nonconventional, and arable species were evaluated for their tolerance to sodicity and successful and most suitable species were identified for agroforestry.

Planting Methods for Sodic/Alkali Soils

As discussed earlier, alkali soils have a compact hard sub-surface layer or a caliche (calcite) bed (of nodulated or amorphous $CaCO_3$) in lower depths, which imposes physical impediment to root penetration/development. The viable planting technique should ensure efficient utilization of rainwater and leaching of reaction products after interaction of amendments and help to root development in the soil profile, soil structural improvement for increased water retention to encourage rapid root penetration in the vertical rather than horizontal direction, and minimize direct sodium toxicity hazards. Keeping this view, several attempts were made to develop suitable techniques for planting trees on such lands. In the past, planting methods like pits and trenches of various shapes and sizes were used for raising trees in alkali soils with some intrinsic drawbacks and the difficulties were overcome through the development of tree planting technique termed as the auger hole planting technique as mentioned earlier in this technique the auger is mounted on a tractor and used for making holes of dimensions 20–25 cm diameter and 1.2–1.8 deep (Fig. 9.2). This technique recognizes that in trees, owing to their deep root systems, management of the root zone by modifying the soil environment to greater soil depths using a limited quantity of amendments, has a vital role to play in terms of success in sapling establishment, cost of plantation, and practical adaptability.

Many forests and fruit tree species can be raised on highly alkali soil (pH > 10) but some of the fruit trees like pomegranate (*Punica granatum*) and bael (*Aegle marmelos*) are unable to tolerate water stagnation during rainy season. Raised and sunken bed technique of agroforestry was developed for such situations (Dagar et al.

Fig. 9.2 Making auger holes in alkali soil to pierce *kankar* pan ($CaCO_3$ layer) for planting tree sapling *source* CSSRI, Karnal

2001a). After refilling the auger holes with soil mixture in a leveled field as mentioned in the above-mentioned technique, the auger holes are marked with sticks. Parallel bunds, each of 1–2 m height and 1–2 m width, are then constructed leaving 4–5 m space between them taking soil from interspaces. The seedlings are raised on middle of bund at marked places and small rings are made around seedlings for initial irrigation. The interspaces can be cultivated growing water-loving crops such as *kallar* grass (*Leptochloa fusca*) or rice (salt-tolerant variety like CSR 10, CSR 13, CSR 30, CSR 36) during rainy season and Egyptian clover (*Trifolium alexandrinum*) or wheat (vars. KRL 1–4, KRL 19, KRL 210, KRL 213) during winter.

Afforestation of Sodic Soils

During the last three decades, numbers of studies were conducted in India to establish positive benefits of the recent planting techniques and to identify most suitable species for alkali soils. Chaturvedi (1984) reported that a 30 % reduction in biomass was observed at pH 10 as compared to pH 7 in tree species like *Acacia nilotica, Terminalia arjuna* and *Pongamia pinnata* whereas at pH 9.5 in *Eucalyptus tereticornis*, *Casuarina equisetifolia* and *Acacia nilotica* could grow well in soil with ESP 30.6, whereas *Pongamia pinnata* and *Dalbergia sissoo* survived only up to pH 9.5 and ESP of 15.2 (Yadav and Singh 1986). Based on the performance of tree saplings planted in soils of different pH (7–12), relative tolerance was reported (Singh et al. 1987) in the order: *Prosopis juliflora* > *A. nilotica* > *Haplophragma adenophyllum* > *Albizia lebbeck* > *Syzygium cuminii.*

In one experiment on high pH (10.3) soil, traditional pit method was compared with auger hole technique (Gill 1985, Gill and Abrol 1993) for establishing *Eucalyptus tereticornis* and *Acacia nilotica* plantations. The results obtained after 6 years of plantation indicated no mortality in both the species when planted with auger hole of 10–15 cm diameter and 120–180 cm depth and confirmed the superiority of the technique to the traditional laborious pit method. Afforestation has tremendous potential for soil amelioration and carbon sequestration not only in aboveground C biomass but also root C biomass in deeper soil depths. *Acacia nilotica* and *Eucalyptus tereticornis* plantations were to have a considerable ameliorative effect on soil properties when planted in alkali soil. Both pH and salinity (EC 1:2) were found to be reduced (Fig. 9.3) significantly. Reduction was greater in the surface soil than at depth. The soil organic carbon content increased to about double the initial value for *Eucalyptus* but tripled under *Acacia* plantation. This was largely due to greater litter accumulation in the *Acacia* stand, the intrinsically higher rates of decomposition and the higher C: N ratio for the litter in *Acacia.* Das and Itnal (1994) reported that organic carbon content was about double in agri-horticultural and agroforesry systems as compared to sole cropping.

Another trial was conducted by Singh et al. (1989) by planting *Prosopis juliflora* on a highly alkali soil. The planting methods compared were auger hole (15 cm diameter, 90 cm deep), pit (30 × 30 × 30 cm), and trench planting (30 cm wide and 30 cm deep dug across the plot). A uniform dose of gypsum at 3 kg per plant was mixed in the dugout soil and refilled in auger holes, pits, and trenches. Later, Singh (1994) extended the site suitability studies to 20 alkali soil sites in Ganga-Yamuna Doab. Barren alkali soils, represented by Natric Comborthids and

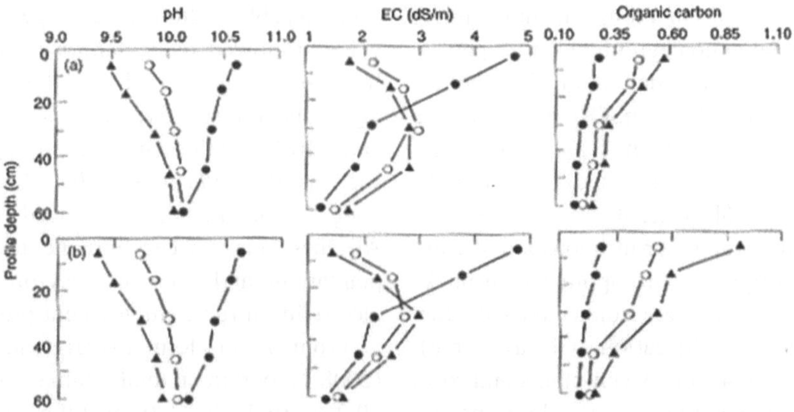

Fig. 9.3 Changes in pH (in 1:2 suspension), electrical conductivity EC (in 1:2 suspension), and percentage organic carbon with soil depth under **a** *Eucalyptus tereticorrais* and **b** *Acacia nilotica* plantations before planting (● = 1st year), after 2 years growth (○), and after 4 years (▲). Modified from Gill and Abrol (1993)

Calciorthids, had pH of 10–10.5, ESP of 60–95, and ECe between 1 and 43 dS m^{-1} and constituted mainly by carbonates and biocarbonates of sodium. *Prosopis juliflora* was found to grow in typic Natrutalfs with a maximum pH of 10.0, ESP up to 60, and ECe < 3 dS m^{-1} in the rooting zone, even though the top 44 cm of soil had pH 10.3, ESP of 70, and ECe at 12 dS m^{-1}. Other planted species failed to grow in this soil. Along with some natural vegetation, *P. juliflora* and *A. nilotica* established well on typic Natrustalfs with an average pH of 9.5, ESP 50, and ECe 10 dS m^{-1}. Other species like *Dalbergia sissoo, Pongamia pinnata, Albizia lebbeck, Terminalia arjuna, Butea monosperma, Capparis decidua* and *Salvadora persica* did well on soil with a pH < 9.1, ESP up to 44 and ECe up to 7.5 dS m^{-1}. The growth of all these species except *P. juliflora* was arrested with a *kankar* pan within 80 cm soil depth. In one study, Singh et al. (1993) observed that 6-year-old plantations of *Prosopis juliflora*, *Acacia nilotica*, *Casuarina equisetifolia*, and *Eucalyptus tereticornis* a biomass (firewood + small timber) of about 26, 21, 19, and 15 t ha^{-1} per annum, respectively, was obtained (Table 9.3) when planted with auger hole technique.

Many workers (Chhabra et al. 1987; Singh et al. 1988a, b, 1991, 1997; Dagar et al. 2001; Singh and Dagar 2005; Singh et al. 2008) conducted experiments with different combinations and doses of amendments applied in auger holes and concluded that original soil mixed up with 3–5 kg gypsum (50 % of gypsum requirement in the auger hole) and 8–10 kg farm yard manure (FYM) in each auger hole (to meet the nutrient requirement of sapling) is most suitable for alkali soils of high pH (up to 10). Application of small dose of nitrogen in the auger hole filling mixture and its regular application every year

Table 9.3 Growth performance and biomass production by four trees raised on high pH soil

Growth parameters	*Prosopis juliflora*	*Acacia nilotica*	*Casuarina equisetifolia*	*Eucalyptus tereticornis*
Height (m)	12.9	11.6	14.5	14.9
DBH (cm)	12.5	13.6	12.0	11.0
Bole weight (kg $tree^{-1}$)	112.6	85.4	84.2	65.6
Branches + leaves (kg $tree^{-1}$)	43.2	43.8	28.4	23.5

Source Singh et al. (1993)

thereafter (25 g both in monsoon and winter) proved beneficial in nonleguminous tree species. Alkali soils are usually deficit in zinc; hence, application of about 25 g $ZnSO_4$ per auger hole is essential. Rings of about 1 m diameter are made around the auger holes and 7 to 9-month-old saplings of salt-tolerant tree species are planted just after rains during monsoon season. Two to three irrigations are applied with buckets. Later on, the rings are connected with furrow channels for irrigation (if available). Spacing of 4 m × 4 m between plants and rows is ideal. For fuelwood plantation, close spacing of 2 m × 2 m may be kept. After 2–3 years, the alternate trees may be harvested for fuelwood. For agroforestry purposes, wider spaces between rows help to grow arable crops in interspaces.

In a well-conducted site-specific field study at Saraswati in semi-arid Haryana, out of 30 tree species planted in highly alkali soil (pH of profile 10.1–10.6), only three species *Prosopis juliflora* (Fig. 9.4), *Acacia nilotica,* and *Tamarix articulata* were found economically suitable with biomass production of 51, 70, and 93 t ha^{-1}, respectively, in 7 years (Dagar et al. 2001; Singh and Dagar 2005). *Tamarix articulata* ameliorated the soil by inducing the maximum reduction of exchangeable sodium percentage (ESP) and pH values followed by *P. juliflora* and *A. nilotica*. Increase in organic carbon in the surface 15 cm layer under respective species was 0.23, 0.26, and 0.10 %. At the same site, species of *Prosopis* such as *P. juliflora, P. alba, P. articulata, P. levigata*, and *P. nigra* produced high biomass. All these can successfully be used as energy plantations and even in gassy-fires to generate electricity in rural employment programmes.

From a long-term experiment, Singh et al. (2008) reported a total biomass ranging from 19.2 to 56.5 t ha^{-1} from different species after 10 years of plantation in high sodic soil of pH 10.6 in Uttar Pradesh (Fig. 9.5). These tree species improved soil in terms of reduction of pH and exchangeable sodium percentage (ESP), and increase in organic carbon significantly (Table 9.4). When harvested after 14 years of plantation, maximum biomass production was achieved in *Eucalyptus teretcornis, Acacia nilotica, Prosopis juliflora* and *Casuarina equisetifolia* giving 231, 217, 208, and 197 kg bole weight per plant, respectively, whereas *Prosopis alba, Pithecellobium dulce, Terminalia arjuna, Pongamia pinnata, Azhadirachta indica* and *Cassia siamea* provided relatively lower

Fig. 9.4 Plantation (*Prosopis juliflora*) in highly sodic soil at Saraswati Range Forest site (The original barren sodic land is in front) *source* Singh and Dagar (2005)

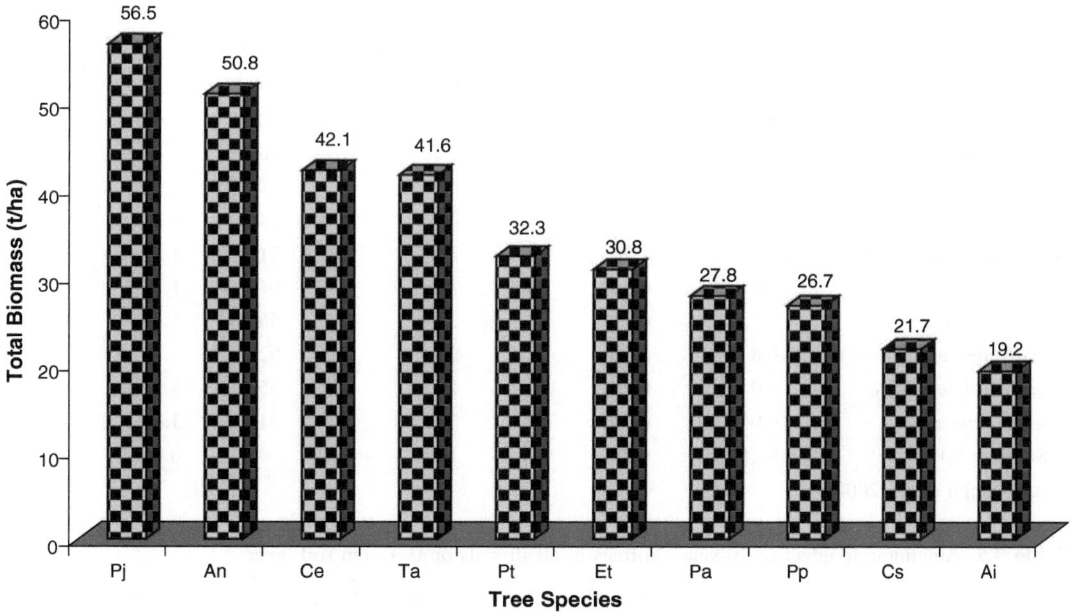

Fig. 9.5 Above-ground air-dried biomass (t ha^{-1}) of different trees grown in sodic soil Pj *Prosopis juliflora*, An *Acacia nilotica*, Ce *Casuarina equisetifolia*, Ta *Terminalia arjuna*, Pd *Pithecellobium dulce*, Et *Eucalyptus tereticornis*, Pa *Prosopis alba*, Pp *Pongamia pinnata*, Cs *Cassia siamea*, Ai *Azadirachta indica* (*Source* Singh et al. (2008))

bole weight of 133, 100, 97, 84, 83, and 52 kg per plant, respectively.

Singh and Gill (1992) reported that 20-year-old plantations of *P. juliflora, Acacia nilotica, Eucalyptus tereticornis, Albizia lebbeck* and *Terminalia arjuna* ameliorated the alkali soil through litter and root decomposition (Table 9.5) to the extent that arable crops could be grown successfully on this soil.

While evaluating different species on alkali soils it was observed that there was heterogeneity in soil as this is a well-known fact that alkali soils have patches of different pH in the same field. At the same time, there was also difference in the growth of trees of the same species because large interspecific and intraspecific differences are found in many genera. In species such as *Eucalyptus camaldulensis* and *E. occidentalis* intraspecific variation is so wide that to a tolerance classification at species level appears to be dubious. Researchers have exploited this intraspecific variation and selected individual plants with high salt tolerance from different areas. These selections have been propagated vegetatively through micropropagation or cuttings. In India, plantations established with nursery seedlings from available seed sources showed very high genetic variation and poor productivity; hence, a well-thought strategy with clear-cut objectives of developing genetically improved, uniform, fast growing, high yielding, disease-resistant, and locality-specific *Eucalyptus* clones was finalized at ITC Bhadrachalam Paper Boards Ltd., Andhra Pradesh in 1987 and several clones were tested at 82 different locations. A comparison between plantation from seed source and clonal stock is shown in Fig. 9.6.

From 1987 to 1995, 6,185 farmers in 1,138 villages were assisted to promote 7,441 ha of *Eucalyptus* plantations with 17.4 million seedlings. More than 613 selections of candidate plus trees (CPTs) of *Eucalyptus tereticornis, E. camaldulensis*, and Mysore Gum (*Eucalyptus* Hybrids—*E. botrryoides* x *E. tereticornis* and *E. robusta* x *E. tereticornis* from Mysore) have been cloned and are being evaluated scientifically at a number of locations. On the basis of their performance, 89 clones have been identified (Piare Lal et al. 1997; Piare Lal 2006) which are fast growing and disease resistant with productivity

Table 9.4 Ameliorative effects (after 10 years) of different species

Tree species	pH (1:2)	EC_2 (dS m^{-1})	OC (g kg^{-1})	ESP	Bulk density (t m^3)
Initial	10.6	1.43	0.8	89	1.57
Terminalia arjuna	9.8	0.39	3.5	60	1.47
Azadirachta indica	9.8	0.33	2.7	56	1.48
Prosopis juliflora	9.5	0.30	4.3	51	1.32
Pongamia pinnata	9.7	0.61	4.0	54	1.36
Casuarina equisetifolia	10.0	1.26	3.6	71	1.21
Prosopis alba	9.9	0.63	3.3	64	1.37
Acacia nilotica	9.7	0.77	3.5	56	1.29
Eucalyptus tereticornis	9.8	0.86	2.4	62	1.38
Pithecellobium dulce	9.9	0.70	2.7	65	1.25
Cassia siamea	10.0	0.69	2.6	71	1.46
LSD (p = 0.05)	0.26	0.31	1.6	4	0.05

Source Singh et al. (2008)

Table 9.5 Ameliorating effect of 20-year-old trees in alkali soils on 0–15 cm soil layer

Species	pH_2	EC_2 (dS m^{-1})	Organic C (%)	Available (kg ha^{-1})	
				P	K
Acacia nilotica	8.4	0.25	0.85	59	499
Eucalyptus tereticornis	8.5	0.44	0.66	33	359
Prosopis juliflora	7.5	0.51	0.93	111	702
Terminalia arjuna	7.9	0.32	0.86	68	410
Albizia lebbeck	7.9	0.32	0.62	43	387
Initial	10.2	1.18	0.22	28	278

Source Singh and Gill (1992)

ranging from 12 to 58 m^3 ha^{-1} $year^{-1}$ under nonirrigated conditions compared to 6–10 m^3 $ha^{-1}year^{-1}$ from seed-originated plantations.

After seeing the progress of these trials, attempts were made by the Forest Department of Haryana and some clones were introduced at Seonthi Research Station in Kurukshetra District from Bhadrachalam in 1993. The Forest Department encouraged farmers of Ambala district in 1999 under Social Forestry Programme by planting clonal *Eucalyptus* in alkali soil (pH up to 9.6) after bringing from Bhadrachalam, Andhra Pradesh. Some progressive farmers planted three clones (C-3, C-7, and C-10) along with seeded *Eucalyptus* under rain-fed condition. The performance of these plantations showed that the farmers could easily get 29.5 t ha^{-1} $year^{-1}$ wood biomass. They earned net profit of INR 52,738 (1US $ = 45 INR) ha^{-1} $year^{-1}$ (Forest News 2005). The results clearly indicated that cloned *Eucalyptus* plantations have more survival, height, and girth as compared to seed-origin plantations (Fig. 9.6). Moreover, the coefficient of variation in all these parameters in cloned plantations is very less as compared to seeded showing the uniformity in these plantations. In addition to this, the crown of clonal plants is thin which facilitates more sunlight to the ground crops. Moreover, these have been developed from superior genotypes with better adaptability, faster growth, higher yield, good bole form, and resistance to major known diseases like *Cylindrocladium* blight and wind damage. Thus, the planting of cloned *Eucalyptus* may be encouraged in agroforestry plantations. The adoptation reflects a great success story of closed plantations on farmers' fields on wider scale.

Further, it has been observed that fruit trees are comparatively less tolerant than some forest trees. Singh and Singh (1990) observed that

Fig. 9.6 *Eucalyptus (Photo Courtsy Dr Piare Lal)* plantations; *Left* Raised from saplings of seed origin; *Right* From cloned saplings

Emblica officinalis, Carissa carandas, Ziziphus mauritiana, Syzygium cuminii, Grewia asiatica, Psidium guajava, Aegle marmelos, and *Vitis vinifera* when grown on alkali soil could produce 6.0–20.5 t ha^{-1} fruits at different pH (Table 9.6).

Out of 10 fruit tree species tested on highly alkali soil (pH > 10) using different soil amendments (Singh et al. 1997; Dagar et al. 2001a; Singh and Dagar 2005), *Ziziphus mauritiana, Syzygium cuminii, Psidium guajava* (Fig. 9.7), *Emblica officinalis* and *Carissa carandas* were found the most successful species showing good growth and also initiated fruit setting after 4–5 years of plantation. After 10 years, these could produce 12–25 t ha^{-1} fruits annually.

On sodic vertisols tree species such as *Prosopis juliflora, Azadirachta indica, Salvadora persica,* and *Acacia nilotica* are found to be highly successful when planted in auger holes. *Cassia siamea* and *Leucaena leucocephala* also performed satisfactorily (Sharma et al. 1992). Species such as Acaciaauriculaeformis, *Dalbergia sissoo, Casuarina equisetifolia, Dendrocalamus strictus* and *Hardwickia binata* did not survive for longer period. Tree species also improved clayey vertisols significantly in terms of reducing soil pH and ESP and increasing soil carbon (Table 9.7).

Based on the evaluation of more than 60 forest and tree species (through series of experimentation on sodic soils), it could be concluded that *Prosopis juliflora* was the best performer for the sodic soils of high pH (> 10) followed by *Tamarix articulata* and *Acacia nilotica*. Species such as *Eucalyptus tereticornis, Terminalia arjuna, Salvadora oleoides, Cordia rothii* and fruit trees (with improved management) such as *Carissa carandus, Emblica officinalis,* and *Psidium guajava* can be grown with great success in moderate alkali soil, preferably at pH around 9.5 or less (Table 9.8).

Silvopastoral Systems

The grazing lands of sodic soils are very poor in forage production under open grazing, but when brought under judicious management these can be explored successfully for sustainable fodder production. Based on a series of long-term experiments (Kumar and Abrol 1983a, b, 1984, 1986; Kumar 1988a, b, 1990, 1998; Singh and Dagar 2005), it was found that Kallar grass (*Leptochloa fusca*) could be rated the most tolerant grass to highly sodic soil and waterlogged conditions as compared to other grasses producing 45 t ha^{-1}

Table 9.6 The fruit yield and optimum sodicity tolerance of some fruit trees when cultivated in alkali soil

Species	Optimum pH of sodicity tolerance	Fruit yield (t ha^{-1})
Emblica officinalis	10.0	20.5
Carissa carandas	10.0	5.2
Ziziphus mauritiana	9.5	15.5
Syzygium cuminii	9.5	16.0
Grewia asiatica	9.2	6.0
Psidium guajava	9.0	12.5
Aegle marmelos	8.5	6.5
Vitis vinifera	9.0	18.3

Source Singh and Singh (1990)

Fig. 9.7 Four-year-old guava (*Psidium guajava*) plantation in highly sodic soil. Source CSSRI, Karnal

green forage without applying any amendment. It produces more biomass in alkali soil than normal soil. It withstands prolonged stagnation of water and also ameliorates soil quickly (Kumar and Abrol 1986). Another interesting grass is Rhodes grass (*Chloris gayana*), which could produce 60–65 t ha^{-1} green biomass and did not show any reduction in biomass up to pH 10, whereas it produced 50 t ha^{-1} forage at pH 10.4 (Kumar 1998). Gutton panic (*Panicum maximum*) produced 60 t ha^{-1} up to pH 9.6 and 45 and 35 t ha^{-1} at pH 10 and 10.4, respectively, and Para grass (*Brachiaria mutica*) produced 90 and 70 t ha^{-1} at pH 9.2 and 9.6, respectively, and 50 and 40 t ha^{-1} at pH 10.0 and 10.4, respectively, showing their potential for sodicity tolerance. *Panicum antidotale, P. laevifolium, P. purpureum*, and *Setaria anceps* were other successful grasses up to soil pH

Table 9.7 Soil properties (0–15 cm soil depth) of alkali black clay soil under different trees after 7 years of plantation

Species	pHs	ECe (dS m^{-1})	ESP	Organic C (%)	Average of N (kg ha^{-1})	Average P_2O_5 (kg ha^{-1})	Hydraulic conductivity
Prosopis juliflora	8.5	1.29	10.2	0.71	263	8.0	0.25
Azadirachta indica	8.5	1.30	14.0	0.62	235	9.6	0.50
Eucalyptus tereticornis	8.2	1.24	20.2	0.67	240	16.8	0.85
Albizia lebbeck	8.5	1.40	20.0	0.57	215	5.6	0.25
Cassia siamea	8.1	1.42	15.6	0.50	235	8.0	0.25
Initial	8.8	4.00	35.0	0.35	185	3.4	Negligible

Source Sharma et al. (1992)

Table 9.8 Relative tolerance of forest and fruit tree species for soil sodicity

Mean pH_2 (0–1.2 m)	Fuelwood/fodder/timber species	Fruit tree species
>10	*Prosopis juliflora, Acacia nilotica, Tamarix articulata*	Not recommended
9.6–10.0	*Eucalyptus tereticornis, Capparis decidua, Pithecellobium dulce, Prosopis alba, P. cineraria, Casuarina equisetifolia*[a], *Salvadora persica, S. oleoides, Terminalia arjuna*	*Carissa carandas, Psidium guajava, Ziziphus mauritiana, Emblica officinalis*[a]
9.1–9.5	*Cordia rothii, Albizia lebbeck, Cassia siamea, Pongamia pinnata, Sesbania sesban, Parkinsonia aculeata, Dalbergia sissoo, Kigelia pinnata, Butea monosperma*	*Punica granatum*[b], *Phoenix dactylifera, Achrus zapota*[a], *Tamarindus indica*[a], *Syzygium cuminii, Feronia limonia*
8.2-9.0	*Grevillea robusta, Azadirachta indica, Melia azedarach, Leucaena leucocephala, Hardwickia binata, Moringa oliefera, Populus deltoides, Tectona grandis*	*Grewia asiatica, Aegle marmelos*[b], *Prunus persica, Pyrus communis, Mangifera indica, Morus alba, Ficus* spp, *Sapindus laurifolius, Vitis vinifera*

[a] (Frost sensitive)
[b] Does not stand water stagnation, may be raised on bunds
Compiled from various sources

9.6. These grasses can be grown successfully with most promising tree species such as *Prosopis juliflora, Acacia nilotica, Tamarix articulata, Casuarina equisetifolia* (susceptible for frost), *Terminalia arjuna*, and *Pongamia pinnata*.

Based on 4 years of study, it was found that on an average 15.6 t ha^{-1} forage of *Leptochloa fusca* could be obtained with *Prosopis juliflora*,, 16.2 t ha^{-1} with *Acacia nilotica*, 17.0 t ha^{-1} with *Dalbergia sissoo*, and 17.4 t ha^{-1} with *Casuarina equisetifolia*. However, among trees *P. juliflora* and *A. nilotica* performed the best followed by *Dalbergia sissoo* and *Casuarina equisetifolia* (Singh and Dagar 2005). Singh et al. (1988, 1989, 1991), Dagar et al. (2001b) and Singh and Dagar (2005) evaluated several tree and grass species for their performance in highly sodic soil and mesquite (*Prosopis juliflora*) and Kallar grass silvopastoral practice were adjudged the most promising for firewood and forage production and also for soil amelioration. *Leptochloa fusca* grown with *P. juliflora* produced 46.5 t ha^{-1} green forage in 15 cuttings over 50-month period without application of any fertilizer or other amendment (Singh et al.

Table 9.9 Biomass production in 6 years by *Prosopis juliflora* and Kallar grass (*Leptochloa fusca*) under different spacing on high pH (10.0–10.2) soils

Spacing	Biomass (t ha^{-1})			
	Prosopis juliflora			Kallar grass
	Lopped	Harvested	Total	
2 m× 2 m	49.1	112.2	161.3	55.6
3 m× 3 m	31.6	55.2	86.8	68.7
4 m× 4 m	25.0	36.1	61.1	80.9

Source Singh et al. (1993)

1988, 1989a, b, 1991, 1993) and *P. juliflora* yielded 161.3 t ha^{-1} air-dried firewood in 6 years when planted at 2 × 2 m^2 spacing besides 55.6 t ha^{-1} *Leptochloa fusca* grass forage (Table 9.9). This system also ameliorated soil to greater extent in terms of reducing soil pH and increasing organic carbon and nutrients (Table 9.10). An associative nitrogen-fixing bacterium, *Azoarcus*, occurs as an endophyte in the roots of *Kallar* grass (*L. fusca*)-a pioneer species of alkali soils that yields 9–12 t ha^{-1} of dry biomass without application of any nitrogen fertilizer, nearly half of the plant N of 90–120 kg ha^{-1} is derived from associative fixation (Malik and Zafar 1984; Malik et al. 1986) and helps the plants survive in adverse habitats. Growth of native nonsymbiotic bacteria is improved by applying amendments (Rao 1998). Symbiotic nitrogen fixation by *Rhizobium* has been extensively investigated in salt-affected soils (Rao and Ghai 1995) and their survival is not a problem as they have considerable tolerance to high pH.

This system improved the soil to such an extent that less tolerant but more palatable fodder species such as Persian clover (*Trifolium resupinatum*), Egyptian clover (*T. alexandrinum*), Lucerene (*Medicago sativa*), and Sweet clover (*Melilotus denticulata*) could be grown under mesquite trees after 52 months producing 23.1, 21.3, 10.3, and 8.0 t ha^{-1} forage, respectively (Singh et al. 1993). The proposed model is shown below:

Sodic land (pH > 10)

↓

Prosopis juliflora + *Leptochloa fusca* grass

↓ 5 years

Replacement of *L. fusca* with *Trifolium resupinatum/Melilotus parviflora/Medicago sativa*

↓ 10 years

Reclaimed land fit for growing almost all crops as intercrops between tree rows.

Grewal et al. (1987) developed a silvopastoral system for rainwater conservation and production of fuel and forage from alkali lands by planting *Acacia nilotica*, *Eucalyptus tereticornis* and *Parkinsonia aculeata* trees on ridges and establishing *kallar* grass (*L. fusca*) in the trenches between ridges. This system conserved rainwater during monsoon, which in turn increased the biomass of trees and intercrops of grasses. In addition to firewood and forage

Table 9.10 Effect of *P. juliflora—L. fusca* silvopastoral system on soil properties (0–15 cm soil depth) of an alkali land

Soil properties	Initial	After 6 years of planting
pH_2	10.3	8.9
EC_2 (dS m^{-1})	2.2	0.36
Organic C (%)	0.18	0.58
Available N (kg ha^{-1})	79	165
Available P (kg ha^{-1})	35	30
Available K (kg ha^{-1})	543	486

Source Singh et al. (1993)

production, this system was found useful in checking runoff and soil loss.

In one silvopastoral experiment conducted by Kaur et al. (2002a, b) on highly sodic soil (pH > 10), the extent of storage of carbon in aboveground parts of the tree + *Desmostachya bipinnata* system (t ha^{-1}) was: 4.95, 6.03, and 14.80 in *Acacia nilotica* + *D. bipinnata*, *Dalbergia sissoo* + *D. bipinnata*, and *Prosopis juliflora* + *D. bipinnata*, respectively, accounting for 66–80 % of total carbon content of the vegetation. The total carbon storage in the tree + *Desmostachya* systems ranged from 6.80 to 18.55 t C ha^{-1} across the treatments. Carbon content in total plant biomass was 1.44 t C ha^{-1} and 12.32 t C ha^{-1} in case of *Dalbergia sissoo* + *Sporobolus marginatus* and *Prosopis juliflora* + *S. marginatus*, respectively (Table 9.11). The amount of total carbon input through net primary production in the trees + *D. bipinnata* systems (t ha^{-1}yr^{-1}) was: 2.81 (*A. nilotica* + *D. bipinnata*), 5.37 (*D. sisoo* + *D. bipinnata*), and 6.50 (*P. juliflora* + *D. bipinnata*). At the same site, Kaur et al. (2002a) also observed a significant relationship between microbial biomass carbon and plant biomass carbon (r = 0.92) as well as the flux of carbon in net primary productivity (r = 0.92). Nitrogen mineralization rates were found greater in silvopastoral systems compared to sole grass stand. Soil organic matter was linearly related to microbial biomass carbon, soil N, and nitrogen mineralization rates (r = 0.95–0.98, $p < 0.01$). Therefore, silvopastoral systems were found to be promising for the highly sodic soils for improving the fertility and carbon sequestration.

In the same study, soil microbial biomass carbon was measured by using the fumigation extraction technique and nitrogen mineralization rate using aerobic incubation method. The microbial biomass carbon in the soil under grasses (*D. bipinnata* and *S. marginatus*) was low. In silvopastoral systems, microbial biomass carbon increased due to the increase in the carbon content in the soil–plant system. A significant relationship was found between microbial biomass carbon and plant biomass carbon (r = 0.83) as well as the flux of carbon in net primary productivity (r = 0.92). Nitrogen mineralization rates were found greater in silvopastoral systems compared to "grass only" systems. Soil organic matter was linearly related to microbial biomass carbon, soil N, and nitrogen mineralization rates (r = 0.95 to 0.98, $p < 0.01$).

The continuing shortage of fuelwood and manures has sparked a renewed interest in woody perennials as a source of biomass and nitrogen. Green-matter production of 17 accessions of *Sesbania* averaged 26 t ha^{-1} (6 t DM) at 54 days of growth, N uptake was 154 kg ha^{-1} and N fixation was 105–150 kg ha^{-1} (Rao and Gill 1993). At 100 days after sowing, green stem and leaf production in semi-reclaimed alkali soil was obtained to be 21.5 and 9.4 t ha^{-1} with a biofertilizer value in leaf and upper tender stems of 125, 5, 81, and 12 kg ha^{-1}, respectively of N, P, K, and S (Rao and Gill 1995). *Sesbania* is also a useful source of firewood and at 200 days after sowing dry stem yield was 20 t ha^{-1} with a calorific value of 4,730 kcal kg^{-1}. Another useful short-duration legume Pigeon pea (*Cajanus cajan*) yielded 9.1 t ha^{-1} dry woody stem and 2 t ha^{-1} litterfall, which led to the recycling of 39.5, 2.1, 7.3, and 2.1 kg ha^{-1} of N, P, K, and S, respectively, to the benefit of next crop in rotation and N fixation in the growing season was 115 kg ha^{-1} (Rao and Gill 1995). Of the various species of perennial *Sesbania*, *S. sesban* had the highest biomass production and nitrogen-fixing ability. The N accumulation in aerial

Table 9.11 Carbon content (t ha^{-1}) of *Acacia nilotica* (An), *Dalbergia sissoo* (Ds), and *Prosopis juliflora* (Pj) along with *Desmostachya bipinnata* (Db) and *Sporobolus marginatus* (Sm) in silvopastoral systems on a sodic soil

Plant component	An + Db	An + Sm	Ds + Db	Ds + Sm	Pj + Db	Pj + Sm
Tree foliage, branches & bole	4.95	-	6.03	0.33	14.80	9.28
Coarse & fine roots	1.48	-	2.06	0.11	3.66	2.80
Grasses	0.37	1.18	1.01	1.00	0.09	0.24

'-'denotes nonsurvival of *A. nilotica* due to high pH
Source Kaur et al. (2002a)

and root parts was 180 and 41 g per tree, amounting to 449 and 102 kg ha^{-1}. In high density plantation managed by coppicing for 6 years, dry biomass production was 25–35 kg ha^{-1}year^{-1} and nitrogen fixation was nearly 350 kg ha^{-1}year^{-1} in the first 3 years and 170–240 kg ha^{-1}year^{-1} in next 3 years (Rao et al. 1990). In a tree-legume (4 years)—cereal crops (6 years) sequential agroforestry system, N fixation by *Sesbania sesban* in the legume phase was 260–330 kg ha^{-1}year^{-1} (Table 9.12). Soil nitrogen enrichment in the 0–60 cm soil layer was 388 kg ha^{-1} and in cereal phase, rice and wheat were grown without application of nitrogen. Total N uptake in six crops each of rice and wheat was higher by 185 kg ha^{-1} in the plots in which *Sesbania* was grown earlier vis-à-vis those maintained fallow. The maximum residual effects were observed in the first 2 years; rice could be sustained at 5.3 t ha^{-1} in subsequent years whereas wheat yields at 1.9 t ha^{-1} were low in comparision with urea fertilization. Even after 6 years of cropping without addition of nitrogen, organic C, available N, and microbial activities were higher in the plots in which *Sesbania* was grown.

The salty soils of black soil zone (saline/sodic vertisols) are generally either contemporary or of secondary origin. The contemporary salty soils exists in the topographic situation having poor drainage conditions. However, the soils that have become sodic due to unjudicious use of irrigation water can be encountered in the irrigation command area. In 14 years of plantation it was found that *P. juliflora* and *Azadirachta indica* were most successful species for these soils. Among grasses, *Aeluropus lagopoides, Leptochloa fusca, Brachiaria mutica, Chloris gayana, Dichanthium annulatum, Bothriochloa pertusa* and species of *Eragrostis, Sporobolus,* and *Panicum* are most successful and form suitable silvopastoral system.

In another experiment on alkaline vertisol it was found that after 7 years of plantations of *P. juliflora* and *Azadirachta indica* forming silvopastoral system with Kallar grass soil pH, ECe, and ESP reduced from 8.8, 4 dS m^{-1} and 35 to 8.5, 1.29 dS m^{-1} and 10, respectively, under *Prosopis*-based system and these values reduced to 8.5, 1.3 dS m^{-1}, and 14, respectively under *Azadirachta* system. The experiments conducted in sodic vertisols with ESP 40 growing grasses like *Leptochloa fusca, Brachiaria mutica,* and *Vetiveria zizanioides*, showed that all these grasses performed well and the forage biomass increased during second year. The uptake of sodium by *L. fusca* was highest followed by *B. mutica* at every stage of cutting. During 3 years, these grasses could remove 144.8, 200.0, and 63.5 kg ha^{-1} sodium from soil, respectively (AICRP 2000–2004).

A large proportion of salt-affected lands (particularly in Indian subcontinent) does not belong to individual farmers, but is either government land or in the custody of village *Panchayats*. Reclamation of such lands for crop production is not feasible because of common property rights. Raising suitable trees and grasses would appear to be a promising use of these lands. As mentioned earlier, the most promising tree species for highly alkali soils such as *Prosopis juliflora, Acacia nilotica,* and *Tamarix articulata* blended with highly salt-

Table 9.12 Nitrogen balance (kg ha^{-1}) in *Sesbania sesban* grown at selected spacings in an alkaline soil

Inter-row spacing (m)→ N (kg ha^{-1}) ↓	0.5	1.0	2.0	3.0
Initial soil N	2,986	3,006	3,026	3,132
Tree biomass N	444	336	237	168
Litter N	247	2,28	243	220
Final soil N	3,221	3,487	3,421	3,574
N balance	+876	+995	+825	+780
N fixation kg ha^{-1}yr^{-1}	292	332	275	260

Source Rao (1998)

tolerant and high biomass producing grass species like *Leptochloa fusca, Brachiaria mutica, Chloris gayana,* and species of *Sporobolus* and *Panicum* form ideal silvopastoral system. The grasses can be managed through "cut and carry" system between the interspaces of trees. Thus, we find that for highly sodic soils as well as alkaline vertisols silvopastoral systems (especially Kallar grass-*Prosopis* based) have shown promise in terms of biomass production as well as soil amelioration.

Performance of Forest and Fruit Trees-Based Cropping Systems (Agri-Silvicultural Systems)

In this land use system, forest or fruit trees are raised in wider spaces (row-to-row 5–6 m, at times even more and plant-to-plant 4–5 m) and the arable crops are cultivated in the interspaces on high pH soils. In one trial, Egyptian clover (*Trifolium alexandrinum*), wheat, rice, onion, and garlic were grown successfully for 3 years in the interspaces of fruit trees *Carissa carandas, Punica granatum, Emblica officinalis, Psidium guajava, Syzygium cuminii* and *Ziziphus mauritiana*; and 10.6–16.7 t ha^{-1} forage from *Leptochloa fusca* grass, 1.6–3.0 t ha^{-1} grains from wheat, 1.8–3.4 t ha^{-1} onion bulb, and 2.3–4.1 t ha^{-1} garlic were harvested (Tomar et al. 2004) showing that during establishment of fruit trees, suitable arable crops can successfully be harvested from the interspaces of trees. As shown earlier, many forests and fruit tree species can be raised in alkali soil (pH up to 9.8) but some of these like pomegranate (*Punica granatum*) and Bael (*Aegle marmelos*) are unable to tolerate water stagnation during rainy season which may be cultivated on raised bunds. Dagar et al. (2001a) planted pomegranate on bunds and water-loving kallar grass (*Leptochloa fusca*) and salt-tolerant rice (variety CSR-30) in sunken beds during rainy season. In winter season, Egyptian clover (*Trifolium alexandrinum*) and wheat (var. KRL 1–4) could be grown in sunken beds. Results showed that on an average, 4.3–4.9 t ha^{-1} rice and 1.2–1.4 t ha^{-1} wheat were obtained. In the second rotation, 21.3–36.8 t ha^{-1} fresh forage of Kallar grass and 44.9– 47.8 t ha^{-1} fresh forage of Egyptian clover were obtained. There was no yield reduction due to plantations at initial stage of growth. Another advantage was that after 2 years, soil amelioration in terms of reduction in soil pH and increase in organic matter and nitrogen contents was significant.

Many progressive farmers of Indo-Gangetic plains like to grow fast-growing trees and intercrops on partial reclaimed soil. To get more income, they reclaim the soil by applying amendment like gypsum before sowing crops. Because of the establishment of several pulpwood industries in Yamunanagar area in Haryana, large number of farmers grow *Populus* and *Eucalyptus* on their farm especially in rice–wheat rotation cropping system both in the field (Fig. 9.8) and as boundary plantations.

Fig. 9.8 Commercial agroforestry plantations in Indo-Gangetic plains *Left Eucalyptus* with wheat; *Right Populus* with wheat (Leaves shedded during winter) (Photo courtsy Dr Pyare Lal)

Singh et al. (1995) evaluated this agroforestry approach on a moderately alkali soil (pH 9.2) in irrigated condition of Haryana by planting three commercial trees namely poplar (*Populus deltoides*), eucalyptus (*Eucalyptus tereticornis*), and kikar (*Acacia nilotica*) in association with rice–wheat, rice–Egyptian clover, pigeonpea/sorghum–mustard rotations, and sole trees and sole crops as control. Results showed that intercrops of Egyptian clover, rice, wheat, and mustard can successfully be grown along with these trees during the initial 3 years (Table 9.13). Later on, these crops may be replaced with shade-loving crops such as turmeric. These intercrops help *Populus* and *Eucalyptus* grow faster but adversely affect the growth of low water demanding trees like *Acacia*. Soil amelioration measured in terms of decrease in pH and improvement in organic carbon and available N, P, and K contents followed the order: *Acacia*-based system > *Populus*-based system > *Eucalyptus*-based system > sole crops (Table 9.14). The benefit: cost ratio was highest (3.30) in case of poplar with rice–wheat followed by poplar with rice–Egyptian clover (2.95), and the lowest (1.76) in *Acacia* with rice–Egyptian clover (Table 9.15). Among trees alone, poplar was most profitable followed by *Acacia* and *Eucalyptus*. Growing trees along with crops should not be viewed only as a better and economically viable in terms of food, fodder, timber, and firewood production system, but also as a promising option to maintain long-term sustainability and also a practical solution for sequestering C in the soil and mitigating climate change.

Singh et al. (1995), Dagar et al. (1995, 2004) and Dagar and Singh (2003, 2004) evaluated the performance of arable crops in the interspaces of several forest and fruit trees such as *Tectona grandis, Ailanthus excelsa, Casuarina equisetifolia*, and *Tamarindus indica* in irrigated system on reclaimed alkali soil and concluded that from irrigation and fertilizer application to crops all the trees were benefitted showing better growth and during initial years of establishment normal crop yield was obtained without any yield reduction but during later stages due to larger canopy there was drastic yield reduction in almost all the crops grown with all the tree species except that the remunerative yield of potato could be obtained under partial shade of *Casuarina* and there was no significant yield reduction as compared to when cultivated in open rather there was no risk of frost under canopy.

Table 9.13 Performance in terms of yield (t ha^{-1}) of different crop rotations with three commercial tree species when grown on partial reclaimed alkali soil for 5 years

Plantations → Crop rotation ↓	*Eucalyptus tereticornis*	*Acacia nilotica*	*Populus daltoides*	Control (without plantation)
Rice–Egyptian clover/cowpea-Egyptian clover Rice (G)	14.4	12.5	11.8	21.6
Egyptian clover (F)	239.7	212.7	234.6	389.1
Cowpea (F)	18.0	4.6	2.5	45.0
Rice–wheat/guinea grass–oats Rice (G)	13.4	11.8	10.3	21.0
Wheat (G)	9.0	8.1	8.4	16.0
Guinea grass (F)	18.1	12.3	2.5	30.0
Oats (F)	23.8	24.0	25.6	42.0
Pigeon pea–mustard/turmeric Pigeon pea (G)	0.7	0.8	0.2	0.7
Mustard (G)	2.3	1.8	2.0	4.0
Sorghum (F)	18.3	8.6	22.8	50.0
Turmeric (R)	22.3	5.9	8.3	22.1

Depictions *G* grain, *F* fodder, *R* rhizome
Source Singh et al. (1997)

Table 9.14 Changes in soil properties (0–30 cm) in 5 years under different tree–crop combinations

Cropping system	Organic carbon (%)	Available nitrogen (kg ha^{-1})
Sole crop	+0.07	+10
Eucalyptus tereticornis based	+0.12	+21
Acacia nilotica based	+0.20	+31
Populus deltoides based	+0.17	+25

Source: Singh et al. (1997)

Table 9.15 Benefit:cost (B:C) ratio, net present worth (NPW), and pay-back period (PBP) of various land uses in three plantations and different intercrops

Land use	B:C	NPW (Indian Rupees)	PBP (Months)
Eucalyptus	1.99	13,618	72
Eucalyptus + rice + Egyptian clover	2.23	48,797	27
Eucalyptus + rice + wheat	2.06	38,820	24
Acacia	2.02	22,569	72
Acacia + rice + Egyptian clover	1.76	31,033	24
Acacia + rice + wheat	1.80	29,347	21
Poplar	2.38	15,807	66
Poplar + rice + Egyptian clover	2.95	80,668	24
Poplar + rice + wheat	3.30	81,804	27
Rice + Egyptian clover	2.39	53,724	–
Rice + wheat	2.79	49,007	–

Source Singh et al. (1997)

In one experiment conducted in Gujarat on alkaline vertisol having ESP values of 25, 40, and 60 it was found that the fruit trees gooseberry (*Emblica officinalis*) and *ber* (*Ziziphus mauritiana*) were the most successful for these soils followed by sapota (*Achrus zapota*). Through series of experiments conducted on raised and sunken beds, it was concluded that both forest tree species such as *Azadirachta indica* and fruit trees like pomegranate (*Punica granatum*), Jamun (*Syzygium cuminii*), and goose berry (*Emblica officinalis*) could successfully be grown on raised bunds and rain-fed rice could be grown during rainy season in sunken beds (CSSRI 2002–2003 to 2012–2013). Seed spices such as fennel (*Foeniculum vulgare*), dill (*Anethum greveolens, A. sowa*), and cumin (*Cuminum cyminum*) and oil crops like sesame (*Sesamum indicum*) could be grown successfully in sunken beds. Castor (*Ricinus communis*) is another successful species for these soils. This technique helps in moisture conservation as in sunken beds rain water accumulates and moisture is retained for a longer period in clayey soil.

Many of the medicinal and aromatic underexplored crops are in great demand for both internal requirements and export. But since these crops are nonconventional in nature, it is not always feasible to produce these on fertile land, which is generally used for arable crops. The marginal lands, specifically the salt lands, where profitable returns are not possible from arable crops, can successfully be utilized for the cultivation of these high value crops with marginal inputs. Results of several experiments conducted by Dagar et al. (2004, 2006a, b) clearly indicated that aromatic grasses such as palmarosa (*Cymbopogon martini*) and lemon grass (*C. flexuosus*) could successfully be grown on moderate alkali soils up to pH 9.2, while vetiver (*Vetiveria zizanioides*) which withstands both high pH and stagnation of water could successfully be grown without significant yield reduction on highly alkali soils. Anwar et al. (1996)

reported safe limit of sodicity tolerace in terms of pH and exchangeable sodium percentage (ESP) to be 9.5 and 55 for both palmarosa and vetiver; 9.0 and 50 for lemon grass; and 10.0 and 55 for Jamrosa (*Cymbopogon khasans*). They also reported pH 9.5 to be a safe limit for German chamomile (*Matricaria chamomilla*) and periwinkle (*Catharanthus roseus*), while safe limit for Rye (*Secale cereale*) for ergot (*Claviceps purpurea*) is reported to be pH 10. Medicinal psyllium (*Plantago ovata*) produced 1.47–1.58 t ha^{-1} unhusked grain at pH 9.2 and 1.03–1.12 t ha^{-1} at pH 9.6 showing its potential at moderate alkali soil (Dagar et al. 2006). *Matricaria chamomilla, Catharanthus roseus*, and *Chrysanthemum indicum* were other interesting medicinal and flower-yielding plants which could be grown on moderate alkali soil (Dagar et al. 2009). All these crops can be blended suitably as intercrops in agroforestry systems on moderate alkali soils.

Agroforestry for Saline Soils

The saline soils suffer from excessive concentration of salts, high water table often leading to water logging and occurrence of poor-quality underground waters in many areas. Poor root zone aeration caused by high water table (water logging) and excess presence of salts, which operate simultaneously, impairs success of plantations on such soils. The planting techniques should be such that salt concentration in the root zone remains at a low level and the plants are able to escape the adverse affects of high salinity.

Developing Suitable Planting Techniques

Through a series of experiments, techniques of plantations on waterlogged saline soils were developed. To provide better aeration and to avoid excessive salinization, planting on high ridges was often considered beneficial for establishing tree plantations on waterlogged saline soils where the salinity is usually more in the surface layers and the same decreases with depth down the water table. On the contrary, the soil moisture is minimum near the surface and maximum in the capillary fringe of water table. Therefore, to encash the advantage of low salinity and better soil moisture regimes in sub-surface layers, Tomar and Gupta (1984–1994) tried the sub-surface planting of saplings (at a depth of 30 cm below surface) and compared it with ridge planting (40 cm high). Substantially, higher salts accumulated in the ridges that resulted in poor survival and sapling growth. It was observed that the greater the surface area of the ridge, the more salts accumulated in the surface 1 m root zone of ridge planted trees. Difficulty of conserving rainwater on the ridge tops and the presence of salts causing higher susceptibility to soil erosion were the other disadvantages encountered with ridge planting. In contrast, under the sub-surface planting method, roots were encountering a milder saline transmission zone and were meeting most of their water requirement from the phreatic zone.

The performance of trees was better when planted with sub-surface method but need for spot irrigation was the main problem. This method was then improved upon by planting the saplings in the sole of furrow (60 cm wide and 20 cm deep), which was subsequently used for irrigating the tree saplings. Tomar et al. (1998) conducted a series of long-term experiments in semi-arid regions (average rainfall 630 mm). The soils were sandy loam (Hyperthermic comborthids) containing high concentrations of chlorides and sulfates of sodium, calcium, and magnesium. The soil pH of entire profile was 7.2 and ECe ranged from 25 to 80 dS m^{-1} (average 36.4 dS m^{-1}) in upper 30 cm layer and gradually decreased with depth. Groundwater was highly saline (EC 30 dS m^{-1}) and remained close to ground surface during rainy season. The mean salinity of groundwater fluctuated and was highest (46 dS m^{-1}) during summer and minimum (2 dS m^{-1}) during rainy season.

During these studies, the sub-surface and furrow planting methods were compared with traditional ridge-trench method (Fig. 9.9) and

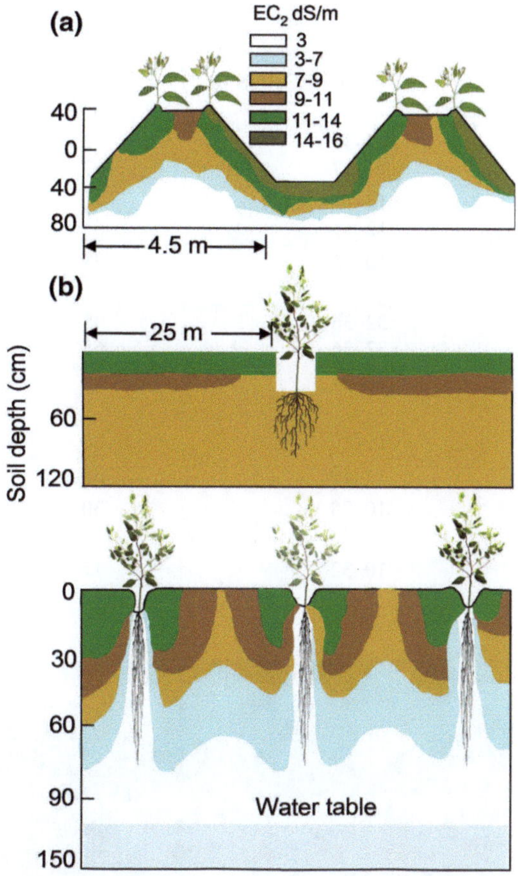

Fig. 9.9 Salt distribution patterns under **a** Ridge-trench, **b** Sub-surface and **c** Furrow planting methods in waterlogged saline soils

more than three dozen woody perennial species were planted under different methods of planting. In the furrow technique, a tractor-driven furrow maker was used to create about 60 cm wide and 20 cm deep furrows. The saplings of a tree species were planted at the base of the furrow. These furrows were subsequently used for irrigating the tree saplings. Establishment of saplings with furrow method was better than sub-surface method of planting. In addition to reducing the water application costs and increasing uniformity in water application, downward and lateral fluxes of water and salts from these furrows helped to create zones of favorable low salinity below their bases, especially when low-salinity irrigation water was used. Creation of such low "salt-niches" favored the establishment of young tree seedlings. With the furrow planting technique, salt concentrations were kept lower in the rooting zone of trees, such that the trees were able to escape the adverse effects of high salinity. Moreover, the furrow system seems more viable than the other techniques from a practical point of view for undertaking large-scale plantation of trees.

Performance of Trees Species and Agroforestry Systems

Singh and Yadav (1985a, b) and Yadav and Singh (1986) observed that *Casuarina equisetifolia* was most tolerant to salinity and was able to survive at ECe 32.5 dS m^{-1}, followed by *Acacia nilotica* and *Pongamia pinnata*. Gupta et al. (1987a, b) reported that salinity for 50 % reduction in the growth of *A. nilotica* and *Eucalyptus camaldulensis* was 5 dS m^{-1} in clay soil, but they grew satisfactorily at ECe 10 dS m^{-1} in sandy soil. They also observed a significant reduction in dry plant weight at ECe 2.5 dS m^{-1} in *Leucaena leucocephala* and *Peltophorum pterocarpum* and at ECe 5 dS m^{-1} in *E. tereticornis* and *Albizia lebbeck*; and at ECe 7 dS m^{-1} in *A. nilotica*. The data on biomass production after 9 years of plantation (Tomar et al. 1998) showed that *P. juliflora* and *Casuarina glauca* was highest (98 and 96 t ha^{-1}), followed by *Acacia nilotica* (52–67 t ha^{-1}) and *A. tortilis* (41 t ha^{-1}) when planted with subsurface or furrow technique showing their potential for saline waterlogged soils (Table 9.16).

On the basis of performance of trees for 6–9 years after planting in saline waterlogged soils (Tomar and Gupta 1984–1994, Tomar and Minhas 1998, Tomar et al. 1994, 1998) it was found that species like *P. juliflora, Tamarix articulata, T. traupii, Acacia farnesiana*, *Parkinsonia aculeata*, *Salvadora persica,* and *S. oleoides* were most tolerant to waterlogged saline soil and could be raised successfully up to salinity levels of ECe 30–40 dS m^{-1}; and species like *A. nilotica, A. tortilis, A. pennatula,*

Table 9.16 Biomass estimation of trees after 9 years of planting on saline soils

Tree species	Method of planting	Range of soil salinity ECe at 0–120 cm depth (dS m^{-1})	Range of water table salinity ECe (dS m^{-1})	Estimated biomass (t ha^{-1})
Acacia nilotica	Subsurface	10.6–25.3	27–33	52
	furrow	11.1–21.0	17–27	67
A. tortilis	Subsurface	6.8–28.1	12–33	41
	ridge	19.7–29.1	12–33	6
Eucalyptus camaldulensis	Furrow	10.0–17.9	10–35	28
Prosopis juliflora	Subsurface	10.3–24.0	32–36	98
	ridge	23.5–57.5	32–36	65
Casuarina equisetifolia	Furrow	5.6–20.7	10–31	28
C. glauca	Furrow	6.5–33.9	12–19	96
C.obesa	Furrow	9.0–19.5	12–19	38
Leucaena leucocephala	Subsurface	6.9–23.9	10–25	30
Tamarix sp.	Furrow	8.2–21.3	10–32	12

Source Tomar et al. (1998)

Casuarina glauca, C. obesa, C. equisetifolia, Callistemon lanceolatus, Eucalyptus camaldulensis, Feronia limonia, Leucaena leucocephala and *Ziziphus mauritiana* could be grown on sites with ECe 10–20 dS m^{-1}.

Casuarina glauca was found as an interesting species which withstood prolonged waterlogging and regenerated naturally as many seedlings were found under its canopy (Fig. 9.10). Other species such as *Casuarina cunninghamiana, Eucalyptus tereticornis, Terminalia arjuna, Albizia carbaea, Dalbergia sissoo, Emblica officinalis, Guazuma ulmifolia, Punica granatum, Pongama pinnata, Samanea saman, Acacia catechu, Syzygium cuminii,* and *Tamarindus indica* could be grown satisfactorily only at ECe < 10 dS m^{-1}. Based on the salinity level at which growth of species was satisfactory, tree species have been grouped (Tomar et al. 1998) into highly tolerant, tolerant, and moderately tolerant categories (Table 9.17).

Fig. 9.10 *Casuarina glauca* on highly waterlogged saline soil (Under canopy are regenerated seedlings)

Silvopastoral Systems

In waterlogged saline areas, many grasses such as *Leptochloa fusca*, species of *Aeluropus, Eragrostis, Sporobolus, Chloris, Panicum, Brachiaria* etc. can successfully be cultivated along with salt-tolerant trees and salty bushes such as *Atriplex, Kochia* and *Salvadora* constituting a viable and sustainable silvopastoral

Table 9.17 Relative salt tolerance by different tree species established with saline water on saline soils

Range of soil ECe ($dS\ m^{-1}$)	Tree species
>35 (highly tolerant)	*Prosopis juliflora, Salvadora persica, S. oleoides, Tamarix ericoides, T. troupii*
25–35 (tolerant)	*Tamarix articulata, Acacia farnesiana, Parkinsonia aculeata*
15–25 (moderately tolerant)	*Casuarina (glauca, obesa, equiselifolia), Acacia tortilis, A. nilotica, Callistemon lanceolatus, Pongamia pinnata, Eucalyptus camaldulensis, Crescentia alata, Albizia lebbeck*
10–15 (less tolerant)	*Casuarina cunninghamiana, Eucalyptus tereticornis, Acacia catechu, A. ampliceps, A. eburnea, A. leucophloea, Terminalia arjuna, Samanea saman, Albizia procera, Borassus flabellifer, Prosopis cineraria, Azadirachta indica, Dendrocalamus strictus, Butea monosperma, Cassia siamea, Feronia limonia, Leucaena leucocephala, Tamarindus indica, Guazuma ulmifolia, Ailanthus excelsa, Dichrostachys cinerea, Balanites roxburghii, Maytenus emarginata, Dalbergia sissoo, Salix babylonica*

Source Tomar et al. (1998)

system to sustain live stock productivity. As advocated earlier, *Leptochloa fusca* grass was found to have special advantages in terms of forage production from stagnant waters and having no ill effects on animal health and playing role in soil amelioration. *Aeluropus lagopoides, Sporobolus helvolus, Cynodon dactylon, Brachiaria ramosa, Paspalum* spp., *Echinochloa colonum, E. crusgalli, Dichanthium annulatum, Digitaria ciliaris, Vetiveria zizanioides*, and *Eragrostis* sp. are important grasses which are tolerant to both salinity and stagnation of water and can successfully be grown in silvopastoral systems on these habitats. Species of *Ziziphus, Atriplex, Kochia, Suaeda, Salsola, Haloxylon* and *Salvadora* are prominent forage shrubs of saline regions and relished by camel, sheep and goats (Dagar 2003, 2005b).

Now a days, in search for potential halophytic crops, work is progressing in number of countries and a number of potential halophytic genera have been identified which include *Acacia, Arthrocnemum, Atriplex, Avicennia, Batis, Bruguera, Cassia, Casuarina, Ceriops, Chloris, Coccoloba, Cressa, Crithmum, Distichlis, Grindelia, Juncus, Kochia, Kosteletzkya, Leptochloa, Limonium, Lumnitzera, Maireana, Pongamia, Panicum, Porterasia, Prosopis, Rhizophora, Salicornia, Salvadora, Simmondsia, Sonneratia, Spergularia, Sporobolus, Suaeda, Taxodium, Thinopyrum, Xylocarpus, Ziziphus*, and *Zostera* to name a few. In India, species of *Phragmites, Rumex, Polygonum, Typha, Coix, Brachiaria, Paspalum, Echinochloa, Panicum, Scirpus, Cyperus, Saccharum*, and *Vetiveria* are among the predominant herbaceous/grasses and species of *Salicornia, Suaeda, Haloxylon, Salsola, Tamarix*, and *Ipomoea* are prominent shrubs or undershrubs found in waterlogged saline situations (Dagar 2003, Dagar and Singh 2007). *Paspalum vaginatum* has an amazing ability to thrive in wet salty areas. *Leptochloa fusca, Brachiaria mutica*, and species of *Paspalum* are excellent fodder grasses which can be cultivated under waterlogged situations in the Indian subcontinent. *Juncus rigidus* and *J. acutus* can successfully be explored for paper and fiber making. *Vetiveria zizanioides*, a tall aromatic grass of waterlogged areas, may be propagated from rootstocks both for fodder and aromatic oil from its roots.

Most of the area in Rann of Kutchh along Gujarat is highly saline and because of low rainfall and high evapotranspiration, the problem becomes more severe. In many areas, natural salt is prepared in evaporation saltpans. *Prosopis juliflora, Salvadora persica, Tamarix articulata, T. troupii* and many halophytes are found growing naturally in these areas with stunted growth. A silvopastoral system may be developed incorporating suitable salt-tolerant forages such as species of *Atriplex*, *Kochia indica*,

Fig. 9.11 *Salvadora persica* on highly saline vertisol

Aeluropus lagopoides, Dichanthium annulatum, Leptochloa fusca, and *Sporobolus helvolus*. High value trees such as *Salvadora persica, Pongamia pinnata* and *Terminalia catappa* and firewood trees like *P. juliflora, Acacia nilotica*, and *Casuarina glauca* may be raised in furrows and above-mentioned grasses in interspaces. Oil-yielding *Salicornia begonia* is being grown at many places as industrial crop. Oil-yielding saltbush *Salvadora persica* performed well both in dry as well as waterlogged situations in saline soils (Fig. 9.11). In a study, *S. persica* based silvopastoral system was developed with *Leptochloa fusca, Eragrostis* sp., and *Dichanthium annulatum* forage grasses on clay loam saline vertisol (clay 40 %, silt 31 %, sand 29 %; pH ranging from 7.2 to 8.9; ECe from 25–70 dS m^{-1}) in Gujarat. The underground water was 0.5–2 m from surface with EC_{iw} ranging from 55 to 60 dS m^{-1}. Based on growth pattern in terms of height, canopy spread, and seed yield, a planting density of 4 m × 4 m was found as optimum for *S. persica*. During fourth year, the seed yield of *Salvadora persica* ranged form 1.84 to 2.65 t ha^{-1} with oil contents ranging from 576 to 868 kg ha^{-1} (Table 9.18) at different salinity levels (Rao et al. 2003). These grasses, (*L. fusca, D. annulatum*, and *Eragrostis* sp.) when planted on 45 cm high ridges, could produce 3.17, 1.85, and 1.09 t ha^{-1} forage, respectively. When planted in furrows, these could yield 3.75, 1.76, and 0.54 t ha^{-1}, respectively, showing their potential for these highly degraded lands.

Multienterprise/Integrated Farming Systems

In India, marginal and small categories of farmers, representing more than 86 % of farm families with holding size below 1.2 ha, are living in risk-prone diverse production conditions. At times, the land is quite degraded due to sodicity or salinity problems. Small and fragmented land holdings do not allow these farmers to have better independent farm resources. To fulfill the basic needs of households including food (cereals, pulses, oil seeds, milk, fruit, honey, fish, meat, etc.) for human consumption, feed and fodder for cattle, fuel and fiber, a well-focused attention toward Integrated Farming System (IFS) research is warranted. Scattered experiments based on IFS approach have been carried out in the country over the years (Balusamy et al. 2003; Singh et al. 2008, 2011; Gill et al. 2009) but the findings of these activities could not be converted into recommendations and failed to reach the real stakeholders. This fact was realized by both the Planning Commission and ICAR; and Project Directorate of Cropping System was renamed as Project Directorate of Farming System Research (PDFSR) with changed mandate in 2009. Research findings of a project on "Development of an integrated farming system model for small land holders of western plain zones of Uttar Pradesh" carried out during 2004–2010, revealed that IFS approach applied on 1.5 ha irrigated land, besides fulfilling all the requirements of seven members household food and fodder demand inclusive cost of production, could create an additional average annual savings of INR 47,000 in the first 4 years of its establishment and more than INR 50,000 in subsequent years (Singh et al. 2011). This saving could assist the family to meet other liabilities including health, education, and social customs improving the livelihood of small farm holders.

Table 9.18 Seed and oil yield of *Salvadora persica* (during different period) grown on saline black soil of different salinity

Soil salinity (dS m^{-1})	Seed yield (t ha^{-1})				Seed oil (%) content			
	First year	Second year	Third year	Fourth year	First year	Second year	Third year	Fourth year
25–35	1.08	1.27	2.41	2.65	32.3	32.8	32.6	32.7
35–45	0.84	1.02	1.35	1.72	32.3	32.2	31.8	31.6
45–55	0.60	0.88	1.29	1.58	31.3	30.6	30.1	30.1
55–65	0.38	0.74	1.27	1.40	29.7	29.8	29.3	29.8
Mean	0.72	0.98	1.58	1.84	31.4	31.9	31.0	31.1

Source Rao et al. (2003)

In this system, out of 1.5 ha area, 0.72 ha was used for cultivation of cereals, pulses, oilseeds, potato, flowers, and sugarcane. An area of 0.22 ha was allotted for a multistoried unit of horticulture containing a mixed plantation of mango, guava, pear, citrus, papaya, and banana. In the interspace, a number of short-duration vegetables and fodder crops were grown. Along the boundary, *Carissa carandas* was planted as hedge which gave additional fruits.The third component was dairy having buffalo and cow as milch animals. To ascertain the supply of green fodder round the year, fodder crops were rotated in 0.32 ha with other field crops. Other components included apiary (a unit of 10 bee boxes), pisciculture (mix of rohu, katla, mirgal, grass carp, and common carp) in 0.1 ha fish pond, poultry, vermicomposting, and goat unit (with 15 females and one male).

One study was initiated in 2006 at Central Soil Salinity Research Institute (CSSRI) on semi-reclaimed sodic land as a model for 2 ha land with interdisciplinary approach integrating various components (Fig. 9.12) which could be divided into two broader categories, i.e., crop component-1.8 ha (rice–wheat–green gram-0.2 ha, maize–wheat–green gram-0.2 ha, soyabean–maize-0.2 ha, pigeon pea–mustard–fodder maize-0.2 ha); fodder production -0.4 ha; horticulture/fruits production-0.2 ha, vegetables-0.2 ha, floriculture 0.2 ha); and subsidiary component 0.2 ha (fish production-0.2 ha, dairy, fruits, and vegetables on pond dykes, ducks and poultry, and bee keeping). The crop component gives income usually on half yearly basis when the crops are harvested, while the subsidiary component generates income on regular basis.

For the initial 5 years, the system was maintained by the scientists of the Institute. The net annual income from crop components was INR 1,15,844 and from subsidiary components INR 1,52,164 with a total of 2,68,007 having net saving of INR 734 per day.

When the same system is being operated in farmers' participatory mode, the gross and net income in 5 months was INR 1,78,888 and 51,155, respectively, with a daily net saving of INR 341 (CSSRI 2007–2012). The difference is in the maintenance of the system. On another pond constructed on alkali soil, various species of carp fish (*Catla catla, Labeo rohita, Cirrhinus mrigala, Hypopthalmichthys molitrix, Ctenopharyngodon idella*, and *Cyprinus carpio*) were cultivated in pond producing a fish biomass of 4.5 t ha^{-1} $year^{-1}$; and dykes were used for growing vegetables and fruits. The overall B: C ratio of the farming system was 5.5 (CSSRI 2011–2012).

Further, about 0.35 Mha of the land in Sharda Sahayak Canal Command is affected by waterlogging and sodicity. Traditional gypsum-based sodic land reclamation technology is not sustainable for reclaiming waterlogged sodic soils. In seepage prone areas of sodic soil (Fig. 9.13), pH generally decreases with increasing depth of soil. The initial soil pH at the site before digging out fish pond was 10.1, 9.7, 9.5, 9.2, 8.9, and 8.8 at 0–15, 15–30, 30–45, 45–60, 60–90, and

Fig. 9.12 Components of one multienterprise farming (agroforestry) system on partially reclaimed alkali soil at *source* CSSRI, Karnal

Fig. 9.13 Waterlogging in sodic soils (*Left*) due to seepage in Sharda Sahayak Canal Command areas of Uttar Pradesh (*Right*) *source* CSSRI, Karnal

90–120 cm depth, respectively. It is obvious that lower depths of soil have low pH. Thus, if soil having low pH is prevalent below 1–2 m depth, the land modification with IFS approach can successfully be employed for bringing waterlogged sodic lands back to cultivation.

Keeping this in view, a fishpond-based IFS model was developed over an area of 1 ha. A fish pond was dug up to a depth of 1.75 m and the soil was spread on adjacent sodic field. Due to mixing of low pH soil from lower depth, the pH of surface 0–15 cm soil was reduced to 9.1. *Eucalyptus* and banana were planted on bunds of the pond and are performing well. Fruit trees such as goose berry (*Emblica officinalis*) and guava (*Psidium guajava*) can also be planted on bunds of fish pond. In this particular case rice, wheat, sorghum, mustard, garlic and onion could be grown successfully on embankment and elevated field beds. Net return from the system was INR 34,694 with B: C ratio 4.45 (CSSRI 2011–2012).

Agroforestry for Combating Waterlogging (Biodrainage)

Introduction of canal irrigation in arid and semi-arid regions without provision of adequate drainage causes rise in groundwater table leading to waterlogging due to seepage and secondary salinization. Presently, about one-third of the world's irrigated area faces the threat of water logging, about 60 Mha is already waterlogged and 20 Mha is salt-affected (Heuperman et al. 2002). In India, the total degraded land due to waterlogging is 6.41 Mha out of which 1.66 Mha is due to surface ponding and 4.75 Mha is under sub-surface waterlogging (Maji et al. 2010). For the reclamation of waterlogged saline soils, the conventional technique is sub-surface drainage which is relatively expensive and generates harmful drainage effluents. A viable alternative of the above technique could be biodrainage, which is "pumping of excess soil water by deep-rooted plants using bioenergy." The root systems of trees intercept saturated zone or unsaturated capillary fringe above water table to control the shallow water table. These plants are known as *phreatophytes.*

Biodrainage is economical because it requires only initial investment for planting the vegetation, and when established, the system provides economic returns by means of fodder, wood, or fiber harvested and additionally sequesters carbon in the timber and soil. Reliance on capability of vegetation to reduce water table has been reported promising both in India (Chhabra and Thakur 1998; Naik and Manjunath 2000; Jeet Ram et al. 2011; Jena et al. 2011; Roy Chowdhury et al. 2011) as well as in other countries (Stirzaker et al. 1999; Zangs1999; Bhutta and Choudhry 2000). Fast-growing plants such as cloned *Eucalyptus* can successfully be grown on ridge particularly in areas where salinity is low. The impact of block plantations of *Eucalyptus tereticornis* on reclamation of waterlogged areas was tested and found effective at the Indira Gandhi Nahar Project (IGNP) site in Rajasthan and Dhob-Bhali research plot in Haryana (Heuperman et al. 2002; Jeet Ram et al. 2007). On these sites it was established that the transect of trees such as species of *Eucalyptus, Acacia, Populus, Prosopis, Casuarina, Pongamia, Terminalia, Syzygium, Dalbergia,* etc. when planted along canals successfully checked seepage and helped in mitigating waterlogging. During the studies conducted in IGNP area (Kapoor and Denecke 2001; Heuperman et al. 2002), groundwater under the tree plantation was reported to fall by 15.7 m over a period of 6 years. At 100 m from the edge of the plantation, the level of the groundwater was about 9 m higher than at the edge, with a draw down of 6.7 m. The higher groundwater level further away from the plantation edge is apparently the result of recharge from irrigation of areas under cultivation. Through these observations, Heuperman et al. (2002) concluded that the plantations act like groundwater pumps, pumping water at the rate of 34,460 $m^3 year^{-1}$ or 3.93 m^3 $h^{-1} ha^{-1}$ of plantation and the water used by plantations in the IGNP command was 3,446 mm $year^{-1}$, which was about 1.4 Class A pan. They further stated that the drawn down of the groundwater table under the plantations could even be 15 m or even more. No abnormal increase in salinity levels of soils and groundwater was observed under these plantations. Jena et al. (2011) planted *Acacia mangium* and *Casuarina equisetifolium* with intercropping of pineapple, turmeric, and arrowroot was taken successfully in Khurda district of Orissa coast. The depth to premonsoon water table changed from 0.5 m to 1.67 m after one year of plantation and to 2.20 in next year and to 3.20 during third year due to biodrainage. *Acacia* was better performer than *Casuarina*. Roy Chowdhury et al. (2011) also summarized the role of plantations (*Eucalyptus* and *Casuarina*) for reclamation of waterlogged situations in Deltaic Orissa which has been dealt in detail under agroforestry systems of coastal regions in Chap. 7.

In a controlled lysimeter study, conducted by Chhabra and Thakur (1998), it was concluded that *Eucalyptus tereticornis* could biodrain 2,880, 5,499, 5,518, and 5,148 mm of water in the first, second, third, and fourth year, respectively, from nonsaline groundwater and a water table depth of 1.5 m. The amount of water biodrained was more at 1.5 m as compared to 1 and 2 m water table depths because of maximum lateral roots in that zone. Further, the biodrainage capacity of trees was significantly affected by the salinity of the groundwater; however, even at a salinity of 12 dS m^{-1}, the plants biodrained 53 % of that under nonsaline conditions. In these experiments, the *Eucalyptus* plants could control water table rise up to 1.95, 3.48, 3.76, and 3.64 m in first, second, third, and fourth year, respectively; while in similar situations, the bamboo (*Bambusa arundinacea*) could control water table rise up to 1.09, 1.86, 2.46, and 2.96 m in first, second, third, and fourth year of growth, respectively. The secondary salinity developed in the root zone, up to 45 cm depth, did not exceed 4 dS m^{-1} even at a water table depth of 1 m with salinity of 12 dS m^{-1}.

To find out the optimum density of *Eucalyptus* for utilization of sewage water of municipal areas for production of wood biomass, an experiment was initiated at CSSRI Karnal in October 2000 by planting saplings in Nelder's competition wheel with 10 concentric rings (8 observation rings at radii 1.98, 3.30, 4.62, 6.47, 9.06, 12.68, 17.75, and 24.85 m and two guard rings at 0.6 and 31.95 m) of 18 plants each. The wheels were irrigated with sewage water following climatological approach, i.e., applying irrigation (D_{iw} 7.5 cm) when the cumulative open pan values (CPE) equal D_{iw} (or D_{iw}/CPE = 1.0). The observations were presented for very high density VHD (6,530 stems per ha), high density HD (1993 stems per ha), moderate density MD (517 stems per ha), and low density LD (162 stems per ha). Transpiration rate was measured using Sap Flow Sensors (Dynamax Flow 32) based on principal of thermodynamics. When data averaged for the period August 2002 to January 2003, transpiration rates for sewage-irrigated plantations was computed to be 23.0 and 17.3 l per day per plant at LD and MD densities. The average transpiration rates in 3-year-old plantations from May to December 2003 were found to be 29.5, 19.8, and 14.4 liters per day per plant in low, optimum, and high density plantations, respectively, which comes to be 189, 339, and 945 mm during this period. In August to December 2005 (5-year-old plantations), the transpiration values increased to 56.5, 30.7, and 18.9 liters per day per plant in respective densities, after 6 years, annual total consumptive use of water was 2,200 mm in high density and 1,300 mm in low density, which is quite reasonable amount of water (CSSRI 2002–2003 to 2007–2008).

The studies conducted by Jeet Ram et al. (2007, 2008) indicated that a block plantation of 18-year-old *Eucalyptus tereticornis* (2.6 ha with 300 trees) could influence the lowering of water table up to a distance of more than 730 m from the edge of plantation in the adjacent crop fields. In one experiment, Jeet Ram et al. (2011) constructed parallel ridges in the north–south direction on the bunds of agricultural waterlogged fields in Haryana. Ridge-to-ridge distance was 66 m. Each ridge was about 1 m high, 2.6 m wide at the base, and 2.0 m wide at the top. Plantations of cloned *Eucalyptus* (clone C-7) were raised as strip plantations (Fig. 9.14). Every strip-plantation contained two rows of plants. The row-to-row and plant-to-plant distance was 1 m, resulting in a total of 1,440 plants in 4.8 ha with a density of 300 plants per ha. The area under strip plantations was about 4 % and the rest 96 % was available for raising

Fig. 9.14 Three-year-old *Eucalyptus* planted in paired strips to lower down the water table in the agricultural field in canal commond area in Haryana

agricultural crops, thereby making it an agroforestry model for biodrainage.

The groundwater table was measured through observation wells installed in two parallel transects which were installed perpendicular to the strip plantations. Distance between the two transects was 60 m. The transpiration rate was measured using thermal dissipation probes (Dynamax, USA-make). In this experiment, it was observed that the groundwater table underneath the strip plantations was lower (1.61 m) than the groundwater table in the adjacent agricultural fields (1.43 m) resulting in a drawdown of 0.18 m by 2-year-old strip plantations. The total drawdown of groundwater table during a period of 3 years was 0.85 m (Fig. 9.15).

Further, the roots of 5.4-year-old trees of *Eucalyptus* penetrated in the soil profile up to a depth of 3.30 m. Therefore, the upper 0.50 m portion of the roots was above the ground level (in the bunds) and the remaining 2.80 m portion below the ground level where the soil was totally wet. It indicated that the roots have reached in the zone of capillary fringe located above the groundwater table for the absorption of groundwater and the strip plantations of clonal *Eucalyptus* were working as biopumps. The average above ground oven dry biomass was 99.9 kg tree^{-1}, of which 92.1 kg (92.3 %) was of timber (poles), 3.3 kg (3.3 %) of fuelwood, and 4.5 kg (4.5 %) of twigs and leaves resulting in 24.0 t ha^{-1} total oven dry above ground biomass of 240 surviving trees. The average below ground oven dry biomass of roots was 37 kg tree^{-1} (about 37.1 % of the average above ground oven dry biomass) resulting in 8.9 t ha^{-1} total oven dry below ground biomass of 240 surviving trees ha^{-1}. Thus, the total above and below oven dry biomass of 240 surviving trees was 32.6 t ha^{-1} (Table 9.19). The carbon percent in the oven dry biomass was 47.0 % in timber, 43.5 % in fuelwood, 43.9 % in twigs and leaves, and 48.0 % in the roots. Therefore, the weight of carbon in 5.4-year-old 240 surviving trees ha^{-1} of clonal *E. tereticornis* was 10.4 t ha^{-1} in timber, 0.3 t ha^{-1} in fuelwood, 0.5 t ha^{-1} in twigs and leaves, and 4.3 t ha^{-1} in roots resulting in a total carbon content of 15.5 t ha^{-1}, which was equivalent to 56.7 t ha^{-1} of CO_2 (Table 9.20).

The average rate of transpiration (measured by sap flow technique) of groundwater by the clonal *E. tereticornis* ranged from (liters day^{-1} tree^{-1}) 44.5 to 56.3 in May, 30.5 to 34.0 in July, 24.1 to 28.3 in October, and 14.8 to 16.2 l day^{-1} tree^{-1}in January. The annual rate of transpiration by 240 trees ha^{-1} was equal to 268 mm per annum. The total decline in the wheat grains yield due to the shading effect of 5-year-old strip plantations of clonal *E. tereticornis* was 11.7 % (0.29 t ha^{-1}). But, in spite of decline in wheat grains yield due to the adverse effect of strip plantations, the wheat grains yield was 2.15 t ha^{-1} as compared to 0.64 t ha^{-1} in the nearby untreated fields without plantation. Therefore, the wheat grains yield research plot was 3.36 times the yield in the nearby untreated fields. The farmers earned INR 72,000 ha^{-1} at a rotation of 5 years and 4 months resulting in a benefit–cost ratio of 3.5:1 at 12 % discount rate of interest. Further, the stools of young felled trees of *E. tereticornis* gave excellent coppice shoots. Therefore, there is no need to artificially regenerate the felled area. The only operation to be carried out is the singling of coppice shoots

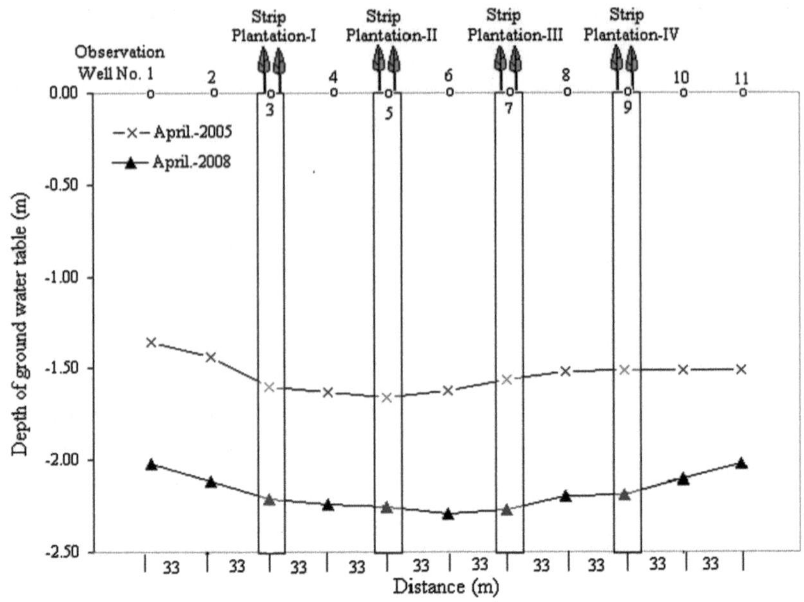

Fig. 9.15 Comparision of drawdown of water table by cloned *Eucalyptus* in a crop fields and under plantations of 3 years old (X) and 5 years old (▲)

Table 9.19 Fresh and dry biomass of 5.4-year-old trees of clonal *E. tereticornis* in biodrainage plantation

Tree components	Fresh biomass		Dry biomass	
	kg tree^{-1}	t ha^{-1}	kg tree^{-1}	t ha^{-1}
Timber	131.6	31.6	92.1	22.1
Fuelwood	4.6	1.1	3.3	0.8
Twigs and leaves	6.4	1.5	4.5	1.1
Total above ground biomass	142.7	34.2	99.9	24.0
Roots	51.4	12.3	37.0	8.9
Grand total	194.1	46.6	136.9	32.9

Source Jeet Ram et al. (2011)

Table 9.20 Carbon and CO_2 sequestered by 5.4-years-old trees of clonal *E. tereticornis*

Tree components	Oven dry biomass (t ha^{-1})	Carbon (%)	Carbon content (t ha^{-1})	CO_2 content (t ha^{-1})
Timber	22.1	47.0	10.39	38.10
Fuelwood	0.8	43.5	0.34	1.25
Twigs and leaves	1.1	43.9	0.47	1.74
Roots	8.9	48.0	4.26	15.63
Total	32.8	182.4	15.47	56.72

Source Jeet Ram et al. (2011)

so that the retained shoots can get maximum growing space. In this way, about 3–4 rotations of coppice crop (each of 5–6 years) can be taken by incurring negligible expenditure on singling of coppice shoots. Thus, the benefit–cost ratio of next 3–4 rotations of coppice crop of *E. tereticornis*, each of 5–6 years, would be much higher due to zero cost of raising and negligible cost of maintenance.

Agroforestry survey, involving 411 farmers of 15 villages belonging to 11 development blocks of 6 waterlogged districts, was conducted to access the availability of waterlogged areas for raising biodrainage plantations. The response of the farmers was very encouraging as 99 % farmers showed their willingness. The total agricultural land owned by these farmers was 1,972 ha, of which 1,877 ha was waterlogged out of which about 79 % of the land was offered for planting biodrainage plantations. Government of Haryana is very keen in mitigating the climate change as is evident from the fact that it is implementing the first small-scale afforestation Clean Development Mechanism (CDM) Project in the world registered with the United Nations Framework Convention on Climate Change under the Kyoto Protocol. Keeping in view the increasing demand of the farmers, the Haryana Government raised biodrainage plantations on 5,000 ha of farmers' agricultural waterlogged fields during the financial year 2008–2010 (Jeet Ram, personal communication).

It is now established fact that trees play very important role in controlling the waterlogging due to seepage along canals. The trees are planted along the canals at close space (1–2 m) in lines. These intercept the seepage and control water logging and salinity. The companion halophytes such as species of *Atriplex, Suaeda, Haloxylon, Kochia,* and *Salsola* and grasses such as *L. fusca, B. mutica, V. zizanioides* and species of *Paspalum* will cover the soil and check upward flux of salts which would follow if water evaporates directly from the soil surface. Considering the local agroclimatic and socioeconomic factors, appropriate trees and companion grasses/shrubs/crops can be chosen. When trees are raised in wider rows, the interspaces may be planted with forage grasses. The grasses in combination with trees are more efficient than the trees alone.

In one experiment, conducted at Gangawati in Karnataka on saline vertisol involving tree species (*Hardwickia binata, Sesbania grandiflora, Acacia nilotica, Dalbergia sissoo, Casuarina equisetifolia* and *Azadirachta indica*) in combination with grasses such as hybrid napier it was found that *A. nilotica* intercepted highest incoming seepage (86.4 %) from canal as compared to control (without trees) followed by *D. sissoo* with 84 % interception (Tomar and Patil 1998). The interception was directly correlated with canopy spread. The trees when planted with napier grass in interspaces showed more efficiency in controlling seepage (Table 9.21) showing the importance of silvopastoral system in such areas.

Among other species *Terminalia arjuna, Pongamia pinnata, Eucalyptus camaldulensis, E. tereticornis* and *Syzygium cuminii* among trees and *Leptochloa fusca, Phragmites australis, Dichanthium annulatum, D. varicosum, Brachiaria mutica* and *Paspalum* spp. among grasses have credibility for such situation.

Table 9.21 Relative efficiency of tree species in controlling canal seepage along 12-m-long strip

Treatment	Seepage (m^3 day^{-1})		Increase in interception (%)		Canopy width (cm)	
	Tree alone	Tree + napier grass	Tree alone	Tree + napier grass	Tree alone	Tree + napier grass
Control (no tree)	0.90	0.90	0.0	0.0	–	–
Azadirachta indica	0.36	0.29	60.0	68.0	102.8	89.9
Hardwickia binata	0.87	0.25	68.0	72.0	92.5	105.98
Casuarina equisetifolia	0.25	0.25	72.0	72.0	316.2	308.1
Sesbania grandiflora	0.25	0.25	72.0	76.0	160.0	149.9
Dalbergia sissoo	0.14	0.18	84.0	80.0	389.7	314.0
Acacia nilotica	0.12	0.12	96.4	86.5	421.8	379.2

Source Tomar and Patil (1998)

Agroforestry Suited to Saline Irrigation

Farmers besides irrigating arable crops also utilize salty waters for cultivation of fruit trees such as date palm (*Phoenix dactylifera*), *ber* (*Ziziphus mauritiana*), guava (*Psidium guajava*), goose berry (*Emblica officinalis*), etc. on coarse-textured saline soil having ECe from 15–20 dS m^{-1}. The rehabilitation of degraded drylands is limited to two possibilities: (i) the exploitation of plants native to arid environments, and (ii) devising efficient cropping systems and techniques for using limited saline groundwater resources judiciously. In the past, efforts toward utilization of saline waters were mainly aimed at enhancing the production of annual arable crops and notion of irrigated forestry or fruit trees, growing forages and other nonconventional high value crops was considered to be less attractive leading to poor economic production. But recent research efforts have shown that agroforestry has greater potential for utilizing degraded lands using saline water for establishment of trees and profitable and sustainable agroforestry systems.

Planting Methods with Saline Irrigation

The traditional approach for sustaining the use of saline water is to irrigate more frequently and provide for leaching requirements (Ayers and Westcot 1985). Nevertheless, such practices demand for application of additional quantities of saline water and thereby also result in the enhancement of salt loads of soils. These approaches were advocated for shallow-rooted crop plants in arid environments mainly because the added salts could be pushed beyond the rooting zone. But in deep-rooted tree plantations, the additional salts going into the soil through enhanced frequency of irrigations during their establishment may rather aggravate the problem as these are likely to persist within their expanding rooting zones and may subsequently hinder the growth of trees. Therefore, irrigation with saline waters should aim to create favorable niches for the better establishment of saplings and also eliminate the over salinity buildup. This could be achieved by irrigating only the limited area under furrows planted with tree saplings (Minhas et al. 1997a, b; Tomar et al. 1994, 2003b; Dagar et al. 2008). In this technique, furrows (15–20 cm deep and 50–60 cm wide) are created at 4–5 m intervals with a tractor-drawn furrow maker. Auger holes (0.2 m diameter and 1.2 m deep) are dug at the sill of these furrows spaced at 2–4 m intervals depending upon the space to be kept between plants. These are refilled with the mixture of original soil plus 8 kg of farmyard manure, 30 g superphosphate, 15 g zinc sulfate, and 15 g of iron sulfate. Six to nine-month-old tree saplings are transplanted during rainy season (July–August) at sites where auger holes are dug. The irrigation with saline water is given in furrows only. The technique is known as subsurface planting and furrow irrigation system (SPFIS).

The irrigation may be provided for the initial 3 years (4–6 times in a year) and thereafter, plantations may be irrigated once during the winter only. Salt storage in soil profile may increase during irrigation period but the added salts get distributed in soil profile as a consequence of seasonal concentration of rainfall during monsoons and some episodic events of rainfall during the following years. Along with forest and fruit trees, arable annual crops such as barley, wheat, clusterbean, pearl millet, mustard grasses such as *Leptochloa fusca, Chloris gayana, Agropyron elongatum,* etc. and medicinal plants such as *Plantago ovata, Aloe vera, Withania somnifera* and *Cassia angustifolia* could be cultivated successfully (Minhas et al. 1997a, b; Tomar and Minhas 2002, 2004a, b; Tomar et al. 2003a, b, 2005, 2010; Dagar et al. 2006a, b, 2008; Dagar 2012).

Performance of Forest Trees on Degraded Land Using Underground Saline Water

With increasing demands of food, forage, fuelwood, timber, and other necessities for ever-increasing population and limited availability of good quality water the saline water irrigation is now considered as an imperative necessity for the sustainable agricultural development, which includes the use of saline groundwater, saline drainage water, and sewage wastewater for irrigation. Agroforestry is sustainable option for utilizing poor quality waters. A long-term field trial with 31 tree species was conducted over 20 years on calcareous soils in a semi-arid region (annual rainfall about 350 mm) of northwest India using furrow method of irrigation.

The saplings were established irrigating with saline water (EC 8–10 dS m^{-1}) for the initial 3 years (4–6 times in a year) and thereafter plants were irrigated once in a year during winter up to the age of 8 years. Most of the tree species (except *Syzygium cuminii, Bauhinia variegata,* and *Crescentia alata*) showed quite high survival rate (71–100 %) during the first 3 years. Ranking in order of survival, growth, and biomass yield showed that *Tamarix articulata, Acacia nilotica, Prosopis juliflora, Eucalyptus tereticornis, Acacia tortilis* and *Cassia siamea* were most successful species (Tomar et al. 2003b). After 8 years of planting, the highest shoot biomass was harvested (Fig. 9.16) from *Tamarix articulata* (71.9 t ha^{-1}) followed by *Acacia nilotica* (23.4 t ha^{-1}), *P. juliflora* (20.2 t ha^{-1}), and *Eucalyptus tereticornis* (14.8 t ha^{-1}). After 16 years of growth, these trees produced 206, 197, 110, and 57 t ha^{-1} biomass, respectively. *Cassia siamea, Acacia tortilis*, and *Azadirachta indica* produced 94, 87, and 67 t ha^{-1} biomass because of their higher survival rate. Based on several experiments, a cafeteria of tree species has been prepared (Table 9.22) which depicts very promising and promising performers when established with saline irrigation.

The soil organic carbon content under these plantations was more than 3.5 g kg^{-1}. Litter fall from the most of tree species resulted in an improvement of organic carbon content of the underlying soils. The prominent species where considerable enhancements in organic carbon contents (>0.4 % in upper 30 cm layer) were observed included: *Acacia tortilis, Cassia siamea*, and *Prosopis juliflora*. There was substantial increase in organic carbon in soil which

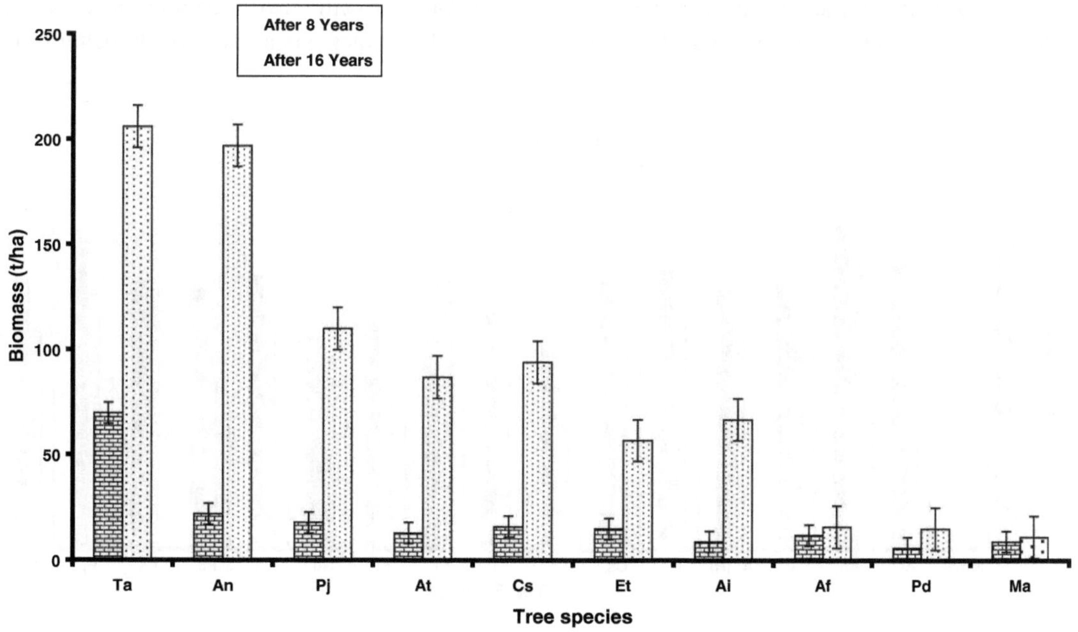

Fig. 9.16 Biomass of some selected individual trees of successful species after 8 and 16 years of growth (Af *Acacia farnesiana*, An *A. nilotica*, At *A. tortilis*, Ai *Azadirachta indica*, Cs *Cassia siamea*, Et *Eucalyptus tereticornis*, Ma *Melia azedarach*, Pd *Pithecelobium dulce*, Pj *Prosopis juliflora*, Ta *Tamarix articulata*)

Table 9.22 Performance of forest and fruit tree species with saline water irrigation (ECiw 10 dS m^{-1})

Performance rating	Forest tree species	Fruit tree species
Very promising	*Tamarix articulata, Azadirachta indica, Acacia nilotica, A. tortilis, A. farnesiana, Cassia siamea, Faidherbia albida, Eucalyptus tereticornis,, Prosopis juliflora, P. cineraria, Pithecellobium dulce, Salvadora persica, S. oleoides*	*Carissa carandas, Aegle marmelos, Cordia rothii, Phoenix dactylifera, Feronia limonia*
Promising	*Melia azedarach, Cassia fistula, Acacia auriculaeformis, Bauhinia variegata, Cassia glauca, C. javanica, Crescentia alata, Pongamia pinnata, Tecomella undulata*	*Ziziphus mauritiana, Vitis vinifera, Syzygium cuminii, Emblica officinalis, Psidium guajava, Punica granatum*

Source Tomar et al. (2003a, b)

was more with increase in age due to more contribution by litter and soil root decomposition (Fig. 9.17).

While studying water storage due to saline irrigation for establishment of trees in dry regions computations showed that out of the total water added (832 mm rain plus 324 mm irrigation) during 3 years, only 17–277 mm was the additional water stored in soils for future extraction by tree saplings when the supplemental irrigation was discontinued. The water storage at this stage ranged between 24.2–26.0 cm/3.0 m soil under some of the better extracting species like *Acacia nilotica, Azadirachta indica* and *Eucalyptus tereticornis* whereas it got improved to 48.3–51.0 cm under poor performing species like *Casuarina equisetifolia* and *Cassia javanica*. The water use efficiency in dry regions is very important parameter. Estimates (Tomar et al. 2003b) show that some of the tree species like *Tamarix articulata, Acacia nilotica*, and *Prosopis juliflora* are very efficient water users with their water use efficiency (WUE) being more than 35 kg cm^{-1} of water (Table 9.23).

The most of the vigorous tree species continued to extract higher amounts of water during their later growth also. Keeping in view the seasonal changes vis-à-vis growth patterns of

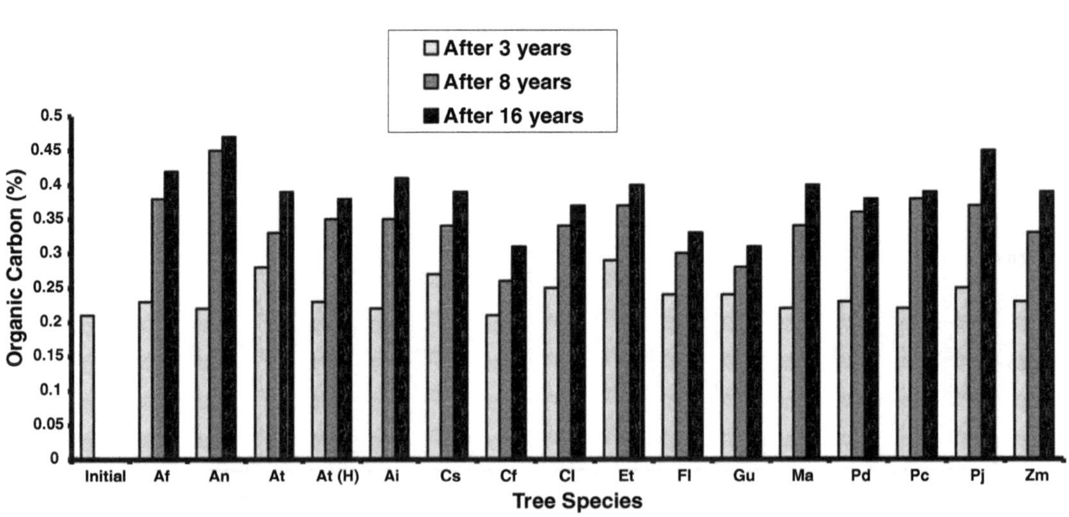

Fig. 9.17 Soil organic carbon status under different trees species at different intervals (Af-*Acacia farnesiana*, An *A. nilotica*, At *A. tortilis*, At(H) *A. tortilis* (hybrid), Ai *Azadirachta indica*, Cs *Cassia siamea*, Cf *C.fistula*, Cl *Callistemon lanceolata*, Et *Eucalyptus tereticornis*, Fl Feronia limonia, Gu *Guazuma ulmifolia*, Ma *Melia azedarach*, Pd *Pithecellobium dulce*, Pc *Prosopis cineraria*, Pj *P.juliflora*, Zm *Ziziphus mauritiana*]

Table 9.23 Water extract and water use efficiency (WUE) of different tree species (as calculated between April, 1994 to January, 1998)

Species	Mean water extraction (mm day^{-1})	WUE (kg cm^{-1})
Acacia farnesiana	1.79	20.5
A. nilotica	3.23	45.0
A. tortilis (hybrid)	3.41	25.7
A. tortilis	3.52	23.0
Azadirachta indica	3.47	13.7
Casuarina equiselifolia	3.34	0.3
Cassia javanica	3.51	6.6
C. siamea	3.33	21.1
Eucalyptus tereticornis	3.42	23.2
Guazuma ulmifolia	3.52	4.5
Melia azedarach	3.40	21.7
Moringa oleifera	3.39	0.9
Pithecellobium dulce	3.41	11.9
Prosopis cineraria	3.56	1.5
P. juliflora	3.46	37.6
Ziziphus mauritiana	3.45	5.3

Source Tomar et al. (2003b)

tree species affecting water extractions, the annual data were partitioned into post-monsoon and monsoon periods. Tree species like *Feronia limonia, Acacia farnesiana, Cassia siamea, Azadirachta indica, Prosopis juliflora, P. cineraria* and *Callistemon lanceolatus* were showing higher water extractions during monsoon season whereas species like *Cassia fistula, Guazuma ulmifolia, Acacia nilotica, Acacia tortilis* (hybrid), *Pithecellobium dulce*, and *Moringa oleifera* extracted more water during the post-monsoon periods. *Guazuma ulmifolia, Acacia tortilis, P. cineraria, P. juliflora, Feronia limonia*, and *Cassia fistula* showed an overall low water extraction but continued to extract soil moisture through out the year. In general, the soil water extraction by different tree species matched well with their growth behavior.

Cultivation of Arable Crops with Fruit Trees Using Underground Saline Water

Till recently, arable crops used to be preferred for cultivation using saline water for irrigation. In recent years as is evident from above-mentioned results that tree-based agricultural systems can successfully be developed through utilizing saline water for irrigation. Fruit trees such as *Ziziphus mauritiana, Carissa carandas, Feronia limonia, Emblica officinalis*, and *Aegle marmelos* could be established irrigating with saline water up to EC 10 dS m^{-1} and intercrops in wider spaces between rows (5 m) such as pearl millet, cluster bean, and barley could be cultivated with success (Fig. 9.18) applying one or two irrigations (Tables 9.24, 9.25). This is very much sustainable and economically viable agroforestry system for calcareous degraded soils. *Karonda* (*Carissa carandas*) started bearing fruits just after 2 years (75 % bushes started bearing) yielding about 1 t ha^{-1} fruits. During the third year, it could produce 1.2–1.8 t ha^{-1} fruits. Karonda–barley is quite sustainable agroforestry system with providing limited saline irrigation for dry region. Goose berry (*Emblica officinalis*) started bearing after fouth year but it was found to be sensitive to frost. There is an initial problem of establishment when planted from grafted seedlings. In-situ grafting may be more successful method of plantation. Bael (*Aegle marmelos*) also started bearing after fourth year and produced fruits up to 3 t ha^{-1} (Dagar et al. 2008).

Fig. 9.18 Performance of arable crops with fruit trees with saline irrigation (*Left*): Barley with Karonda (*Carisa carandas*) (*Right*): Clusterbean with goose berry (*Emblica officinalis*)

In arid regions, *Ber* (*Ziziphus mauritiana*) has performed very well when irrigated with saline water of EC_{iw} 8 dS m^{-1} using drip irrigation. The results in terms of fruit yield indicated that the fruit yield was maximum at 0.6 PE of best available water-BAW (64.6 kg per tree) followed by 0.6 PE of saline water of EC_{iw} 8 dS m^{-1} (56.7 kg per tree) and this yield was better than all other treatments including 0.8 and 0.4 PE of BAW (AICRP 2010–2012) showing its potential for dry region.

Silvopastoral Systems Established with Saline Irrigation

On calcareous sandy loam soils, silvopastoral systems involving fodder trees from the above-mentioned forest trees and several local and introduced grasses may be developed with great success which may sustain the live stock and livelihood of the resource poor and landless farmers of dry regions. In tree plantations when protected from grazing local grasses predominated by *Cenchrus ciliaris, C. setigerus* and *Dactyloclenium sindicum* covered the interspaces. These could produce 3–4 t ha^{-1} forage and if irrigated once with saline water during summer green forage could be available during dry period when there is scarcity of green forage. The results of one trial conducted by Tomar et al. (2003a) on perennial grasses irrigating with saline water (EC 10 dS m^{-1}) at different frequency showed that *Panicum laevifolium* and *P. maximum* were most suitable species producing annually 11–17 t ha^{-1} dry forage (Table 9.26). About 25–30 % of total forage was also available during lean period of summer when most of the people become nomadic along with their cattle. The water use efficiency was also highest in these two species. These grasses along with native *Cenchris setigerus* and *S. ciliaris* can successfully be grown with trees mentioned above and one irrigation with saline water which is always available during summer can produce reasonably good biomass for live stock in dry regions.

Medicinal and Arromatic Plants with Saline Irrigation

Salt tolerant and low water requiring medicinal and aromatic plants can provide viable alternative to effectively utilize the degraded calcareous lands. Tomar and Minhas (2002, 2004a, b); Tomar et al. (2005, 2010) and Dagar et al. (2008, 2009, 2013) in series of experiments evaluated the performance of aromatic and medicinal plants and winter annual flowers under saline irrigation (EC 8–10 dS m^{-1}) in isolation as well as in partial shade of trees. Among the species tested for medicinal value, the most promising was psyllium (*Plantago ovata*) with average seed yield of 1,050 kg ha^{-1} and did not show any adverse impact when compared with canal water irrigation. When different frequencies of irrigation were compared using water of low salinity (ECiw 4.0 dS m^{-1}), high salinity (EC 8.6 dS m^{-1}), and providing irrigation with waters of low and high salinity alternately, the average unhusked seed

Table 9.24 Mean grain and straw yield (t ha^{-1}) of intercrops (mean of 2 years) along with three fruit trees during initial establishment of trees

Fruit trees	Crops	T_1		T_2		T_3		T_4	
		Grain	Straw	Grain	Straw	Grain	Straw	Grain	Straw
Karonda	Pearl millet	2.45	11.2	2.51	11.28	2.35	11.25	2.18	10.65
	Cluster bean	0.85	2.63	0.91	2.96	0.79	2.51	0.7	2.46
Anola	Pearl millet	2.5	10.98	2.64	11.32	2.25	11	1.98	10.48
	Cluster bean	0.96	3.31	1.03	3.43	0.95	3.2	0.91	2.86
Bael	Pearl millet	2	10.45	2.25	11.05	1.84	10	1.72	9.85
	Cluster bean	0.92	3	0.97	3.31	0.9	2.95	0.86	2.74
LSD ($p = 0.05$)									
		Grain				Straw			
		Pearl millet	Cluster bean			Pearl millet	Cluster bean		
Between fruit trees (A)		0.134	0.073			NS	0.425		
Between treatments (B)		0.097	NS			NS	0.189		
Interactions (A) x (B)		NS	NS			NS	NS		

Treatments depict as: Planting methods T_1 Traditional pit method irrigated with water of low salinity (ECiw 5–6 dS m^{-1}), T_2 Planted in furrows and irrigated with water of low salinity, T_3 planted in furrows and irrigated alternatively with low and high salinity, T_4 planted in furrows and irrigated with water of high salinity (ECiw 10–12 dS m^{-1})

yield was found to be 1,102, 885, and 1,159 kg ha^{-1}, respectively, showing significant advantage when the crop was irrigated alternately with water of low and high salinity. There was increase in yield with increase of frequency of irrigation. Among eight varieties, the best performance was shown by variety JI-4 followed by Sel-10, Niharika, HI-5, GI-2, GI-1, local, and HI-34, in descending order (Tomar et al. 2005, 2010). Psyllium did not show any yield reduction with *Acacia nilotica* trees even at later stages showing its suitability for partial shade tolerance.

Lemon grass (*Cymbopogon flexuosus*) was also found promising crop with saline irrigation (Dagar et al. 2013). The average fresh foliage yield was found to be 12.0–13.0, 6.7–8.3, and 9.1–9.8 t ha^{-1}, respectively, when irrigated with water of low salinity (ECiw 4.0 dS m^{-1}), high salinity (ECiw 8.6 dS m^{-1}), and alternately with two waters. There was increase in yield with increase of frequency of irrigation particularly during second year. Furrow planting was a superior planting method than other methods including the flat planting method. Among the cultivars tested, RRL-16 and OD-58 showed better performance followed by Praman and Krishna (Dagar et al. 2013). The overall results indicated the possibilities of raising Lemongrass on degraded calcareous soil using saline water up to EC 8.6 dS m^{-1} (Tomar and Minhas 2004a; Dagar et al. 2008, 2013) without build up of soil salinity if normal rainfall occurs once in 3–4 years. There was no impact on quality of oil due to salinity. The aromatic grasses such as vetiver, lemon grass and palmarosa, when irrigated with saline water (EC 8.5 dS m^{-1}) could produce on an average 90.9, 10.4, and 24.3 t ha^{-1} dry biomass, respectively (Tomar and

Table 9.25 Grain and straw yield (t ha^{-1}) of cluster bean and barley with three plantations after three years

Fruit tree	Treatment	*Yield of* cluster bean		Yield of barley	
		Grain	Straw	Grain	Straw
Karonda	T_1	0.88	1.46	3.58	3.88
	T_2	0.86	1.38	3.47	3.97
	T_3	0.81	1.27	3.45	3.71
	T_4	0.76	1.15	3.10	3.32
Anwla	T_1	0.79	1.29	4.19	3.40
	T_2	0.81	1.33	3.63	3.83
	T_3	0.76	1.24	3.24	3.34
	T_4	0.69	1.18	2.87	3.00
Bael	T_1	0.75	1.23	3.27	3.45
	T_2	0.71	1.21	3.22	3.35
	T_3	0.67	1.06	2.73	2.86
	T_4	0.63	1.02	2.52	2.64
LSD (p = 0.05)					
Factor A (species)		0.13	NS	0.12	0.17
Factor B (treatment)		0.02	0.11	0.14	0.15
Interaction (A x B)		NS	NS	0.24	0.26

Treatments depict as: Planting methods T_1 Traditional pit method irrigated with water of low salinity (ECiw 5–6 dS m^{-1}), T_2 Planted in furrows and irrigated with water of low salinity, T_3 planted in furrows and irrigated alternatively with low and high salinity, T_4 planted in furrows and irrigated with water of high salinity (ECiw 10–12 dS m^{-1})

Table 9.26 Gross dry-matter yield (Average of 3 years) and water use efficiency of different grasses when cultivated with saline water

Species	Yield (t ha^{-1}) at Diw/CPE			Water use efficiency (kg ha^{-1} cm)		
	0.2	0.4	0.8	0.2	0.4	0.8
Brachiaria mutica	9.54	12.15	11.72	26.7	12.15	11.72
Cenchrus setigerus	4.64	4.57	4.38	12.9	4.57	4.38
Cynodon dactylon	8.91	9.23	10.20	25.0	9.23	10.20
Panicum antidotale	9.34	11.41	11.77	26.2	11.41	11.70
P. coloratum	6.95	10.29	8.93	19.5	10.29	8.93
P. laevifolium	13.49	16.85	16.88	37.8	16.85	16.88
P. maximum (C)	10.87	13.04	12.72	30.5	13.04	12.72
P.maximum (wild)	14.00	14.72	13.72	39.3	14.72	13.72
P. virgatum	9.95	12.10	11.36	27.9	12.10	11.36
Mean	9.74	11.60	11.30	27.3	11.60	11.30

Diw/CPE = irrigation water/cumulative pan evaporation
Source Tomar et al. (2003a)

Minhas 2004a). Different cultivars of vetiver could produce 72.6–78.7 t ha^{-1} shoot biomass and 1.12–1.71 t ha^{-1} root biomass. The roots are used to extract aromatic oil.

Aloe barbadensis was also equally tolerant and could produce 18 t ha^{-1} fresh leaves under partial shade. *Ocimum sanctum* could produce 910 kg ha^{-1} dry shoot biomass. In a separate trial, dill (*Anethum gravelens*), taramira (*Eruca sativa*), and castor (*Ricinus communis*) could produce 931, 965, and 3,535 kg seeds per ha, respectively, when provided with three irrigation of saline water of EC 10 dS m^{-1} (Dagar et al. 2008). *Cassia senna* and *Lepidium sativum* could also be cultivated successfully irrigating with saline water up to 8 dS m^{-1}. Ornamental and medicinal flowers such as *Chrysanthemum, Calendula,* and *Matricaria* were also found successful when cultivated irrigating with water of EC up to 5 dS m^{-1}. These species could yield 13.2, 4.7, and 3.5 t ha^{-1}, respectively, fresh flowers in a season when irrigated with good quality water available at site. A few irrigations with good quality water for establishment will increase the yield of flowers significantly. All these high value crops could successfully be grown as intercrops with forest and fruit trees at least during the initial years of establishment (Dagar 2009; Dagar et al. 2006a, b, 2008, 2009, 2013).

Afforestation of Coastal Saline Areas

Agroforestry systems including afforestation, biodrainage plantations on saline and waterlogged situations have been dealt in Chap. 7.

Other Management Practices

As stated earlier, saline and sodic soils are found in arid, semi-arid and hot and dry subhumid regions, where about 70–80 % of the total rainfall is usually received during the months of July–September (particularly in Indian subcontinent). Usually the rains are erratic and periods following them are dry. Though the transplanting of tree saplings is mostly completed during rainy season, the saplings may suffer for want of water during the post-monsoon period especially the summers. Young saplings of many species except a few may also suffer from frost injuries during winters in northern regions of India. Therefore, provision of supplemental irrigation for early establishment and growth and avoiding frost injuries of saplings is very crucial. Like crops, trees also need irrigations to meet their evapotranspiration demands. The experiments on irrigation requirements (Gupta et al. 1995, Tomar et al. 2003b, Singh and Dagar 2005) clearly indicated that plantations need regular irrigation after planting the saplings especially during summer (2 irrigations) and winter (1 irrigation) at least for the initial 3 years. Higher plant population may necessitate further extending the support period.

Irrigation practices for trees on saline and alkali soils should further ensure that there is no waterlogging and aeration problem, optimize the availability of moisture in the root zone, and minimize the secondary salinization of the root zone. Therefore, sub-furrow planting and furrow irrigation method has proved ideal. It may be pointed that when water table is within 50 cm during rainy season or within 2 m during later part of the year, irrigation may be avoided. This is because capillary rise of water toward surface will suffice irrigation needs of saplings. Irrigation under situations of high water table may preferably be applied for leaching the salts, when accumulated in excessive amounts.

Afforestation programs where good quality water is not sufficient can only succeed if the poor quality groundwater aquifers are developed and utilized judiciously for irrigating the saplings. Using furrow technique of irrigation, saline waters of ECiw up to 12 dS m^{-1} in sandy loam and loam soils can successfully be utilized. In finer-texture soils (clay loam), saline waters of ECiw up to 6–8 dS m^{-1} can be used with some reduction in tree growth. Use of water having residual alkalinity renders the soils alkali. The severity of problem depends mainly upon the quality and quantity of water used. Since smaller irrigation quantities suffice for the growth of tree saplings using furrow method of irrigation, alkali waters are unlikely to cause any serious problem unless residual sodium carbonate (RSC) is excessively high (>10 meq l^{-1}). Under such situations, a gypsum bag may be kept in irrigation channel to neutralize RSC in the water.

To give proper shape to the plants and to get clear boll, annual pruning preferably during winter is required. The trees like *Acacia nilotica* and *Prosopis juliflora*, which have more number of branches require regular annual pruning. The lower one-third part must be pruned but not more than that. The trees like *Eucalyptus* do not require intensive pruning. The pruned biomass may be used as fuelwood. Lopping of selected *Acacia* trees at 16 and 42 months of growth gave 2.5 and 7.2 t ha^{-1} fuelwood, respectively. The tree plantation program should also include measures for protection against wild or stray animals. The protection may be with barbed wire fencing or live fencing of spiny bushes or unpalatable hedges. Environmentally safe animal repellents like "Ro-Pel" are now available, which may be sprayed on saplings.

Environmental and Socioeconomic Benefits from Agroforestry

With ever-increasing human and cattle population, we require increased production from the nonproductive salt-affected lands which need to be used for producing food, forages, timber, fuelwood, oil seeds, medicine, and other minor products. In the scenario of scarcity of good quality water for irrigation, the use of poor quality waters, particularly in the arid and semi-arid regions is unavoidable. The judicious use of saline waters for conventional and nonconventional salt-tolerant crops through sustainable agroforestry systems is a viable option. Many halophytes provide high biomass and high protein or mineral levels with outstanding ability to a wide range of environmental stresses. Aronson (1989) identified more than 1,600 salt-tolerant plants of coastal and inland desert environments to have an economic potential. More than 1,000 plant species of economic importance are found on Indian saline environment (Dagar and Singh 2007). Many of these species find utilization as food, firewood, timber, forage, medicine, and minor products. When grown on salty lands, many species act as bioameliorative and soil acts as sink for carbon sequestration. In agroforestry systems, these not only provide food security but also create employment and improve microclimate and general environment and help in biodiversity conservation.

Le Houe'rou (1986), Sankary (1986), Aronson (1989), NAS (1990), Dagar et al. (1991, 1993), Lieth and Al Masoom (1993), Jaradat (2003), Dagar (1995, 2003, 2005, 2006), Al-sharhan et al. (2003) and Dagar and Singh (2007) have generated valuable information regarding the potential economic value of salt-tolerant plants as food, forage, edible oil, fuelwood, liquid fuel, pulp and fiber, bioactive derivatives, medicine, landscape, and ornamental. With the increasing world population and the need for increased crop production, the nonproductive salt-affected lands and saline water (for irrigation) have to be used for producing nonconventional crops of economic value. Many halophytes are already being used for a variety of purposes by traditional societies as well as through scientific technologies by different stakeholders for high economic yield. Some halophytes have agricultural or industrial potential and their cultivation is also considered as a biological method for soil desalination and reclamation. A brief account of potential salt-tolerant plants particularly as agroforestry crops is given here.

Contribution of Agroforestry Crops to Food, Fodder and Fuelwood Production

Food

Biosaline agroforestry provides food in many ways, especially in areas where traditional agriculture cannot be profitably practiced. Appropriate halophytes can be domesticated (NAS 1990) as agroforestry crops and their seeds, fruits, roots, tubers, or foliage can be used directly or indirectly as food. Dagar and Dagar (1999) and Dagar and Singh (1999) listed several food plants in wild being consumed by the aborigines of the Bay Islands any many of these such as fruits of *Pandanus*, radicles of mangroves, tubers of Dioscoreas, nuts of coconut are consumed as stable food. There is ample evidence (O'Leary 1993, Dagar 2003, Dagar and Singh 2007) that little-known seed-bearing halophytes are potential candidate for germplasm resources for the development of salt-tolerant agroforestry crops in the Middle East and Central Asia. It has been reported (Aronson 1989) that almost 50 species of seed-bearing halophytes are potential source of grains and oil. Pearl millet (*Pennisetum typhoides*) grows well on sand dunes and tolerates EC of 27–37 dS m^{-1} for irrigation (NAS 1990) and yields 1.6 and 6.5 t ha^{-1} grains and fodder, respectively, and can be grown successfully as intercrop with many fruit trees of dry region (Dagar et al. 2008).

Many species of *Acacia* produce seed that are rich in nutrients, with higher energy, protein and fat contents. For example, *A. aneura* is potential oil crop (37 % fat), whereas *A. dictyphleba* is considered potential grain crop with 26.8 % protein content (Turnbull 1986, Goodchild and McMeniman 1987). Wild water chestnut (*Eleocharis dulcis*) produces tubers that can be cooked or pounded to meal (Glenn and O' Leary 1985). Similarly, the roots and stems of saltwort (*Batis maritima*) can be used for food as green leaves and the plant produces up to 17 t ha^{-1} of dry biomass using seawater for irrigation (NAS 1990).

Dagar et al. (1993) and Dagar and Dagar (1991, 1999) have listed many wild coastal plants of tidal zone used in one or the other form by the aborigines of Andaman-Nicobar Islands as food items. Among conventional crops, beet root (*Beta vulgaris*) and date palm (*Phoenix dactylifera*) are well known for their food value and these can be grown successfully irrigating with saline water and salt-tolerant rice varieties (e.g., CSR10, CSR 27, CSR 30, CSR 36) can be cultivated on sodic soil of high pH (up to 10). Fruit bearing gooseberry (*Emblica officinalis*), Karonda (*Carissa carandas*), Ber (*Ziziphus mauritiana*), and Bael (*Aegle marmelos*) withstand drought as well as salinity. These can be cultivated with success irrigating with water up to 12 dS m^{-1}. These along with guava (*Psidium guajava*) and *Syzygium cuminii* could be grown on highly alkali soil (pH up to 9.8) with application of amendments (gypsum) in auger holes. *Ziziphus nummularia, Lycium* spp., *Santalum acuminatum,* and *Coccoloba uvifera* are other potential fruit genetic resources.

Salicornia bigelovii and *S. herbacea* have been evaluated as a source of vegetable oil and the cake as animal feed (Riley and Abdal 1993, Dagar 2003). Samphire (*Salicornia bigelovii*) yields about 28.2 % seed oil, 31.2 % protein, 5.3 % fiber, and 5.5 % ash from seeds. Its oil resembles to Saffola oil and also considered for cosmetic and pharmaceutical industries. Straw and cake are used as forage and considered suitable for paper pulp. It can be irrigated with water of sea salinity and has been successfully grown in some coastal areas of Gujarat. Under seawater irrigation, *Salicornia* has been reported to produce 20 t ha^{-1} plant biomass, out of which, 2 t ha^{-1} as oilseed (NAS 1990). It withstands high salinity both of soil and water. Raw fruits of Kair (*Capparis decidua*) tree are used for pickles and possess medicinal value. It grows naturally on both saline and sodic soils and can be domesticated using saline water after preparing nursery from its rootstocks, seeds, and also stem cuttings and then transplanting. The coastal badam (*Terminalia catappa*), salt bush *Salvadora persica* and species of *Pandanus* are known for their oils of industrial application. Fruits of *Pandanus* are staple food for coastal population of Bay Islands and many species are found natural growing in tidal zone which can be improved and domesticated successfully in coastal areas. Palmirah palm (*Borassus flabellifer*), widely used for toddy, jaggery, vinegar, beverage, juice for sugar, and edible radicles and fruits is found widely distributed all along Andhra coast. It needs to be genetically improved for wider cultivation. The use of *Suaeda maritima* in *papar* industry in Rajasthan & Gujarat is well known. The young leaves and shoots of *Chenopodium album*, species of *Amaranthus*, *Portulaca oleracea*, *Sesuvium portulacastrum* and many others are used as vegetable and salad in many parts of India and other countries. Many of these are even cultivated as agroforestry crops. Many seed spices such as coriander, fenugreek, dill, celery, cumin etc. are moderately tolerant to saline irrigation (up to EC 10 dS m^{-1}) and are cultivated as agroforestry crops in dry regions.

Forage

Halophytes have been used as forage in arid and semi-arid areas for millennia. The value of certain salt-tolerant shrubs and grasses has been recognized by their incorporation in pasture-improvement programmes in many salt-affected regions throughout the world. There have been recent advances in selecting species with high biomass and protein levels in combination with their ability to survive a wide range of environmental conditions, including salinity (NAS 1990; Ulery et al. 1998; Barrett-Lennard 2003; Jaradat 2003; Dagar 2003, 2005a, b, 2006a, b, 2009). Halophytes suitable for forage production either as native stands or established as agroforestry crops include the herbaceous species: *Elymus elongatus, Leptochloa fusca, Aeluropus lagopoides, Hedysarum carnosum, Cynodon dactylon* vars *hirsutissimum* and *villosum, Puccinellia ciliata*, species of *Panicum, Paspalum, Sporobolus, Chloris, Brachiaria, Eragrostis*, etc.; the forage shrubs: *Haloxylon persicum, H. aphyllum, H. salicornicum, Kochia indica, Maireana brievifolia* and many Atriplex species (*amnicola, atacamensis, barclayana, canescens, cinerea, glauca, halomus, isatidea,lentiformis, nummularia, paludosa, polycarpa, semibaccata, undulata*); and plenty of tree genera such as *Acacia, Prosopis, Ficus, Balanites, Ailanthus, Ziziphus, Cordia, Salvadora, Capparis, Azadirachta, Feronia, Albizia, Cassia*, etc. are traditional fodder resources.

In many coastal areas where mangroves occur sporadically and there is scarcity of fodder, the foliage of many mangroves (*Avicennia, Ceriops, Bruguiera, Rhizophora, Kandelia*, etc.) and their associated species such as species of *Terminalia, Pongamia, Salvadora, Ficus* and many others, are used as forage for cattle, goats, and camel. Among grasses *Distichlis spicata, Paspalum vaginatum, Sporobolus virginicus, Pennisetum clandestinum*, and *Chloris gayana* have tolerance to waterlogging and low to moderate salinity. Mixed stands of these and other halophytic grasses may be managed for production of silage.

Grazing lands of arid and semi-arid regions are poor in yield. One site in northwestern India with average yield of 0.85 t ha^{-1} when protected from grazing could produce 2.4 t ha^{-1} forage and the same site when brought under cultivated grasses and irrigated with saline water (ECiw 8–10 dS m^{-1}) could produce up to 22 t ha^{-1} dry forage. Even in the lean period when people are forced to lead nomadic life along with their herds of cattle, sufficient forage (30 % of total) was available from these perennial grasses. Species of *Salicornia, Chenopodium, Kochia, Atriplex, Salsola, Suaeda, Trianthema, Amaranthus, Portulaca, Tribulus*, and *Alhagi* along with several grasses such as *Leptochloa fusca, Aeluropus lagopoides, Cynodon dactylon, Dactyloctenium sindicum, Paspalum vaginatum, Sporobolus airoides, S. marginatus, Chloris gayana, Echinochloa crus-gali, E. colonum, Eragrostis tenella, Dichanthium annulatum, Brachiaria mutica, Bothriochloa pertusa* and many others are commonly used as forages from alkali and saline grazing areas (Dagar 2006). As described earlier, many of these forages can be cultivated successfully on degraded salt-affected soils or in drought prone areas irrigating with saline water, where other arable crops cannot be grown.

Kallar grass (*Leptochloa fusca*) has gained attention in India and Pakistan as fodder grass for saline and highly sodic soils (pH as high as 10.4) producing more than 45 t ha^{-1} green forage. Biomass production by several grasses has already been discussed in this Chapter. Species of *Sporobolus, Brachiaria, Panicum, Dactyloctenium*, and *Eragrostis* are tolerant to high salinity. *Aeluropus lagopoides* is a dominant grass in entire Rann of Kutchh and is used for grazing. The grazing lands of sodic soils are poor in forage production under open grazing but when brought under judicious management these can be explored successfully for sustainable forage production.

Production of Fuelwood

Often fuelwood is obtained from salt-tolerant trees and shrubs, which include species of *Prosopis, Acacia, Tamarix, Casuarina, Eucalyptus, Parkinsonia, Capparis, Sesbania*, and *Salvadora*. In coastal areas, the mangroves and their associates are widely used for fuelwood and timber leading to deforestation of these habitats. Species of *Rhizophora, Bruguiera, Ceriops, Kandelia, Avicennia, Sonneratia, Xylocarpus, Heritiera*, and *Excoecaria* are excellent fuelwood and are also used for making charcoal. *Avicennia marina* is one of the most widely distributed species throughout the mangal stands of the world and tolerates biotic stress. Many species have been planted successfully in coastal regions of Goa and Tamil Nadu. *Casuarina equisetifolia* has been widely planted along Orissa coast. *Beta vulgaris* and *Nypa fruticans* have been identified as potential source of liquid fuels while species such as *Jatropha curcas, Pongamia pinnata*, and *Euphorbia antisyphilitica* are among potential diesel–fuel plants and these can be grown successfully on degraded lands with saline irrigation. The energy yield (in the form of biogas) from *Leptochloa fusca* has been estimated at 15×10^6 kcal ha^{-1} (Jaradat 2003). At present, every year about 1.2 million tons of tree-borne nonedible seed oil is produced in the country (GoI 2008) and there is huge potential to increase this figure by planting these perennial plants on degraded including salt-affected lands.

Among fuelwood trees and bushes, *Acacia (ampliceps, linarioides, tortilis, nilotica, senegal, africana, auriculaeformis, saligna, ligulata), Azadirachta indica, Casuarina equisetifolia, C. glauca, Cajanus cajan, Commiphora riparis, Cassia siamea, Eucalyptus camaldulensis, E. tereticornis, Haloxylon aphyllum, Parkinsonia aculeata, Prosopis (alba, juliflora, tamarugo, chilensis, pallida, stephanian), Tamarix articulata, Salvadora persica* and *Ziziphus* are important.

Aromatic, Medicinal, and Other Uses

The global demand for herbal products is not only large, but also growing. The largest markets for medicinal and aromatic plants besides India are China, France, Germany, Italy, Japan, Spain, UK, and USA. Due to ever-increasing population, we have an optimum pressure on arable lands for cultivation of foodcrops; therefore, utilization of degraded wastelands including salt-affected soils is a viable option for cultivation of medicinal & aromatic plants.

There are many reports on the medicinal uses of halophytes (CSIR 1986; Dagar 1995, 2003; Dagar and Dagar 1991, 1999; Dagar and Singh 1999; Dagar et al. 1991, 1993, 2004a, b, 2005, 2006a, b; Dagar and Singh 2007). Among mangroves and their associate species, leaves of *Acrostichum aureum* are applied on wounds and boils; *Acanthus ilicifolius* and *A. volubilis* are used in rheumatism, neuralgia, paralysis, asthma, as blood purifier and dressing boils and wounds; *Barringtonia acutangula* for diarrhea, toothache; *B. racemosa* in cough, asthma, diarrhea, jaundice; seed oil of *Calophyllum inophyllum* in rheumatism, skin diseases and leprosy; *Cerbera manghas* in rheumatism; *Cynometra ramiflora* in leprosy, scabies and cutaneous diseases; *Heritiera fomes* to cure piles; *H. littoralis* in diarrhea and dysentery; *Xylocarpus granatum* and *X. moluccensis* in dysentery and breast tumors; *Terminalia catappa* in cutaneous diseases; *Sonneratia caseolaris* as vermifuge; *Thespesia populnea* in stomach trouble; and *Suaeda fruticosa* is used as emetic and to cure sores on camel back (Dagar and Dagar 1991, 1999; Dagar and Singh 1999, 2007).

Among other important salt-tolerant plants found as wild or cultivated in agroforestry systems, *Achyranthes aspera* is used in asthma, renal dropsis; *Adhatoda vasica* in asthmatic problems; *Aloe barbadensis* in piles, rheumatism, boils, and stomach problems; *Azadirachta indica* in skin infections, eczema, and ulcers; *Balanites roxburghii* in whooping cough, skin troubles, source of diosgenin for synthesis of several steroidal drugs; *Calotropis procera* in skin diseases, tumors, piles, and as abortifacient and anticoagulant; *Capparis decidua* to cure cough, asthma, inflammations, cardiac troubles and biliousness; alkaloids of *Catharanthus roseus* are used for leukemia and blood pressure; *Citrullus colocynthis* in jaundice, rheumatism, and urinary diseases; *Clerodendrum inerme* in malarial fever as substitute to quinine; *Cressa cretica* as tonic, aphordisiac, and stomachic; *Jatropha curcas* in diarrhea, toothache, piles, rheumatism and skin diseases; *Kochia indica* as cardial stimulant; *Pandanus odoratissimus* in leprosy, scabies, diseases of heart, and oil as antispasmodic; *Pongamia pinnata* in diarrhea, cough, leprosy, gonorrhea, and rheumatic pains; *Ricinus communis* in boils, sores, and lumbago; *Salsola baryosma* posseses anthelmintic, emmenagogue, diuretic properties; ash for itch; *Salvadora persica* and *S. oleoides* in cough, rheumatism, suppositories, toothache, and piles; *Solanum surattense* in cough, asthma, sore throat, and rheumatism; *Tamarix articulata* and *T. troupii* in eczema, ulcers, piles, sore throat, diarrhea, liver disorders; *Trianthema portulacastrum* in asthma, amenorrhea, dropsy, rheumatism, liver problems, and as abortifacient; *Tribulus terrestris* as tonic, diuretic, and in painful micturition and calculous affections; grass *Vetiveria zizanioides* in rheumatism, fever, headache, toothache, and as tonic; *Withania somnifera* in asthma, cough; *Ziziphus nummularia* in skin diseases, cold, cough, biliousness; and the herb *Zygophyllum simplex* has cardiac properties and applied in eye diseases (CSIR 1986; Dagar 2003; Dagar and Singh 2007).

Among potential agroforestry salt-tolerant high value crops of medicinal importance include Isabgol (*Plantago ovata*), Periwinkle (*Catharanthus roseus*), Aloe (*Aloe barbadensis*), celery (*Apium graveolens*), chrysanthemum (*C. indicum*), calendula (*C. officinalis*), *Azadirachta indica, Cordia rothii, Salvadora persica, Ricinus communis, Jatropha curcas, J. gossypifolia,* and *Adhatoda vasica*.

The important plants yielding aromatic oil include vetiver (*Vetiveria zizanioides*), palmarosa (*Cymbopogon martini*), and lemon grass (*C. flexuosus*). German chamomile (*Matricaria chamomilla*) is a medicinal and flower-yielding salt-loving herbaceous crop of winter season.

Seed of *Salvadora oleoides* and *S. persica* contain 40–50 % fat and is good source of lauric acid. Purified fat is used for soap and candle making. Among other products of economic value such as medicinal oil from *Terminalia catappa, Calophyllum inophyllum;* perfume from male flowers of *Pandanus spp;* essential oil from *Mentha piperita, M. arvensis* (entire plants), *Anethum graveolens* (fruit), and species of *Ocimum* (entire plant); beverages from mangrove palm *Nypa fruticans*; aromatic resin from *Grindelia camporum, humilis, stricta, latifolia, integrifolia*, and *Larrea tridentata*; aromatic oil similar to sperm whale oil from *Simmondsia chinensis*; rubber from *Chrysothamnus nauseosus* and *Parthenium argentatum*; pulp and fiber from *Phragmites australis, P. karka, Juncus rigidus, J. acutus, Pandanus tectorius, Typha domingensis, T. australis, Hibiscus cannabinus, H. tiliaceous, Urochondra setulosa* and many others; bioactive compounds from *Calophyllum inophyllum, Balanites roxburghii, Salsola baryosma, Catharanthus roseus*, and liquid fuel from *Jatropha curcas, J. gossypifolia, Pongamia pinnata, Ricinus communis* and *Euphorbia antisyphilitica* are worth mentioning.

Many salt-tolerant species can also be used as landscape plants, especially in areas where fresh water is not available for irrigation. These may include trees, shrubs, succulents, ground covers, and lawn grasses. These observations show that there lie promising avenues of research and development on halophytes leading to biosaline agroforestry.

Bioamelioration and Carbon Sequestration

The biomass production and bioamelioration properties of several trees and grasses in system mode have already been discussed above in this Chapter. Reclaiming waterlogged, salt-affected lands which are low in organic carbon, through fast-growing plantations is a useful strategy for carbon sequestration. Increase in soil carbon through plantations may also act as an important carbon sink. The results discussed in biodrainage section show that by raising strip plantations of *Eucalyptus* (4 % of total agricultural area) in waterlogged areas of Haryana, 1.33 million tons additional C would be sequestered per annum. Therefore, apart from lowering the groundwater table, *Eucalyptus* plantation provides additional benefits in terms of carbon sequestration. Available estimates suggest that land use may mitigate additionally from 1 to 2 Gt C per year. Restoration of 1×10^9 ha of soil in the tropics has a potential to sequester C at the rate of 1.5 Pg year^{-1} (Lal and Kimble 2000). Achieving this full carbon mitigation potential will require that we use all land use-related options for carbon sequestration. The present stock of carbon in Indian soils is estimated to 63.19 Pg (Velayutham et al. 2000), which is just 4.2 % of the world and the C carrying capacity of Indian soils is estimated to 85.04 Pg (Dagar and Swarup 2003), therefore, there is a scope of additional C sequestration of 21.85 Pg. Earlier, Gupta and Rao (1994) advocated that grasses and trees when planted on the degraded lands have a potential of sequestering 1.9 Pg in 7 years as against emission of 2.27 Pg during the same period thus may help in slowing down the warming. The estimation proposed by them is toward conservative side. Analyzing the productive potential critically this figure may go higher.

Agroforestry has been recognized as having high potential for sequestering carbon and mitigating climate change (IPCC 2007; Lal 2011; Dagar et al. 2012). On an average, carbon storage by agroforestry land use system has been estimated to be around 9, 21, 50 and 63 t C ha^{-1} in semi-arid, sub-humid, humid, and temperate regions, respectively (Schroeder 1994). In India, agroforestry potential (mainly on degraded lands) carbon sequestration (on average basis) has been estimated to be 25 t ha^{-1} (Sathey and Ravindranath 1998). Singh and Lal (2000) have estimated that by bringing 40 Mha degraded land under forest with 5 t C ha^{-1}, the carbon mitigation of 3.32 Gt would be possible in next 50 years with annual reduction of about 0.072 Gt of carbon. Thus, optimization of biomass production through multispecies stands in agroforestry or sole forestry on degraded lands should be the priority for sustaining the system and continuously improving the environment. Wherever there is a scarcity of good quality water for establishment of tree plantation or silvopastoral system or growing medicinal plants, underground saline water can successfully be used for irrigation.

Control of Salinity and Waterlogging Through Vegetation

Tree planting should be considered in conjunction with other modified farming practices that minimize groundwater recharge, such as changing from bare ground fallowing to minimum tillage cultivation practices and the use of deeper rooted crops/trees or pasture species. Recharge areas are zones of high water infiltration. Trees have two major beneficial effects: (i) interception and evaporation of rainfall and (ii) transpiration of soil water. In waterlogged saline areas, salt-tolerant grasses such as *Leptochloa fusca, Brachiaria mutica, Thinopyrum ponticum,* and *Puccinellia ciliata* can be established. Once established, less-tolerant trees, shrubs, and grass species can be established as the water table falls and salt is leached as a result of recharge control. In some locations, high density tree planting on recharge areas could be in the form of commercial plantations. As discussed earlier under biodrainage that trees play very important role in controlling the waterlogging caused by seepage in canal command areas. Trees such as *Acacia ampleceps, Eucalyptus tereticornis, E. camaldulensis, Casuarina glauca, C. equisetifolia, Terminalia arjuna, Pongamia pinnata,* and *Syzygium cuminii* and grasses such as *Leptochloa fusca, Phragmites australis, Dichanthium annulatum, D. caricosum, Brachiaria mutica,* and *Paspalum* spp. have credibility for lowering down water table in waterlogged areas.

We can safely conclude that biodrainage can effectively contribute strongly in reducing the problems experienced from waterlogging in irrigated and nonirrigated agriculture. At least it is a safe remedy to control seepage all along canals. The problem of salinization associated with rise in water table can effectively be delayed using strips of plantations in semi-arid and arid regions. Undoubtedly, the added advantages of biodrainage and biodisposal systems, depending on the vegetation used, include production of timber, fuelwood, other products such as oil, honey, fruits, fiber etc., contribution to carbon sequestration, diminishing the effects of wind erosion, provision of shade and shelter, function as wind breaks, soil amelioration through litter fall, enhancement of biodiversity, and improvement in general environment.

Conservation of Biodiversity

Biodiversity consists of a hierarchy of definitions from the molecular level through taxa to the landscape level. The United Nations Convention on Biological Diversity (CBD) defines biodiversity as "the variability among living organisms from all sources including inter alia, terrestrial, marine and other aquatic ecosystems and ecological complexes of which they are a part; this included diversity with in species, between species and of ecosystems" (CBD 1992). The issues of biological and genetic diversity management in agroforestry are complex.

Atta-Krah et al. (2004) have assessed the diversity within and between tree species in traditional agroforestry systems. Their assessment shows that although the practice of agroforestry has been a diversity management and conservation system, research in agroforestry over time has emphasized the diversity element; nevertheless, farmers do value diversity and do manage agroforestry from that prospective. As discussed in detail (Chap. 7), homegardens are rich in biodiversity both of higher plants as well as microorganisms. Based on a profiling of various traditional agroforestry systems and research development technologies, a strong case is made for increased species and genetic diversity, at both inter- and intraspecific levels. It has been well recognized that agroforestry can serve to bridge the conflict that often exists between the need for conservation of biodiversity and provision of needs of human society. More attention needs to be given to the need to conserve biodiversity, through promotion of sustainable management and use of our genetic resources.

While dealing with biodiversity of saline habitats, Dagar (2000, 2003) mentioned that the inland highly salty lands are either devoid of any vegetation or support very meager cover and only some restricted species are found in highly inland saline regions. Contrary to inland saline habitats, the coastal tidal areas are rich in biodiversity. The mangrove ecosystems support several interesting animals such as saltwater crocodiles, turtles, water monitor lizards, snakes, wild pigs, monkeys, deer, even tigers (in Sunderbans), several indigenous and migratory birds, skippers, crabs, mollusks, insects, crustaceous, and microflora and fauna species. These are nursery and breeding ground and source of food for variety of animals and microorganisms. The aerial roots, pneumatophores and branches serve as shelter and food for large number of organisms including many beautiful orchids, ferns, lichens, mosses, jungermanniales, algae, fungi, bacteria, ciliates, nematodes, and amphipods which colonize and take part in complex food chain of this unique ecosystem. But unfortunately, all these habitats are now under tremendous anthropogenic pressure. Many species of this ecosystem are at the verge of extinction. As a result of the unsustainable exploitation of these habitats, many species are dwindling at a fast rate and the entire biodiversity is at stake. Certainly, the afforestation of these habitats will improve the biodiversity.

Afforestation of inland salty soils not only supports a variety of wild life such as hare, deer, blue bull, and birds but also improves the microbial status of soil. In a recent study it was found that in a silvopastoral system microbial biomass carbon increased tremendously. There was a significant relationship between microbial biomass carbon and plant biomass carbon ($r = 0.83$) as well as the flux of carbon ($r = 0.92$) in net primary productivity (Kaur et al. 2002b).

While reviewing biological amelioration of salt-affected soils, Rao (1998) concluded that due to litterfall and root decomposition, the microbial activities in soil improved a lot. Biological reclamation of salty lands is based on increasing soil carbon levels by addition of organic materials. By growing tolerant tree and grass/crop species, we also increase microbial activities. Growing of nitrogen-fixing trees and green manuring crops, we not only ameliorate soil but also improve the microbial biodiversity. Various species of *Rhizobium, Azotobacter*, and vesicular carbuncular mycorrhiza (VAM) fungi are found associated with nitrogen-fixing trees. In salty soils, organic matter application/addition or inoculation of crops and trees with tolerant *Rhizobium* strains or VAM helps plants in tolerating stress through their nutrition. Rao and Ghai (1995), Rao and Pathak (1996) and Rao (1998) reported several evidences where cultivation of nitrogen-fixing trees/crops helped in amelioration of salt-affected soils and microbial diversity conservation.

Improvement in Microclimate and General Environment

No doubt, the agroforestry systems ameliorate the salty lands by addition of organic matter in the form of litter and decomposition of fine roots in the soil. The water infiltration rates of soil also improve to a greater extent. The mean air temperature under canopy is lowered during summer and increased during winter. In a study, Gill et al. (1990) and Gill and Abrol (1993) found that under *Acacia nilotica* canopy, the mean air temperature was lowered by 2–5 °C during summer and increased by 2–4 °C during winter (Fig. 9.19). Similar results were found under *Eucalyptus* plantation but the magnitude of the modifying effect was less because of less spread of canopy. The influence of the two canopies on the thermal regime of the soil at 5 cm and 15 cm depth was also of the modifying type. Fluctuations were more marked for soil temperature recorded at 5 cm depth than at 20 cm depth in open. Effects of the plantations during summer were of blanketing type and of sheltering type during winter. Modification in thermal characteristics due to tree plantations, thus, owes to their sheltering effect by cutting insulation and by their blanketing effect in restricting outflow of heat thereby keeping the ground warmer during winter and cooler in the summer.

Plantations also improve water intake in soils and thereby aid in conservation of rainwater which otherwise is lost as runoff. Alkali soils, otherwise poorly permeable, when planted with trees, the infiltration rate increased and thus the runoff decreased (Gill et al. 1990, Gill and Abrol 1993). Trees because of their deeper roots can meet their water requirement from the deeper horizons helping themselves in overcoming drought. A tree plantation on ridge of alkali soils was found to conserve whole of the annual precipitation. Land use of alkali soils as silviculture to reduce runoff was also reported by Kamra et al. (1987). The percentage of runoff due to fallow, pastoral (*Leptochloa fusca* grass), and silvicultural (*Acacia nilotica*) land use was found to be 87, 62, and 44 of a rainfall event. The revegetation of alkali soils with tree and grass species helped in reducing the runoff volume and peak runoff rate significantly as compared to fallow alkali soils. As discussed earlier, the general root environment under plant canopy is congenial for microbial population. Thus, through agroforestry systems, we obtain not only food, timber, fuel, and forage from trees and crops and ameliorate soil but the systems also improve the microclimate and general environment.

Food Securities and Employment Generation

The global Hunger Index (GHI), worldwide improved from 19.8 % in 1990 to 15.1 % in 2010. With the GHI at 24.1 %, India ranks to

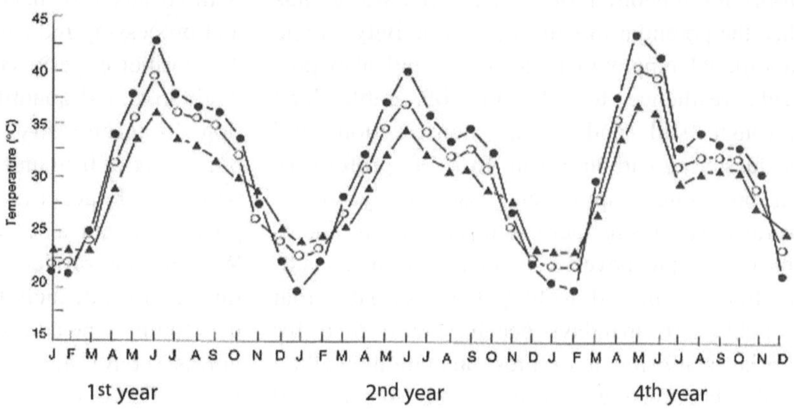

Fig. 9.19 Changes in atmospheric temperature in the bare fallow (●) and underneath canopies of *Eucalyptus tereticornis* (○) and *Acacia nilotica* (▲) *source* Gill and Abrol (1993)

67th position in the world, at an alarming high level. As per the UN report, India accounts for 42 % of world's underweight and about 40 % of the undernourished children (Rai 2013). Energy, iron, and vitamin A remain the major nutrient deficiencies in the semi-arid tropics. Due to population pressure, our forest and other natural resources including soil are dwindling at fast rate. Only nitrogen was deficient in our soil about 50 years ago. Now as many as 15 plant nutrients, both micro and macro, are deficient in major parts of the country. Some of the major nutrients like potash and phosphate are being imported. Therefore, focus on organic crop production, rejuvenation of old orchards, and high density plantations of fruit crops, rehabilitation of degraded (including salt-affected and waterlogged) lands, production and certification of quality planting material, popularization of tissue culture cultivation, establishment of improved storage facilities, value addition for enhanced income, establishment of suitable marketing chain, and farmer friendly agriculture policy ensuring food security and employment generation must get the outmost priority. In recent past, agroforestry practices have been paid more attention and hold a promising solution in the developing countries to help in securing food production, while protecting a safe environment and conserving biodiversity. Agroforestry will have close links with social forestry, particularly in degraded lands, to meet the needs of small holders and landless farmers.

In India, development of agroforestry-based technologies in harmony with biotechnological tools for rehabilitation of 121 Mha wastelands has the potential to enhance productivity, create additional employment avenues, and also provide resilience to 142 Mha of arable land through soil and water conservation and sequestering carbon in soil. Role of agroforestry in providing employment, particularly in the rural areas where there is often serious unemployment and poverty, is an important consideration. Abrol and Joshi (1984) estimated that roughly 216 mandays per ha are needed for initial establishment of forest plantation on alkali soils. For raising and managing *Acacia* and *Eucalyptus* for 7 years, approximately 1,092 and 940 mandays are needed, respectively.

Further, the watershed management technology (where agroforestry is major component), besides having the potential to reduce runoff from 42 to 14 % and soil loss from 11 to 2 t ha^{-1} is generating 215 mandays per ha per annum during implementation phase and 20 mandays per ha per annum in post-implementation phase (Rai 2005a, b). Thus, through implementation of improved agroforestry technology (including value addition to agroforest products), we stand a chance to capitalize on the natural resources on sustainable basis to eradicate poverty and nutrient imbalance. Agricultural produces in India including that from horticulture, dairy, fisheries, and animal husbandry amounting to more than 600 M tones in general is perishable. At present, post-harvest losses in India (so is true for many other developing countries) are estimated to be about 25 % and only 2 % is processed. The value addition is only 7 % of agro-produce as compared to 23 % in China, 30 % in Thailand, 70 % in Brazil, and 80 % in Malaysia.

Thus, the value addition and processing in the area of production is of paramount importance to convert the losses into earnings. The food and bioprocess engineers will always be needed to manage biological and food processes, improve process efficiency, food and worker safety, and reduce environmental degradation. Another important aspect is creation of suitable markets at the site of production. At present, the agricultural produce from production catchments is taken without any post-harvest processing to markets and processing industries located in urban areas. The farmer experiences the produce losses both qualitative and quantities during this transportation. In agroforestry, products presentation and packaging of fruit and vegetables-based products will not only generate many-fold income to rural population but also will generate employment. We need the production of fruit and vegetables, milk and meat, including poultry and fish products (with value addition) in accordance with the norms set for the consumers of the developed world. The developing countries like India will be able to diversify their agriculture and garner larger share in exports particularly in the scenario of a new global trade regime.

Gender Issues

There is increasing feminisation of agriculture in families with small and marginal holdings, particularly due to the out-migration of men. The problem is more acute in hilly areas of India like Uttranchal, Jharkhand, and North East States. In most of the developing countries (both in Africa and Asia), the women is deprived of property right. Besides sharing the labor in agriculture, she has responsibility of running the household, bringing the fuelwood from forest and cooking the food for the entire family. In many regions, they extract fruit juice, rear cattle, and prepare the country liquor. They have to earn for paying the fee for the school going children. Creating facilities of value addition to agroforestry products, employment can be generated for rural women. Garrity (2004) reviewed the role of agroforestry in the advancement of women. He stated that 60–80 % of the farmers in the developing world are women and in these countries rural women grow and harvest most of the staple crops that feed their families. In India, agriculture sector employs about 80 % of all economically active women and about 48 % of self-employed farmers are women. There are 75 million women engaged in dairying as against 15 million men and 20 million in animal husbandry as compared to 1.5 million men (Nagarajan et al. 2005). Women possess detailed knowledge of uses of plant and plant products for food, medicine, and animal feed. Women farmers possess a unique knowledge of plant products such as preparing pickles, milk products, fish processing and farming, cattle rearing, and handle most of the related works.

Women have trouble in diversifying their crops because they have difficulty obtaining the credit and land needed to shift to nontraditional exports. In India, out of nearly 44 million Kisan (farmer) Credit Cards issued till 2005, < 10 % were issued to women (Swaminathan 2005a, b). The proposed conferment of land rights to women will help to redress this distressing situation. Much more needs to be done to understand the kinds of traditional and nontraditional agroforestry products that are accessible to women, and to get research attention focused on them. This also applies to value-added processing activities and marketing. Trees are a medium for long-term investment on the farm. Thus, the propensity to cultivate them is particular sensitive to property rights. Policy research in agroforestry must continue to strengthen our understanding of these linkages. We need to assist in identifying the means by which womens' land rights can be made more secure to enhance the intensification of farming in general and the acceleration of tree cultivation in particular.

Efforts have been initiated in the recent past both by the governments and NGOs to incorporate gender issues in the development agenda to ensure womens' full and equitable participation agricultural (including agroforestry) development programmes. However, as stated above, statistics still indicate that these efforts have not been sufficient enough to bridge gender inequalities. Despite the key role of women in crop husbandry, live stock rearing, fisheries, forestry, and post-harvest technologies (particularly in value addition), those incharge of formulating packages of technologies, services, and public policies for rural areas have often tended to neglect the productive role of women. They have traditionally been discriminated in their access to productive resources and have been denied ownership of land, cattle, trees, harvest, and shelter and above all in access to credit and marketing facilities for their economic activities. It is thus, essential to develop strategies and mechanism to improve women's access to agricultural support services.

Other Social and Environmental Issues

Agroforesty is the major source of fuelwood, food, fodder, timber, tannin, gum, medicinal herbs, and a wide range of minor products. Biofuels are emerging as a renewable and eco-friendly source of energy which could help in enhancing the self-sufficiency in energy and minimizing dependence of a nation on imported fossil fuels. The feedstocks identified are molasses for the production of ethanol and tree-born nonedible oil-seed crops like *Jatropha* and *Pongamia* for the production of biodiesel. These plants have advantage as these can be grown on degraded lands. The National Policy on Biofuels, released in 2009, foresees biofuels as a potential means to stimulate rural development and generate employment opportunities, as well as aspires to reap environmental and economic benefits arising out of their large-scale use (GoI 2009). Every year around 1.2 million tons of tree-born nonedible seed oil is produced in the country (GoI 2008). Raju et al. (2012) have given insight to the potential, policy, and emerging paradigms to biofuels in India.

From adopting agroforestry especially on degraded lands, not only the individual farmer is benefited but the entire community around the area is benefited in many ways. The soil and water conservation measures by creating the bunds cross the nalas, fields, and ripping along the contour result in checking the soil erosion and harvesting rain water in situ. Creating the ponds for fish culture helps in meeting the water requirement of cattle and wild life population in surrounding areas. The grass harvested by local people from the afforested areas helps many families especially landless farmers to rear cattle and sustain their families. The trees influence the temperature extremes and help in percolation of rainwater into loose and porous soils (recharging of sub-soil water). The livefencing helps in protection of crop fields. Several medicinal herbs are collected by the local practitioners who sustain their families by subscribing plant-based drugs. Recently, the attention is being paid toward commercial forestry, raising block plantation of commercial trees, and also trees yielding biodiesel such as *Jatropha curcas, Pongamia pinnata, Ricinus communis*, etc. This approach will change the economical scenario by reducing the import of fossil fuels. As discussed in earlier sections, trees play a vital role both in lowering down water table (in water-logged areas) and also recharging the groundwater in dry regions where water table is falling drastically. By adopting agroforestry practices, we shall be able to diversify the cropping pattern when more production will be obtained per unit of water available. Adopting biosaline agroforestry, the nomadic behavior of large population will be checked in dry regions. We need change in our policies tilting toward agroforestry.

References

Abrol IP, Bhumbla DR (1978) Some comments on terminology relating to salt-affected soils. In: Proceedings of the sub-commission on dryland saline-seep control, 11th congress international soc. of soil science, Edmonton, pp 6.1–6.27

Abrol IP, Joshi PK (1984) Economic viability of reclamation of alkali lands with special reference to agriculture and forestry. In: Sharma K (ed) Economics of wastelands development. SPWD, New Delhi, pp 19–30

AICRP (2000–2004) Biennial reports of the all India coordinated research project on management of salt-affected soils and use of saline water in agriculture. CSSRI, Karnal, p 199

AICRP (2010–2012) Biennial reports of the all India coordinated research project on management of salt-affected soils and use of saline water in agriculture. CSSRI, Karnal, p 188

Alsharhan AS, Wood WW, Goudie AS, Fowler A, Abdellatif EM (eds) (2003) Desertification in the third millennium. Swets & Zeitlinger Publishers, Lisse, p 489

Anwar M, Patra DD, Singh DV (1996) Influence of soil sodicity on growth, oil yield and nutrient accumulation in vetiver (*Vetiveria zizanoides*). Ann Arid Zone 35(1):49

Aronson JA (1989) Haloph: a database of salt tolerant plants of world. Office of arid land studies. University of Arizona, Tucson, p 77

Atta-Krah K, Kindt R, Skilton JN, Amaral W (2004) Managing biological and genetic diversity in tropical groforestry vol. I. In: Nair PKR, Rao MR, Buck LE (eds) Advances in agroforestry: new vistas in agroforestry-a compendiumn for the 1st world congress of agroforestry. Kluwer Academic Publishers, Dardrecht, pp 183–194

Ayers RS, Westcot DW (1985) Water quality for agriculture. Irrigation & Drainage paper No. 29. FAO, Rome, p 173

Balusamy M, Shanughum PM, Bhaskaran R (2003) Mixed farming-an ideal farming. Intensive Agric 41(11–12):20–25

Barrett-Lennard EG (2003) Salt land pastures in India: a practice guide. State of Western Australia, Department of Agriculture, CISRO, CRC for Plant-based Management of Dryland Salinity Australia. pp 176

Bhumbla DR, Abrol IP (1979) Saline and sodic soils. In: Soil and rice symposium, (20–23 Sept, 1977), IRRI, Los Banos, Laguna, Phillipines, pp719–738

Bhutta MN, Choudhry MR (2000) Biological control of waterlogging: Biodrainage and other alternatives to drainage. In: Proceedings Proc. 8th ICID international drainage workshop on role of drainage and challenges in 21st century held during January 31–February 4, 2000 at New Delhi, India

Bridges EM, Oldeman LR (1999) Global assessment of human-induced soil degradation. Arid Soil Res Rehabil 13:319–325

CBD (1992) Convention on biological diversity (CBI) Article 2, use of terms (http://www.biodiv.org/convention/articles.asp?Lg=0&a=cbd.02)

Chaturvedi AN (1984) Firewood crops in areas of brackish water. Indian For 110(4):364–366

Chhabra R, Abrol IP, Chawla K (1987) Tolerance and productivity of Sesbania species for fuelwood production in sodic soil. In: Proceedings of international symposium on afforestation of salt affected soils, vol. 2, CSSRI, Karnal, India, pp 33–47

Chhabra R, Thakur NP (1998) Lysimeter study on the use of biodrainage to control waterlogging and secondary salinisation in (canal) irrigated arid/semi-arid environment. Irrig Drainage Syst 12:265–288

CSIR (1986) The useful plants of India. CSIR, New Delhi

CSSRI (2002–2003 to 2012–2013) Annual reports of Central Soil Salinity Research Institute, Karnal, India

Dagar HS, Dagar JC (1991) Plant folk medicines among the Nicobarese of Katchal Island India. Econ Bot 45:114–119

Dagar JC (1982) Some ecological aspects of mangrove vegetation of Andaman and Nicobar Island. Sylvatrop,Philippines For Res J 7:177–216

Dagar JC (1991) Agroforestry practices for coastal Bay Islands. J Indian Soc Coastal Agril Res 9(1&2):433–439

Dagar JC (1994) Strategies for sustainable use of island resources of India. J Indian Soc Coastal Agril Res 12(1&2):187–191

Dagar JC (1995) Ecology of halophytic vegetation in India: a review. Int J Ecol Environ Sci 21:273–296

Dagar JC (1996) Rehabilation of coastal saline lands by planting with suitable species. J Indian Soc Coastal Agril Res 14(1&2):185–191

Dagar JC (2000) Agroforestry systems for coastal and island regions. Indian J Agrofor 2:59–74

Dagar JC (2003) Biodiversity of Indian saline habitats and management and utilization of high salinity tolerant plants with industrial application for rehabilitation of saline areas. In: Alsharhan AA, Wood WW, Gouie AS, Fowler A, Abdellatif EM (eds) Desertification in the third millennium. Swets and Zeitlinger Publishers, Lisse, pp 151–172

Dagar JC (2005a) Salinity resources in India: an overview. Bull Natl Instit Ecol 15:69–80

Dagar JC (2005b) Ecology, management and utilization of halophytes. Bull Nat Instit Ecol 15:81–97

Dagar JC (2006) Halophytic forages for salty soils in deserts. In: Pathak PS, Kundu SS (eds) Livestock feeding strategies for dry regions. International Book Distributing Co, Lucknow, pp 193–223

Dagar JC (2008) Indian mangrove: status, management and their plausible benefits for livelihood security. J India Soc Coastal Agril Res 26(2):121–128

Dagar JC (2009) Opportunities for alternate land uses in salty and water scarcity areas. Int J Ecol Environ Sci 35(1):53–66

Dagar JC (2012) Utilization of degarded lands/habitats and poor quality water for livelihood security and mitigating climate change. Indian J Agrofor 14(1):1–16

Dagar JC, Dagar HS (1999) Ethnobotany of aborigines of Andaman-Nicobar Islands. Surya International Publications, Dehra Dun, p 203

Dagar JC, Bhagwan Hari, Kumar Y (2004a) Effect on growth performance and biochemical contents of *Salvadora persica* when irrigated with water of different salinity. Indian J Plant Physiol 9(3):234–238

Dagar JC, Kumar Y, Tomar OS (2006a) Cultivation of medicinal Isabgol (*Plantago ovata*) in alkali soils in semiarid regions of northern India. Land Degrad Dev 17:275–283

Dagar JC, Mongia AD, Bandyopadhyay AK (1991) Mangroves of Andaman & Nicobar Islands. Oxford & IBH Publishing Co. Ltd., New Delhi, p 166

Dagar JC, Sharma HB, Shukla YK (2001a) Raised and sunken bed technique for agroforestry on alkali soils of northwest India. Land Degrad Dev 12:107–118

Dagar JC, Singh AK, Singh R, Arunachalum A (2012) Climate change vis-à-vis Indian agriculture. Annal Agric Res New Ser 33(4):189–203

Dagar JC, Singh G (1998) Vegetation ecology of salty soils. In: Tyagi NK, Minhas PS (eds) Agricultural salinity management in India. CSSRI, Karnal, pp 221–241

Dagar JC, Singh G (2003) Pasture production in degraded and problematic soil status and prospects. In: Jakhmola RC, Jain RK (eds) Sustainable animal production. Pointer Publishers, Jaipur, pp 94–119

Dagar JC, Singh G (2004) Evaluation of crops in agroforestry with *Casuarina equisetifolia* (Linn.) plantations. Indian J Agrofor 3(1):47–50

Dagar JC, Singh G (2007) Biodiversity of saline and waterlogged environments: documentation, utilization and management. National Biodiversity Authority, Chennai, p 76

Dagar JC, Singh G, Singh NT (1995) Evaluation of crops in agroforestry with teak (*Tectona grandis* Linn-f.), maharukh (*Ailanthus excelsa* Roxb.) and tamarind (*Tamarindus indica* Linn.) on reclaimed salt-affected soil. J Trop For Sci 7(4):623–634

Dagar JC, Singh G, Singh NT (2001b) Evaluation of forest and fruit trees used for rehabilitation of semiarid alkali/sodic soils in India. Arid Land Res Manage 15:115–133

Dagar JC, Singh NT (1999) Plant resources of Andaman & Nicobar Islands. Bishen Singh Mahendra Pal Singh Publishers, Dehra Dun, p 985

Dagar JC, Singh NT, Mongia AD (1993) Characteristics of mangrove soils and vegetation of Bay Islands in India. In: Lieth H & Al Masoom A (eds) Towards the rational use of high salinity tolerant plants vol. 1. Kluwer Academic Publishers, The Netherlands, pp 59–80

Dagar JC, Swarup A (2003) Potential of afforestation and agroforestry in carbon sequestration for mitigating climate changes. In: Pathak PS, Ram N (eds) Agroforestry: potentials and opportunities. Agrobios and Indian Society of agroforestry, pp 43–65

Dagar JC, Tomar OS, Kumar Y, Bhagwan H, Yadav RK, Tyagi NK (2006b) Performance of some under-explored crops under saline irrigation in a semi-arid climate in northwest India. Land Degrad Dev 17:285–299

Dagar JC, Tomar OS, Kumar Y, Yadav RK (2004b) Growing three aromatic grasses in different alkali soils in semi-arid regions of northern India. Land Degrad Dev 15:143–151

Dagar JC, Tomar OS, Minhas PS, Kumar M (2013) Lemon grass productivity as affected by salinity of irrigation water, planting methods and fertilizer doses on a calcareous soil in a semi-arid region of northwest India. Indian J Agric Sci 83(7):734–738

Dagar JC, Tomar OS, Minhas PS, Singh G, Ram J (2008) Dryland biosaline agriculture -Hisar experience. Techanical Bulletin 6: CSSRI, Karnal, pp 28

Dagar JC, Rao Gururaja G, Shukla YK, Sharma HB (2009) Performance of three flower-yielding plants in different sodic soils. Indian J Hortic 66(3):404–409

Das SK, Itnal CJ (1994) Capability based land use systems: role in diversifying dryland agriculture. In: Soil management for sustainable agriculture in dryland area. Bull Indian Soc Soil Sci 16:92–100

FAO/AGL (2000) Extent and causes of salt-affected soils in participating countries. FAO/AGL-Global network on integrated soil management for sustainable use of salt-affected lands (http://www.fao.org/ag/agl/agll/spush/topic2.htm)

Forest News (2005) Planting of clonal *Eucalyptus*. For News-For Dept Haryana 2(2):5–6

Garrity DP (2004) Agroforestry and the achievements of the Millennium development. goals. In: Nair PKR, Rao MR & Buck LE (eds) Advances in Agroforesty Vol. 1; New Vistas in Agroforestry—a compendium for the 1st world congress of agroforestry. Kluwer Academic Publishers, Dordrecht, pp 5–17

Gill, HS (1985) Studies on the evaluation of selected tree species for their tolerance to sodicity and mechanical impedance in a highly sodic soil with particular reference to root growth behaviour. Ph.D. Thesis, Kurukshetra University, Kuruskhetra, India

Gill HS, Abrol IP (1993) Afforestation and amelioration of salt-affected soils in India. In: Davidson N, Galloway R (eds) Productive use of saline land. ACIAR proceedings No.42, Canberra, pp 23–27

Gill HS, Rao DLN, Abrol IP (1990) Litter yield and its quality in plantation of *Leucaena leucocephala* and *Sesbania sesban* in alkali soils: In: Pathak PS, Deb Roy R, Singh P (eds) Multipurpose tree species for agroforestry systems. Proceedings of National workshop, range management society of India. Jhansi, pp 187–196

Gill MS, Singh JP, Gangwar KS (2009) Integrated farming system and agriculture sustainability. Indian J Agron 54(2):128–139

Glenn EP, O'Leary JW (1985) Productivity and irrigation requirements of halophytes grown with seawater in the Sonoran desert. J Arid Environ 9:81–91

GoI (2008) Agricultural statistics at glance. Ministry of Agriculture Government of India (GoI), New Delhi

GoI (2009) National policy on biofuels. Ministry of New and Renewable Energy Government of India (GoI), New Delhi

Goodchild AV, McMeniman NP (1987) Nutritive value of *Acacia* foliage and pods for animal feeding. In: Turnbull JW (ed), Australian acacias in developing counties. ACIAR proceedings, no. 16, Canberra, Australia, pp 101–106

Greig G (1883) Notes on experimental plantations on Usar. Indian For 9:454–457

Grewal SS, Abrol IP, Singh OP (1987) Rain water management for establishing agroforestry on alkali soils. Indian J Agric Sci 57:30–37

Gupta GN, Homan S, Prasad KG (1987a) Salt tolerance of selected tree species. J Trop Forest 3:217–226

Gupta GN, Homan S, Prasad KG (1987b) Salt tolerance of selected tree species at seedling stage. Indian For 112:101–113

Gupta RK, Abrol IP (1990) salt affected soils: their reclamation and management for crop production. In: Lal R, Stewart BA (eds) Advances in soil science soil degradation III. Springer, New York, pp 223–288

Gupta RK, Rao DLN (1994) Potential of wasterlands for sequestering carbon by restoration. Curr Sci 66:378–380

Gupta RK, Tomar OS, Minhas PS (1995) Managing salt affected soils and waters for afforestation, bulletin 7/95. CSSRI, Karnal, p 23

Heuperman AF, Kapoor AS, Denecke HW (2002) Biodrainage: principles, experiences and applications. International, programme for technology & research in irrigation and drainage IPTRID Secretariat. FAO, Rome, p 78

IPCC (2007) Climate change: climate change impacts, adaptation and vulnerability. Summary for policy-makers. Inter-governmental Panel on Climate Change (IPCC)

Jaradat AA (2003) Halophytes for sustainable biosaline farming systems in the Middle East. In: Alsharhan AS, Wood WW, Goudie AS, Fowler A, Abdellatif EM (eds) Desertification in the third millennium. Swets & Zeitlinger Publishers, Lisse, pp 187–204

Jeet Ram, Dagar JC, Lal K, Singh G, Toky OP, Tanwar RS, Dar SR, Mukesh Kumar (2011) Biodrainage to combat water logging, increase farm productivity and sequester carbon in canal command area of north-west India. Curr Sci 100(11):1673–1680

Jeet Ram, Dagar JC, Singh Gurubachan, Lal Khajanchi, Tanwar VS, Shoeran SS, Kaledhonkar MJ, Dar SR, Kumar Mukesh (2008) Biodrainage: eco-friendly technique for combating waterlogging & salinity. Techanical Bulletin 9: CSSRI, Karnal, pp 24

Jeet Ram, Garg VK, Toky OP, Minhas PS, Tomar OS, Dagar JC, Kamra SK (2007) Bio-drainage potential of *Eucalyptus tereticornis* for reclamation of shallow water table areas in north-west India. Agrofor Syst 69:147–165

Jena SK, Sahoo N, Roy Chowdhury S, Mohanty RK, Kundu DK, Behera MS, Patil DU, Kumar A (2011) Reclamation of coastal waterlogged wasteland through biodrainage. J Indian Soc Coastal Agric Res 29(2):57–62

Kamra SK, Rao KVGK, Dhruva Narayana VV (1987) Runoff response of alkali watersheds to different land use treatments. In: Proeedings of the international Symposium on afforestation of salt affected soils, vol. 3. CSSRI, Karnal, India, pp 105–119

Kapoor AS, Denecke HW (2001) Biodraiange and biodisposal: the Rajasthan experience. In: GRID, IPTRID's Network Magazine No. 17

Kaur B, Gupta SR, Singh (2002a) Bioamelioration of sodic soil by silvopastoral systems in northwestern India. Agrofor Syst 54:13–20

Kaur B, Gupta SR, Singh (2002b) Carbon storage and nitrogen cycling in slivi-pastoral systems on a sodic soil in northwestern India. Agrofor Syst 54:21–29

Khan MAW, Yadav JSP (1962) Characteristics and afforestation problems of saline alkali soils. Indian For 88(4):259–271

Kogo M, Kaminura D, Miyagi T (1986) Research for rehabilitation of mangroves in truck Islands. In: Mangroves of Asia & the pacific-status and monument. technical report. UNDP/4NESCO regional Mangrove project, New Delhi, pp 419–439

Kumar A (1988a) Long term forage yields of five tropical grasses on an extremely sodic soil and resultant soil amelioration. Expl Agric 24:89–96

Kumar A (1988b) Performance of forage grasses in saline soils. Indian J Agron 33:26–30

Kumar A (1990) Effect of gypsum compared with that of grasses on the yield of forage crops on a highly sodic soil. Exp Agric 24:89–96

Kumar A (1998) Forage in slat-affected soils. In: Tyagi NK, Minhas PS (eds) Agricultural salinity ssmanagement in India. CSSRI, Karnal, pp 489–498

Kumar A, Abrol IP (1983a) Effect of periods of submergence on the performance of selected grasses. Indian J Agric Sci 53:694–698

Kumar A, Abrol IP (1983b) Effect of gypsum on tropical grasses grown in normal and extremely sodic soil. Exp Agric 19:167–177

Kumar A, Abrol IP (1986) Grasses on Alkali Soils. Bulletin No. 11, CSSRI, Karnal, India

Lal R (2011) Sequestering carbon in soils of agro-ecosystems. Food Policy 36:533–539

Lal R, Kimble JM (2000) Tropical ecosystems and the global C cycle. In: Advances in soil science: global climate change & tropical ecosystems. CRC Press Roca Raton, pp 3–32

Leather JW (1897) Reh. Agricul Ledger 7:129–137

Le Houe'rou HN (1986) Salt-tolerant plants of economic value in the Middle East basin. Res Revegitation 5:319–341

Leith H, AL Masoom AA (eds) (1993) Towards the rational use of high salinity tolerant plants, vol. 1&2. Tasks for vegetation science 27&28, Kluwer Academic Publishers, The Netherlands, pp 521–447

Lieth H, Lieth A (1993) Seawater irrigation studies in the thirsted Arab Emirates. In: Lieth H, Al Masoom A (eds) Towards the rational use of high salinity tolerant plants, vol. 1. Kluwer Academic Publishers, The Netherlands, pp 1–10

Maji AK, Reddy GPO, Sarkar D (2010) Degraded and wastelands of India: status and spatial distribution. Indian Council of Agricultural Sciences, New Delhi, p 158

Malik KA, Aslam Z, Naqvi M (1986) Kallar grass-a plant of saline soils. Nuclear Institute for Agriculture and Biology, Faisalabad

Malik KA, Zafar Y (1984) Use of ^{15}N isotop dilution technique for quantification of associative BNF in kallar grass. In: Veegar C, Newton WE (eds) Advances in nitrogen fixation research. Nijhoff Junk, The Hague, p 58

Mandal AK, Sharma RC, Singh G, Dagar JC (2010) Computerized database on salt affected soils in India. Technical Bulletin No.2/2010. Central Soil Salinity Research Institute, Karnal, p 28

Mass EV (1985) Crop tolerance to saline sprinkling waters. Plant Soil 89:273–284

Mass EV(1990) Crop salt tolerance. In: Tanji KK (ed) Agricultural salinity assessment and management. ASCE Manu Rep Eng Practices 71

Minhas PS (1998) Use of poor quality of waters. In: Singh GB, Sharma BR (eds) 50 Years of natural resource management. ICAR, New Delhi, pp 327–346

Minhas PS, Gupta RK (1992) Quality of irrigation water: assessment and management. ICAR, New Delhi, p 1223

Minhas PS, Sharma OP (2003) Management of soil salinity and alkalinity problems in India. J Crop Prod 7:181–230

Minhas PS, Singh YP, Tomar OS, Gupta Raj K, Gupta RK (1997a) Saline water irrigation for establishment of furrow-planted trees in northwest India. Agrofor Syst 35:177–186

Minhas PS, Singh YP, Tomar OS, Gupta Raj K, Gupta RK (1997b) Effect of saline irrigation and its schedules on growth, biomass production and water use by *Acacia nilotica* and *Dalbergia sissoo* in a highly calcareous soil. J Arid Environ 36:181–192

Moreland WH (1901) Reh. Agricult Ledger 13:415–463

Nagarajan S, Ramamurthy VV, Prasanna BM (2005) Farm research: eucation and policy dimensions. In: Ram N (ed) The Hindu survey of Indian agriculture

2005. Kasturi & Sons Ltd., National Press, Kasturi Building Chennai, pp 28–34

Naik GR, Manjunath GTS (2000) Project (UKP) areas through biodrainage concept. In: Proceedings of the 8th ICID international drainage workshop on role of drainage and challenges in 21st century held during January 31-February 4, 2000 at New Delhi, India

NAS (1990) Saline agriculture: salt-tolerant plants for developing countries. National Academy of Sciences, Washington, DC, 143 pp

O'Leary JW (1993) Growth and physiology of *Salicornia bigelovii* Torr. Int J Plant Sci 156:197–205

Oldeman RL, Hakkeling RTA, Sombroek WG (1991) World map of the status of human-induced soil degradation. International Soil Reference & Information Centre, Wageningen

Oliver EE (1881) Report on Reh. Swamp and Drainage of the Western Jamuna Canal Districts, Public Works Department Press, Lahore

Pande GC (1967) Afforestation of Usar lands. In: Proceedings of 11th all India silviculture conference, FRI, Dehra Dun

Pasternak D (1982) Biosaline research in Israel: alternative solutions to a limited water supply. In: San Pietro A (ed) Biosaline research: a look to the future. Plenum, New York, pp 39–57

Piare L (2006) Private sector forestry research: a success story from India. In: Chauhan R, Chauhan SK, Gill SS, Khajuria HN (eds) Production technology and management of agroforestry models. Agrotech Publishing House, Udaipur, pp 398–411

Piare L, Kulkarni HD, Srinivas K, Venkatesh KR, Kumar PS (1997) Genetically improved clonal planting stock of Eucalyptus—a success story from India. Indian For 123(12):1117–1138

Rai M (2005a) Employment generation through entrepreneurship development in agriculture. Financing Agric 37(2):25–29

Rai M (2005b) Policy structure: internalising synergies. In: Ram N (ed) The Hindu survey of Indian agriculture 2005. Kasturi & Sons Ltd, National Press, Chennai, pp 20–27

Rai M (2013) Challenging agriculture in the changing times Dr AG Joshi Memorial Lecture. National Academy of Agricultural Sciences, New Delhi, p 22

Raju SS, Parappurathu S, Chand R, Joshi PK, Kumar P, Msangi S (2012) Biofuels in India: potential, policy and emerging paradigms, policy paper 27. National Centre for Agricultural Economics and Policy Research, New Delhi, p 90

Rao DLN (1998) Biological amelioration of salt-affected soils. In: Subba Rao NS, Donimergues YR (eds) Micobial Interactions in agriculture and forestry, vol. I. Oxford & IBH Publ. Co., New Delhi, pp 219–238

Rao DLN, Ghai SK (1995) Predicting nitrogen fixation and N accumulation in field grown annual *Sesbania* spp Proc. Indian Natl Sci Acad (B) 61:57–62

Rao DLN, Gill HS (1993) Nitrogen fixation, biomass production and nutrient uptake by annual *Sesbania* species in the alkaline soils. Biol Fertil Soils 15:73–78

Rao DLN, Gill HS (1995) Biomass and biofertiliser production by *Sesbania cannabina* in alkaline soils. Bioresources Technol 53:169–172

Rao DLN, Gill HS, Abrol IP (1990) Regional experience with perennial Sesbania in India. In: Macklin B, Evans DO (eds) Proceedings of Workshop on Perennial Sesbania Species in Agroforestry Systems held at ICRAF, Nairobi. NFTA Special Publication, Hawaii, USA, pp 189–198

Rao DLN, Pathak H (1996) Ameliorative influence of organic matter on biological activity of salt-affected soils. Arid Soil Res Rehabil 10:311–319

Rao Gururaja G, Nayak AK, Chinchmalatpure AR (2003) *Salvadora persica*: a life support species for salt affected black soils. Technical Bulletin 1/2003. CSSRI, Karnal, p 44

Rengasamy P, Greene RSB, Ford GW, Mehanni AH (1984) Identification of dispersive behaviour and the management of Red-brown Earths. Aust J Soil Res 22:413–431

Rhoades JD, Kandiah A, Mashali AM (1992) The use of saline waters for crop production. FAO irrigation & drainage paper 48. Food & Agriculture Organisation, Rome

Riley JJ, Abdal M (1993) Preliminary evaluation of Salicornia production and utilization in Kuwait. In: Lieth H, Masoom A Al (eds) Towards the rational use of high salinity tolerant plants, vol 2. Kluwer Academic Publishers, The Netherlands, pp 319–329

Roy Chowdhury S, Kumar A, Brahmanand PS, Ghosh S, Mohanty RK, Jena SK, Sahoo N, Panda GS (2011) Application of biodrainage for reclamation of waterlogged situations in deltaic Orissa. Research bulletin 53. Directorate of water Management, Bhubaneswar, p 32

Ryan JG, Spencer DC (2001) Future challenges and opportunities for agricultural R &D in the Semi-Arid Tropics. ICRISAT, Patancheru 83

Sandhu SS, Abrol IP (1981) Growth responses of *Eucalyptus tereticornis* and *Acacia nilotica* to selected cultural treatments in a highly sodic soil. Indian J Agric Sci 51:437–443

Sankary MN (1986) Species distribution and growth in salt-affected land in Syria. Reclam Revegetation Res 5:125–142

Sathey JA, Ravindranath NH (1998) Climate change mitigation in the energy and forestry sectors of developing countries. Ann Rev Energy Environs 23:387–437

Schroeder P (1994) Carbon storage benefits of agroforestry systems. Agrofor Syst 27:89–97

Sen DN, Rajpurohit KS, Wissing FW (1982) Survey and adaptive biology of halophytes in western Rajasthan, India. In: Sen DN, Rajpurohit KS (eds) Contribution to the ecology of halophytes. tasks for vegetation Series Vol. 2. Dr. W. Junk Publishers, The Hauge, pp 61–78

Sharma OP, Verma SK, Kool YM, Desnonki SC (1992) Progress report of thr project evaluation of tree plantations for the control of salinity and water table. Jawaharlal Nehru Krishi Vishava Vidyalaya, Indores, p 26

Singh G, Abrol IP, Cheema SS (1989a) Effects of spacing and lopping on mesquite (*Prosopis juliflora*)—Karnal grass (*Leptochloa fusca*). Expt Agric 25:401–408

Singh G, Dagar JC (1998) Agroforestry in salt affected soils. In: Tyagi NK, Minhas PS (eds) Agricultural salinity management in India. CSSRI, Karnal, pp 473–487

Singh G, Dagar JC (2005) Greening sodic soils: Bichhian model. Technical bulletin No.2/2005. Cental Soil Salinity Research Institute, Karnal, p 51

Singh G, Dagar JC, Singh NT (1997) Growing fruit trees in a highly alkali soils—a case study. Land Degrad Dev 8:257–268

Singh G, Gill HS (1992) Ameliorative effect of tree species on characteristic of sodic soils at Karnal. Indian J Agri Sci 62:142–146

Singh G, Singh NT, Dagar JC, Singh H, Sharma VP (1995) An evaluation of agriculture, forestry and agroforestry practices in moderately alkali soil in north-western India. Agrofor Syst 37:279–295

Singh G, Singh NT, Tomar OS (1993) Agroforestry in salt-affected soils. Technical bulletin No.17, CSSRI, Karnal, India, 65 pp

Singh G, Abrol IP, Cheema SS (1988) Agroforestry on alkali soil-effect of planting methods and amenments on initial growth, biomass accumulation and chemical composition of mesquite (*Prosopis juliflora* (SW)DC with inter-space planted with and without Karnal grass (*Diplachne fusca* (Linn) P. Beauv.). Agrofor Syst 7:135–160

Singh G, Abrol IP, Cheema SS (1989b) Effects of gypsum application on mesquite (*Prosopis juliflora*) and soil properties is an abandoned sodic soil. Forest Ecol Manage 29:1–14

Singh G, Gill HS, Abrol IP, Cheema SS (1991) Forage yield, mineral composition, nutrient cycling and ameliorating effects of Karnal grass (*Leptochloa fusca*) grown with *Prosopis juliflora* in a highly alkali soil. Field Crops Res 26:45–55

Singh IS, Singh RK (1990) Fruits as an component of agroforstry. In: agroforstry: present status and scope for future development in farming systems. Narendra Dev University of Agriculture & Technology, Faizabad, India

Singh JP, Gangwar B, Pandey DK, Kochewad SA (2011) Integrated Farming system model for small farm holders of western plain zone of Uttar Pradesh. Project Directorate for Farming Systems Research, Modipuram 58

Singh K (1994) Site suitability and tolerance limits of trees, shrubs and grasses on sodic soils of Ganga-Yamuna Doab. Indian For 120:225–235

Singh K, Yadav JSP (1985a) Growth response and cationic uptake of *Eucalyptus hybrid* and varying levels of soil salinity and sodicity. Indian For 111:1123–1135

Singh K, Yadav JSP (1985b) Growth response of shisham (*Dalbergia sissoo*) at varying levels of soil salinity and sodicity. Curr Agric 9:43–47

Singh NT (1998) Historical perspective. In: Tyagi NK, Minhas PS (eds) Agricultural Salinity Management in India. Central Soil Salinity Research Institute, Karnal, pp 9–19

Singh UN, Bhatt DN, Yadav JSP (1987) Growth and biomass production of certain forest species as influenced by varying pH levels. In: Proceedings of international symposium on afforestation of salt affected soils vol. 2, CSSRI, Karnal, India, pp 51–62

Singh YP, Sharma DK, Singh G, Nayak AK, Mishra VK, Singh R (2008) Alternate land use management for sodic soils. CSSRI Technical Bull No. 2/2008. CSSRI, Karnal, p 16

SSSA (1987) Glossary of soil science terms. Technology Committee, Soil Science Society of America, WisconsinSS

Stirzaker RJ, Cook FJ, Knight J (1999) Where to plant trees on cropland for control of dryland salinity: some approximate solution. J Agric Water Manage 39:115–133

Sumner ME, Rengaswamy P, Naidu R (1998) Sodic soils: a reappraisal sodic soils. In: Sumner ME, Naidu R (eds) Distribution, properties management and environmental consequences. Oxford Uniersity Press, New York, pp 3–19

Swaminathan MS (2005a) India's greatest living industry: hundred years later. Financing Agric 37(2):3–24

Swaminathan MS (2005b) Sustainable farming: three-pronged strategy needed. In: Ram N (ed) The Hindu survey of Indian agriculture 2005. Kasturi & Sons Ltd. National Press, Kasturi Building Chennai, pp 8–15

Szabolcs I (1979) Review of research on salt-affected sols (with a bibliography compiled by Varallyay G). Natural Resources Research XV, UNESCO, Paris

Szabolcs I (1989) Salt-affected soils. CRC Press, Inc. Boca Raton, Florida, p 274

Tomar OS, Dagar JC, Minhas PS (2010) Evaluation of sowing methods, irrigation schedules, chemical fertilizer doses and varieties *of Plantago ovata* Forsk to rehabilitate degraded calcareous lands irrigated with saline water in dry regions of north western India. Arid Land Res Manage 24:133–151

Tomar OS, Gupta RK (1984–1994) Performance of forest tree species in saline soils, annual reports. CSSRI, Karnal

Tomar OS, Gupta RK, Dagar JC (1998) Afforestation techniques and evaluation of different tree species for waterlogged saline soils in semiarid tropics. Arid Soil Res Rehabil 12(4):301–316

Tomar OS, Minhas PS (1998) Afforestation of salt-affected soils. In: Tyagi NK, Minhas PS (eds) Agricultural salinity management in India. CSSRI, Karnal, pp 453–471

Tomar OS, Minhas PS (2002) Performance of some ornamental winter annual flowering species under saline irrigation. Indian J Horticul 59(2):201–206

Tomar OS, Dagar JC, Singh YP (2004) Forest and fruit trees for alkali soils. Indian Farming 53(11):44–47

Tomar OS, Minhas PS (2004a) Relative performance of aromatic grasses under saline irrigation. Indian J Agron 49(3):207–208

Tomar OS, Minhas PS (2004b) Performance of medicinal plant species under saline irrigation. Indian J Agron 49(3):209–211

Tomar OS, Minhas PS, Dagar JC (2005) Isabgol (*Plantago ovata* Forsk.): a potential crop for saline irrigation and moderate alkali soils. Technical bulletin No. 1/2005. CSSRI, Karnal, p 17

Tomar OS, Minhas PS, Gupta RK (1994) Potential for afforestation of waterlogged saline soils. In: Singh P, Pathak PS and Roy MM (eds) Agroforestry systems for degraded lands vol. 1. Oxford & IBH Pub. Co. Pvt. Ltd., New Delhi, India, pp 111–120

Tomar OS, Minhas PS, Sharma VK, Gupta RK (2003b) Response of nine forage grasses to saline irrigation and its schedules in a semi-arid climate of north-west India. J Arid Environ 55:533–544

Tomar OS, Minhas PS, Sharma VK, Singh YP, Gupta RK (2003b) Performance of 31 tree species and soil condition in a plantation established with saline irrigation. For Ecol Manage 177:333–346

Tomar OS, Patil SG (1998) Alternate land uses. In: Minhas PS, Sharma OP, Patil SG (eds) 25 years of research on management of salt-affected soils and use of saline water in agriculture. CSSRI, Karnal, pp 189–202

Turnbull JW (1986) Australian acacia in developing countries. ACIAR Proc. N. 16, Canberra, Australia

Ulery AL, Teed JA, Van Genuchten M-Th, Shannon MC (1998) SALTDATA: a database of plant yield response to salinity. Agronomental J 90:556–562

Untawale AG (1993) Development of an intertidal mangrove nursery and afforestation techniques along the Indian Coast. In: Lieth H, Al Masoom A (eds) Towards the rational use of high salinity tolerant plants vol. 1, Kluwer Academic Publishers, The Netherlands, pp 371–378

USSL Staff (1954) Diagnosis and improvement of saline and alkali soils. USDA Hand Book No. 60, United State Salinity Laboratory Staff, Department of Agricultural, Washington DC

Vannucci M (1989) The Mangroves and US. A synthesis of insight. Indian Association of the Advancement of Science, New Delhi

Velayutham M, Pal DK, Bhattacharya T (2000) Organic carbon stock in soils of India. In: Lal R, Kimble JM, Stewart BA (eds) Advances in soil science—global climate change and tropical ecosystems. CRC Press, Boca Raton, pp 71–95

Yadav JSP (1972) Improvement of saline and alkali soils through biological methods. Indian For 101:385–395

Yadav JSP, Bhumbla DR, Sharma OP (1975) Performance of certain forest species on a saline sodic soil. In: Proceedings of international symposium on new development in the field of salt affected soils of subcommission of salt affected soils. International Society of Soil Science at Cairon, pp 683–690

Yadav JSP, Singh K (1986) Response of *Casuarina equisetifolia* to soil salinity and sodicity. J Indian Soc Coastal Agricul Res 4:1–8

Yensen NP, Bedell JL (1993) Consideration for the selection, adoption and application of halophyte crops to highly saline desert environments as exemplified by the long-term development of cereal and forage cultivars of *Distichlis* spp (Poaceae). In: Lieth H, Masoom A Al (eds) Towards the rational use of high salinity tolerant plants, vol 2. Kluwer Academic Publishers, The Netherlands, pp 305–313

Yensen NP, Bedell JL, Yensen SB (1988) The history of the development od *Distichlis* cultivars for grain forage and sad. In: Proceedings of the Mexican national conference on halophytes. Ciudad Obregon, Sonora, Mexico

Yensen NP, Yensen SB, Weber CW (1985) a review of Distchlis spp for production and nutritional value. In:Whitehead EE, Hutchenson CF Timmermann BN, Varady RG (eds) Arid land today and tomorrow. Westview Press, Boulder, Co. pp 809–822

Zang L, Dawes WR, Slavich PG, Meyer WS, Thorburn PJ, Smith DJ, Walker GR (1999) Growth and ground water uptake responses of Lucerne to changes in ground water levels and salinity: lysimeter, isotope and modeling studies. J Agric Water Manage 39:265–282

Wetland-Based Agroforestry Systems: Balancing Between Carbon Sink and Source

10

A. Arunachalam, D. Balasubramanian, K. Arunachalam, J. C. Dagar, and B. Mohan Kumar

Abstract

Wetlands of India, estimated to be 58.2 million ha, are important repositories of aquatic biodiversity. The diverse ecoclimatic regimes extant in the country resulted in a variety of wetland systems ranging from high altitude cold desert wetland to hot and humid wetlands in coastal zones with its diverse flora and fauna. These ecosystems provide immense services and commodities to humanity. Wetlands perform numerous valuable functions such as recycle nutrients, purify water, attenuate floods, maintain stream flow, recharge ground water, and also serve in providing livelihood to local people in terms of fish, drinking water, fodder, fuel, and environmental services. With rapidly expanding human population, wetlands of India are threatened and facing severe anthropogenic pressures. There is obviously much ground to be covered in our conservation efforts of wetlands. Various agencies at local and government level need to join hands in making these viable, functional, and sustainable. Being diversified farming systems, agroforestry opportunities are abundant in rehabilitation of wetland systems. The nutrient-rich riparian zone provides a suitable site for harnessing the ecosystem services of tree-based farming in the flood plains and in the ecologically fragile hilly region. Ecologically, wetland use as a component in agroforestry may be more acceptable in areas which are facing frequent/seasonal or permanently flooding. It is envisaged that wetland agroforestry can alleviate poverty by making substantial contribution toward local economy in terms of fish and agricultural production.

A. Arunachalam (✉) · D. Balasubramanian ·
J. C. Dagar · B. Mohan Kumar
Division of Natural Resource Management, Indian Council of Agricultural Research, Krishi Anusandhan Bhawan II, Pusa, New Delhi, 110012, India
e-mail: arun70@gmail.com

K. Arunachalam
School of Environment and Natural Resources, Doon University, Dehradun, Uttarakhand 248001, India

J. C. Dagar et al. (eds.), *Agroforestry Systems in India: Livelihood Security & Ecosystem Services*, Advances in Agroforestry 10,
DOI: 10.1007/978-81-322-1662-9_10, © Springer India 2014

Introduction

A wetland is a land area that is saturated with water, either permanently or seasonally, such that it takes on the characteristics of a distinct ecosystem (Wikipedia- State of Florida, DoEP 2011). Primarily, the factor that distinguishes wetlands from other land forms or water bodies is the characteristic vegetation that is adapted to its unique soil conditions, which are hydric and support aquatic plants (Butler 2010). Wetlands include lands that are permanently or temporally flooded and many different habitats such as ponds, marshes, swamps, bogs, fans, and peat lands. The water found in wetlands can be salt water, fresh water, or brackish. Wetlands function as 'ecotones', transitions between different habitats, and have characteristics of both aquatic and terrestrial ecosystems. Thus, the biological resources tend to be highly productive. In India, wetlands are reported to occupy 58.2 million ha (Mha) (Prasad et al. 2002) under various categories (Table 10.1) which is about 17.7 % of the total geographic area. But, under National Wetland Conservation and Management Programme (NWCMP) only 3.02 Mha area covering 94 sites in different states are under wetlands (Table 10.2), out of which only 677 thousand ha area has been considered of International importance covering 25 (Ramsar) sites (MoEF 2007).

Mostly the wetlands are distributed across the floodplains and high hills. While 80 % human in the country depend on agrarian economy, still agroforestry in its diverse form become significant both at subsistence and semi-subsistence level (Asian Development Bank 2005). At systematic level, agroforestry opportunities are abundant in wetland systems. The nutrient-rich riparian zone provides a suitable site for harnessing the ecosystem services of tree-based farming in the flood plains and in the ecologically fragile hilly region. Ecologically, in wetland land use system agroforestry as a component may be more acceptable in areas which are facing frequent/seasonal or permanently flooding. It is envisaged that wetland agroforestry can alleviate poverty by making substantial contribution toward local economy in terms of fish and several agricultural products (Fig. 10.1). Nonetheless, the wetlands are continuing to be threatened owing to reveal the factors that are both ecological and anthropogenic.

Agricultural Perspective

India, with its annual rainfall of over 130 cm, varied topography and climate regimes support and sustain diverse and unique wetland habitats. Natural wetlands in India consists of the high altitude Himalayan lakes, followed by wetlands situated in the flood plains of the major river systems; saline and temporary wetlands of the arid and semi-arid regions; coastal wetlands such as lagoons, backwaters, and estuaries; mangrove swamps; coral reefs and marine wetlands, etc. In fact, with the exception of bogs, fens, and typical salt marshes, Indian wetlands

Table 10.1 Extent of wetlands in India

Wetlands in India	Area (million ha)
Area under wet paddy cultivation	40.9
Area suitable for fish culture	3.6
Area under capture fisheries	2.9
Mangroves	0.4
Estuaries	3.9
Backwaters	3.5
Impoundments	3.0
Total area	58.2

Source Parikh and Parikh (1999)

Table 10.2 State wise distribution of wetlands under National Wetland Conservation and Management Programme (NWCMP)

State/UT	Number of wetlands	Name of wetland	Area (ha)
Andhra Pradesh	1	Kolleru	90,100
Assam	2	Deepor Beel, Urpad Beel	4,504
Bihar	3	Kabar, Barilla, Kusheshwar Asthan	11,490
Chandigarh	1	Sukhna	148
Gujarat	8	Nalsarovar, Great Rann of Kachh, thol Bird Sanctuary, Khijadiya Bird Sanctuary, Little Rann of Kacch, Pariej, Wadhwana, Nanikakrad	12,70,875
Haryana	2	Sultanpur, Bhindavas	288
Himachal Pradesh	5	Renuka, Pong Dam, Chandratal, Kewalsar, Khajjiar	15,736
Jammu & Kashmir	7	Wular, Tsomoriri, Tisgul Tso & Chisul Marshes, Hokersar, Mansar-Surinsar, Ranjitsagar, Pangong Tso	1,17,325
Jharkhand	2	Udhwa, Tilaiya Dam	98,965
Karnataka	7	Magadhi, Gudavi Bird Sanctuary, Bonal, Hidkal & Ghataprabha, Heggeri, Ranganthittu, KG Koppa	4,250
Kerala	5	Ashtamudi, Sasthamkotta, Kottuli, Kadulandi, Vembnad Kol	2,13,229
Madhya Pradesh	12	Barna, Yashwant Sagar, Wetland of Ken River, National Chambal Sanctuary, Ghatigaon, Ratapani, Denwa Tawa, Kanha Tiger Riserve, Sakhyasagar, Dihaila, Govindsagar	3,59,814
Maharashtra	3	Ujni, Jayakawadi, Nalganga	40,298
Manipur	1	Loktak	26,600
Mizoram	2	Tamdil, Palak	285
Orissa	4	Chilika, kuanria, Kanjia, Daha	1,22,580
Punjab	3	Harike, Ropar, Kanjali	5648
Rajasthan	1	Sambhar	24,000
Sikkim	6	Khechuperi, Holy Lake, Tamze, Tembao Wetland Complex, Gurudokmar, Tsomgo	164
Tamil Nadu	3	Point Callimere, Kaliveli, Pallaikarni	46,283
Tripura	1	Rudrasagar	240
Uttar Pradesh	9	Nawabganj, Sandi, Lakh Bahoshi, Samaspur, Alwara, semarai Lake-Nagaria Lake Complex, Keetham Lake, Shekha, Saman Bird Sanctuary & sasai Nawar Complex	12,083
Uttaranchal	1	Ban Ganga Jhimli Tal	800
West Bengal	5	East Kolkata Wetlands, Sunderbans, Ahiron Beel, Rasik Beel, Santragahi	5,53,090
Total	94	Total wetlands (94)	30,18,795

Source MoEF (2007)

cover the whole range of the ecosystem types found. The various reservoirs, shallow ponds, and numerous tanks support wetland biodiversity and add to the countries wetland wealth. It is estimated that wetland alone support 20 % of the known range of biodiversity in India (Deepa and Ramachandra 1999, Prasad et al. 2002). The total number of aquatic plant species exceeds 1200 and a partial list of animals and avifauna has been given by Gopal (1995). This biodiversity may help in improving agricultural production including fish production having trees on boundaries of wetlands, which contribute toward shade, shelter, and nutrition as litter.

Fig. 10.1 Few wetland based agroforestry systems *Top* On the brink/slope of wetland, *Middle* fish pond surrounded by trees, *Bottom* On the bank (lowlying area) of wetland

Out of about 3.56 Mha of inland wetlands, north-eastern states and West Bengal cover about 370,000 ha and these areas fall under Indo-Burma global mega biodiversity hotspot and harbor large number of wetlands. Major wetland types are rivers, streams, lakes, ponds, waterlogged areas, ox-bow lakes, high altitude lakes, and reservoirs. Wetlands of this region are traditionally used for agriculture and fisheries since time immemorial.

Apart from the fishes, which are major produces from the wetlands, significant number of edible plant species such as *Alocasia cucullata*, *Euryale ferox*, *Hedychium coronarium*,

Lemanea australis, and *Nelumbo nucifera* are also collected and sold in the local market.

Conversions of wetlands to agricultural land and for shifting cultivation have been projected as a severe environmental problem in the north-eastern region (Balasubramanian et al. 2011, Arunachalam et al. 2002). Recent reports do indicate that the natural wetlands are the largest sources of green house gases (GHGs) such as CO_2, CH_4, and N_2O that significantly contribute to global warming. Further, their loss and degradation of wetlands can result in releases of large amounts of GHGs into the atmosphere, negating gains made from emission reductions. It has been hypothesized that the wetland ecosystems could be a better carbon C sink if managed properly by plantation of multipurpose trees in and around the wetlands (Fig. 10.2). In brief eco-restoration of degraded wetlands could help protect their carbon stores and improve their ability to sequester more C in future. Contrarily, it may also increase the amount of methane released due to anaerobic microbial activity that is pronounced when water level rises. Feasibility analysis of wetland agroforestry as one of the C balancing technique, prescribes it as an alternative farming system in the fragile flood plain ecosystems, based on a few socio-eco-biological values.

Carbon Sequestration

In tree-based systems, carbon sequestration is a dynamic process and can be divided into phases. At establishment phase, many systems are likely to be source of GHGs. These follow a quick accumulation phase and a maturation period when tons of carbon are stored in boles, stems, and roots of trees and in soils. At the end of rotation period, when the trees are harvested and land returned to cropping (sequential systems), part of carbon will

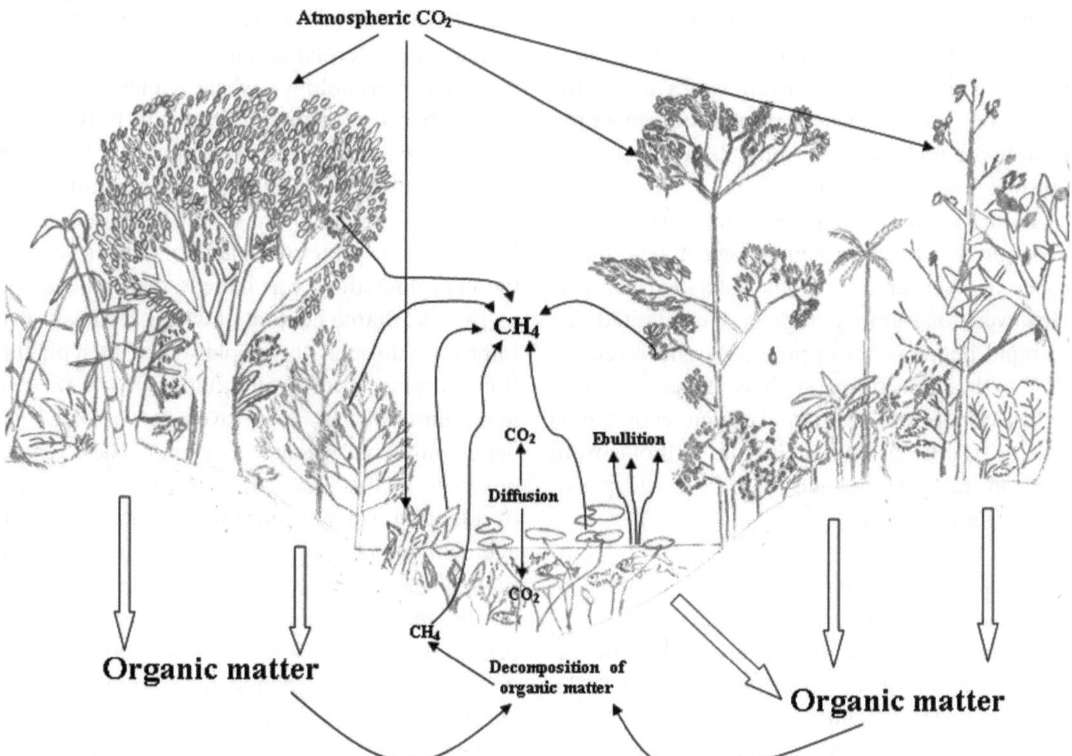

Fig. 10.2 Hypothetical structure of wetland agroforestry system and C balancing function

be released back to the atmosphere (Dixon 1995). In case of simultaneous systems like hedgerow intercropping, silvopastoral systems and agri-silviculture systems; fate of carbon will be different. Therefore, effective sequestration can only be considered if there is a positive net carbon balance from an initial stock after a few decades (Feller et al. 2001). In fact, practice of agroforestry system in and around the natural wetlands may be unique, which could lead to the irreversible alteration in the major ecological functions of wetlands (Box 1).

There are several mechanisms that influence the carbon storage capacity of a wetland ecosystem. For instance, wetland trees and other plants convert atmospheric carbon dioxide into biomass through the mechanism of photosynthesis. Hence carbon may be temporarily stored in wetlands as trees and plants and the living material which feed upon them, and detritus including fallen plants and animals which feed upon them. Many wetland plants are known to use atmospheric CO_2 as the C source, and their death/decay and ultimate settlement at a wetland bottom can have profound effect on C sequestration. This mechanism even varies along the latitudinal gradient where growth of vegetation is slow for high latitude wetlands with less sun, nutrient, and colder temperature.

Further, carbon-rich sediments are trapped and stored that are brought along floods, hurricanes or even drained from watershed sources. However, long-term storage is often limited due to rapid decomposition processes and rerelease of C to the atmosphere, such as in case of paddy fields. Hence, wetlands are dynamic ecosystems where significant quantities of C, both from wetland and non-wetland sources, could possibly be sequestered.

Box 1.

Poplar-based wetland agroforestry system in China: a case study (Fang et al. 2005).

Poplars are the major tree component of traditional agroforestry systems throughout the south temperate central area of China which includes all or portions of Jiangsu, Anhui, Zhejiang, Hubei, Henan, Shandong, and Shanxi provinces, an area of 600,000 km^2. In order to develop a poplar-crop inter-planting pattern in floodplain wetland areas, that is, economically viable, environmentally sound, technically workable, and socially compatible in wetland plain areas new poplar-crop inter-planting patterns were designed using the principle of edge effects and established in 1992 (Fang et al. 2005). The study site located at Hanyuan Forestry Farm, Baoying County, Jiangsu Province, P. R. China (33 °08 °N, 119 °19 °E), and a part of the Lixia River wetland experiences a warm temperate climate with an average growing season of 229 frost-free days.

The poplar-crop interplantations modified the microclimate by decreasing solar radiation, lowering air temperature, and enhancing relative humidity. Overall, the reduction of total solar radiation and illumination intensity in the interplantations ranged from 3.9 to 36.2 %, while relative humidity increased from 2.5 to 3.9 %, depending on phenological phases. The lower air temperature and higher relative humidity in the poplar-crop interplantations enabled the growth and development of winter wheat. Various poplar-crop interplantations have a great impact on plant morphology, wheat yields, and crop quality because of the modification in microclimate. The most dramatic impact on wheat quality between poplar-crop inter-plantation and agricultural monoculture was in protein content. For instance, protein content of wheat seed in the interplantations was increased by an average of 18.3 %. Starch content of wheat seed was also improved through the poplar-crop inter-plantation, varying from a minimum of 10.9 % to a maximum of 15.0 % (average of 12.1 %) depending on plant spacing. The results observed in this study highlight the importance of adopting a poplar-crop interplantations management system to improve crop quality, even though the system reduces some crop yields. The results indicated that the profits of the interplantations per unit are 20–47 % more than that of the agricultural monoculture. Hence, poplar agroforestry is a more profitable option for farmers at current prices, provided that the correct poplar-crop inter-planting patterns and cultivars are used.

Scope of Poplar in India

Poplar is a very prominent taxonomical group of tree species in agro (plantation) forestry in India. It occurs in natural forests also. However, its population in natural stands is small and is gradually declining. Bulk of the plantations is composed of *Populus deltoides*, an exotic species. Indigenous species of poplar occur in the Himalayan region in northern part of India. Other exotic species like *P. xeuramericana* and *P. xberolinensis* were introduced in India in 1950. Clones of *P. canescens, P.maximowiczii, P. trichocarpa, P. simonii, P. szechuanica, P. yunnanensis,* etc., were introduced in the subsequent years. *P. deltoides* performed better than all other exotic poplars in the plains of North India, and relegated most other exotic poplars to the status of anonymity in India. In the hills, *P.yunnanensis* and *P. xeuramericana* 'Robusta' proved better than other species.

Adoption of poplar (*Populus deltoides*) agroforestry in northern India, however, is occurring in areas where land and water are already intensively used and managed for agricultural production. Poplar-based agroforestry plantations in Saharanpur and Yamunanagar districts of northwestern India produce considerable amount of biomass, which helps the farmers in generating income. Poplar-based boundary and agrisilviculture systems account for 99–304 t ha^{-1} CO_2 assimilation at the rotation period of 7 years in the two districts (Rizvi et al. 2010). The potentials of promising poplar agroforestry are yet to be unleashed in mainstreaming livelihood security through intercropping.

The Mangroves

Mangroves are salt tolerant plants that occur between 32 °N and 38 °S latitudes (Ajai et al. 2013). According to a trend analysis conducted on available data by FAO (2007), 15.2 Mha of mangroves are estimated to exist worldwide as on the 2005, down from 18.8 Mha in 1980. Mangroves in India account for about 5 %of the world's mangrove vegetation and are spread over an area of about 4662 km^2 along the coastal States/Union Territories of the country (FSI 2011). Mangroves provide suitable opportunity for practicing tree-based farming systems in coastal areas due to their immense beneficial properties in both ecological and economical point of view (Box 2).

Box 2

Mangrove-based farming System for Coastal areas: case study in Kosrae (Conroy and Fares 2011, Drew et al. 2005).

On the island of Kosrae (Federated States of Micronesia), freshwater-forested wetlands dominated by *Terminalia carolinensis* are often found just upslope from mangrove forests, which appear to be hydrologically connected to them. Many of these *Terminalia* forests have been converted into agroforests (Conroy and Fares 2011, Drew et al. 2005). Kosrae's wetland agroforests have been in existence for approximately 1,350–1,500 years (Athens et al. 1996). In this type of agroforestry systems in coastal freshwater wetlands, with *Terminalia carolinensis* as a dominant overstory species, swamp taro (*Cyrtosperma chamissonis*), banana, and breadfruit are cultivated as main agricultural crops. Cultivation of *C. chamissonis* does not seem to alter the hydrological condition of the land, so that natural wetland conditions are maintained (Chimner and Ewel 2004, 2005). A survey of 10 % of the households on Kosrae showed that 89 % owned some *Terminalia* land. Most grew taro, bananas, and sugar cane, either in or immediately adjacent to *Terminalia* forests. Most owned canoes constructed of *Terminalia* logs, and nearly half had harvested trees from these forests during the past years: 64 % to clear land for agricultural purposes, 36 % for building canoes, and 31 % for other uses. *Terminalia* forests provided over $3.1 million worth of goods to Kosraeans, primarily from agricultural production (Drew et al. 2005). Approximately, 2/3 of those surveyed understood that *Terminalia* grows best in a wetland setting. Most thought that *Terminalia* forests provide erosion protection and improve water quality. However, very few were cognizant of the ecological links between *Terminalia* and mangrove forests. Kosraeans attached little importance to the fact that *Terminalia* is endemic to the eastern Caroline Islands. If human dependence on these wetlands increases, the integrity of *Terminalia* forests, as well as adjacent mangrove forests, could be at risk.

In most of the north-eastern India and coastal Kerala, one could see the cultivation of Taro (*Colocasia esculenta*) in wetland areas and buffer zone of fish ponds. It is suggested that root crops like Taro which grow well in wetland areas could be worthwhile in wetland agroforestry system along with the other tree crops. Coastal almond (*Terminalia catappa*), *Calophyllum inophyllum,* and species of *Pandanus* are very common behind mangroves in Andamans and *Nypa fruticans* along creeks of Andamans and Sunderbans; may be explored as commercial plantations. Cashew nut (*Anacardium occidentale*) *Pandanus* and *Casuarina* are cultivated along beaches and back spaces along Orissa coast.

Mangroves also play a significant role in sequestering of carbon and reducing greenhouse gases (Patil et al. 2012). The below-ground content of mangroves is 4–18 times higher than the carbon content of tropical rainforests (Tateda 2005). This indicates that positive action in mangrove conservation and rehabilitation would contribute immensely to sequestration of CO_2. The higher rates of carbon sequestration in salt marshes, soils of tidal salt, and lower methane emissions make coastal wetlands more valuable carbon sinks than other ecosystems in a warmer world. Dagar (2003, 2008) reported 38 species of exclusive mangroves and more than 180 species of associates of mangroves (including climbers, epiphytes, ferns, etc.) from India. In India, mangroves are mostly dominated by genera like *Aegiceras,Avicennia, Bruguiera, Ceriops, Excoecaria,Heritiera,Lumnitzera, Rhizophora, Sonneratia,* and *Xylocarpus*. *Nypa fruticans*, a mangrove palm is frequent in Sunderbans and Andamans only. Mangrove trees have been the source of firewood in India since ancient time. Because of the high specific gravity of wood, the species of *Rhizophora*, *Kandelia*, *Ceriops,* and *Bruguiera* are preferred for firewood. *Heritiera fomes* is used for boat building, while *Avicennia* and *Rhizophora* are used for brick-burning. Species of *Bruguiera* are used to make poles. All mangrove species are rich in tannin. Honey collection from the mangrove forest is a promising business in India. It has been estimated that Sundarbans mangrove alone produce 111 tons of honey annually. Honey collected from *Cynometra ramiflora* and *Aegialitis rotundifolia* has a good market value and is in demand. *Avicennia* spp., *Phoenix paludosa*, and *Sonneratia caseolaris* are used for human consumption and as cattle feed. *Nypa fruticans* is tapped for an alcoholic drink. Leaves of *Nypa* palm are used for thatching of roofs. The uses of mangrove and associate species have been dealt in detail by Dagar (2003, 2008), Dagar et al. (1991, 1993), and Dagar and Singh (2007). Further, mangroves can possibly provide a gene bank for cultivating salt tolerant species of crops.

Backwater Farming in Kerala

The backwaters (or '*Kaayals*' in Malayalam) in Kerala are a chain of brackish lagoons and lakes lying parallel to the Arabian Sea coast (also known as the Malabar Coast). The network includes five large lakes linked by canals, both manmade and natural, fed by 38 rivers, and extending virtually half the length of Kerala state. It is estimated that the total area of these backwaters has shrunk from 440 km^2 in 1968 to 350 km^2 due to the population and extended agricultural activities (Namboothiry 2000). The nutrient-rich upland area of backwaters with perfectly balanced tropical climate is very conducive for practicing tree-based farming system. The upland backwater area with expansion of mangrove forests provides suitability for growing paddy fields and coconut (*Cocos nucifera*) based agroforestry systems which are practiced traditionally by the upland communities. The coconut plantations contain a carbon stock of about 30–70 t ha^{-1} depending on the agro-climatic and soil conditions (Ranasinghe 2012).

Cultivation of Colocasia (*Colocasia esculenta*) under the coconut trees are major economic activities in backwater area. Fruit crops

Table 10.3 Potential indigenous plant species for wetland agroforestry system based on their socio-ecological properties

Plant Species	Ecological/Agroclimatic Zone	Selection Restraints/Criteria		C Stock (t ha^{-1}) (Reference)
		Ecological Services[a]	Economic Services	
Albizia procera	TSEF; TMDF; DTF; NSBLF	Y, H	Y, L	0.95 (Ullah and Al-Amin 2012)
Alnus nepalensis	A; B; C	Y, H	Y, L	30.20 (Ranabhat et al. 2008)
Areca catechu	K; Q; R; S	N	Y, H	NA
Artocarpus heterophyllus	HT; HST; NT	N	Y, H	2.26 (Alamgir and Al-Amin 2007)
Aquilaria malaccensis	Subtropical; Sub-temperate	Y, L	Y, H	NA
Bambusa balcooa	Tropical; Subtropical; HST	Y, H	Y, H	2.22 (Nath and Das 2011)
Bambusa pseudopallida	Tropical; HT; HST	Y, H	Y, H	NA
Bombax ceiba	Tropical; Subtropical	Y, H	Y, H	0.09(Ullah and Al-Amin 2012)
Cassia fistula	Tropical; Subtropical	Y, H	Y, M	0.64(Ullah and Al-Amin 2012)
Castaniopsis indica	Tropical; Subtropical; Sub-temperate	Y, H	Y, M	0.05(Ullah and Al-Amin 2012)
Dalbergia latifolia	Tropical	Y, M	Y, L	NA
Dendrocalamus strictus	Tropical; Subtropical; Temperate	Y, H	Y, H	NA
Duabanga grandiflora	Tropical; Subtropical;	Y, M	Y, M	0.07(Ullah and Al-Amin 2012)
Eucalyptus tereticornis	Tropical; Subtropical; Temperate	Y, H	Y, M	98.27 (Devi el al. 2013)
Grevillea robusta	Warm temperate; Subtropical	Y, H	Y, H	NA
Livistona jenkinsiana	TEF; NSBLF	Y, L	Y, H	NA
Mangifera indica	Tropical; Warm temperate	N	Y, H	0.03 (Chavan and Rasal 2011)
Phyllostachys assamica	Tropical; Temperate; Assam Plains	Y, H	Y, H	NA
Piper nigrum	Tropical; Pantropical; Temperate	Y, M	Y, H	NA
Populus deltoides	Temperate; Warm temperate; Tropical	Y, H	Y, H	31.95 (Rizvi et al. 2010)

TSEF – Tropical Semi-Evergreen Forests; *TMDF* – Tropical Moist Deciduous Forests; *DTF* – Dry Tropical Forests; *NSBLF* – Northern Subtropical Broad-Leaved Forests; *TEF* – Tropical Evergreen Forests; *HT* – Humid Tropical; *HST* – Humid Subtropical; *NT* – Near Tropical

A, B, C Western, Central and Eastern Himalayas respectively. *K, Q, R, S* – Deccan Plateau, Bengal and Assam Plains, Western Coastal Plains, Eastern Coastal Plains respectively.

Y – Yes; *N* – No; *L* – Low; *M* – Medium; *H* – High, *NA* – Not Available.

[a] Based on the assumption of C sequestration potential of particular plant species.

like banana, mango, jackfruit, and pineapple also grown in this system. Plantation crops like cocoa and coffee are also cultivated for cash generation activities. Paddy-cum-fish cultivation (locally known as '*Pokkali*') practice is a unique traditional farming system in backwater area of Kerala. The rice crop cultivated is highly resistance to salinity and very productive without addition of manure/fertilizers.

The Way Forward

Appropriate choice of plant species is necessary for successful wetland agroforestry system. While selecting tree species for the wetland agroforestry, the factors such as adaptation to highly fluctuating flooding condition, water level, salinity, nutrient enrichment, etc. are important Planting trees for agroforestry around wetlands may help to extract and remove excess nutrients. Trees may be even more effective than some forage species, by using more extensive and deeper root systems to recapture deep leached nutrients. A major concern with growing trees in this part of the landscape is the risk of flood damage. This depends on the risk of a flood event, the duration of flooding and soil saturation, the flood tolerance of the tree species, and the expected harvest interval of the tree crop. A few potential trees for wetland agroforestry system with both ecological (mainly C storage) and economic importance have been identified (Table 10.3). A thorough surveying of woody perennials that could be managed under short-rotation forests is warranted for diversifying the species specific to different agro-climatic condition with intricate work plan, such tree-based system in wetland could yield substantial economic return while reducing management costs. Further, carbon loss from the wetlands could also be significantly balanced by the trees which capture and fix the atmospheric carbon in the form of biomass and return to the soil subsequently by decomposition. Thus, wetland agroforestry provides tremendous potential for carbon storage by balancing C between its sink and source, and help to achieve sustainable livelihoods in source rich and wet northeast India.

References

Ajai BA, Chauhan HB, Sarma KS, Bhattacharya S, Ashutosh S, Pandey CN, Thangaradjou T, Gnanppazham L, Selvam V, Nayak SR (2013) Mangrove inventory of India at community level. Nat Acad Sci Lett 36(1):67–77

Alamgir M, Al-Amin M (2007) Organic carbon storage in trees within different geopositions of chittagong (south) forest division, Bangladesh. J For Res 18(3):174–180

Arunachalam A, Khan ML, Arunachalam K (2002) Balancing traditional Jhum cultivation with modern agroforestry in eastern Himalaya-A biodiversity hot spot. Curr Sci 83:117–118

Asian Development Bank (2005) Federated states of Micronesia economic report toward a self sustainable economy. Asian Development Bank, Manila, The Philippines, Pacific Studies Series. http://www.adb.org/sites/default/files/pub/2005/fsm-economic-report.pdf Accessed on 19 Feb 2013

Athens JS, Ward JV, Murakami GM (1996) Development of agroforest on a Micronesian high island: prehistoric Kosraean agriculture. Antiquity 70:834–846

Balasubramanian D, Arunachalam A, Arunachalam K, Das AK (2011) Nutrient accumulation pattern of *Eichhornia crassipes* Mart. (Solms.) in natural wetlands with different trophic condition. In: Malik DS, Kumar S, Bharti U (eds) Water pollution and management. Biotech Books, New Delhi, pp 30–56

Gopal B (1995) Biodiversity in freshwater ecosystems including wetlands, biodiversity and conservation in India. A status report, vol 4. Zoological Survey of India, Calcutta

Butler S (ed) (2010) Macquarie concise dictionary, 5th edn. Macquarie Dictionary Publishers Ltd., Sydney

Chavan BL, Rasal GB (2011) Potentiality of carbon sequestration in six year ages young plant from university campus of Aurangabad. Global J Res Eng 11(7):15–20

Chimner RA, Ewel KC (2004) Differences in carbon fluxes between forested and cultivated Micronesian tropical peatlands. Wetland Ecol Manage 12:419–427

Chimner RA, Ewel KC (2005) A tropical freshwater wetland: II. Production, decomposition and peat formation. Wetland Ecol Manage 13:671–684

Conroy NK, Fares A (2011) A snapshot of agroforestry in *Terminalia carolinensis* wetlands in Kosrae, Fed States Micronesia 41(2):177–195

Dagar JC (2003) Biodiversity of Indian saline habitats and management and utilization of high salinity tolerant plants with industrial application for rehabilitation of saline areas. In: Alsharhan AA, Wood WW,

Gouie AS, Fowler A, Abdellatif EM (eds) Desertification in the third millennium. Swets and Zeitlinger Publishers, Lisse, pp 151–172

Dagar JC (2008) Indian mangroves : Status, management and their plausible benefits for livelihood security. J India Soc Coast Agric Res 26(2):121–128

Dagar JC, Mongia AD, Bandyopadhyay AK (1991) Mangroves of Andaman & Nicobar Islands. Oxford & IBH Publishing Co. Ltd., New Delhi, p 166

Dagar JC, Singh G (2007) Biodiversity of Saline and Waterlogged Environments: Documentation, Utilization and Management. NBA Scientific Bulletin Number - 9, National Biodiversity Authority, Tamil Nadu, India, p 78

Dagar JC, Singh NT, Mongia AD (1993) Characteristics of mangrove soils and vegetation of bay islands in India. In: Lieth H, Al Masoom A (eds) Towards the rational use of high salinity tolerant plants, vol 1, Kluwer Academic Publishers, The Netherlands, pp 59–80

Deepa RS, Ramachandra TV (1999) Impact of Urbanization in the interconnectivity of wetlands. In: Proceedings of National Symposium on Remote Sensing Applications for Natural Resources: Retrospective and perspective, Indian Society of Remote Sensing, Bangalore

Devi B, Bhardwaj DR, Panwar P, Pal S, Gupta NK, Thakur CL (2013) Carbon allocation, sequestration and carbon dioxide mitigation under plantation forests of north western Himalaya, India. Ann For Res 56(1): 123–135

Dixon RK (1995) Agroforestry systems: sources or sinks of greenhouse gases. Agrofor Syst 31:99–116

DoEP (2011) Glossary. Department of Environmental Protection, state of Florida (http:www.dep.state.fl.us/evergladesforever/about/glossary.html)

Drew WM, Ewel KC, Naylor RL, Sigrah A (2005) A tropical freshwater wetland: III. Direct use values and other goods and services. Wetlands Ecol Manage 13:685–693

Forestry Paper 153 (2007) The worlds mangroves 1980–2005: a thematic study prepared in the framework of the global forest resources assessment 2005. FAO, Rome, p 77

Fang S, Xu X, Yu X, Li Z (2005) Poplar in wetland agroforestry: a case study of ecological benefits, site productivity, and economics. Wetlands Ecol Manage 13:93–104

Feller C, Albrecht A, Blanchart E, Cabidoche YM, ChevallierT, Hartmann C, Eschenbrenner V, Larre-Larrouy MC, Ndandou JF (2001) Soil carbon sequestration in tropicalareas: general considerations and analysis of some edaphicdeterminants for lesser antilles soils. Nutr Cycl Agroecosyst 61:19–31

FSI (2011) India state of forest report 2011. forest survey of India, ministry of environment and forests, government of India, Dehra Dun, pp 286

MoEF (2007) Conservation of wetlands in India: a profile. Ministry of environment and forests, government of India, New Delhi, pp56

Namboothiry SS (2000) Romancing with Kerala Backwaters. http://pib.nic.in/feature/feyr2000/fjun2000/f220620002.html Accessed 02 February 2013

Nath AJ, Das AK (2011) Carbon storage and sequestration in bamboo-based smallholder homegardens of Barak valley, Assam. Curr Sci 100(2):229–233

Patil V, Singh A, Naik N, Seema U, Sawant B (2012) Carbon sequestration in mangroves ecosystems. J Environ Res Dev 7(1A):576–583

Parikh J, Parikh K (1999) Sustainable Wetland. Environmental Governance – 2, Indira Gandhi Institute of Development Research, Mumbai

Prasad SN, Ramachandra TV, Ahalya N, Sengupta T, kumar A, Tiwari AK, Vijayan VS, Vijayan L (2002) Conservation of wetlands in India–a review. Trop Ecol 43:173–186

Ranabhat S, Awasthi KD, Malla R (2008) Carbon sequestration potential of *Alnus nepalensis* in the mid hill of Nepal: a case study from Kaski district. BankoJanakari 18(2):3–9

Ranasinghe S (2012) Potential impacts of climate change on coconut. http://climatenet.blogspot.in/2012/07/potential-impacts-of-climate-change-on.html. Accessed 02 Feb 2013

Rizvi RH, Dhyani SK, Yadav RS, Singh R (2011) Biomass production and carbon stock of poplar agroforestry systems in Yamunanagar and Saharanpur districts of northwestern India. Curr Sci 100(5):736–742

Tateda Y (2005) Estimation of CO_2 sequestration rate by mangrove ecosystem, http://criepi.denken.or.jp/en/publications/annual/2005/05juten16.pdf. Accessed on 02 Feb 2013

Ullah MR, Al-Amin M (2012) Above- and below-ground carbon stock estimation in a natural forest of Bangladesh. J For Sci 58(8):372–379

Agroforestry in India and its Potential for Ecosystem Services

11

S. K. Dhyani and A. K. Handa

Abstract

India has a long tradition of agroforestry and many different types of indigenous agroforestry systems are found in different parts of the country. Most of these systems are location specific, and information on them is mostly anecdotal. The situation is now changing and a lot of research inputs have been at place since the existence of All India Coordinated Project on Agroforestry and establishment of National Centre for Agroforestry at Jhansi. Many technologies for rehabilitation of degraded lands including salt-affected and eroded lands have been developed and benefits of these technologies are reaching to the stake holders. Some of these issues and ecosystem services rendered through agroforestry systems, which are otherwise not highlighted, are discussed in this chapter. The need of separate agroforestry policy at place has been felt.

Introduction

Forests played crucial role in the human evolution. Man learnt the art of domesticating plants and animals after leaving the hunting and gathering habit. Man's desire to live in co-existence in a community created settled agriculture. Increase in human and live stock population necessitated acquiring more and more land under cultivation to meet the ever increasing demand for food, fodder, vegetables, fuel wood, timber, medicines, etc. Further, demographic pressure has forced man to seek unconventional methods of agriculture to utilize land to the maximum extent. Therefore, in the quest of optimizing productivity, the multitier system came into existence. The origin of agroforestry practices in India, i.e., growing trees with food crops, grasses, and other components is believed to have started during *Vedic* era, though agroforestry as a science evolved in recent years. Agroforestry as is now understood as a science of designing and developing integrated self sustainable land management systems which involve introduction and/or retention of woody components such as trees, shrubs, bamboos, canes, palms along with agricultural crops including pasture/animals, simultaneously or sequentially on the same unit of land and time, to satisfy the ecological as well as socio-economic needs of the people.

However, referring to the recent literature, agroforestry combines agriculture and forestry technologies to create more integrated, diverse,

S. K. Dhyani (✉) · A. K. Handa
National Research Centre for Agroforestry, Jhansi, 284003, UP, India
e-mail:
shivkdhyani@gmail.comskdhyani@nrcaf.ernet.in

J. C. Dagar et al. (eds.), *Agroforestry Systems in India: Livelihood Security & Ecosystem Services*, Advances in Agroforestry 10,
DOI: 10.1007/978-81-322-1662-9_11, © Springer India 2014

productive, profitable, healthy, and sustainable land use systems. Some important agroforestry practices include: agrisilviculture, agrihorticulture, silvopasture, alley cropping, etc. According to the Association for Temperate Agroforestry—"Agroforestry practices are *intentional* combinations of trees with crops and/or livestock that involve *intensive* management of the *interactions* between the components as an *integrated* agroecosystem." These key characteristics are the essence of agroforestry and are what distinguish it from other farming or forestry practices. To be called agroforestry, a land use practice must satisfy all of these criteria:

- *Intentional*: Combination of trees, crops, and/or animals are intentionally designed and managed as a whole unit, rather than as individual elements that may occur in close proximity but are controlled separately.
- *Intensive*: Agroforestry practices are intensively managed to maintain their productive and protective functions; these practices often involve annual operations such as cultivation and fertilization.
- *Interactive*: Agroforestry management seeks to actively manipulate the biological and physical interactions between tree, crop, and animal components. The goal is to enhance the production of more than one harvestable component at a time, while also providing conservation benefits.
- *Integrated*: The tree, crop, and/or livestock components are structurally and functionally combined into a single, integrated management unit. Integration may be horizontal or vertical, and above or below ground. Such interaction utilizes more of the productive capacity of the land and helps balance economic production with resource conservation.

Traditional Agroforestry Systems

Scanty information is available on the historical aspect of agroforestry in the country, however, it is more in the form of anecdotes and narration without much backing from authentic literature. Though some information is cited under Introduction of this publication a periodic historic information is included under this section. The following text is based on the published papers such as Tejwani (1994), Pathak and Dagar (2000), Dhyani et al. (2006, 2009a, b, 2011), and Kareemulla et al. (2007, 2009).

Agroforestry During Vedic Era

Man was basically a food collector during *Vedic* era, who gradually acquired the knowledge of plant species yielding food, fruits, vegetables, medicines, etc. The origin of agroforestry practice, i.e., growing trees with food crops and grasses believed to have occurred during the Vedic and Pre-Vedic period in India. The God's prime *Rishi* and *Muni* and *Guru* were residing in forest and derived food in the form of vegetables, dry fruits, sweetening medium, medicine, and gum besides fuel from the forest. Further, they developed *Ashrams* and started growing fruit trees like mango, jackfruit, guava, mahua, and others along with flowers, vegetables and *Kutoo* (food grain) to meet their needs. Man gradually developed settled farming with the passage of time. However, there were many who moved with receding forests, entirely depending on forest for food and shelter. They came to be known as tribal people who cleared some forest area and started growing agricultural crops without any input such as fertilizer and water. They cultivated agricultural crop for 3–5 years thereafter the practice kept on rotation. This practice was coined as shifting cultivation and became popular in Madhya Pradesh, Orissa, Andhra Pradesh, Bihar, North Eastern States, and Andaman Nicobar Islands.

Agroforestry During Epic Era

The period of Indian history during which two great epics (Ramayana and Mahabharata) were composed is known as epic era. Ramayana was composed by great poet Maharishi Valmiki. In this epic, he illustrated that *Rishi* and *Muni* were used to grow mango, banana, guava, sweet potato, barley, and flowers. They mostly utilized the food grains, fruits, and vegetables grown on the land of *Ashram*. Maharishi Valmiki had vividly illustrated the arrangement of trees such as Ashok, Neem, Pipal, Bargad in outer ring, jack fruits, mango, banana, and guava in middle ring and flowers namely, Champa, Chameli, Juhi, Bela, etc., in inner ring and *sarover* (water-tank) in the centre in *Ashok vatika*. This clearly indicated that trees, fruit trees and flower plants grown in association in the *vatika* (garden), which could be considered an agroforestry practice. Mahabharat was composed by Maharishi Ved Vyas. The agroforestry practices were improved during this period as compared to Ramayana period. The great teacher Dronacharya not only raised trees, fruit trees, and vegetables, but also raised *Sanwa* and barley in the orchard, which was used for food and worship of God. Rearing of silkworm was a popular profession during that period. It is evident from these two epics that agroforestry practices such as agrisilviculture and agrihorticulture were in operation in their primitive state during epic era.

Agroforestry During Medieval Period

The agriculture was in its primitive state during the medieval period. However, many agroforestry practices existed. Example includes planting trees for shade, shelter, boundary demarcation by planting trees, multipurpose trees, nitrogen-fixing trees, timber trees, fruits, nuts, food, fuel, and fodder yielding trees. Flowers and fruits of mahua, bamboos, sal seeds, and others were being consumed by people. Alcohol was also produced from trees such as palms and mahua. The tribal dances performed around these trees are symbolic of the intimate ethnic bias to protect trees. Trees had considerable socio, ethnic, and religious backing.

Agroforestry During Recent Past

In the recent times especially for the last 3–5 centuries, spectacular improvements were made in tree plantation. The most important agroforestry practice is known from the Kangeyam tract of Tamil Nadu where *Acacia leucophloea* + *Cenchrus setigerus* in silvopasture system was perfected. Similarly in ravines of Yamuna and Chambal, trees, shrubs, and bamboos with grasses were planted for rearing the milk producing Jamunapari breed of goats and sheep. Scattered trees such as Khejri (*Prosopis cineraria*) and Mehndi (*Lawsonia alba*) in association with pearl millet, sorghum, and chilly were grown in the semi-arid area of Tamil Nadu. Tree plantations continued as demarcation and control against wind erosion throughout the country. Plantation of Khejri trees for various uses on farm was a common practice in Rajasthan. In the northeast, large cardamom (*Amomum subalatum*) was introduced in Alder (*Alnus nephalensis*) forests in Sikkim Himalayas by the local farmers more than 200 years ago. Now the Alder-based agroforestry system is covering more than 48,000 ha area in northeastern states (Dhyani 1998). Deliberately growing of trees like *Grewia, Celtis* etc., on field bunds and systematic plantation of shade trees for tea and coffee plantation are other common examples of agroforestry practices in the hilly areas. In the coastal areas, *Casuarina equisetifolia* and other trees were grown in association with crops on farmlands for cash and small timber. Live hedges were common as an agroforestry practice in which Mehndi, *Agave sisalana, Euphorbia* species were commonly raised. In paddy sowing areas, *Pongamia pinnata* and *Sesbania grandiflora* were grown, lopped annually and their leaves applied to fields as green manure. Application of green manure to paddy fields also was common in Madhya Pradesh and Uttar Pradesh. In Western Ghats, *Terminalia* leaves were harvested, spread on land burnt and then paddy and millets were sown. Multistoried homesteads/home gardens were in existence in Kerala, Karnataka, and Tamil Nadu as an important agroforestry practice. This agroforestry practice is still followed in aforesaid states to fulfill the needs of fodder, fodder, fuel wood, fruit, vegetable, medicines, etc (Kumar 2003).

Some of the traditional systems such as shifting cultivation, taungya cultivation, home gardens, trees on boundaries, and block plantations are important and have been discussed in detailed in the regional chapters where these are prevalent.

Agroforestry Research in India

In order to meet the increasing demand of our fast growing population, we require the production of food grain and fuel wood for human consumption and green and dry fodder for livestock to the tune of 250, 350, and 2,085 million ton, respectively besides 75 million m^3 of timber. Under these circumstances, there is no alternative but to follow the integration of trees with agricultural crop, which is known as agroforestry. Agroforestry research is more than 100 years old in India beginning with introduction of shade trees in tea gardens, but organized research was initiated during the second half of the last century in few ICAR Institutes namely Central Arid Zone Research Institute, Jodhpur, Central Soil and Water Conservation Research and Training Institute, Dehradun, Indian Grassland and Fodder Research Institute, Jhansi, as fuel-fodder plantations on degraded lands or establishing the silvopastoral systems.

India has been at the forefront since organized research in agroforestry started worldwide. In 1979 at Imphal, the Ist Seminar on Agroforestry was organized by the Indian Council of Agricultural Research, New Delhi to accumulate and compile the data on the research and development of agroforestry in India. The All India Coordinated Research Project (AICRP) on Agroforestry with 20 centers all over country was launched in 1983 to implement the major recommendations of the aforesaid seminar. Further, National Research Centre for Agroforestry was established in 1988 at Jhansi to accelerate the basic, strategic, and applied research in agroforestry. At the initial phase, agroforestry research also got financial support from different international agencies primarily USAID, UNDP, and others, which helped in strengthening infrastructure and human resource component. At present, there are 37 centers under All India Coordinated Research Project on Agroforestry with project coordinating unit at National Research Center for Agroforestry, Jhansi. These centers represent almost all the agro-climates of the country. In addition to ICAR, Indian Council of Forestry Research and Education (ICFRE) and its Institutes, private institutions and NGOs such as WIMCO, ITC, BAIF, IFFDC, West Coast Paper Mills Ltd., Hindustan Paper Mills Ltd, and National Tree Growers Cooperatives are also engaged in research and promotion of agroforestry in the country.

The major thrust of agroforestry research in the beginning was on the following areas:

- Diagnostic survey and appraisal of existing agroforestry practices.
- Collection and evaluation of promising fuel wood, fodder and timber yielding tree species/ cultivars.
- Studies on management practices of agroforestry systems.
- Development of wastelands, watersheds, and community lands through agroforestry interventions.
- Integration of livestock, fishery, apiculture, lac, etc., as component of agroforestry.

The research and developmental efforts undertaken during the last more than two decades have brought results, which have clearly demonstrated the potential of agroforestry for resource conservation, improvement of environmental quality, rehabilitation of degraded lands, and providing multiple outputs to meet the day–to-day demand of the rural population.

The diagnostic survey and appraisal of existing agroforestry practices in the country revealed that there are a number of agroforestry practices prevalent in different agro-ecological zones (Chauhan and Dhyani 1989, 1994; Singh and Dhyani 1996; Pathak et al. 2000; Sharda et al. 2001). The review of agroforestry system research indicates that agroforestry systems or practices are widely based on nature and arrangement of the components and ecological or socioeconomic criteria. In the documented agroforestry practices, trees serve as wind breaks and shelterbelts, delineate boundaries, and provide shade, ornamentation, and seclusion around homesteads. They supply not only poles, stakes, timber, and fuel, but also cash crops, fodder, fruits and nuts, dyes, gums, resins, fiber, and medicines. Fodder and food trees provide balanced diets during dry seasons, when other foods are scarce. Trees, with their deep rooting systems, consume moisture and nutrients from higher depth in the soil than arable and pastoral crops, and thus there is least competition among the different components. Thus, the value of trees on farmland may considerably exceed that offered by woodlands and plantations (Singh 1987).

The second major aspect of the research endeavors under agroforestry was collection and evaluation of promising tree species/cultivars of fuel, fodder, and small timber. Quality germplasm of important agroforestry trees has been collected and evaluated in arboretum established by different Centres of the AICRP (Dhyani et al. 2008; Solanki 2006). About 184 promising tree species have been determined based on growth performance trials at these centers. The results indicate that the safest choice of agroforestry species have come from the native vegetation, which has a history of adaptation to local precipitation and other climatic regimes. There are a number of tree-crop combinations, which in turn reflect the differences in the climates and soil fertility of various regions in the country. The examples of major trees in agroforestry practices include: *Grewia optiva*, *Ulmus wallichiana*, *Morus alba*, and *Robina pseudoacacia* in western Himalayan region; *Acacia auriculaeformis, Alnus nepalensis*, Bamboos, *Parasarianthes falcataria* and *Gmelina arborea* for north-eastern Himalayan region; Poplars, Eucalyptus, *Acacia* and *Dalbergia sisoo* in Indo-Gangetic region; *Dalbergia sissoo, Acacia tortilis, A. nilotica, Ailanthus excelsa, Prosopis cineraria, Leucaena leucocephala* and *Azadirachta indica* for arid and semi arid regions; *Acacia nilotica, Prosopis cineraria* and *Zizyphus* in western India; *Tectona grandis*, *Tamarindus indica*, Para rubber (*Hevea brasilensis*) and cashew nuts (*Anacardium occidentale*) in southern region; *Albizia* spp. *Gmelina arborea, Gliricidia, Acacia auriculaeformis, A mangium* for humid and sub-humid regions; and *Artocarpus*, *Pongamia*, *Casuarina equisetifolia, Grevillea robusta* and bamboos in coastal and island regions. The efforts made so far has created voluminous database, which is a great strength for planning agroforestry research strategy. The information collected may be utilized for creating local and regional timber volume tables. The Planning Commission (2001) has also identified six species for major thrust in agroforestry and Joint Forest Management (JFM) programs in the country. The species are *Acacia nilotica*, Bamboos, *Casuarina equisetifolia, Eucalyptus* spp., *Populus deltoides* and *Prosopis cineraria.*

On the basis of identification and evaluation of promising trees and D&D survey of the existing systems, agroforestry interventions were initiated in different agro-climatic regions. Primarily these systems are put in different categories such as agrisilviculture, agri-horticulture, agri-horti-silviculture, hortipastoral, silvopastoral, and specialized systems. Among these systems, agrisilviculture followed by agrihorticlture, are the most prominent being practiced and advocated for the majority of the agro-climatic zones. Home gardens, block plantation, energy plantation, shelterbelts, and shifting cultivation are some of the specialized agroforestry systems developed by the research institutions with the knowledge gained by documenting and analyzing the traditional systems. However, the agroforestry systems developed and recommended by these institutions are location specific, besides in many situations these systems are not looked favorably by the farmers due to wide variations in the choice of species, crop preferences, etc. Also some of the systems developed have not yet completed a full rotation for the tree component, therefore, biomass production and returns are only on the basis of extrapolation. The stakeholders also require full package of practices, choice of crops with trees in temporal sequence and information on insect-pest-disease and effect of aberrant weather conditions, which are now of common occurrence.

Agroforestry initiatives in the country has greatest role in creating awareness among the common man for the protection and preservation of trees, bushes, and grasses. It has been realized that destruction and degradation of forest resources may have detrimental effect on soil, water and hence on human and animal life. In fact, Stern (2006) points out that reducing deforestation and forest degradation would be one of the most cost-effective mitigation approaches. Unfortunately, most forest soils are not suitable for agriculture and quickly became unproductive. The felling of trees for commercial and domestic wood products is mostly unregulated and beyond the forests' ability to replenish itself. Similarly, the grazing of livestock in forested areas is often beyond their carrying capacity. Going by the potential for economic exploitation, it would appear that 90 % of the forests are performing critical functions of protecting fragile watersheds and are not fit for commercial exploitation. As a shift in the National Forest Policy of India harvesting from forests has now practically been banned with social benefits mainly flowing from the protective and environmental functions of the forest apart from meeting the subsistence needs of the communities living close to the forests.

Agroforestry and Ecosystem Services

Owing to increase in population of human and cattle, there is increasing demand of food as well as fodder in the country due to large human and livestock population. Besides, there is little scope to increase food production by increasing the area under cultivation. Hence food production is to be increased from the land already under cultivation or from land not conventionally considered arable. A management system therefore needs to be devised that is capable of producing food from marginal agricultural land and is also capable of maintaining and improving the quality of producing environment. Agroforestry has both productive and protective potential, and it can play an important role in enhancing the productivity of our lands to meet the demand of ever-growing human and livestock population.

In the country, about 7.45 million ha area has been planted with different types of agroforestry plantations (Dhyani et al. 2006). Besides, 25.72 million ha area has been covered under various types of tree plantations viz. agroforestry, social forestry/farm forestry. Future prospects for expansion of agroforestry in different agro ecological regions of India exist in 10 million ha irrigated and 18 million ha rainfed areas (Planning Commission 2001). Further, 15 million ha degraded forest could be developed through joint forest management (JFM). The increase in area under agroforestry and trees outside forests is expected to minimize current deficit in supply of fuel wood (334.5 million tons) and timber (16.0 million cum).

Systemic research in agroforestry has clearly shown that it can contribute significantly to meet the deficits of fuel, fodder, timber, accelerate economic growth, and help in poverty alleviation, women empowerment, and livelihood support in several ways. A major role for agroforestry is now emerging in the domain of ecosystem or environmental services such as biodiversity conservation, watershed protection, carbon sequestration, and mitigating climate change effects besides in livelihood security and employment generation. The major environmental functions of agroforestry can be enumerated as-

- Control of soil degradation and rehabilitation of problem soils
- Control of desertification
- Flood and drought moderation
- Reduction in the pollution of groundwater resulting from high inputs of fertilizers
- Increasing biodiversity in the farming system and watershed scale
- Increasing food security and thereby reduce pressure on land resources
- Checking deforestation and its associated impact on environment
- Reducing pressure on forests through on-farm supply of fuelwood, fodder and other forest products
- Reduction in the build-up of atmospheric carbon dioxide and other greenhouse gases and mitigating adverse effects.

Agroforestry is playing the greatest role in maintaining the resource base and increasing overall productivity of agriculture, thus helping in building climatic resilient agriculture.

Agroforestry for Food and Nutritional Security

The country's food production has increased many folds since independence but recent improvements in food supply have been insufficient to fulfill the nutritional needs of the average person in an ever increasing population of the country. Agroforestry with appropriate tree-crop/legume combination is one option in this regard. The different agroforestry systems provide the desired diversification options to increase the food security of the country and act as a shield against the poor production during drought and other stress conditions. The agroforestry also provides nutritional security because of diverse production systems which include fruit, vegetables, legumes, oilseed crops, medicinal, and aromatic plants in addition to normal food crops grown by the farmers. With the rapid growth of urbanization and economic growth in the country, farming community have witnessed unprecedented opportunities for moving beyond subsistence farming to supplying products needed by urban population. Agroforestry products such as timber, fruit, food, fibre, fodder, medicine, and others are progressively meeting the subsistence needs of households and providing the platform for greater and sustained productivity. Thus, agroforestry systems offer opportunities to farmers for diversifying their income and to increase farm production. Research results from different agro-climatic regions of the country show that financial returns generated from agroforestry systems vary greatly but are generally much higher than returns from continuous unfertilized food crops. The higher returns associated with agroforestry can translate into improved household nutrition and health, particularly when women control the income. Agroforestry has proven as an important tool for crop diversification. By virtue of diversity of the components of the agroforestry systems like fruits, vegetables, nutritional security to the communities could be ensured. There are ample evidences to show that the overall (biomass) productivity, soil fertility improvement, soil conservation, nutrient cycling, microclimate improvement, carbon sequestration potential of an agroforestry system is generally greater than that of an annual system. In agroforestry, the potentially higher productivity could be due to the capture of more growth resources, e.g., light or water or due to improved soil fertility. The best example is of poplar (*Populus deltoides*)—a popular species in agroforestry system in the Upper Indo-Gangetic region. Poplar was a best choice as it was fast growing, compatible with wheat and other crops and has industrial use. Therefore, poplar-(*Populus* spp.) based agroforestry in northern India made rapid strides. At present, there are 70 million poplar trees in the agricultural fields of the upper Gangetic region producing 10.40 million m^3 of industrial wood. Woodlots of other fast growing trees such as *Eucalyptus* spp., *Leucaena leucocephala*, *Casuarina equisetifolia*, *Acacia mangium*, *A. auriculaeformis*, *Ailanthus*, teak, and *Melia dubia* are also becoming increasingly popular among the farmers in several parts of the country due to their great market potential. Genetically improved clonal planting stock of eucalypts, poplars and acacias has transformed the productivity and profitability of plantations. Average yields from such clonal plantations are 20–25 times higher compared to the average productivity of forests in India. Almost 50 million plants of improved *Eucalyptus* are being planted every year.

Agroforestry for Fodder Production

It is evident from discussion in several chapters that trees and shrubs often contribute substantial amount of leaf fodder in all the agro-climatic regions. In arid and semi-arid and hill regions fodder scarcity is more especially during lean period when fodder is collected through lopping/pruning of trees, popularly known as top feed. The leaf fodder yield depends on species, initial age, lopping intensity, and interval as well as agro-climatic conditions. Fodder from trees is mainly available from two parts viz., leaf twigs and pods.

This forage is usually rich in proteins, vitamins, and minerals like calcium. They are however in general low in phosphorus and crude fiber (Singh 1990). Such top feeds species play an important role in human food security through their function as animal feed resource. The importance of top feeds increases with the severity of drought and progression of drought season.

Other uses such as for live fencing are complimentary, as they encourage cultivation of the species and increase the availability of feed. Silvopastoral system is the most appropriate land use system for degraded lands. The top feeds are also considered very important in vegetation stabilization and sustained productivity of rangelands (Dhyani 2003). They also play an important role as windbreaks and by providing shade for the grazing animal. The important ones are *Prosopis cineraria, Albizzia lebbeck. Acacia* spp., *Leucaena leucophloea, Dalbergia sissoo, Ailanthus excelsa, Azadirachta indica, Acacia leucophoela*, etc., for the arid and semi-arid region and *Grewia optiva, Morus alba, Celtis australis, Albizia,* Oaks, species of *Ficus,* etc., for the hilly regions. Tree fodder provides enough nutrients and can serve very well as a green fodder supplement. Besides providing green fodder, such leaves are also conserved in the form of hay and silage to supplement feed during scarcity periods. Bamikole et al. (2003) reported that feed intake, weight gain, digestibility, and nutrient utilization can be enhanced by feeding *Ficus religiosa* in mixture with *Panicum maximum*, and it can be used in diet mixtures up to 75 % of dry matter fed. Dagar et al. (2001) reported that for silvopastoral system on alkali soils *Prosopis juliflora, Acacia nilotica* and *Tamarix articulata* are the most promising trees and *Leptochloa fusca, Chloris gayana* and *Brachiaria mutica* most suitable grasses. *L. fusca* in association with *P. juliflora* produced 46.5 t ha^{-1} green fodders over a period of four years without applying any amendments and fertilizer. More details are discussed in Chap. 9

Agroforestry for Biofuel Production and Energy Security

India ranks 6th in the world in terms of energy demand accounting for 3.5 % of worlds energy demand since the beginning of twenty first century. The energy demand is expected to grow at 4.8 %. A large part of India's population mostly in rural areas, does not have access to the conventional source of energies. Further the Indian scenario of the increasing gap between demand and domestically produced petroleum is a matter of serious concern. In this connection, fuels of biological origin have drawn a great deal of attention during the last two decades. Biofuels are renewable liquid fuels coming from biological raw materials and has proven to be good substitute for oil in the transportation sector as such biofuels are gaining worldwide acceptance as a solution for problems of environmental degradation, energy security, restricting imports, rural employment and agricultural economy. The potential tree borne oilseeds (TBOs) holding promise for biofuel are *Jatropha curcas*, *Pongamia pinnata*, *Simarouba*, *Azadirachta indica*, *Madhuca* spp., etc. These biofuel species can be grown successfully under different agroforestry systems. There is need to identify the genetically superior germplasm of these biofuel species for higher seed yield and oil content (Dhyani et al. 2011). At present, the germplasm of *Jatropha* and *Pongamia* is under multilocation trials to identify the superior germplasm. The promotion of the use of oils could also provide a poverty alleviation option in the rural areas. Farmers can use vacant, waste, and marginally used land for growing such trees and benefit from the annual produce, which will add as their income. With the increased green cover, the environment will also benefit greatly. The use of oils is also CO_2 neutral, which would mitigate greenhouse effect. But the economics and viability of the Jatropha plantation and biofuel production are still at initial stage and will be governed by international market prices of crude oil as well as government policies.

Agroforestry for Energy Plantation

The main biomass energy sources in rural areas which are being used in the households, include wood (from forest, croplands, and homesteads), cow dung, and crop biomass. Among the sources 70–80 % energy comes through biomass from trees and shrubs. Due to the agroforestry initiatives, large amount of woods are now being produced from outside the conventional forestlands. Small landholdings and marginal farmers, through short rotation forestry and agroforestry practices are now providing the bulk of country's domestically produced timber products. Ravindranathan et al. (1997) reported for a Karnataka village that 79 % of all the energy used came mainly from trees and shrubs. *Prosopis juliflora* due to high calorific value of over 5,000 kcal is the major source of fuel for the boilers of the power generation plants in Andhra Pradesh (the other materials are rice husk, cotton stalks, other wood, etc.). About INR 700–1,300 per ton is the price offered for *P. juliflora* wood at factory gate depending on the season and moisture content. An estimated 0.51 million ha area is considered under *P. juliflora*. Even if 25 % of this area is utilized for power generation leaving the rest 75 % for fuel and charcoal, the bioenergy potential works out to 1,000 MW. Similarly, other fast growing and high biomass producing trees/shrubs like *Leucaena, Jatropha*, and *Gliricidia* can be used for running the biomass based power plants. Thus, a total of 5,000 MW power could be produced from the biomass sources from trees under moderate conditions, which meets almost one-third of the ultimate potential of 16,000 MW from biomass. R&D can help in enhancing productivity and assisting the power plants in captive plantation management on degraded lands. Promoting bioenergy through *P. juliflora* also encourages tremendous employment generation to the tune of 6.34 million mandays and 7.03 million woman days for fuel making in Tamil Nadu alone.

The fuel wood potential of indigenous (*Acacia nilotica, Azadirachta indica, Casuarina equisetifolia, Dalbergia sissoo, Prosopis cineraria,* and *Ziziphus mauritiana*) and exotics (*Acacia auriculaeformis, A. tortilis, Eucalyptus camaldulensis* and *E. tereticornis*) trees was studied by Puri et al. (1994). The calorific value ranges from 18.7 to 20.8 MJ $kg^{-1}$for indigenous tree species and 17.3 to19.3 MJ kg^{-1} for exotics. Pathak (2002) opined that species such as *C. equisetifolia, Prosopis juliflora, Leuceana leucocephala* and *Calliandra calothyrsus* have become prominent due to their potential for providing wood energy at the highest efficiency, shorter rotation and also their high adaptability to diverse habitats and climates.

Agroforestry for Soil Conservation and Amelioration

Agroforestry plays a key role in keeping the soil resource productive, which is one of the major sustainability issues. Closely spaced trees on slopes reduce soil erosion by water through two main processes: first as a physical barrier of stems, low branches, superficial roots, and leaf litter against running water and secondly as sites where water infiltrates faster because of generally better soil structure under trees than on adjacent land. Agroforestry played a major role in the recent past in rehabilitation of wasteland such as desert and lands that have been degraded by salinization and ravines, gullies and other forms of water and wind erosion hazards. These aspects have been discussed in detail in earlier chapters.

Agroforestry systems on arable lands envisage growing of trees and woody perennials on terrace risers, terrace edges, field bunds, as intercrops and as alley cropping in the shape of hedge row plantation. Integrating trees on the fields act as natural sump for nutrients from deeper layers of soil, add bio-fertilizer, conserve moisture, and enhance productivity of the system. The alley cropping with leguminous trees such as Subabul (*Leucaena leucocephala*) has been most widely used on field bunds for producing mulch material for moisture conservation and nutrient recycling. Alley cropping with *Leucaena leucocephala* was effective for erosion control on sloping lands up to 30 %. Reduction in crop yield could be minimized by shifting the management of trees to contour hedge rows. The sediment deposition along the hedge and tree rows increased considerably with consequent reduction in soil loss. Improvement in the organic matter status of the soil can result in an increased activity of the favourable micro-organisms in the root zone. In addition to the nutrient relations, such micro-organisms may also produce growth-promoting substances through desirable interaction and result in better growth of plant species. Inclusion of trees and woody perennials on farm lands can, in the long run, result in marked improvements in the physical conditions of the soil, e.g., its permeability, water-holding capacity, aggregate stability, and soil-temperature regimes. Although these improvements may be slow, their net effect is a better soil medium for plant growth. Experimental evidences give a very clear picture about agroforestry system that increased soil organic carbon and available nutrients than growing sole tree or sole crop. An increase in organic carbon, available N, P, and K content in Khejri based silvopastoral system over no-Khejri soil, advocating retention/plantation of Khejri tree in pasture land to get higher fodder production and to meet requirement of food, fodder, fuel, and small timber is one such example. Similarly, an increase in soil organic carbon status of surface soil under *Acacia nilotica* + *Sacchram munja* and under *Acacia nilotica* + *Eulaliopsis binata* after 5 years was observed. It was found that *Acacia nilotica* + *Eulaliopsis binata* are conservative but more productive and less competitive with trees and suitable for eco-friendly conservation and rehabilitation of degraded lands of Shiwalik foot hills of sub-tropical northern India. Rehabilitation of degraded forests is possible through afforestation by adopting integrated land use planning with soil and water conservation measures on watershed basis. NRCAF observed that in agrisilviculture growing of *Albizia procera* with different pruning regimes, the organic carbon of the soil increased by 13–16 % from their initial values under different pruning regimes, which was five to six times higher than growing of either sole tree or sole crop.

Agroforestry systems have been developed using local resources and conservation-based measures in the North Eastern Hill (NEH) region. Suitable alternate land use systems involving agriculture, horticulture, forestry, and agroforestry have been designed with the support of local natural resources for almost identical hydrological behaviour as under the natural system. The model land use suggests utilizing slopes below 50 % toward lower foothills and valley lands for agricultural crops and pisciculture, middle slopes between 50 and 100 % for horticulture and top slopes over 100 % for forestry/silvopastoral establishment. Under agri-horti-silvopastoral systems, the reduction in runoff was 99 % and in soil loss 98 %. Combining fine-root system of grasses and legumes, such as *Stylosanthes guyanensis*, *Panicum maximum*, *Setaria,* etc., and deep-root system of fodder trees, such as alder (*Alnus nepalensis*) in a silvopastoral system stabilizes terrace risers and provides multiple outputs. In-depth evaluation of soil chemical properties of traditional agroforestry system in northeastern region indicated a spectacular increase in soil pH, organic-C, exchangeable Ca, Mg, K, and build up of available P (Bray's P_2-P) under different agroforestry practices (AFP) within 10–15 years of practice. The exchangeable Al, potential cause of infertility of these lands disappeared completely within 10–15 years of agroforestry practice. Therefore, the agroforestry practices were found to have built in dynamism for the restoration of soil fertility and sustained yield (Singh et al. 1994a, b). Similar results were obtained when multipurpose trees were evaluated in an extremely P-deficient acid Alfisol in Meghalaya (Dhyani et al. 1994).

The use of trees as shelterbelts in areas that experience high wind or sand movement is well-established example of microclimate improvement that resulted in improved yields. Increased agricultural production due to windbreaks and shelterbelts in India has been well demonstrated. Establishment of micro-shelterbelts in arable lands, by planting tall and fast-growing plant species such as castor bean on the windward side, and shorter crop such as vegetables in the leeward side of tall plants helped to increase the yield of lady's finger by 41 % and of cowpea by 21 % over the control. In general, the use of shelterbelts brought about a 50 % reduction in the magnitude of wind erosion. In studies carried out in an agroforestry system, *Acacia tortilis* (7-year old) and guar crop at Jodhpur indicated that relative humidity recorded beneath the tree canopy during the active cropping season of guar was found to be 7 % more than in the open. This will, in turn, help for better growth of the crops.

Agroforestry practices have been developed for arable lands and nonarable degraded lands, boulder laden riverbed land, torrent control, landslide and landslip stabilization, abandoned mine-spoil area rehabilitation, and as an alternative to shifting cultivation. Also, agroforestry systems have proven their efficacy in prevention of droughts, reclamation of waterlogged areas, flood control, rehabilitation of wastelands, ravine reclamation, sea erosion control, control of desertification and mine-spoil rehabilitation, and treatment of saline and alkaline lands. Agricultural use of salt-affected lands and water resources increase due to increasing demands of food and fodder is yet another example where agroforestry played a great role in enhancing the productivity of land and also address the environmental issues. Removal of salts from the soil surface is neither possible nor practical; therefore attempts have been made to minimize adverse effect of salts on crop by developing agro techniques. Central Soil Salinity Research Institute, Karnal (Haryana) has developed appropriate package of agro-techniques for crop production in salt-affected soils. The Institute also developed special planting techniques for sodic and saline soil for better establishment and growth of multipurpose trees. The technique ensures more than 80 % tree survival even after 10 years in highly alkali soil. See the details in Chap. 9.

Bio-Diversity Conservation Through Agroforestry

It is established that India is having rich vegetation with good bio-diversity. According to the Government of India report, biological diversity is estimated to be over 45,000 plant species and 810,000 animal species, representing 7 % of the world flora and 6.5 % of the world fauna, respectively. The UN Convention on Biological Diversity calls for conservation of the biological diversity, sustainable use of its components and fair and equitable sharing of benefits arising out of the utilization of genetic resources. Agroforestry innovations contribute to bio-diversity conservation through integrated conservation-development approach. Forest degradation has caused immense losses to the bio-diversity, which can be conserved through agroforestry by adopting a strategy of conservation through use. The biodiversity shall help in the development or improvement of new varieties or populations. It will further help in enhancing the availability of improved planting material, which is a key to the increase the productivity and production at farm level. Swaminathan (1983) has pointed out that biodiversity is the feed stock for a climate resilient agriculture. Agroforestry with components like trees, agricultural crops, grasses, livestock, etc., provides all kinds of life support system. Trees in agro-ecosystems in Rajasthan and Uttarakhand have been found to support threatened cavity nesting birds and offer forage and habitat to many species of birds. The traditional society of coastal belts and tropics of the country practicing homegardens and sacred groves help in biodiversity conservation. In a majority of the traditional villages, trees have been planted and dedicated to different gods/goddesses or have stated as abodes of spirit, making them sacred. Frequently, species selected by the local people for social significance might turn out to be of ecological significance. It helps ensure sustainability, stability, and productivity of production systems in spite of the level of complexity of the ecosystem in which it occurs, and in the final analysis, it contributes to social welfare of the population through its involvement to poverty mitigation and sustainable food security.

Carbon Sequestration Potential of Agroforestry

Agroforestry has importance as a carbon sequestration strategy because of carbon storage potential in its multiple plant species and soil as well as its applicability in agricultural lands and in reforestation. The potential seems to be substantial; average carbon storage by agroforestry practices has been estimated as 9, 21, 50, and 63 t C ha^{-1} in semiarid, subhumid, humid, and temperate regions (Schroeder 1994). For smallholder, agroforestry systems in the tropics, potential carbon sequestration rate ranges from 1.5 to 3.5 t C ha^{-1} yr^{-1} (Montagnini and Nair 2004). Agroforestry can also have an indirect benefit on carbon sequestration when it helps to decrease pressure on natural forests, which are the largest sinks of terrestrial carbon. Another indirect avenue of carbon sequestration is through the use of agroforestry technologies for soil conservation, which could enhance carbon storage in trees and soils. Carbon compounds are sequestered or accumulated by plants to build their structure and maintain their physiological process. The energy captured in the molecular bonds of carbon compounds generally present between 2 and 4 % of the radiation absorbed by the tree canopy. Stem wood growth often accounts for less than 20 % of the dry matter produced in a year, the rest being used by foliage most of which is shed during leaf fall which is an important pathway for the flow of organic matter and energy from the canopy to the soil. Only green plant can assimilate carbon on the earth. The analysis of carbon stocks from various parts of the world showed that significant quantities could be removed from the atmosphere over the next 50 years if agroforestry systems are implemented on a global scale. Carbon storage depends on several factors including climatic, edaphic, and socio-economic conditions. Perennial systems like home gardens and agroforestry can store and conserve considerable amounts of carbon in living biomass and also in wood products. For increasing the carbon sequestration, potential of agroforestry systems practices such as conservation of biomass and soil carbon in existing sinks; improved lopping and harvesting practices; improved efficiency of wood processing; fire protection and more effective use of burning in both forest and agricultural systems; increased use of biofuels; increased conversion of wood biomass into durable wood products needs to be exploited to their maximum potential. Agroforestry practices such as agrisilviculture or agrihorticulture systems for food and wood/fruit production; boundary and contour planting for wind and soil protection; silvopasture system for fodder production as well as soil and water conservation; complex agroforestry systems, viz., multistrata tree gardens, home gardens, agrisilvohorticulture and horti-silvopasture systems for food, fruits, and fodder especially in hill and mountain regions and coastal areas and biofuel plantations are suitable for sequestering atmospheric carbon and act as the potential sinks for sequestering surplus carbon from the atmosphere.

In India, evidence is now emerging that agroforestry systems are promising land use system to increase and conserve aboveground and soil C stocks to mitigate climate changes (Dhyani et al. 2009b). The average potential of agroforestry has been estimated to be 25 t C ha^{-1} over 96 million ha^{-1} (Sathaye and Ravindranath 1998). In this way, the total potential of agroforestry in India to store C is about 2,400 million tons. Although there is variation in the estimation of area under agroforestry and C stock but there is good indication of agroforestry for gaining popularity for mitigating climate change because desired tree cover can only be achieved including tree in farm field/bunds. The C storage capacity varied from region to region and also depends upon the growth and nature of tree species involved in the system.

Agroforestry for Livelihood Security and Employment Opportunities

Agroforestry systems due to diverse options and products provide opportunities for employment generation in rural areas. Dhyani et al. (2003, 2005) have highlighted the role of agroforestry products and environmental services to meet the subsistence needs of low income households and providing a platform for greater and sustained livelihood of the society. Increased supply of wood in the market has triggered a substantial increase in the number of small-scale industries dealing with wood and wood based products in the near past. Such industries have promoted agroforestry and contributed significantly to increasing area under farm forestry. Recognizing agroforestry as a viable venture, many business corporations, limited companies such as ITC, WIMCO, West Coast Paper Mills Ltd., Hindustan paper Mills Ltd., financial institutes such as IFFCO have entered into the business and initiated agroforestry activities in collaboration with farmers on a large scale. Besides the existing agroforestry practices, there is a tremendous potential for employment generation with improved agroforestry systems to the tune of 943 million person days annually from the 25.4 million ha of agroforestry area. Dhyani and Sharda (2005); Dhyani et al. (2005) have indicated the potential of agroforestry for rural development and employment generation to the tune of 5.763 million human days per year from Indian Himalayas alone.

Sericulture is being practiced in different parts of the country since time immemorial. On the basis of climate, edaphic conditions, host plants, and insect species, four types of sericulture is practiced in India viz. (i) mulberry, (ii) tasar—[a] tropical tasar, [b] oak tasar, (iii) eri, and (iv) muga. Dhyani et al. (1996) successfully developed and demonstrated mulberry and muga sericulture based agroforestry systems for the north-eastern hill region. Sericulture with fruit plants and grass model was highly preferred by farmers, followed by sericulture with field (uplands) crops. Now, tasar sericulture is being promoted in different districts of Bundelkhand region. For this, tasar sericulture insect *Antheria mylata* is being reared on Arjun (*Terminalia arjuna*) which is common tree species of the forest area in this area. Arjun can also be cultivated under dense plantation (2,800 plants per ha at 2 m × 2 m spacing) and intensive management for economic tasar cultivation. The plantation is ready for rearing tasar silkworm within 3–4 years. As per the information of Sericulture Directorate, Uttar Pradesh, one ha dense plantation of Arjun can generate an income of INR 30 to 50 thousand from tasar cultivation. There is scope of tasar sericulture under agroforestry. To promote livelihood opportunities for the farmers, NRCAF introduced lac based agroforestry system for the semi-arid Bundelkhand region on Palas (*Butea monosperma*) and Ber trees which are very common in this region. Success of lac cultivation in *katki* crop (rainy season) was observed in the region. The preliminary results indicate good possibility to promote livelihood through lac cultivation in the region. Similarly, there is scope for augmenting income of farmers by collecting gum and resins from trees. NRCAF identified suitable trees for gum and resin in different agro-climatic regions for development under agroforestry.

However, to achieve the above potential, it will require appropriate research interventions, adequate investment, suitable extension strategies, harvest process technology development of new products and market infrastructure. It will also need the development of mechanism to reward the rural poor for the environmental services such as biodiversity conservation, watershed protection and carbon sequestration that they provide to society. Above and all, it will require to develop a forward looking National Agroforestry Policy.

Agroforestry Policies

Agroforestry growth and development is influenced by various policies of the economy like credit, trade, taxation, power, transport, etc. These policies impact the sector either directly or indirectly besides the core forest and agriculture policies. The forest policies have a larger bearing on the agroforestry policies in the Indian context. Since forests are in the concurrent list of the constitution, both the union and state governments are empowered to amend the forest laws. Hence, the scope for conflicting policies exists. Historically trees in India have been managed and regulated by customary law. The British exploited the commercial value of Indian forests by establishing rigid control. A systematic forest policy was initiated in 1855. This was followed by the Government Forests Act of 1865 and Indian Forest Act 1878, which classified forests into reserve forests, protected forests, and village forests. The National Forest Policy of 1894 encouraged diversion of forestlands for agriculture, which suited the growing population. Independent India has given two forest policies. The National Forest Policy 1952 gave thrust on having one-third geographical area under forests with a target of 60 % forest cover in hills and 20 % in plains. The National Commission on Agriculture (NCA 1976) had recommended commercialization of forests to meet the needs of defense forces and industry. The involvement of corporate bodies in providing technical knowhow to farmers who took up farm forestry and the momentum given by the social forestry projects were very conducive for tree plantation outside the forest areas. The revised National Forest Policy, 1988 encourages private sector tree plantations on other than forestlands. The Joint Forest Management (JFM) programmes were initiated in the country during 1990 as a sequel to the revised NFP of 1988. The 73rd amendment to the constitution and the recommendations of Bhuria committee, 1996 have given the *Panchayats* wide powers of control over the natural resources including land and forest produce. Subsequently, state governments issued enabling resolutions one by one. As of now, all the state governments in India have issued resolutions in this regard. The MOEF in 1999 initiated National Forestry Action Plan (NFAP) and National Forestry Research Plan (NFRP), which may have some influence on agroforestry development. For promoting agroforestry, some of the states took innovative policy decisions. However, there are many bottlenecks in the marketing of agroforestry produce like inadequate infrastructure, lack of awareness on final use of the produce, exploitation by the middlemen, etc. The tree felling and transit rules as applicable to trees on private lands unique to each of the states have a direct bearing on the progress of agroforestry. There is wide variation in the scope of the felling and transit rules across states. Realizing the magnitude of the problem the MOEF has circulated guidelines to states in December 2004 for relaxing the felling and transit regulations for tree grown on non-forest laws. The guidelines are primarily aimed at achieving the goal of one-third-tree cover in the country. Responding to the guidelines, some states enacted suitable laws but many of them such as Kerala, Odisha, and others retracted instantly to prevent large scale felling. Much more need to be done, say empowering the *Panchayats* to permit the felling and transit of less unrestricted, i.e., category B species.

Agroforestry promotion for realizing the true potential requires efforts from all the concerned like forest and agriculture departments, research and financial institutions, corporate sector, *panchayats*, NGOs, and farmers. The policy thrust must be on creating infrastructure, ensuring quality planting material, plugging the vacuum of extension network, favourable pricing mechanism including support price system, encouraging public–private partnerships in all spheres of agroforestry among others. Recently, an Agroforestry Policy Initiative was jointly organized by the Indian Council of Agricultural Research (ICAR) and The World Agroforestry Centre (ICRAF) at New Delhi on June seventeenth, 2011 to prepare a road map for development of a National Agroforestry Policy. In XIIth Plan it is expected that some concrete proposals for launching a National Mission on Agroforestry and National Agroforestry Policy will be initiated by the MoEF and MoA.

Success Stories

Poplar Based System

The technology has been developed by AICRP on Agroforestry centre at PAU, Ludhiana. Punjab is an agricultural state and the state has only 5.57 % area under the forest (2.81 lakh ha). The farmers of Punjab have taken to tree planting on their private lands in a big way by raising *Eucalyptus* and poplar plantations on a large tract. The major production of poplar wood in the state is consumed by plywood industries. In Punjab, 75 units (big and small) are manufacturing plywood and plyboard. The plywood industry in Punjab has expanded by almost 50 % from 49 units in 1993 and 90 units in 2008. Owing to short duration and fast growing habits, poplar has become an important agroforestry tree species in Punjab. It offers opportunities for increasing both the income of farmers and the area under tree growth. Farmers have been practicing the integrated cultivation of agricultural crops with fast growing trees to meet their requirements of firewood, fodder, small timber for agricultural and construction purposes. Farmers have earned good returns from the sale of these farm grown trees and thus total area under poplar increased many fold. The central zone has come up as poplar farming belt as 68 % tree growers are practicing poplar based agroforestry system. Large percentage of poplar growing farmers are planting this species in riverine belts of Sutlej and Beas, as the growth of poplars in *'bet'* area is 50–60 % more as compared to table lands. Poplar growers are quite limited in the south-west zone and sub-montane zone due to regular and high irrigation requirements of this tree species which is scanty in these zones. Poplars are grown on agricultural lands either in single rows along field or farm boundaries or even in block plantation at a spacing of 5 × 4 or 7 × 3 m along with annual agricultural crops such as cash crops like sugarcane, turmeric, wheat, pearl millet, oats, maize, mustard, vegetables, and several other fodder crops oats, fodder (berseem), etc., throughout the rotation. Generally farmers do not grow *'kharif'* season crops owing to dense shade of poplars during later years (after 4 years). Poplar + wheat + fodder based agroforestry system is a great success in Punjab. This system is economically viable and much more profitable in comparison to sole cropping system (wheat + rice) commonly followed by majority of farmers. The comparative economics of poplar-based agroforestry model in block and boundary plantation revealed that these were 2.8 and 1.6 times more profitable respectively than rice–wheat rotation. In addition, this system has played an important role in the upliftment of social status of farmers, generated employment and became one of the viable options for crop diversification. The impact of the success resulted in

- Area under tree cover increased in the State
- Productivity per unit area and per unit time increased
- Overall income in terms of money increased per unit area and per unit time over the existing cropping system
- Employment generation at primary, secondary, and tertiary levels
- Social upliftment of small and marginal farmers by adopting boundary plantation.

Sapota/Mango-Teak Based Agroforestry Systems

A multicomponent agroforestry system with sapota (*Achrus zapota*) as base crop, teak (*Tectona grandis*) in the sapota line and agricultural crop in the interspaces was developed by AICRP on Agroforestry, Dharwad Centre for high rainfall areas/areas with irrigation facility of Karnataka. Broad spacing between trees provides an opportunity to intercrop in the initial few years. Under Technology Extension Project on Agroforestry, demonstrations were taken up in 1996 in the village Kyarakoppa, Dharwad, Karnataka. Sapota was planted at a recommended spacing of 10 × 10 m, tree rows being across the slope. Three teak plants were planted at distance of 3–2 to 2–3 m in between two sapota trees. Field crops viz. horse gram, jowar and bajra were grown in the interspaces of sapota + teak alleys. Sapota crop served as an insurance against failure of field crops. Teak trees act as a fixed deposit to the farmers to come in handy during adverse situations. The system generated employment throughout the year. The same technology is adopted by other farmers with a modification viz. Sapota being replaced by Mango. Fruit yields in sapota and mango started from the year 2003 and 2004, respectively. Presently, the sapota tree is yielding 30–40 kg per plant which accounts to be INR 22,000–25,000 per ha and from mango is yielding 30–50 kg per plant which accounts to INR 36,000–60,000 per ha. The income generated from field crop in both the cases is about INR 2,500–3,500 per ha. Growing of field crop was stopped from 2007 due to closure of canopy. The value estimation of teak reveals that each teak pole fetches INR 120. Adoption of Sapota/Mango–Teak based agroforestry model improved microclimatic conditions in the plantation. Under the system, employment generated was to an extent of 180 man days per year. The socio-economic status of the farmers improved and farmers are earning on an average of INR 23,500 ha^{-1} y^{-1} with sapota and INR 48,000 ha^{-1} y^{-1} with mango as against INR 3,000 ha^{-1} y^{-1} during initial period on the same land.

Continuous Contour Trench: A Phule Technique

Shri. Genuji Bhimaji Kurkute a person from middle level farmers family made sincere and continuous effort for the development of Jogaldara hilly area of Jachakwadi of Akola Tahasil in Ahmednagar District of Maharashtra State and succeed to change the whole scenario of Gajanan valley of Jogaldara hill. He purchased 35 acre of mountains hilly land of Sahaydri ranges at Akole and Sangmner Taluka in Ahmednagar District of Maharashtra, locally known as Jogaldara hill. After that he contacted the Scientists of Mahatma Phule Krishi Vidyapeeth, Rahuri, and discussed regarding the development of Jogaldara hill by adopting soil conservation techniques for development of huge hilly, undulated area by tree plantation, intercropping, suitable crops, supplemental irrigation system, etc., through continuous contour trenches (CCT) technique developed by AICRP on Agroforestry centre at MPKV, Rahuri.

Before implementing the development work, he got surveyed all the 35 ha area. After survey of near about 272 feet gradient of hill, he adopted CCT techniques for plantation of different fruit trees. Initially, pomegranate (Var. Mridula and Bhagava) fruit trees were planted on 18 ha hilly area, thereafter, custard apple (Balanagar), mango (Keshar), and gooseberry (Krishna, Kanchan, Narendra, etc.) were planted on 12 ha area The plantation of pomegranate was done on 3 × 3 m spacing and used only organic manure in the plantation. For supplementary irrigation during summer season was very much essential. He constructed one water tank of 180 thousand liters capacity and purchase drip irrigation system in consultation with Jain Irrigation by submitting project proposal to 'National Horticulture Board' for financial assistance in the name of "Gajanan Valley High Tech Project." Thereafter, he has established three bore wells at suitable locations and all these three bore wells were interlinked with each other. The capacity of each of these bore well are 12.5, 7.5, and 5 HP. These three bore wells continuously lifting water, for collection in storage water tank. Shri Kurkute emphasized the benefits of CCT techniques in getting water throughout the year. The main, submain, and laterals of drip irrigation system was attached to the main storage water tank, from where the water is lifted at the peak point of hill through 15 hp electric pump. He got 20 kg average yield of Pomegranate (rate INR 30–35 per kg i.e. INR 600–700 per tree). Accordingly from one hectare area he received near about INR 500–600 thousand. Similarly, from Custard apple he got INR 400–500 thousand per hectare. Shri. Kurkute had obtained INR 4 million Bank loan, within

6/7 year he refunded up to INR three million to bank and till to date only INR 1million is outstanding. According to his experience due to fluctuation in market rate and expenditure on labour the fruit farming with proper utilization of CCT techniques is far better than vegetable farming. Now the Jogaldara hill became a visiting spot to the students from the Agriculture Colleges and Farmers from the neighboring villages/ Districts. Shri Kurkute made a mile stone by adopting technology developed by Mahatma Phule Krishi Vidyapeeth, Rahuri and put forth an example that, how middle level farmers can change the whole scenarios of such mountainous, undulating, hilly region of our country.

There are similar success stories of farmers adopting agroforestry in different agro-climatic regions.

Model watershed project on natural resource management through agroforestry interventions at Garhkundar, Teekamgarh, Madhya Pradesh

One watershed management program has been initiated by NRC for Agroforestry, Jhansi in Garhkundar-Dabar, district Tikamgarh in Madhya Pradesh of Bundelkhand region to demonstrate agroforestry technologies in participatory mode. The impact of integrated watershed management interventions, viz. soil and water conservation measures, agroforestry development, crop demonstrations with improved package of practices, plantation, and human resource development on natural resource conservation and livelihood security in Garhkundar–Dabar watershed was assessed for last 5 years. Soil and water conservation measures generated 25,000 m^3 water storage capacity, reduced number of dry wells from 2 to 86 %, increased average available water column depth in wells from 0.88 to 4.36 m and enhanced water availability to round-the-year from 4 to 5 months during the study period. Runoff per unit area and soil loss from treated watershed was 46 and 42.2 % lower than the untreated watershed respectively in 2009. Average productivity and crop intensity of major crops, increased by 26 and 119.5 %, respectively in 2009–2010 as compared to 2005–2006. The fodder availability increased by 208 % and within four years, watershed became a fodder secure area with fodder surplus of 1.992 t y^{-1} $animal^{-1}$ as compared to (−) 0.569 t y^{-1} $animal^{-1}$ in 2005–2006. The increased direct and indirect employment opportunities in watershed reduced migration to 9 % in 2009–2010 from 29 % in 2007–2008 (Palsaniya et al., 2012). This clearly indicates that watershed management through agroforestry interventions is the only way out for sustainable management of natural resources and to support livelihood in the semi-arid and arid region.

Research Gaps and Way Forward

- Inadequate understanding of the causes and processes of biophysical and physio-biochemical issues related to productivity and resource sharing under agroforestry system.
- Lack of knowledge of genetic potential, breeding behavior, and inheritance pattern of economic traits of important woody perennials.
- Paucity of methods to assess the tangible and intangible benefits of agroforestry.
- Limited commercialization of potential technologies.
- Lack of independent agroforestry Policy.

Conclusions

The retrospective and critical perusal of the agroforestry research conducted during the last three decades exhibits its wide spectral potential in sustenance of agriculture as these systems provide food, fodder, fruit, vegetables, fuel wood, timber, medicines, fiber, etc., from the same piece of land at a time which not only fulfill the demand of people but also elevate their socioeconomic status and standard of life.

Table 11.1 Potential agroforestry systems for different agroclimatic zones of the country

Agro-climatic zone	Agroforestry system	States	Tree component	Crop/grass	Net income $year^{-1}$	Potential area (ha)
Western himalayas	Silvopasture (RF)	Himachal Pradesh	*Grewia optiva*	*Setaria* spp.	18,670	228,487
		Uttar Pradesh	*Morus alba*	*Setaria* spp.	19,732	228,487
	Agrihorticulture	Uttrakhand	*Prunus persica*	Maize, soybean	4,360	265,114
Eastern Himalayas	Agrisilviculture	Arunachal pradesh	*Anthocephalus cadamba*	Paddy	13,880	68,737
		Assam, manipur,				
		Meghalaya, Mizoram,				
		Nagaland, Tripura,				
		West bengal, sikkim				
Lower gangetic plains	Agrisilviculture	West bengal	*Eucalyptus*	Paddy	10,603	66,730
Middle gangetic plains	Agrisilviculture	Bihar,	*Populus deltoides*	Sugarcane-wheat	24,281	161,929
	Agrisilviculture	Uttar Pradesh	*Eucalyptus*	Rice–wheat	16,124	202,411
	Agrisilviculture		*Dalbergia sissoo*	Sesamum	11,600	242,893
Transgangetic plains	Agrihorticulture	Uttar pradesh	*Emblica officinalis*	Black gram/green gram	13,108	372,984
Upper gangetic plains	Agrisilviculture	Haryana, Punjab	*Populus deltoides*	Wheat, bajra fodder	37,125	240,427
Eastern plateau and hills	Agrisilviculture	Delhi, Bihar	*Gmelina arborea*	Paddy, linseed	26,402	185,772
	Silviculture	Jharkhand, MP, Orrisa	Bamboos		6,700	553,460
		West Bengal				
Central plateau and hills	Agrihorticulture (Irrigated)	Madhya pradesh,	*Psidium gujava*	Bengal gram	11,700	177,911
	Agrihorticulture (RF)	Rajasthan, Uttar pradesh	*Emblica officinalis*	Black gram/Green gram	13,108	133,433
	Silvipasture (RF- degraded lands)	Maharashtra	*Albizia amara, L.leucocephala, D.cinerea*	*C. fulvus, S. hamata, S. scabra*	6,095	926,606
	TBOs (RF)		*Jatropha curcas*	–	3,741	308,869
Western and Plateau Hills	Agrihortisilviculture (Irrigated)	Karnataka	*Tectona grandis*	Paddy, maize	80,613	371,378
		Madhya Pradesh	*Achras zapota*			

(continued)

Table 11.1 (continued)

Agro-climatic zone	Agroforestry system	States	Tree component	Crop/grass	Net income year^{-1}	Potential area (ha)
Southern plateau and hills	Agrisilviculture (RF)	Andhra Pradesh	*Eucalyptus*	Cotton	25,605	115,982
	Agrisilviculture (Irrigated)	Karnataka	*Eucalyptus*	Chillies	38,695	463,928
		Tamil Nadu				
	Block plantation (RF)		*Leucaena leucocephala*	–	12,810	231,964
	Block plantation (RF)		*Eucalyptus*	–	32,666	231,964
	Agrihorticulture		*Tamarindus indica*	Chilli	16,126	115,982
	TBOs		*Pongamia pinnata*	–	4,000	60,017
East coast plains and hills	Agrisilviculture (RF)	Andhra Pradesh	*Ailanthus excelsa*	Cow pea	13,237	61,103
		Tamil Nadu				
		Pondicherry				
West coast plains and hills	Agrisilviculture (RF)	Kerala	*Acacia auriculiformis*	Black pepper	114,240	44,629
		Maharashtra				
	Agrihorticulture (RF)	Tamil Nadu	*Artocarpus heterophyllus*	Black pepper	97,440	44,629
	Agrisilviculture (RF)	Goa, Karnataka	*Acacia auriculiformis*	Paddy	21,032	44,629
	Agrisilviculture		*Casurina equisetifolia*	Paddy	24,968	22,315
Gujarat coast plains and hills	Agrisilviculture	Gujarat	*Azadirachta indica*	Cow pea	10,896	98,313
		Dadra& N. Haveli	*Ailanthus excelsa*	Green gram	6,025	98,313
		Daman				
Western dry region	Agrisilviculture	Rajasthan	*Prosopis cineraria*	Pearl millet	30,215	193,768
	TBOs (RF)	Lakshadweep	*Jatropha curcas*	–	3,741	6,058,812
All Islands	Agrihorticulture		*Cocos nucifera*	Paddy	15,433	1

As per perspective plans (based on supply and demand for timber, fuel, and fodder), 338,068 ha of poplars (@ 160 t ha^{-1} productivity at 7 years rotation)
Assuming availability of another 10 % of community wastelands for silvipasture, the additional fodder production (@ 2.5 t ha^{-1} yr^{-1} at 50 % of research yield) would be 14.75 million t from 5.9 million ha in silvipasture (Dhyani et al. 2009a)

Agroforestry is key path to prosperity for millions of farm families, leading to extra income, employment generation, greater food and nutritional security, and meeting other basic human needs in a sustainable manner. As mitigation strategy to climate change as well as rehabilitation of degraded land, the conversion of unproductive grasslands and crop land to agroforestry is a major opportunity as it helps for carbon sequestration and makes land productive and reduces further soil degradation. By virtue of diversity of the components of the agroforestry systems like food grains, vegetables, fruits, nutritional security to the communities could be ensured. Induction of fodder cultivation under agroforestry land use will ensure production of milk, meat, and animal products and also wide range of food crops, pulses, and oil seeds can meet diverse needs of society. The analysis presented here gives a clear identification of the advances made in understanding and appreciation the potential of agroforestry. Owing to increased supply of wood in the market, there has been a significant increase in the number of factories/industries dealing with wood and wood-based ventures. Such industries have promoted agroforestry (through Poplar, *Eucalyptus, Leucaena*, etc.) and contributed significantly in increasing area under agroforestry. On the whole, in addition to promoting indigenous agroforestry models, it appears that a great deal of research needs to be done to identify short rotation, high value species, which suit the farmers' requirement of planting on marginal lands. It would probably be more realistic to select trees that could provide more cash benefit to farmers through their products, and to accept that in the longer term they will also provide environmental benefits arising from a more complex agro- ecosystem (Table 11.1).

References

Bamikole MA, Babayemi OJ, Arigbede OM, Ikhatua UJ (2003) Nutritive value of *Ficus religiosa* in west African dwarf goats. Anim Feed Sci Technol 105(1):71–79

Chauhan DS, Dhyani SK (1989) Traditional agroforestry practices in N. E. Himalayan region. Indian J Dryland Agric Res Dev. 4(2): 73–81

Chauhan DS, Dhyani SK (1994). Agroforestry-an approach to sustainable development of degraded lands in NEH region of India. In: Punjab S et al (eds) agroforestry systems for degraded lands, vol I. Oxford & IBH Publishing Co. Pvt. Ltd., New Delhi, pp 154–158

Planning Commission (2001) Report of the task force on greening India for livelihood security and sustainable development. Planning Commission, GoI, New Delhi, p 254

Dagar JC, Sharma HB, Shukla YK (2001) Raised and sunken bed technique for agroforestry on alkali soils of northwest India. Land Degrad Dev 12:107–118

Dhyani SK (1998) Tribal alders-an agroforestry system in the northeastern hills of India. Agrofor Today (ICRAF) 10(4):14–15

Dhyani SK (2003) Role of watershed management in improving forage production. In: Jakhmola RC, Jain RK (eds) Sustainable animal production. Pointer Publishers, Jaipur, pp 173–207

Dhyani SK, Sharda VN (2005) Agroforestry systems as rural development options for the Indian Himalayas. J Tree Sci 24(1):1–19

Dhyani SK, Singh BP, Chauhan DS, Prasad RN (1994) Evaluation of MPTs for agroforestry system to ameliorate infertility of degraded acid alfisols on sloppy lands. In: Punjab S, Pathak PS, Roy MM (eds.) Agroforestry systems for degraded lands, vol I. Oxford & IBH Publishing Co.Pvt. Ltd., New Delhi, pp 241–247

Dhyani SK, Chauhan DS, Kumar D, Kushwaha RV, Lepcha ST (1996) Sericulture based agroforestry systems for hilly areas of north-east India. Agrofor Syst 34:1–12

Dhyani SK, Sharda VN, Sharma AR (2003) Agroforestry for water resources conservation; issues, challenges and strategies. In: Pathak PS, Ram Newaj (eds) Agrofor potentials opportunities. Agribios (India) and ISA, pp 9–30

Dhyani SK, Sharda VN, Samra JS (2005) Agroforestry for sustainable management of soil, water and environmental quality; looking back to think ahead. Range Manag Agrofor 26(1):71–83

Dhyani SK, Kareemulla K, Handa AK (2006) Agroforestry potential and returns across agroclimatic zones. NRCAF, Jhansi

Dhyani SK, Handa AK, Uma (2008) Twenty five years of agroforestry networking (1983–2008). NRCAF, Jhansi, p. 102

Dhyani SK, Kareemulla K, Ajit, Handa AK (2009a). Agroforestry potential and scope for development across agro-climatic zones in India. Indian J. Forest 32:181–190

Dhyani SK, Ram N, Sharma AR (2009b). Agroforestry: its relation with agronomy, challenges and opportunities. Indian J. Agron 54(3):70–87

Dhyani SK, Kumar RV, Ahlawat SP (2011) *Jatropha curcas* L.: A potential biodiesel crop and its current R&D status. Indian J Agric Sci 81(4):295–308

Kareemulla K, Dhyani SK, Rizvi RH, Yadav RS, Ram M (2007) Agroforestry for rural development: a cooperative experience. NRCAF, Jhansi 91

Kareemulla K, Rhizvi RH, Yadav RS, Munna R, Dhyani SK (2009) Agroforestry for rural development-cooperatives as viable institutions. Agriculture situation in India. pp 61–67

Kumar BM (2003) Homegardens as a livelihood security system in the humid tropics with special reference to Kerala. In: Pathak PS, Ram N (eds) Agroforestry: potentials and opportunities, Agrobios, Jodhpur, pp 121–142

Montagnini F, Nair PKR (2004) Carbon sequestration: an under exploited environmental benefit of agroforestry systems. Agrofor Syst 61:281–295

Palsaniya DR, Singh R, Tewari RK, Yadav RS, Dhyani SK (2012) Integrated watershed management for natural resource conservation and livelihood security in semi arid tropics of India. Indian J. Agric. Sciw 82(3):241–247

Pathak PS (2002) Common pool degraded lands: technological and institutional options. In: Marothia DK (ed) Common Pool Resources. Concept Publishing Company, New Delhi, pp 402–433

Pathak PS, Dagar JC (2000). Traditional agroforestry sytems: an overview. In: Soni P, Srivastava AK (eds) Landmarks of botany in India, Suriya International publisher, Dehradun, pp 26–49

Pathak PS, Pateria NM, Solanki KR (2000) Agroforestry systems in India: a diagnosis and design approach. NRC for Agroforestry. ICAR. pp 166

Puri S, Singh S, Bhushan B (1994) Evaluation of fuel wood quality of indigenous and exotic tree species of India's semiarid region. Agrofor Syst 26:123–130

Ravindranathan NH, Chanakya HN, Kerr JM (1997) An ecosystem approach for analysis of biomass of energy In: Ker JM, Marothia DK, Singh K, Ramaswamy C. and Bentley WR (eds.) Natural resource economics theory and application in India. Oxford & IBH, New Delhi. pp 156–163

Sathaye JA, Ravindranath NH (1998) Climate change mitigation in the energy and forestry sectors of developing countries. Annu Rev Energy Env 23:387–437

Schroeder P (1994) Carbon storage benefits of agroforestry. Agrofor Syst 27:89–97

Sharda VN, Dhyani SK, Sharma AR (2001). Role of agroforestry in disaster management. Submitted to Chairman, HPC (DM Plan), DOA&C, GoI, New Delhi

Singh GB (1987) Agroforestry in the Indian subcontinent: past, present and future. In: Steppler HA, Nair PKR (eds) Agroforestry–a decade of development. International council for research in agroforestry nairobi, Kenya, pp 117–138

Singh P (1990) Agrisilviculture systems in India. In: Devendra C (ed) Shrubs and tree fodder for farm animals. IDRC, Ottawa (Canada), pp 186–195

Singh BP, Dhyani SK (1996) An appraisal of indigenous land use systems in Meghalaya for developing sustainable rainfed hill agriculture, In: Kohli RK, Arya KS (eds) Proceedings. IUFRO-DNAES International meet on resource inventory techniques to support agroforestry and environment activities, (eds) DNAES, Chandigarh, Oct. 1–3, 1996, pp 61–72

Singh BP, Dhyani SK, Chauhan DS, Prasad RN (1994a) Effect of multipurpose tree species on chemical properties of an acid Alfisols in Meghalaya. Indian J Agric Sci 65(5):345–349

Singh BP, Dhyani SK, Prasad RN (1994b) Traditional agroforestry system and their soil productivity on degraded alfisols/ultisols in hilly terrain. In: Panjab Singh, Pathak PS, Roy MM (eds) Agrofor Sys Degraded Lands. Oxford & IBH Publishing Co. Pvt. Ltd., New Delhi, Vol. I, p. 205–214

Solanki KR (2006) Agroforestry- opportunities and challenges. In: Chauhan SK, Gill SS, Khajuria HN, Chauhan R (eds) Production technology and management of agroforestry models. Agrotech Publ, Academy, pp 14–15

Stern N (2006) Stern review on the economics of climate change. HM Treasury, London

Swaminathan MS (1983).Genetic conservation: microbes to man. Presidential address to the 15th International Congress on Genetics. In: Genetics: new frontiers, Vol. 1. Oxford & IBH Publishing Co., New Delhi, India

Tejwani KG (1994) Agroforestry in India. Oxford and IBH, New Delhi 233 pp

Agroforestry Policy Issues and Challenges

12

A. K. Singh and S. K. Dhyani

Abstract

Agroforestry growth and development is influenced by various policies of the economy like credit, trade, taxation, power, transport, etc. These policies impact the sector either directly or indirectly besides the core forest and agriculture policies which have a larger bearing on the agroforestry policies in the Indian context. However, agriculture is a State Subject, whereas Forestry is in the concurrent list. As per the rules of business, agroforestry research is with the Indian Council of Agricultural Research (ICAR) but its development, extension is not allotted to any ministry or departments. There is a need for better Center—State coordination on agroforestry. A more effective coordinating mechanism between Forest and Agriculture Departments both at the national and state levels is required for which establishment of an apex coordinating body or institutional support mechanism at inter-ministerial level to take policy decisions on agroforestry is need of the hour.

Introduction

Agroforestry is an important land use system, particularly, in the present day scenario, when the availability of land has decreased tremendously and the population is increasing at a fast pace. It has been envisaged that agroforestry programs cannot succeed in isolation. They have to be people's programs. Participation of local people at every stage, viz., planning, execution, selection of area, choice of species, management, harvesting, distribution of produce and benefits, etc., is essential. People's participation should be encouraged owing to the fact that people in the villages know their needs and interests better than government officials working at block, district, and state levels. The main reasons for the non-adoption agroforestry programs to the desired extent have been the lack of participation of people and their representatives at the planning and execution stages at the local level in the absence of separate policy for Agroforestry.

A. K. Singh
Natural Resource Management Division, ICAR, New Delhi, 110012, India

S. K. Dhyani (✉)
National Research Centre for Agroforestry, Jhansi, 284003, Uttar Pradesh, India
e-mail: skdhyani@nrcaf.ernet.inshivkdhyani@gmail.com

J. C. Dagar et al. (eds.), *Agroforestry Systems in India: Livelihood Security & Ecosystem Services*, Advances in Agroforestry 10,
DOI: 10.1007/978-81-322-1662-9_12, © Springer India 2014

Need of Agroforestry Policies

Agroforestry growth and development is influenced by various policies of the economy like credit, trade, taxation, power, transport, etc. (Kareemulla et al. 2007). These policies impact the sector either directly or indirectly besides the core forest policies which have a larger bearing on the agroforestry policies in the Indian context. Historically, trees in India have been managed and regulated by customary law. The 73rd Amendment to the Constitution and the recommendations of Bhuria Committee in 1996 have given the *Panchayats* wide powers of control over the natural resources including land and forest produce. For promoting agroforestry, some of the states took innovative policy decisions, for instance, Andhra Pradesh abolished the agricultural cess that was collected on *Subabul* (*Leucaena leucocephala*) in the *Mandis* (markets). This encouraged increased production of this species.

The tree felling and transit rules, as applicable to trees on private lands, are unique to each of the State and have a direct bearing on the progress of agroforestry as there is wide variation in the scope of the felling and transit rules across states. Realizing the magnitude of the problem, the Ministry of Environment and Forests (MOEF) circulated guidelines to states in December 2004 for relaxing the felling and transit regulations for trees grown on non-forest lands. However, much more need to be done, like empowering the *Panchayats* to permit the felling and transit of less unrestricted, i.e., category B species.

Agroforestry based products have a significant potential for providing employment opportunities as well as a range of products such as fuelwood, edibles, fodder, pulpwood, building materials, and medicines. Trees play an important role in contributing toward sustainable livelihoods of rural poor (Dhyani et al. 2005, 2007). The rural population depends on agroforestry for at least part of their subsistence and cash/livelihoods, which they earn from fuelwood, fodder, poles, and a range of other products such as fruits, flowers, and medicinal plants. Trees not only provide socio-economic development but also have a special role in the ethos of the people of India. A number of tree species are revered as sacred trees and sacred groves and are found all over the country. The great poet Kalidas had penned verses in praise of the fragrance of *siris* (*Albizia lebbeck*) and *kadam* (*Anthocephalus cadamba*). India has a long tradition of growing trees on farms and around homesteads. Such traditions have a positive impact on the ecological, economical, and social well-being of the people (Kumar et al. 2012).

The report of the "Task Force on Greening of India for Livelihood Security and Sustainable Development 2001" (Planning Commission 2001) provides valuable data on agroforestry models for different agro-climatic zones with associated costs and benefits. This report gives the potential of wood production according to land capability class. In northwest India, yield is typically over 40 m^3 $ha^{-1}annum^{-1}$ because of good soil and availability of irrigation. If forest productivity is to be increased, only improved seed and planting material should be used. In fact, subsidized or free supply of seedlings indirectly encourages low productivity. In northwest India, productive clones of both poplar (*Populus*) and *Eucalyptus* have been identified and are being propagated on a large scale. The ITC clones of *Eucalyptus* are also being tested against Punjab selections and grown in different regions. Punjab has identified superior clones of *shisham* (*Dalbergia sissoo*), and research to discover better clones is on. Seed orchards of important plantation species have been established. The potential for growing teak (*Tectona grandis*) in farm forestry systems is also high. Profitable agroforestry models are being developed through painstaking research.

Agroforestry is now providing powerful technological and policy innovations that are rapidly spreading in Africa, Asia, Latin America, and more recently, in several developed countries. It is one key path to prosperity for poor people suffering from hunger, malnutrition, poverty, and the deterioration of the environment in the areas that have been bypassed by the "Green Revolution." By far, the vast majority of trees are grown under agroforestry systems. Agroforestry offers a better livelihood strategy to farmers due to relatively low input costs, flexible labor requirements, higher profitability, diversity of income sources, and more bargaining power at the marketing stage than with annual crops. Incorporating trees into farming systems leads to greater prosperity at the farm level. Trees provide farmers with marketable products—such as lumber, building poles, firewood, animal fodder, fruits, and medicines—all of which earn extra income. In northwest India, in the states of Punjab, Haryana, Uttarakhand, and Uttar Pradesh, tree-crop combinations have the potential to give better economic returns than trees or crops alone. In these areas, clonal poplar and eucalypts are fast emerging as the preferred species for agroforestry. Of late, teak and *Melia composita* are also being seriously considered. Farm and agroforestry have been accepted as a viable alternative for diversification of agriculture in these states. Further gain is possible by integrating trees with wood-based industries through its multiplier effect.

Agroforestry Opportunities

There are immense agroforestry options particularly to reclaim degraded habitats and as a component in multi-enterprise farming systems. Some of the opportunities may include;

- Agroforestry practices are intertwined in the various developmental programs/schemes of Govt. of India, e.g., watershed development projects, etc. (Dhyani et al. 2006, 2009).
- Forestry represents second largest land use in India after agriculture but contributes merely 1 % in GDP. It reflects the dormant potential of the sector.
- Degraded wastelands have potential to change fortunes of small and marginal farmers and rural India through promotion of agroforestry.
- By converting these wastelands into productive and renewable agroforestry resources, the challenges of livelihood, energy, and environment can be met besides unleashing the economic growth in the region.
- One of the major contributions of agroforestry for the economy is the livelihood impact both in terms of income and employment generation.
- The potential of agroforestry in livelihood creation is the most important factor in country like India.
- Besides the existing agroforestry practices, there is a tremendous potential for employment generation with improved agroforestry systems to the tune of 943 million person days per annum.
- Every ha of plantation in wood-based industries creates about 450 man days of employment, thus, 30 million ha has the potential to create approximately 15,000 million man days of employment besides creating job opportunities in wood-based value chain.
- The social benefits of agroforestry outweigh many other sectors. Agroforestry is an instrument of climate change mitigation.
- Bio-remediation for improvement of the soil and environmental quality is the added benefit (Dhyani 2012).
- By greening, even 10 million ha of degraded land, nearly 500 million Mg of carbon dioxide can be sequestered.
- The National Mission of Green India envisages the afforestation of 6 million ha to enhance forest and tree cover to 33 %.
- Achieving national goal of tree cover will also help in narrowing down the gap between demand and supply of tree products.
- In the scenario of climate change, agroforestry-based farming systems are most sustainable and easy to be adopted in extreme climatic conditions.

Key Policy and Other Issues Affecting Adoption of Agroforestry by Farmers

Farmers in India have made significant contribution toward production of wood and other produces through agroforestry. However, there are a number of constraints, which inhibit the growth and development of agroforestry to its full potential. The major constraints are:

- Long gestation period of trees along with market uncertainties.
- Difficulties in getting long term bank credit and insurance for plantations.
- Absence of regulated timber markets for transparent trading.
- Non-availability of genetically improved/certified forest reproductive material and absence of efficient extension services.
- Over regulation often restricting access to markets for farmer-grown timber and tree products, partly because of rules intended to curb illegal logging from natural forests and government plantations.
- Inadequate wood-based enterprises affecting farmers income.
- Lack of strategic directions (supply–demand forecasts) for agroforestry linking potential production with market demand.
- Practice of supply of forest produce at concessional rate to industry.
- Marketing and price support system to ensure right prices and smoothen market fluctuations.
- Unfavorable export and import policy of the Government.

Government of India's policies should address the genuine concerns of tree growers. The important policy issues are:

- Strengthening agroforestry research and extension services to develop profitable and replicable agroforestry models. All research organizations of Central Government and the States, universities as well as the private sectors may be involved in this important endeavor.
- Promotion of capital flows to agroforestry products for marketing and processing businesses.
- Promotion of investments in the sub-sector by making institutional funds available to the growers on concessional terms.
- Certification of FRM and registration of nurseries to facilitate availability of quality planting material.
- Operationalization of the strategy for multi-stakeholder partnerships involving forest departments, communities, and the investors for undertaking plantations on degraded forests, wastelands, and private lands.
- Grant of incentives for growth and development of farm forestry and simplification of rules for timber transit and felling of trees for farm-grown timbers.
- Compensation for environmental services to farmers growing trees.
- PPP in commercialization of agroforestry technologies.
- Developing networks for transferring proven technologies to the farmers.
- Developing decision support systems for replication of successful agroforestry systems/practices.
- Exploring unexploited and under-exploited species of high economic value, such as medicinal, aromatic, oil-yielding plants, etc.
- Local Panchayati Raj Institutions (PRI) to be involved in harvesting and marketing.
- Policies on inter-state transport of agroforestry needs to be liberalized.
- Practice of supply of forest produce at concessional rate to industry needs to be reviewed.
- Public investment in generating high quality material for reproduction to ensure availability of quality planting material.
- No free distribution of material.
- Awareness creation campaigns at the level as in case of other agricultural enterprises.
- Space for agroforestry in various inter-sectoral panels/committees.

Programs on Agroforestry

There is no specific scheme or program to promote agroforestry in the country. However, there are number of schemes of Government of India, State governments, and other organizations where agroforestry is recognized as a component and being promoted as such. Some of these schemes and programs include, Integrated Watershed Management Programme (IWMP), Mahatma Gandhi National Rural Employment Guarantee Act (NREGA) under Ministry of Rural Development, Soil Conservation in the Catchment of River Valley Project and Flood Prone River, National Horticulture Mission,

National Bamboo Mission, and National Mission on Medicinal Plants (DAC, MoA) to name a few. Many states and organizations in the country have also brought a number of innovative schemes for plantations and agroforestry. WIMCO initiated poplar plantation in early 1980s with buy back scheme with the farmers and it became one of the most successful programs linking farmers with industries. Now, ITC is promoting agroforestry at small and marginal farmers' fields in Andhra Pradesh. It has promoted more than 100 thousand ha under *Eucalyptus, Leucaena,* and *Casuarina*. Similarly, under Public–private sector linkages, ITC Bhadrachalam Paper Board Ltd. also promoted these three species, West Coast Paper Mills Ltd. promoted *Acacia mangium x Acacia auriculaeformis* hybrid and Hindustan Paper Mills Ltd. promoted bamboos in farmers' field. In recent years, BAIF (www.baif.org.in) in association with NABARD through its innovative Wadi (a horti-forestry orchard raised by tribal family on sloping uplands) program implemented in about 60,000 ha in Gujarat, Maharashtra, and Karnataka promoted agroforestry.

New Initiatives for Agroforestry

There are number of schemes and programs being initiated by various agencies in the country. Several initiatives have been taken by Government of India to encourage crop diversification in the earlier 'green revolution' states. Punjab wants to bring an additional area of 200 thousand ha under agroforestry to its present 130 thousand ha as crop diversification strategy. Haryana and Uttar Pradesh are also planning on similar lines. At Government of India level, a newly launched Green India Mission (http://moef.nic.in), 3.0 million ha of degraded lands and fallows are to be brought under agro-/ social forestry within 10 years. Agroforestry will receive further boost as a component of the National Mission for Sustainable Agriculture (DAC, Ministry of Agriculture, Government of India) which is being launched in the current Plan. However, at present there is no convergence and there is a lack of coordination amongst the schemes/departments and no implementation frame work including institutional mechanisms exists at the Center, state, and district levels to pursue agroforestry in a systematic manner.

In order to promote agroforestry, it will require appropriate research interventions, adequate investment, suitable extension strategies, providing incentives to agroforestry, removing legal barriers in felling trees by farmers on their fields, transportation and marketing of agroforestry produce, harvest process technology development of new products and market infrastructure and above all a forward looking Agroforestry Policy to address these issues. Recently, a Committee constituted by Ministry of Environment and Forests (MoEF 2013) analyzed existing State regulations governing felling and transit of trees grown on private lands and found that there are wide variations in the rules and regulations related to felling of trees and transportation of felled timber across various States especially in a particular region, impacting inter-state boundary movement of agroforestry produces. Certain agroforestry species which are common in a particular State are exempted from transit regulations, but the same species are not so common in adjoining States and are subjected to transit regulation. There is also lack of unified approach for such agroforestry species even in adjoining districts within a State. Farmers grow trees primarily for commercial reasons and are often discouraged due to lack of uniformity in approach and desired level of clarity in regulations. Generally, fast growing short rotation tree species like *Eucalyptus, Populus, Casuarina, Ailanthus, Melia, Albizia, Acacia auriculaeformis, Gmelina*, Kadam (*Anthocephalus cadamba*), and Bamboo (*Bambusa* spp.) are preferred by farmers for obvious reasons. Most of these species are exempted under

Transit Rules in some States. Further, agroforestry has generally progressed in Northern states like Punjab, Haryana, Uttar Pradesh and Gujarat, etc., which have relatively low forest cover coupled with high productive tracts giving quick return on short rotation crops. Besides, farmers in different parts of the country are practicing a variety of agroforestry models and making significant contributions toward production of wood and other produces (Dhyani et al. 2013).

In fact, Indian Council of Agricultural Research (ICAR) organized a workshop on Agroforestry Policy Initiative with all stakeholders at New Delhi on June 17th, 2011. Now, efforts are on to take up this subject further and develop a draft National Policy on Agroforestry. There is also a proposal to launch a National Mission on Agroforestry in the current plan itself.

Conclusions

Agroforestry promotion for realizing the true potential requires efforts from all the concerned like Forest and Agriculture Departments, research and financial institutions, corporate sector, *panchayats*, NGOs, and farmers. The policy thrust must be on creating infrastructure, ensuring quality in planting material, plugging the vacuum of extension network, favorable pricing mechanism including support price system, encouraging public–private partnerships in all spheres of agroforestry among others.

References

Dhyani SK (2012) Agroforestry interventions in India: focus on environmental services and livelihood security. Indian J Agrofor 13(2):1–9

Handa SK, Uma AK (2013) Area under agroforestry in India: an assessment for present status and future perspective. Indian J Agrofor 15(1): 1–8

Dhyani SK, Kareemulla K, Ajit, Handa AK (2009) Agroforestry potential and scope for development across agro-climatic zones in India. Indian J For 32:181–190

Dhyani SK, Kareemulla K, Handa AK (2006) Agroforestry potential and returns across agroclimatic zones. NRCAF, Jhansi

Dhyani SK, Samra JS, Ajit, Handa AK, Uma (2007) Forestry to support increased agricultural production: focus on employment generation and rural development. Agric Econ Res Rev 26(2):179–202

Dhyani SK, Sharda VN, Samra JS (2005) Agroforestry for sustainable management of soil, water and environmental quality; looking back to think ahead. Range Manag Agrofor 26(1):71–83

Kareemulla K, Dhyani SK, Rizvi RH, Yadav RS, Munna R (2007) Agroforestry for rural development: a cooperative experience. NRCAF, Jhansi, pp 91

Kumar BM, Singh AK, Dhyani SK (2012) South Asian agroforestry: traditions, transformations, and prospects. In: Nair PKR, Garrity D (eds) Agroforestry—the future of global land use, advances in agroforestry 9, Springer, Dordrecht, DOI 10.1007/978-94-007-4676-3_19

MoEF (2013) Report of the committee on the regulatory regime regarding felling and transit regulations for tree species grown on non forests/private lands. Ministry of Environment and Forests (MoEF), Government of India (moef.nic.in), New Delhi, pp 42

Planning Commission (2001) Report of the task force on greening india for livelihood security and sustainable development. Planning Commission, New Delhi, p 254

Synthesis

13

J. C. Dagar

Abstract

In India, though agroforestry is an age-old land use system since time immemorial, the organized research was initiated in early 1970s through industry participation in plantation of commercial tree species through Indian Council of Agricultural Research (ICAR) institutes. As follow-up, the All India Coordinated Research Project (AICRP) on Agroforestry was established in 1983 by the ICAR for conducting research work in collaboration with State Agricultural Universities situated in different agro-climatic zones. Since then voluminous data has been generated about both traditional as well as improved agroforestry systems and practices. Now, about 25 million ha area in the country (8.2 % of the total reported geographical area), is reported under agroforestry in both irrigated and rain-fed agriculture which also includes trees outside forests and scattered trees on and off the agricultural fields. In this publication, the various chapters are compiled in such a way that a clear picture of various potential agroforestry systems both traditional and improved found in different agro-ecological regions is presented. Some systems are adopted across a number of climatic regions (for example salt-affected and waterlogged areas); but the problems are of different nature in different regions. To deal with such cases, separate chapters are included to present agroforestry approaches dealing with specific problems. The entire information has been synthesized under heads: progress made (diagnostic survey, evaluation of promising MPTs, rehabilitation of degraded lands, use of poor quality water for agroforestry interventions, muti-enterprise agroforestry models, emerging paradigms of biofuels, climate resilient systems, etc.), lessions learnt, and way forward.

In India, though agroforestry is an age-old land use system since time immemorial, the organized research was initiated in early 1970s through industry participation in plantation of commercial tree species through Indian Council of Agricultural Research (ICAR) institutes.

J. C. Dagar (✉)
Krishi Anusandhan Bhawan–II, Indian Council of Agricultural Research, New Delhi, 110012, India
e-mail: dagarjc@gmail.com

J. C. Dagar et al. (eds.), *Agroforestry Systems in India: Livelihood Security & Ecosystem Services*, Advances in Agroforestry 10,
DOI: 10.1007/978-81-322-1662-9_13,

As follow-up, the All India Coordinated Research Project (AICRP) on Agroforestry was established in 1983 by the ICAR for conducting research work in collaboration with State Agricultural Universities situated in different agro-climatic zones. Since then voluminous data has been generated both about traditional as well as improved agroforestry systems and practices. Now, about 25 million ha area in the country (8.2 % of the total reported geographical area), is reported under agroforestry in both irrigated and rain-fed agriculture which also includes trees outside forests and scattered trees on and off the agricultural fields. In this publication, the various chapters are compiled in such a way that a clear picture of various potential agroforestry systems both traditional and improved found in different agro-ecological regions is presented. Some systems are adopted across a number of climatic regions (for example salt-affected and waterlogged areas); but the problems are of different nature in different regions. To deal with such cases, separate chapters are included to present agroforestry approaches dealing with specific problems. The entire information has been synthesized under heads: progress made (diagnostic survey, evaluation of promising MPTs, rehabilitation of degraded lands, use of poor quality water for agroforestry interventions, muti-enterprise agroforestry models, emerging paradigms of biofuels, climate resilient systems, etc.), lessions learnt, and way forward.

Progress Made

Agroforestry has played a vital role in the Indian economy by way of tangible and intangible benefits. Through agroforestry we meet at least half of the demand of fuel wood, 2/3rd of the small timber and also the subsistence needs of households. It is playing the greatest role in maintaining the resource base and increasing overall productivity especially in arid and semi-arid regions. Changing priority in avenues like biofuels, employment generation, value addition, carbon sequestration, and mitigating climate change are now being focused through agroforestry systems. Industries have taken up fast growing soft-wood trees for commercial exploitations. The Task Force on Greening India (through Planning Commission, the highest policy formulating body of India) has already identified a potential of 10 million ha irrigated and 18 million ha rainfed area that could be developed through agroforestry on a watershed basis. However, it will require appropriate research and policy interventions. Progress made so far in agroforestry research in the country may be summarized as follows:

Diagnostic Survey and Appraisal of Existing Agroforestry Practices

The diagnostic survey and appraisal of existing traditional and improved agroforestry practices/systems prevalent in different agro-ecological zones in the country revealed that there are a number of agroforestry practices\systems, which are widely based on nature and arrangement of the components and have been adopted widely based on ecological or socioeconomic criteria. The tree-based systems provide not only poles, stakes, timber, and fuel, but also cash crops, food, fodder, fruits and nuts, dyes, gums, resins, fiber, and medicines. Fodder and food trees provide balanced diets during dry seasons, when other foods are scarce. They serve as wind breaks and shelterbelts, delineate boundaries, and provide shade, ornamentation, and seclusion around homesteads. Trees, with their deep rooting systems, consume moisture and nutrients from higher depth in the soil than arable and pastoral crops, and thus there is least competition among the different components. Farmers of arid region understand well that crops are not harmed rather benefitted due to retaining/growing of deep rooted leguminous Khejri (*Prosopis cineraria*) trees in wider spaces, which is considered "Tree of Life" in dry ecologies. Thus, the value of trees on farmland may considerably exceed that offered by woodlands and plantations.

The common denominator and scientific foundation of all these systems are the multipurpose trees (MPTs). The success of agroforestry will depend upon the extent to which the productive, protective, and service potential of MPTs is understood and exploited (through research) and realized (through development, experience and extension efforts). In fact there are number of trees with immense uses mentioned in literature (including medicinal uses) found distributed in different agro-climatic regions but all of these cannot find a place as MPTs in a particular agroforestry system. It is very important to know the autecology of the species, its utility, adaptability, and acceptability in the system. For example, coastal almond *Terminalia catappa, Casuarina equisetifolia,* and *Pongamia pinnata* are tolerant to high salinity and found growing in tidal zone but *Terminalia* and *Casuarina* do not withstand frost and cannot be grown on salty soils in northern parts of India while *Pongamia* performs well. At the same time, it is fact that many tree species have potential but most of these do not appear in the research agenda of both agricultural and forestry research organizations. These species are truly under-exploited with vast potential for improvement. Many of these play vital roles in India's farming systems in producing various products and providing a variety of non-monetary services, and are the key components of sustainability and environmental protection. Many of these are traditionally being used as integral parts of farming systems but most of these indigenous trees and shrubs have been totally ignored. This needs a critical review and attempts should be made to identify potential species based on their agro-climatic suitability and adaptability; and research efforts should be made to exploit their utility for better economic gains.

An assessment of voluminous descriptions of many indigenous agroforestry systems, which are mostly qualitative, shows that the systems are extremely site-specific with very few examples of their extrapolatability. A vast majority of the virtues and benefits ascribed to them are anecdotal, i.e., they have not been tested and proven scientifically. It could be argued that the obvious need not to be tested; but quantification and scientific cataloguing of the benefits and values are essential for system improvement.

Collection and Evaluation of Promising Fuel Wood, Fodder, and Timber Yielding Tree Species/Cultivars

The second major aspect of the research endeavors under agroforestry was collection and evaluation of promising tree species/cultivars of fuel, fodder, and small timber. Quality germplasm of important agroforestry trees was collected and evaluated in arboretum established by different Centers of the AICRP. Based on growth performance trials at these centers, about 200 promising tree species have been identified. The results indicate that the safest choice of agroforestry species has come from the native vegetation, which has a history of adaptation to local precipitation and other climatic regimes. There are a number of tree-crop combinations, which in turn reflect the differences in the climates and soil fertility of various regions in the country. The examples of major trees in agroforestry practices include: *Grewia optiva*, *Ulmus wallichiana, Morus alba*, and *Robina pseudoacacia* in western Himalayan (cold) region; *Alnus nepalensis*, Bamboos, *Acacia auriculaeformis, Parasarianthes falcataria*, and *Gmelina arborea* for north-eastern Himalayan region; Poplars (*Populus* spp.), *Eucalyptus, Acacia nilotica* and *Dalbergia sissoo* in Indo-Gangetic region; *Dalbergia sissoo*, *Acacia tortilis*, *A. nilotica*, *Ailanthus excelsa*, *Prosopis cineraria, Leucaena leucocephala*, and *Azadirachta indica* for arid and semi-arid regions; *Acacia nilotica, Prosopis cineraria, Tecomella undulata*, and *Ziziphus* spp. in western India (Rajasthan); *Tectona grandis, Tamarindus indica, Borassus flabellifer, Cocos nucifera*, Para rubber (*Hevea brasilensis*), and cashew nuts (*Anacardium occidentale*) in southern peninsular region; *Albizia* spp., *Gmelina arborea, Gliricidia sepium, Acacia auriculaeformis, A. mangium* for humid and subhumid regions; and *Artocarpus* spp., *Cocos nucifera*, *Pongamia pinnata*, *Ceiba pentandra*, *Casuarina equisetifolia*, *Grevillea robusta*, and bamboos in coastal and island regions. The efforts made so far have created voluminous database, which is a great strength for planning agroforestry research strategy in future. The information collected may be utilized for creating local and regional timber volume tables. There are many potential species found in different agro-ecological regions which are completely under-explored and need to be domesticated and given a place in suitable agroforestry systems.

Rehabilitation of Degraded Habitats

The research and developmental efforts undertaken during the past three decades for rehabilitation of degraded lands have brought results, which have clearly demonstrated the potential of agroforestry for greening the degraded lands including salt-affected and waterlogged areas; mined, ravine and degraded watershed areas; sand dune stabilization; resource conservation; improvement of environmental quality; and providing multiple outputs to meet the day to day demand of the rural population. The development of innovative technologies such as auger-hole technique for establishing trees on highly sodic land; subsurface planting and furrow irrigation technique for waterlogged saline soils and use of saline water for establishment of trees on degraded lands in dry regions; biodrainage for controlling waterlogging; and sand dunes stabilization has been a revolutionary achievement through which large chunk of degraded wastelands has been brought under plantations. Research inputs in improved systems such as alley cropping on sloping lands; sand dune stabilization; and establishment of multi-enterprise farming systems by integrating fruit trees, MPTs, livestock, poultry, fish culture, apiculture, floriculture and food crops have brought encouraging results. The incorporation of water harvesting and use of micro-irrigation technology for irrigation (even using saline water) has added to the success of agroforestry programmes in these areas.

Utilization of Sewage/Poor Quality Waters for Agroforestry Interventions

Irrigation water is one of the most scarce but critical resources for agricultural production in arid and semi-arid areas. Most of the water scarcity areas are usually underlain by aquifers of poor quality. With developing scenario of severe water scarcity and competition from other sectors of economy, it appears axiomatic that agriculture would have to increasingly depend upon marginal and poor quality waters including sewage water. During past three decades technologies have been developed to use saline water for establishing agroforestry systems on degraded lands in dry regions. Forest and fruit trees can successfully be established using saline water in furrows and providing one or two irrigations to low water requiring intercrops. For providing fodder during lean period the high yielding salt-tolerant grasses also grown in interspaces of trees. These systems are highly profitable and sustainable. It has been found that no significant salinity is found to develop with saline irrigation (with water up to EC_{iw} of 10 dS m^{-1}) if there is 1 year of normal rainfall in 3–4 years.

Techniques have also been developed to grow fast growing trees like *Eucalyptus* by utilising sewage water, which otherwise has problem of disposal. In the interspaces crops which do not play role in food chain such as aromatic grasses like lemon grass or flowers may be cultivated.

Developing Integrated (Multi-Enterprise) Farming System Models

In India more than 86 % farm families are of marginal and small categories with holding size below 1.2 ha and are living in risk prone diverse production conditions as small and fragmented land holdings do not allow these farmers to have better independent farm resources. To fulfill the basic needs of households including food (cereals, pulses, oil seeds, milk, fruit, honey, fish, meat, etc.) for human consumption, feed and fodder for cattle, fuel and fiber, and other essential items of routine use; a well-focused attention toward Integrated Farming System (IFS) research is warranted. Based on the scattered experiments carried out in the country over the years, it has been realized at the level of Planning Commission and ICAR that this approach is quite feasible and the Project Directorate of Cropping System was renamed as Project Directorate of Farming System Research (PDFSR) with changed mandate in 2009. Now, multi-enterprise farming system models are being evolved for farmers having less than 2 ha land involving components from small ruminants or cattle, fish culture in small pond, fruit trees, and vegetables on dykes of pond, food and fodder crops, bee keeping, poultry, piggery (north-eastern states), floriculture, mushroom culture, and preparation of vermi-compost. From this model farmers get income throughout the year. There are many incentives on many components such as micro-irrigation and construction of fish pond. Many farmers have gone for organic agriculture and generating their own manure and energy from the left over agricultural waste. This is highly remunerative and climate resilient system which may be popularized widely.

Emerging Paradigms of Biofuels

Energy security has emerged as an important policy issue all over the world. One of the key challenges facing the developing countries including India is how to meet its growing needs and sustain economic growth without contributing to climate change. Among several alternative renewable sources of energy, biofuels have emerged as a most potent source and have been often regarded as "future fuels". Government of India has undertaken several policy measures to augment production and use of biofuels during the past one decade. The launching of National Biofuel Mission (NBM) in the year 2003 under the aegis of the Planning Commission is the frontrunner of such efforts in the country. Special focus has been on phased expansion of area under biofuel feedstock crops like *Jatropha curcas* and *Pongamia pinnata*. The research and development efforts have been focused on plantations, production, and processing technologies of biofuels, as well as maximizing efficiencies of different end-use applications and utilization of by-products. Several experiments related to evaluation of various cultivars of these species for developing site-specific genotypes that can tolerate adverse climatic conditions; studies on genetic diversity, variability, and other biotechnological traits; production and demonstration of quality planting material; clonal cultures to increase productivity; and developing unconventional methods where mycorrhiza application for increasing growth and *Pongamia* productivity have been conducted and desired results are expected. At present biodiesel production in India is at the nascent stage, with about 95 million liters being produced from and *Jatropha* and *Pongamia* and for 10% blending we shall require 12.31 million tons of biodiesel from about 13 million ha area by 2020. Currently, *Jatropha* is being cultivated on about 0.5 million ha area. Other potential trees include *Azadirachta indica, Simarouba glauca* and *Madhuca indica.*

Agroforestry and Industrial Interface

Agroforestry products are either consumed locally or traded conventionally. The market is highly unorganized. Commercial agroforestry in tandem with wood-based industry has shown immense potential as industries are exempted from the falling restrictions. *Populus, Eucalyptus, Casuarina* , and *Leucaena*-based commercial agroforestry plantations could flourish in different regions of the country when promoted by industries. Similarly, tea (*Thea chinensis*), coffee (*Coffea arabica*), cacao (*Theobroma cacao*), red oil palm (*Elaeis guineensis*), and rubber (*Hevea *brasiliensis*) plantations are successful when promoted by industries. In recent years, fruit-based industries coupled with food supply chain and creation of infrastructural facilities have changed marketing of fruits, vegetables, and medicinal, and aromatic plants. Bioenergy is another area where industrial groups can play an important role to meet the requirements for energy as well as for employment generation. Thus, there is vast potential for investments and creation of employment in agroforestry sector through involvement of corporate sector.

Climate Resilient Farming System Research/Agroforestry

During past one decade research initiatives have been undertaken to study the impact of climate change on various farming systems, carbon sequestration, and adaptation and mitigation measures. The Indian Council of Agricultural Research (ICAR) has responded to the challenges of climate change on Indian agriculture by launching a multi-disciplinary and multi-institutional major scheme, National Initiative on Climate Resilient Agriculture (NICRA) during 2011 which is to be continued during XII Five Year Plan. The scheme aims at climate resilient agricultural technologies to address the long-term changes in climate expected in the country and also demonstrate the best bet practices that help farmers cope with current climate variability. It covers all major commodities in agriculture (including agroforestry component) and allied sectors. It has made significant progress both in strategic and technology demonstration. The interventions of technology demonstration component being tried in 100 climatic vulnerable districts of the country like drought and submergence tolerant crop varieties, water harvesting and recycling, soil health enhancement, livestock shelter and feeding, and custom hiring has got tremendous response by all stake holders in the country. Agroforestry/farming systems especially live stock, fishery, fruit trees, and water harvesting components are the major interventions. In strategic research major emphasis is given on identifying and developing climate resilient/stress tolerant plant and animal resources. Attempts are being made to develop GIS and remote sensing technology to quantify agroforestry trees and prepare a map showing agroforestry area. Major study has been initiated to quantify carbon sequestration both in above and below ground vegetation and how to increase carbon sequestration to mitigate climate change, where agroforestry has to play a vital role.

Lessons Learnt

It is well recognized fact that there are many traditional and improved agroforestry systems in India, most of these, however, are location specific and information generated from them is anecdotal. Therefore, their benefits have remained vastly underexploited. A great deal can be accomplished by improving the indigenous systems utilizing the wealth of information available about these systems. But most of the on-going efforts are of the "What" type dealing with component and system management, with too little attention being paid to the "Why" (reason for the observed behavior) and the "How" (process involved) of the results. Therefore, most of the results obtained from such efforts are also largely location specific. In order to improve the efficiency of indigenous systems, as well as to assess the performance of improved technologies, we need to have a systematic procedure to evaluate such systems.

The thrusts of agroforestry efforts are different in different agro-ecological regions depending on the region's characteristics and land-use constraints. Therefore, it should be possible to develop an evaluation matrix of agro-ecological regions versus agroforestry potentials and attributes. Such a matrix could be a framework for evaluating agroforestry systems, as well as a guide for the type of research and nature of data that will need to be generated from agroforestry projects. The matrix with necessary adjustments can be made use of for evaluation of systems within relatively small areas such as sub-regions to larger areas such as states, or zones consisting of several states. The results of such evaluation can be the bases for appropriate extension recommendations for that specific area or region. Productivity, sustainability and adoptability are the major attributes of all the agroforestry systems hence these parameters should be the criteria of evaluation of the agroforestry systems.

In most of the evaluation trials of the fruit tree species for stress tolerance such as salt tolerance various species are planted for comparison but in a particular species different varieties/cultivars vary in their tolerance, hence by planting one cultivar the desired information is incomplete, therefore, different cultivars of a species should be considered while planting. Moreover, in evaluation of tree species the nursery should be prepared from the same seed stock or cloned material. So far very little attention has been paid on nursery techniques and collection and multiplication of quality planting material. Biotechnological tools should be used for improving the tree germplasm as is the case in fruit tree species.

As stated earlier several potential tree species are available in India which can be part of agroforestry systems but no attention has been paid on domestication of potential species. For example, domestication trials with Nauni (*Morinda citrifolia*) in Andamans have proved that this multipurpose tree species is quite fast growing, tolerant to salinity and waterlogging stresses, and its fruit can be used to prepare juice, jam, pickle, and leaves as fodder. Various parts of the tree are constituent of several medicines used by the aborigines. Many wild cultivars of tree species of highly economic importance such as nutmeg (*Myristica* spp.), Rudraksh (*Elaeocarpus* spp.), and many others are available in Andaman-Nicobar Islands and can form the base material for improvement of these species. Similarly, Common Seabuck-Thorn (*Hippophae rhamnoides*) was a brush wood along road side of Cold Desert of Leh up to sometimes back, has become famous as an industrial crop. Its fruit is rich source of vitamin C and is converted into juice, jelly and jam; syrup used in pulmonary complaints; and seed contain semi-drying oil. Many highly useful woody species growing wild are becoming rare due to over-exploitation and need domestication before these are lost from their natural habitats. For example, Phog (*Calligonum polygonoides*) was predominant bush on sand dunes of Thar Desert but now it can be seen only in Bikaner belt as its inflorescence is edible as vegetable and all parts are used in one or the other way. Such plants need to be domesticated and multiplied.

Way Forward

It is worth mentioning here that agroforestry systems are probably the only means for getting the desired tree cover in the country, especially in states that have low forest area. Agroforestry is not only a technique of growing food crops in association with woody perennials and livestock to optimize the use of natural resources but also a source of renewable energy resource and a means of reducing the risk of environmental degradation and climate change. The livelihood security through agroforestry and its potential in meeting basic needs of poor and landless farmers in terms of food, fuel, and fodder and employment generation are well-known. Before formulation of any big program for implementation we need exact area under agroforestry. Now GIS and Remote Sensing methodology has been developed, which should be implemented at country level and exact area under agroforestry trees and fruit orchards must be mapped to formulate future strategy.

By virtue of diversity of the components of the agroforestry systems like food grains, vegetables, and fruits, and the nutritional security to the communities could be ensured. The fodder cultivation under agroforestry land use will ensure production of milk, meat, and animal products. Wide range of food crops, pulses, and oil seeds can meet diverse needs of the society. This can be assured through multi-enterprise farming systems. These systems have to go long way for bringing sustainability and livelihood security among small and marginal stakeholders. These are site-specific and need more research inputs and demonstrations on farmers' fields.

Tree domestication and the commercial processing and marketing of tree products and services, is a new frontier for agroforestry research and development. A major role is also emerging in the domain of environmental services, particularly the development of mechanisms to reward the rural poor for the watershed protection, biodiversity conservation and carbon sequestration that they provide to the society. Agroforestry has the greatest potential to mitigate climate change through adaptation. It is the tested mechanism to tolerate diversified climate through adaptations towards variety of climate aberrations. More research and extension efforts are needed to popularize potential agroforestry systems in the scenario of climate change.

In coastal areas mangroves play a vital role in protection of coast line and providing nutrients to aquatic life, hence require more attention for rehabilitation of coastal areas by planting suitable salt-tolerant mangrove and other associated species. Aquaculture keeping mangroves intact is challenging but more sustainable and viable option in the scenario of climate change.

Further, we need multi-disciplinary and multi-institutional collaborations in the field of value addition. Very good basic and applied research is being carried out in many institutions related to food, beverage and bioactive products but there is no coordination among these researchers with agricultural scientists hence the useful information could not be translated into livelihood security and profitability for poor and landless farmers through their agricultural farms. Despite of good research work in the field we could not institutionalize the agroforestry in the country due to lack of any agroforestry policy. Like ecology and environmental sciences it is an interdisciplinary subject related to soil science, water, agriculture, ecology, environment, animal sciences, fishery, forestry, biochemistry, and social sciences. As stated earlier we need separate agroforestry policy to address all agroforestry related issues including carbon credits for farmers, environmental benefits, biodiversity conservation (including microbial diversity) through agroforestry systems, value addition, capacity building, bye back policy of agroforestry products, contract farming, agroforestry as compulsory subject in school curricula, and vocational courses at university level. No program can be successful until we develop the human resource; therefore, training especially women and knowledge sharing with all stake holders about agroforestry technologies and value added products must get priority. There is no substitution of "demonstration" of a viable and perfect technology. Therefore, agricultural extension scientists must demonstrate the available agroforestry technologies on farmers' fields in different agro-ecological zones. "Farmer First" must be the approach of all concerned with welfare of Indian Agriculture and attempts should be made to create opportunities in agriculture so that the "Youth" is attracted to opt the livelihood from agriculture and there must not be any doubt that to achieve this Goal, going for climate risk free profitable agriculture, and mitigating climate change "Agroforestry" is the only viable option. The country must have the Agroforestry Mission at place.

About the Book

Agroforestry, the word coined in the early 1970s, has made its place in all the developed and developing countries of the world and is now recognized as an important approach to ensuring food security and rebuilding resilient rural environments. India has been an all-time leader in agroforestry. The South and Southeast Asia region comprising India is often described as the cradle of agroforestry. Almost all forms of agroforestry systems exist across India in ecozones ranging from humid tropical lowlands to high-altitude and temperate biomes, and perhumid rainforest zones to parched drylands. The country ranks foremost among the community of nations not only in terms of this enormous diversity and long tradition of the practice of agroforestry, but also in fostering scientific developments in the subject. Agroforestry applies to private agricultural and forest lands and communities that also include highly erodible, flood-prone, economically marginal, and environmentally sensitive lands. The typical situation is agricultural, where trees are added to create desired benefits. Agroforestry allows for the diversification of farm activities and makes better use of environmental resources. Owing to an increase in the population of human and cattle, there is increasing demand for food as well as fodder, particularly in developing countries like India. So far, there is no policy that deals with specifics in agroforestry in India. However, the Indian Council of Agricultural Research has been discussing the scope of having a National Agroforestry Policy at appropriate platforms. However, evolving a policy requires good and reliable datasets from different corners of the country on the subject matter. This synthesis volume containing 13 chapters is an attempt to collate available information in a classified manner into different system ecologies, problems, and solutions, and converging them into a policy support.

J. C. Dagar et al. (eds.), *Agroforestry Systems in India: Livelihood Security & Ecosystem Services*, Advances in Agroforestry 10,
DOI: 10.1007/978-81-322-1662-9, © Springer India 2014

About the Editors

Dr. Jagdish Chander Dagar has been well recognized for his research in the area of agroforestry, both nationally and internationally. He is presently working as Scientist Emeritus in Central Soil Salinity Research Institute, Karnal; previously Dr. Dagar was the Assistant Director General (Agronomy/Agroforestry) at the headquarters of ICAR (2010–2012). His research interest has been in the areas of biosaline agriculture, agroforestry, rehabilitation of degraded lands, management of natural resources, biodrainage, climate change, and sustainable agriculture and policy. He has written several books and has published more than 200 research papers of high repute in international and national level journals. Recognizing the research contributions, Dr. Dagar has been conferred with several awards and honors: Sajjad Memorial Gold Medal, 1973–1974; Hari Om Ashram Trust Award, 2005; Swami Pranavananda Saraswati National Award, 2005; CSSRI Excellence Award on Soil Salinity & Water Management, 2007–2009; Bharat Excellence Award & Gold Medal, 2009; Dr. K.G. Tejwani Award for Excellence in Agroforestry Research & Development, 2010–2011. Further, he is a Fellow of National Academy of Agricultural Sciences and several other professional societies and has also been consulting several national and international agencies such as FAO and CIMMYT. Dr. Dagar has travelled in several countries in different capacities.

Dr. A. Arunachalam is a trained ecologist from the North-Eastern Hill University, Shillong, and has grown professionally through strong pursuit in ecological research that is evident from his 135 research papers, 90 book chapters, and 20 popular articles. He started his service as a Lecturer/Assistant Professor in Forestry, North Eastern Regional Institute of Science & Technology (Arunachal Pradesh) and currently he is working as a Principal Scientist in the headquarters of Indian Council of Agricultural Research (ICAR), Krishi Bhawan, New Delhi. His academic excellence is also evident from the awards/honors that he has received: DST-Visiting Fellowship in Biodiversity, Jawaharlal Nehru Centre for Advanced Study in Science & Technology,

J. C. Dagar et al. (eds.), *Agroforestry Systems in India: Livelihood Security & Ecosystem Services*, Advances in Agroforestry 10,
DOI: 10.1007/978-81-322-1662-9, © Springer India 2014

Jakkur, 1998–1999; DST-BOYSCAST Fellowship in Restoration Ecology, University of Massachusetts, Boston (USA), 2000–2001; Dr. K.G. Tejwani Award for Management of Natural Resources (biennium 2002–2003), Indian Association of Soil and Water Conservationists; Pran Vohra Award in Agriculture and Forestry Sciences (2003–2004), Indian Science Congress Association; Environmentalist of the Year award (2008) and Eminent Scientist of the Year award (2012) by the National Environmental Science Academy. Further, Bharat Excellence Award, RashtriyaVikasRatan Award, RashtriyaSamman Puraskar Award & Gold Medal and Lifetime Achievement Award by several professional organizations; Dr. Arunachalam is a Fellow/Associate several professional societies and is an Editor of *Journal of Agricultural Extension and Rural Development*, *Climate Change* and *Environmental Sustainability* and an Associate Editor, *Indian Journal of Hill Farming*.

Dr. Anil Kumar Singh is currently the Vice-Chancellor of the Rajmata Vijayaraje Scindia Krishi VishwaVidyalaya, Gwalior. Earlier he was the Deputy Director General (NRM) in ICAR—2007–2012; Acting Director, IARI—2005–2006; Project Director, Water Technology Centre—1999–2007. Dr. Singh joined the Agricultural Research Service in 1976 and grew up to Principal Scientist until 1999. His areas of research interest have been soil-water-nutrient-plant-atmosphere interactions, crop growth simulation, irrigation water management, and integrated watershed management. Dr. Singh has been known as a DAAD Fellow at University of Hohenheim, Germany and has served in International Rice Research Institute between 1993 and 1995. He has been a fellow of several professional societies in the country. Dr. Singh is a recipient of several award/honors: National Science Talent Scholarship, 1967–1976; Gold Medal in M.Sc. 1972; 11th D.P. Motiramani Memorial Lecture Award, 2004; IARI Best Teacher Award, 2005; Soil Conservation Society of India Gold Medal, 2006; 33rd Dr. R.V. Tamhane Memorial Lecture Award, 2006; 6th Prof. S.C. Das Memorial Lecture, 2008; 19th Dr. S.P. Raychaudhary Memorial Lecture, 2008; Platinum Jubilee Lecture—Indian Science Congress, 2009; 8th Dr. P.G. Krishna Memorial Oration, 2009; 19th Nehru Memorial Lecture, 2009; Prof. S.K. Mukherjee Commemoration Award, 2009, Indian Science Congress Association; Dr. C. Subramanium Outstanding Teacher Award 2007–2008; 10th Shri Hari Krishna Shastri Memorial Award, 2009; 20th Dr. Daroga Singh Memorial Lecture, 2011; Rajesh Pilot Memorial Award 2011.

Index

J. C. Dagar et al. (eds.), *Agroforestry Systems in India: Livelihood Security & Ecosystem Services*, Advances in Agroforestry 10,
DOI: 10.1007/978-81-322-1662-9, © Springer India 2014

B